Rarefied Gas Dynamics: Theoretical and Computational Techniques

Edited by
E. P. Muntz
University of Southern California
Los Angeles, California

D. P. Weaver
Astronautics Laboratory (AFSC)
Edwards Air Force Base, California

D. H. Campbell
The University of Dayton Research Institute
Astronautics Laboratory (AFSC)
Edwards Air Force Base, California

Volume 118
PROGRESS IN
ASTRONAUTICS AND AERONAUTICS
Martin Summerfield, Series Editor-in-Chief
Princeton Combustion Research Laboratories, Inc.
Monmouth Junction, New Jersey

Technical Papers selected from the Sixteenth International Symposium on Rarefied Gas Dynamics, Pasadena, California, July 10-16, 1988, subsequently revised for this volume.

Published by the American Institute of Aeronautics and Astronautics, Inc., 370 L'Enfant, Promenade, SW, Washington, DC 20024-2518

American Institute of Aeronautics and Astronautics, Inc.
Washington, D.C.

Library of Congress Cataloging in Publication Data

International Symposium on Rarefied Gas Dynamics
  (16th:1988:Pasadena, California)
  Rarefied gas dynamics: theoretical and computational techniques/edited by E.P. Muntz, D.P. Weaver, D.H. Campbell.

  p. cm. – (Progress in astronautics and aeronautics; v. 118)
  "Technical papers selected from the Sixteenth International Symposium on Rarefied Gas Dynamics, Pasadena, California, July 10-16, 1988, Subsequently revised for this volume."
Includes index.
1. Rarefied gas dynamics – Congresses. 2. Gases, Kinetic theory of – Congresses. 3. Aerodynamics – Congresses. I. Muntz, E. Phillip (Eric Phillip), 1934- . II. Weaver, D.P. III. Campbell, D.H. (David H.) IV. Title. V. Series.
TL507.P75   vol. 118   89-14995
[QC168.86]
629.1 s – dc20
[533'.2]
ISBN 0-930403-55-X

Copyright © 1989 by the American Institute of Aeronautics and Astronautics, Inc. All rights reserved. Reproduction or translation of any part of this work beyond that permitted by Sections 107 and 108 of the U.S. Copyright Law without the permission of the copyright owner is unlawful. The code following this statement indicates the copyright owner's consent that copies of articles in this volume may be made for personal or internal use, on condition that the copier pay the per-copy fee ($2.00) plus the per-page fee ($0.50) through the Copyright Clearance Center, Inc., 21 Congress Street, Salem, Mass. 01970. This consent does not extent to other kinds of copying, for which permission requests should be addressed to the publisher. Users should employ the following code when reporting copying from this volume to the Copyright Clearance Center:

0-930403-55-X/89 $2.00 + .50

Progress in Astronautics and Aeronautics

Series Editor-in-Chief

Martin Summerfield
Princeton Combustion Research Laboratories, Inc.

Series Editors

A. Richard Seebass
University of Colorado

Allen E. Fuhs
Carmel, California

Assistant Series Editor

Ruth F. Bryans
Ocala, Florida

Norma J. Brennan
Director, Editorial Department
AIAA

Jeanne Godette
Series Managing Editor
AIAA

Symposium Advisory Committee

J. J. Beenakker (Netherlands)
G. A. Bird (Australia)
V. Boffi (Italy)
C. L. Burndin (UK)
R. H. Cabannes (France)
R. Campargue (France)
C. Cercignani (Italy)
J. B. Fenn (USA)
S. S. Fisher (USA)
W. Fiszdon (Poland)
O. F. Hagena (FRG)
F. C. Hurlbut (USA)
M. Kogan (USSR)

G. Koppenwallner (FRG)
I. Kuscer (Yugoslavia)
E. P. Muntz (USA)
R. Narasimha (India)
H. Oguchi (Japan)
D. C. Pack (UK)
J. L. Potter (USA)
A. K. Rebrov (USSR)
Yu. A. Rijov (USSR)
B. Shizgal (Canada)
Y. Sone (Japan)
J. P. Toennies (FRG)
Y. Yoshizawa (Japan)

Technical Reviewers

K. Aoki
G. Arnold
V. Boffi
J. Brook
R. Caflisch
D. H. Campbell
C. Cercignani
H. K. Cheng
N. Corngold
J. Cross
R. Edwards

D. Erwin
W. Fiszdon
A. Frohn
O. Hagena
L. J. F. Hermans
W. L. Hermina
E. L. Knuth
G. Koppenwallner
K. Koura
J. Kunc
H. Legge
E. P. Muntz

K. Nanbu
D. Nelson
J. Oguchi
J. L. Potter
A. Rebrov
Yu. A. Rijov
J. Scott
A. K. Sreekanth
B. Sturtevant
H. T. Yang

Local Organizing Committee

E. P. Muntz (Chairman)
D. P. Weaver
D. H. Campbell
R. Cattolica
H. K. Cheng
D. Erwin
J. Kunc
M. Orme
B. Sturtevant

Symposium Sponsors

Los Alamos National Laboratory
Strategic Defense Initiative Organization
University of Southern California
U. S. Army Research Office
U. S. Air Force Astronautics Laboratory

Table of Contents

Preface xix

Chapter I. Kinetic Theory 1

Well-Posedness of Initial and Boundary Value Problems for the Boltzmann Equation 3
   R. E. Caflisch, *Courant Institute of Mathematical Sciences, New York University, New York, New York*

Stationary Flows from a Model Boltzmann Equation 15
   Y. Y. Azmy and V. Protopopescu, *Oak Ridge National Laboratory, Oak Ridge, Tennessee*

A Tensor Banach Algebra Approach to Abstract Kinetic Equations 29
   W. Greenberg, *Virginia Polytechnic Institute and State University, Blacksburg, Virginia,* and C. V. M. van der Mee, *University of Delaware, Newark, Delaware*

Singular Solutions of the Nonlinear Boltzmann Equation 39
   J. Polewczak, *Virginia Polytechnic Institute and State University, Blacksburg, Virginia*

Spatially Inhomogeneous Nonlinear Dynamics of a Gas Mixture 48
   V. C. Boffi, *University of Bologna, Bologna, Italy,* and G. Spiga, *University of Bari, Bari, Italy*

Diffusion of a Particle in a Very Rarefied Gas 61
   B. Gaveau, *Université Pierre et Marie Curie, Paris, France,* and M.-A. Gaveau, *Centre d'Etudes Nucléaires de Saclay, Gif-sur-Yvette, France*

Heat Transfer and Temperature Distribution in a Rarefied Gas Between Two Parallel Plates with Different Temperatures: Numerical Analysis of the Boltzmann Equation for a Hard Sphere Molecule ... 70
   T. Ohwada, K. Aoki, and Y. Sone, *Kyoto University, Kyoto, Japan*

Chapter II. Discrete Kinetic Theory 83

Low-Discrepancy Method for the Boltzmann Equation............... 85
    H. Babovsky, F. Gropengiesser, H. Neunzert, J. Struckmeier,
    and B. Wiesen, *University of Kaiserslautern, Kaiserslautern, Federal Republic of Germany*

Investigations of the Motion of Discrete-Velocity Gases 100
    D. Goldstein, B. Sturtevant, and J. E. Broadwell, *California Institute of Technology, Pasadena, California*

Discrete Kinetic Theory with Multiple Collisions:
Plane Six-Velocity Model and Unsteady Couette Flow 118
    E. Longo and R. Monaco, *Politecnico di Torino, Torino, Italy*

Exact Positive (2 + 1)-Dimensional Solutions
to the Discrete Boltzmann Models............................ 131
    H. Cornille, *Physique Théorique CEN-CEA Saclay, Gif-sur Yvette, France*

Initial-Value Problem in Discrete Kinetic Theory.................. 148
    S. Kawashima, *Kyushu University, Fukuoka, Japan,* and H. Cabannes, *Université Pierre et Marie Curie, Paris, France*

Study of a Multispeed Cellular Automaton....................... 155
    B. T. Nadiga, J. E. Broadwell, and B. Sturtevant, *California Institute of Technology, Pasadena, California*

Direct Statistical Simulation Method and Master
Kinetic Equation.. 171
    M. S. Ivanov, S. V. Rogasinsky, and V. Ya. Rudyak, *USSR Academy of Sciences, Novosibirsk, USSR*

Fractal Dimension of Particle Trajectories in Ehrenfest's
Wind-Tree Model 182
    P. Mausbach, *Rheinisch Westfälische Technische Hochschule Aachen, Aachen, Federal Republic of Germany,* and *Eckard Design GmbH, Cologne, Federal Republic of Germany*

Scaling Rules and Time Averaging in Molecular Dynamics
Computations of Transport Properties......................... 194
    I. Greber, *Case Western Reserve University, Cleveland, Ohio,* and H. Wachman, *Massachusetts Institute of Technology, Cambridge, Massachusetts*

Chapter III. Direct Simulations.......................... 209

Perception of Numerical Methods in Rarefied Gasdynamics 211
    G. A. Bird, *University of Sydney, Sydney, New South Wales, Australia*

Comparison of Parallel Algorithms for the Direct Simulation
Monte Carlo Method: Application to Exhaust Plume Flowfields...227
   T. R. Furlani and J. A. Lordi, *Calspan Advanced Technology Center,
Buffalo, New York*

Statistical Fluctuations in Monte Carlo Calculations 245
   I. D. Boyd and J. P. W. Stark, *University of Southampton,
Southampton, England, United Kingdom*

Applicability of the Direct Simulation Monte Carlo Method
in a Body-fitted Coordinate System 258
   T. Shimada, *Nissan Motor Company, Ltd., Tokyo, Japan,* and T. Abe,
Institute of Space and Astronautical Science, Kanagawa, Japan

Validation of MCDS by Comparison of Predicted with Experimental
Velocity Distribution Functions in Rarefied Normal Shocks 271
   G. C. Pham-Van-Diep and D. A. Erwin, *University of Southern California,
Los Angeles, California*

Direct Monte Carlo Calculations on Expansion Wave Structure
Near a Wall..284
   F. Seiler, *Deutsch-Französisches Forschungsinstitut Saint-Louis (ISL),
Saint-Louis, France,* and B. Schmidt, *University of Karlsruhe, Karlsruhe,
Federal Republic of Germany*

Chapter IV. Numerical Techniques......................295

Numerical Analysis of Rarefied Gas Flows
by Finite-Difference Method...................................297
   K. Aoki, *Kyoto University, Kyoto, Japan*

Application of Monte Carlo Methods
to Near-Equilibrium Problems.................................323
   S. M. Yen, *University of Illinois at Urbana-Champaign, Urbana, Illinois*

Direct Numerical Solution of the Boltzmann Equation
for Complex Gas Flow Problems..............................337
   S. M. Yen and K. D. Lee, *University of Illinois at Urbana-Champaign,
Urbana, Illinois*

Advancement of the Method of Direct Numerical Solving
of the Boltzmann Equation....................................343
   F. G. Tcheremissine, *USSR Academy of Sciences, Moscow, USSR*

New Numerical Strategy to Evaluate the Collision Integral
of the Boltzmann Equation....................................359
   Z. Tan, Y.-K. Chen, P. L. Varghese, and J. R. Howell, *The University
of Texas at Austin, Austin, Texas*

Comparison of Burnett, Super-Burnett, and Monte Carlo Solutions
for Hypersonic Shock Structure................................374
   K. A. Fiscko, *U. S. Army* and *Stanford University,* and D. R. Chapman,
   Stanford University, Stanford, California

Density Profiles and Entropy Production in Cylindrical Couette Flow:
Comparison of Generalized Hydrodynamics
and Monte Carlo Results..396
   R. E. Khayat and B. C. Eu, *McGill University, Montreal, Quebec, Canada*

Chapter V.   Flowfields...................................411

Direct Simulation of AFE Forebody and Wake Flow
with Thermal Radiation...413
   J. N. Moss and J. M. Price, *NASA Langley Research Center,*
   Hampton, Virginia

Direct Monte Carlo Simulations of Hypersonic Flows
Past Blunt Bodies..432
   W. Wetzel and H. Oertel, *DFVLR/AVA, Göttingen, Federal*
   Republic of Germany

Direct Simulation of Three-Dimensional Flow
About the AFE Vehicle at High Altitudes 447
   M. C. Celenligil, *Vigyan Research Associates, Inc., Hampton, Virginia,*
   J. N. Moss, *NASA Langley Research Center, Hampton, Virginia,*
   and G. A. Bird, *University of Sydney, Sydney, New South*
   Wales, Australia

Knudsen-Layer Properties for a Conical Afterbody
in Rarefied Hypersonic Flow 462
   G. T. Chrusciel and L. A. Pool, *Lockheed Missiles and Space*
   Company, Inc., Sunnyvale, California

Approximate Calculation of Rarefied Aerodynamic Characteristics
of Convex Axisymmetric Configurations 476
   Y. Xie and Z. Tang, *China Aerodynamics Research and Development*
   Centre, Mianyang, Sichuan, People's Republic of China

Procedure for Estimating Aerodynamics of Three-Dimensional Bodies
in Transitional Flow...484
   J. L. Potter, *Vanderbilt University, Nashville, Tennessee*

Drag and Lift Measurements on Inclined Cones
Using a Magnetic Suspension and Balance........................493
   R. W. Smith and R. G. Lord, *Oxford University, Oxford, England,*
   United Kingdom

**Three-Dimensional Hypersonic Flow Around a Disk
with Angle of Attack** 500
   K. Nanbu, S. Igarashi, and Y. Watanabe, *Tohoku University, Sendai,
Japan,* and H. Legge and G. Koppenwallner, *DFVLR, Göttingen,
Federal Republic of Germany*

**Direct Simulation Monte Carlo Method
of Shock Reflection on a Wedge** 518
   F. Seiler, *Deutsch-Französisches Forschungsinstitut Saint-Louis (ISL),
Saint-Louis, France,* H. Oertel, *DFVLR-AVA, Göttingen, Federal
Republic of Germany,* and B. Schmidt, *University of Karlsruhe,
Karlsruhe, Federal Republic of Germany*

**Interference Effects on the Hypersonic, Rarefied Flow
About a Flat Plate** 532
   R. G. Wilmoth, *NASA Langley Research Center, Hampton, Virginia*

**Numerical Simulation of Supersonic Rarefied Gas Flows
Past a Flat Plate: Effects of the Gas-Surface Interaction Model
on the Flowfield** 552
   C. Cercignani and A. Frezzotti, *Politecnico di Milano, Milano, Italy*

Rarefied Flow Past a Flat Plate at Incidence. 567
   V. K. Dogra, *ViRA, Inc., Hampton, Virginia,* and J. N. Moss and
J. M. Price, *NASA Langley Research Center, Hampton, Virginia*

**Monte Carlo Simulation of Flow into Channel
with Sharp Leading Edge** 582
   M. Yasuhara, Y. Nakamura, and J. Tanaka, *Nagoya University,
Nagoya, Japan*

**Structure of Incipient Triple Point at the Transition
from Regular Reflection to Mach Reflection** 597
   B. Schmidt, *University of Karlsruhe, Karlsruhe, Federal
Republic of Germany*

Author Index for Volume 118 608

List of Series Volumes 609

Other Volumes in the Rarefied Gas Dynamics Series 616

Table of Contents for Companion Volume 116

Preface xix

Chapter I. Rarefied Atmospheres 1

**Nonequilibrium Nature of Ion Distribution Functions
in the High Latitude Auroral Ionosphere** 3
    B. Shizgal, *University of British Columbia, Vancouver,
British Columbia, Canada*, and D. Hubert, *Observatoire de Meudon, Meudon, France*

**VEGA Spacecraft Aerodynamics in the Gas-Dust Rarefied
Atmosphere of Halley's Comet** 23
    Yu. A. Rijov, S. B. Svirschevsky, and K. N. Kuzovkin, *Moscow
Aviation Institute, Moscow, USSR*

**Oscillations of a Tethered Satellite of Small Mass
due to Aerodynamic Drag** 40
    E. M. Shakhov, *USSR Academy of Sciences, Moscow, USSR*

Chapter II. Plasmas 53

Semiclassical Approach to Atomic and Molecular Interactions 55
    J. A. Kunc, *University of Southern California, Los Angeles, California*

**Monte Carlo Simulation of Electron Swarm
in a Strong Magnetic Field** 76
    K. Koura, *National Aerospace Laboratory, Chofu, Tokyo, Japan*

**Collisional Transport in Magnetoplasmas in the Presence
of Differential Rotation** 89
    M. Tessarotto, *Università degli Studi di Trieste, Trieste, Italy*, and
P. J. Catto, *Lodestar Research Corporation, Boulder, Colorado*

**Electron Oscillations, Landau, and Collisional Damping
in a Partially Ionized Plasma** 102
    V. G. Molinari and M. Sumini, *Università di Bologna, Bologna, Italy*,
and B. D. Ganapol, *University of Arizona, Tucson, Arizona*

**Bifurcating Families of Periodic Traveling Waves
in Rarefied Plasmas** 115
    J. P. Holloway and J. J. Dorning, *University of Virginia,
Charlottesville, Virginia*

Chapter III. Atomic Oxygen Generation and Effects 127

**Laboratory Simulations of Energetic Atom Interactions
Occurring in Low Earth Orbit** 129
    G. E. Caledonia, *Physical Sciences, Inc., Andover, Massachusetts*

High-Energy/Intensity CW Atomic Oxygen Beam Source 143
    J. B. Cross and N. C. Blais, *Los Alamos National Laboratory,*
    Los Alamos, New Mexico

Development of Low-Power, High Velocity Atomic
Oxygen Source 156
    J. P. W. Stark and M. A. Kinnersley, *University of Southampton,*
    Southampton, England, United Kingdom

Options for Generating Greater Than 5-eV Atmospheric Species 171
    H. O. Moser, *Kernforschungszentrum Karlsruhe, Karlsruhe, Federal*
    Republic of Germany, and A. Schempp, *University of Frankfurt,*
    Frankfurt, Federal Republic of Germany

Laboratory Results for 5-eV Oxygen Atoms on Selected
Spacecraft Materials 180
    G. W. Sjolander and J. F. Froechtenigt, *Martin Marietta Corporation,*
    Denver, Colorado

Chapter IV.   Plumes 187

Modeling Free Molecular Plume Flow and Impingement
by an Ellipsoidal Distribution Function 189
    H. Legge, *DFVLR, Göttingen, Federal Republic of Germany*

Plume Shape Optimization of Small Attitude Control Thrusters   Concerning Impingement and
Thrust 204
    K. W. Naumann, *Franco-German Research Institute of Saint-Louis (ISL),*
    Saint-Louis, France

Backscatter Contamination Analysis 216
    B. C. Moore, T. S. Mogstad, S. L. Huston, and J. L. Nardacci Jr.,
    McDonnell Douglas Space Systems Company, Huntington Beach, California

Thruster Plume Impingement Forces Measured in a Vacuum
Chamber and Conversion to Real Flight Conditions 226
    A. W. Rogers, *Hughes Aircraft Company, El Segundo, California,*
    J. Allègre and M. Raffin, *Société d'Etudes et de Services pour Souffleries et Installations*
    Aérothermodynamiques (SESSIA), Levallois-Perret, France, and J.-C. Lengrand, *Laboratoire*
    d'Aérothermique du Centre National de la Recherche Scientifique, Meudon, France

Neutralization of a 50-MeV H⁻Beam Using the Ring Nozzle 241
    N. S. Youssef and J. W. Brook, *Grumman Corporation, Bethpage,*
    New York

Chapter V.   Tube Flow 255

Rarefied Gas Flow Through Rectangular Tubes: Experimental
and Numerical Investigation 257
    A. K. Sreekanth and A. Davis, *Indian Institute of Technology,*
    Madras, India

Experimental Investigation of Rarefied Flow Through Tubes
of Various Surface Properties 273
    J. Curtis, *University of Sydney, Sydney, New South Wales, Australia*

Monte Carlo Simulation on Mass Flow Reduction
due to Roughness of a Slit Surface 283
M. Usami, T. Fujimoto, and S. Kato, *Mie University, Kamihama-cho, Tsu-shi, Japan*

Chapter VI. Expansion Flowfields 299

**Translational Nonequilibrium Effects in Expansion Flows
of Argon** 301
D. H. Campbell, *University of Dayton Research Institute, Air Force
Astronautics Laboratory, Edwards Air Force Base, California*

Three-Dimensional Freejet Flow from a Finite Length Slit 312
A. Rosengard, *Commissariat à l'Energie Atomique, Centre d'Etudes
Nucléaires de Saclay, France*

**Modification of the Simons Model for Calculation
of Nonradial Expansion Plumes** 327
I. D. Boyd and J. P. W. Stark, *University of Southampton, Southampton,
England, United Kingdom*

Simulation of Multicomponent Nozzle Flows into a Vacuum 340
D. A. Nelson and Y. C. Doo, *The Aerospace Corporation, El Segundo, California*

**Kinetic Theory Model for the Flow of a Simple Gas
from a Two-Dimensional Nozzle** 352
B. R. Riley, *University of Evansville, Evansville, Indiana,* and
K. W. Scheller, *University of Notre Dame, Notre Dame, Indiana*

**Transient and Steady Inertially Tethered Clouds of Gas
in a Vacuum** 363
T. L. Farnham and E. P. Muntz, *University of Southern California,
Los Angeles, California*

**Radially Directed Underexpanded Jet
from a Ring-Shaped Nozzle** 378
K. Teshima, *Kyoto University of Education, Kyoto, Japan*

Three-Dimensional Structures of Interacting Freejets 391
T. Fujimoto and T. Ni-Imi, *Nagoya University, Furo-cho, Chikusa-ku, Nagoya, Japan*

**Flow of a Freejet into a Circular Orifice
in a Perpendicular Wall** 407
A. M. Bishaev, E. F. Limar, S. P. Popov, and E. M. Shakhov,
USSR Academy of Sciences, Moscow, USSR

Chapter VII. Surface Interactions 417

Particle Surface Interaction in the Orbital Context: A Survey 419
F. C. Hurlbut, *University of California at Berkeley, Berkeley, California*

**Sensitivity of Energy Accommodation Modeling of Rarefied Flow
Over Re-Entry Vehicle Geometries Using DSMC** 451
T. J. Bartel, *Sandia National Laboratories, Albuquerque, New Mexico*

**Determination of Momentum Accommodation
from Satellite Orbits: An Alternative Set of Coefficients** 463
R. Crowther and J. P. W. Stark, *University of Southampton,
Southampton, England, United Kingdom*

Upper Atmosphere Aerodynamics: Gas-Surface Interaction
and Comparison with Wind-Tunnel Experiments 476
  M. Pandolfi and M. G. Zavattaro, *Politecnico di Torino, Torino, Italy*

Nonreciprocity in Noble-Gas Metal-Surface Scattering 487
  K. Bärwinkel and S. Schippers, *University of Osnabrück, Osnabrück,
  Federal Republic of Germany*

Studies of Thermal Accommodation and Conduction
in the Transition Regime 502
  L. B. Thomas, C. L. Krueger, and S. K. Loyalka, *University of Missouri,
  Columbia, Missouri*

Large Rotational Polarization Observed in a Knudsen Flow
of H_2-Isotopes Between LiF Surfaces 517
  L. J. F. Hermans and R. Horne, *Leiden University, Leiden,
  The Netherlands*

Internal State-Dependent Molecule-Surface Interaction
Investigated by Surface Light-Induced Drift 530
  R. W. M. Hoogeveen, R. J. C. Spreeuw, G. J. van der Mee, and L. J. F.
  Hermans, *Leiden University, Leiden, The Netherlands*

Models for Temperature Jumps in Vibrationally
Relaxing Gases 542
  R. Brun, S. Elkeslassy, and I. Chemouni, *Université de Provence-Centre
  Saint Jérôme, Marseille, France*

Variational Calculation of the Slip Coefficient and the Temperature
Jump for Arbitrary Gas-Surface Interactions 553
  C. Cercignani, *Politecnico di Milano, Milano, Italy,* and M. Lampis,
  Università di Udine, Udine, Italy

Author Index for Volume 116 562

List of Series Volumes 563

Other Volumes in the Rarefied Gas Dynamics Series 570

Table of Contents for Companion Volume 117

Preface xvii

Chapter I. Inelastic Collisions 1

Inelastic Collision Models for Monte Carlo
  Simulation Computation 3
    J. K. Harvey, *Imperial College, London, England, United Kingdom*

Null Collision Monte Carlo Method: Gas Mixtures with Internal
  Degrees of Freedom and Chemical Reactions 25
    K. Koura, *National Aerospace Laboratory, Chofu, Tokyo, Japan*

Nitrogen Rotation Relaxation Time Measured in Freejets 40
    A. E. Belikov, G. I. Sukhinin, and R. G. Sharafutdinov, *Siberian Branch of the USSR Academy of Sciences, Novosibirsk, USSR*

Rate Constants for R-T Relaxation of N_2
  in Argon Supersonic Jets 52
    A. E. Belikov, G. I. Sukhinin, and R. G. Sharafutdinov, *Siberian Branch of the USSR Academy of Sciences, Novosibirsk, USSR*

Rotational Relaxation of CO and CO_2 in Freejets
  of Gas Mixtures 68
    T. Kodama, S. Shen, and J. B. Fenn, *Yale University, New Haven, Connecticut*

Diffusion and Energy Transfer in Gases
  Containing Carbon Dioxide 76
    J. R. Ferron, *University of Rochester, Rochester, New York*

Freejet Expansion of Heavy Hydrocarbon Vapor 92
    A. V. Bulgakov, V. G. Prikhodko, A. K. Rebrov, and P. A. Skovorodko, *Siberian Branch of the USSR Academy of Sciences, Novosibirsk, USSR*

Chapter II. Experimental Techniques 105

Optical Diagnostics of Low-Density Flowfields 107
    J. W. L. Lewis, *University of Tennessee Space Institute, Tullahoma, Tennessee*

Electron Beam Flourescence Measurements of Nitric Oxide 133
    R. J. Cattolica, *Sandia National Laboratories, Livermore, California*

Measurements of Freejet Densities by Laser Beam Deviation 140
    J. C. Mombo-Caristan, L. C. Philippe, C. Chidiac, M. Y. Perrin,
      and J. P. Martin, *Laboratoire d'Energetique Moleculaire et Macroscopic Combustion du Centre Nationale de la Recherche Scientifique, Chatenay Malabry, France* and *Ecole Centrale Paris, Chatenay Malabry, France*

**Turbulence Measurement of a Low-Density Supersonic Jet
with a Laser-Induced Fluorescence Method** 149
    M. Masuda, H. Nakamuta, Y. Matsumoto, K. Matsuo, and M. Akazaki,
    Kyushu University, Fukuoka, Japan

**Measurement of Aerodynamic Heat Rates by Infrared
Thermographic Technique at Rarefied Flow Conditions** 157
    J. Allègre, X. Hériard Dubreuilh, and M. Raffin, *Société d'Etudes
    et de Services pour Souffleries et Installations Aérothermodynamiques
    (SESSIA), Meudon, France*

**Experimental Investigation of CO_2 and N_2O Jets
Using Intracavity Laser Scattering** 168
    R. G. Schabram, A. E. Beylich, and E. M. Kudriavtsev, *Stosswellenlabor,
    Technische Hochschule, Aachen, Federal Republic of Germany*

**High-Speed-Ratio Helium Beams: Improving Time-of-Flight
Calibration and Resolution** 187
    R. B. Doak and D. B. Nguyen, *AT&T Bell Laboratories, Murray Hill,
    New Jersey*

Velocity Distribution Function in Nozzle Beams 206
    O. F. Hagena, *Kernforschungszentrum Karlsruhe, Karlsruhe,
    Federal Republic of Germany*

**Cryogenic Pumping Speed for a Freejet
in the Scattering Regime** 218
    J.-Th. Meyer, *DFVLR, Göttingen, Federal Republic of Germany*

**Effectiveness of a Parallel Plate Arrangement
as a Cryogenic Pumping Device** 233
    K. Nanbu, Y. Watanabe, and S. Igarashi, *Tohoku University, Sendai,
    Japan*, and G. Dettleff and G. Koppenwallner, *German Aerospace
    Research Establishment, Göttingen, Federal Republic of Germany*

Chapter III.   Particle and Mixture Flows 245

**Aerodynamic Focusing of Particles and Molecules
in Seeded Supersonic Jets** 247
    J. Fernández de la Mora, J. Rosell-Llompart, and P. Riesco-Chueca,
    Yale University, New Haven, Connecticut

**Experimental Investigations of Aerodynamic Separation of Isotopes
and Gases in a Separation Nozzle Cascade** 278
    P. Bley and H. Hein, *Kernforschungszentrum Karlsruhe, Karlsruhe,
    Federal Republic of Germany*, and J. L. Campos, R. V. Consiglio,
    and J. S. Coelho, *Centro de Desenvolvimento da Tecnologia Nuclear,
    Belo Horizonte, Brazil*

General Principles of the Inertial Gas Mixture Separation 290
    B. L. Paklin and A. K. Rebrov, *Siberian Branch of the USSR Academy
    of Sciences, Novosibirsk, USSR*

Motion of a Knudsen Particle Through a Shock Wave 298
    M. M. R. Williams, *University of Michigan, Ann Arbor, Michigan*

Method of Characteristics Description of Brownian Motion
Far from Equilibrium 311
   P. Riesco-Chueca, R. Fernández-Feria, and J. Fernández de la Mora,
   Yale University, New Haven, Connecticut

Chapter IV.   Clusters 327

Phase-Diagram Considerations of Cluster Formation When Using Nozzle-Beam Sources 329
   E. L. Knuth and W. Li, University of California, Los Angeles, California,
   and J. P. Toennies, Max-Planck-Institut für Strömungsforschung,
   Göttingen, Federal Republic of Germany

Fragmentation of Charged Clusters During Collisions
of Water Clusters with Electrons and Surfaces 335
   A. A. Vostrikov, D. Yu. Dubov, and V. P. Gilyova, Siberian Branch
   of the USSR Academy of Sciences, Novosibirsk, USSR

Homogeneous Condensation in H_2O - Vapor Freejets 354
   C. Dankert and H. Legge, DFVLR Institute for Experimental Fluid
   Mechanics, Göttingen, Federal Republic of Germany

Formation of Ion Clusters in High-Speed Supersaturated
CO_2 Gas Flows 366
   P. J. Wantuck, Los Alamos National Laboratory, Los Alamos, New
   Mexico, and R. H. Krauss and J. E. Scott Jr., University of Virginia,
   Charlottesville, Virginia

MD-Study of Dynamic-Statistic Properties of Small Clusters 381
   S. F. Chekmarev and F. S. Liu, Siberian Branch of the USSR Academy
   of Sciences, Novosibirsk, USSR

Chapter V.   Evaporation and Condensation 401

Angular Distributions of Molecular Flux Effusing from a Cylindrical
Crucible Partially Filled with Liquid 403
   Y. Watanabe, K. Nanbu, and S. Igarashi, Tohoku University, Sendai, Japan

Numerical Studies on Evaporation and Deposition of a Rarefied Gas
in a Closed Chamber 418
   T. Inamuro, Mitsubishi Heavy Industries, Ltd., Yokohama, Japan

Transition Regime Droplet Growth and Evaporation:
An Integrodifferential Variational Approach 434
   J. W. Cipolla Jr., Northeastern University, Boston, Massachusetts,
   and S. K. Loyalka, University of Missouri-Columbia, Columbia, Missouri

Molecular Dynamics Studies on Condensation Process of Argon 439
   T. Sano, Tokai University, Kitakaname, Hiratsuka, Kanagawa, Japan,
   and S. Kotake, University of Tokyo, Hongo, Bunkyo-ku, Tokyo, Japan

Condensation and Evaporation of a Spherical Droplet
in the Near Free Molecule Regime 447
   J. C. Barrett and B. Shizgal, University of British Columbia, Vancouver,
   British Columbia, Canada

**Theoretical and Experimental Investigation of the Strong
Evaporation of Solids** 460
    R. Mager, G. Adomeit, and G. Wortberg, *Rheinisch-Westfälische
Technische Hochschule Aachen, Aachen, Federal Republic of Germany*

**Nonlinear Analysis for Evaporation and Condensation of a Vapor-Gas
Mixture Between the Two Plane Condensed Phases.
Part I: Concentration of Inert Gas $\sim O(1)$** 470
    Y. Onishi, *Tottori University, Tottori, Japan*

**Nonlinear Analysis for Evaporation and Condensation of a Vapor-Gas
Mixture Between the Two Plane Condensed Phases.
Part II: Concentration of Inert Gas $\sim O(Kn)$** 492
    Y. Onishi, *Tottori University, Tottori, Japan*

Author Index for Volume 117 514

List of Series Volumes 515

Other Volumes in the Rarefied Gas Dynamics Series 522

Preface

The 16th International Symposium on Rarefied Gas Dynamics (RGD 16) was held July 10–16, 1988 at the Pasadena Convention Center, Pasadena, CA. As anticipated, the resurging interest in hypersonic flight, along with escalating space operations, has resulted in a marked increase in attention being given to phenomena associated with rarefied gas dynamics. One hundred and seventy-three registrants from thirteen countries attended. Spirited technical exchanges were generated in several areas. The Direct Simulation Monte Carlo technique and topics in discrete kinetic theory techniques were popular subjects. The inclusion of inelastic collisions and chemistry in the DSMC technique drew attention. The Boltzmann Monte Carlo technique was extended to more complex flows; an international users group in this area was formed as a result of the meeting.

Space-related research was discussed at RGD 16 to a greater extent than at previous meetings. As a result of Shuttle glow phenomena and oxygen-atom erosion of spacecraft materials, the subject of energetic collisions of gases with surfaces in low Earth orbit made a strong comeback.

There were 11 excellent invited lectures on a variety of subjects that added significantly to the exchange of ideas at the symposium.

The Symposium Proceedings have, for the first time, been divided into three volumes. A very high percentage of the papers presented at the symposium were submitted for publication (practically all in a timely manner). Because of the large number of high-quality papers, it was deemed appropriate to publish three volumes rather than the usual two. An additional attraction is that the size of each volume is a little more convenient. Papers were initially reviewed for technical content by the session chairmen at the meeting and further reviewed by the proceedings' editors and additional experts as necessary.

In the proceedings of RGD 16, there are 107 contributed papers and 11 extended invited papers dealing with the kinetic theory of gas flows and transport phenomena, external and internal rarefied gas flows, chemical and internal degree-of-freedom relaxation in gas flows, partially ionized plasmas, Monte Carlo simulations of rarefied flows, development of high-speed atmospheric simulators, surface-interaction phenomena, aerosols and clusters, condensation and evaporation phenomena, rocket-plume flows, and experimental techniques for rarefied gas dynamics.

The Rarefied Gas Dynamics Symposia have a well-deserved reputation for being hospitable events. RGD 16 continued this admirable tradition. It is with much appreciation that we acknowledge Jan Muntz's contributions to RGD history along with Jetty and Miller Fong and Noel Corngold.

Many people helped make the symposium a success. Gail Dwinell supervised organizational details for the symposium as well as the pre-symposium correspondence. Eric Muntz was responsible for the computerized registration. Kim Palos, whose services were provided by the University of Dayton Research Institute, assisted with registration and retyped several manuscripts. Jerome Maes also retyped a number of manuscripts for these volumes. Nancy Renick and Marilyn Litvak of the Travel Arrangers of Pasadena were tireless in their efforts to assist the delegates.

E. P. Muntz
D. P. Weaver
D. H. Campbell
May 1989

Chapter 1. Kinetic Theory

Well-Posedness of Initial and Boundary Value Problems for the Boltzmann Equation

R. E. Caflisch*
Courant Institute of Mathematical Sciences, New York University, New York, New York

ABSTRACT

Existence results for the Boltzmann equation are reviewed and discussed. These include the spatially homogeneous problem [Ackeryd (1972), Elimroth (1984)], steady flow outside a planar boundary with prescribed flux on the boundary, [Coron, Golse and Sulem (1988)], weak shock waves [Caflisch and Nicolaenko (1982)], and a recent result on global existence (but not uniqueness) for initial data of arbitrary size [DiPerna and Lions (1988)]. A convergence result for the modified Nanbu (Monte Carlo) method is also presented [Babovsky and Illner (1988)].

I. Introduction

Mathematical analysis of solutions for the nonlinear Boltzmann equation has proved to be difficult because of the nonlinearity, singular spatial dependence, and many degrees of freedom in the equation. The focus of this review paper is on existence theory for the nonlinear Boltzmann equation for a monatomic gas. We shall describe fundamental existence results on the spatially homogeneous problem (Section II), steady flow with boundaries (Section III) and shock waves (Section IV). A convergence result for a Monte Carlo numerical method is presented in Section V. Although not an existence result, this is of basic importance for analysis of computational methods. Finally, in Section VI, we discuss an important recent result of DiPerna and Lions,[1] which establishes

---

Presented as an Invited Paper.
Copyright © 1989 by the American Institute of Aeronautics and Astronautics, Inc. All rights reserved.
*Professor, Mathematics Department.

global existence (but not uniqueness) for solutions of the Boltzmann equation for initial data of any size.

The emphasis in this presentation will be on physical interpretation and scientific significance of the mathematical results and methods rather than on mathematical details. Mathematical analysis requires careful attention to possible degeneracies in the solution, such as blowup, loss of smoothness, loss of energy, or nonequilibration. Even if such degeneracies do not occur, attention to such details may point to possible sources of error and instabilities in numerical methods, and thus aid in the design and evolution of numerical methods.

II. Spatially Homogeneous Boltzmann Equation

The nonlinear Boltzmann equation for a monatomic gas in

$$(\partial_t + \underset{\sim}{\xi} \cdot \partial_x) F = Q(F, F) \tag{1}$$

in which $t \in [0, \infty]$ is time, $\underset{\sim}{x} \in \mathbf{R}^3$ is space, $\underset{\sim}{\xi} \in \mathbf{R}^3$ is velocity, and $F = F(\underset{\sim}{x}, \underset{\sim}{\xi}, t)$ is the distribution function. The collision operator is $Q(F, F)$. The spatially homogeneous Boltzmann equation is

$$\partial_t F = Q(F, F) \tag{2}$$

Initial data

$$F(t = 0) = F_0 \tag{3}$$

should also be posed. Besides existence of solutions, equilibration of the solution as $t \to \infty$ is important. Strong existence and equilibration results have been established by Arkeryd and Elmroth:[4]

Theorem 1.1 : Assume that $F_0 \geq 0$ and

$$\int F_0 (1 + |\underset{\sim}{\xi}|^4 + |\log F_0|) d\underset{\sim}{\xi} < \infty \tag{4}$$

Then, there is a unique solution $F(\underset{\sim}{\xi}, t)$ of Eqs. (2) and (3), with $F \geq 0$, $F \in L^1(\underset{\sim}{\xi})$ for all t. In addition, mass, momentum, and energy are conserved and the H integral is decreasing; i.e., for all $T > 0$,

$$\rho = \int F d\underset{\sim}{\xi} = \int F_0 d\underset{\sim}{\xi}$$

$$\rho \underset{\sim}{u} = \int \underset{\sim}{\xi} F d\underset{\sim}{\xi} = \int \underset{\sim}{\xi} F_0 d\underset{\sim}{\xi} \tag{5}$$

$$3\rho T = \int (\underset{\sim}{\xi} - \underset{\sim}{u})^2 F d\underset{\sim}{\xi} = \int (\underset{\sim}{\xi} - \underset{\sim}{u})^2 F_0 d\underset{\sim}{\xi}$$

INITIAL AND BOUNDARY VALUE PROBLEMS

$$\int F \log F \, d\underline{\xi} \leq \int F_0 \log F_0 \, d\underline{\xi} \tag{6}$$

As $t \to \infty$, F approaches the Maxwellian distribution

$$M = \rho(2\pi T)^{-3/2} \exp(-(\underline{\xi} - \underline{u})^2 / 2T)$$

in $L^1(\underline{\xi})$, i.e.,

$$\int (F - M)^2 \, d\underline{\xi} \to 0 \text{ as } t \to \infty \tag{7}$$

The proof of this result depends on two facts: First for such an existence proof, the solution F is found as a limit of approximate solutions. Convergence of the limit is guaranteed by the following lemma.

<u>Lemma 2.2</u> : If F_n is a sequence of functions in $L^1(\underline{\xi})$ with $\int F_n(1 + |\underline{\xi}|^2 + |\log F_n|) d\underline{\xi} < c$ independent of n, then F_n is in a weakly compact set in L^1; i.e., there is a subsequence F_{n_i} such that for any $\psi \in L^\infty(\underline{\xi})$, $\int \psi F_{n_i} d\underline{\xi} \to \int \psi F d\underline{\xi}$ as $i \to \infty$. The function F is the weak limit of F_{n_i}.

Second, the proof of the asymptotic approach to a Maxwellilan equilibrium relies on clever use of the fact that M mimizes the H integral over all non-negative functions satisfying Eq. (5).

This result, and improvements on it is Arkeryd[5] and Gustaffson[6] provides a fairly satisfactory mathematical analysis of the spatially homogeneous problem. However, one important potential application of this result to analysis of initial layers in the fluid dynamic limit of the Boltzmann equation may require solutions in function spaces that are stronger than L^1.

III. Boundary Value Problems for Steady Flows

Boundary value problems for steady flows are much more difficult than the spatially homogeneous problem. A number of interesting linear and nonlinear boundary value problems have been analyzed, for example, in Ukai and Asano[7,8]. I will discuss only the linearized planar boundary-layer problem with prescribed incoming flux on the boundary.

The planar linearized steady Boltzmann equation is

$$\xi_x \partial_x f = Lf, \quad 0 < x < \infty \tag{8}$$

in which ξ_x is the x component of the velocity, and the boundary condition at $x = 0$ is

$$f(0, \underline{\xi}) = g(\underline{\xi}) \quad \text{for } \xi_x > 0 \tag{9}$$

In Eq. (8), f is the perturbation from a prescribed Maxwellian distribution, i.e., $F = M + f$, and the linearized collision operator is $Lf = 2Q(M,f)$. As $x \to \infty$, we expect f to approach a linearized equilibrium, i.e.

$$f \to (a + \underline{b}\cdot\underline{\xi} + d\xi^2)M \qquad (10)$$

Problem (8-10) has been solved by Coron, Galse and Sulem[9] following earlier work of Bardos, Caflisch, and Nicolaenko[10] and Cercignani[11]. The well-posedness of this problem depends on the Mach number $m = u/c = u/\sqrt{5T/3}$, in which u and T are the velocity and temperature corresponding to the Maxwellian distribution M. In order to describe this dependence, define the following parameters:

$$\lambda_0 = \int (\xi_y/\sqrt{T})(\xi_x - u)fd\underline{\xi}$$

$$\lambda_1 = \int (\xi_z/\sqrt{T})(\xi_x - u)fd\underline{\xi}$$

$$\lambda_2 = \int (\sqrt{5/2} - |\xi|^2/\sqrt{10}T)(\xi_x - u)fd\underline{\xi}$$

$$\lambda_3 = \int (\xi_x/\sqrt{2T} + |\xi|^2/\sqrt{30}T)(\xi_x - u)fd\underline{\xi}$$

$$\lambda_4 = \int (-\xi_x/\sqrt{2T} + |\xi|^2/\sqrt{30}T)(\xi_x - u)fd\underline{\xi} \qquad (11)$$

These parameters are independent of x, and they are linear combinations of the parameters a, b_1, b_2, b_3, and d in Eq. (10). The theorem of Coron, Golse, and Sulem[9] is the following:

<u>Theorem 3.1</u> : For arbitrary g, problem (8-10) has a unique solution f if, in addition, the following parameters are prescribed:

1) For $c \leq u$, no parameters are prescribed
2) For $0 \leq u < c$, λ_4 is arbitrarily prescribed
3) For $-c \leq u < 0$, $\lambda_0, \lambda_1, \lambda_2, \lambda_4$ are arbitrarily prescribed.
4) For $u < -c$, $\lambda_0, \lambda_1, \lambda_2, \lambda_3, \lambda_4$ are arbitrarily prescribed.

The solution f approaches the linearized Maxwellian in formula (10) at an exponential rate. In addition,

5) For $u = c$, $\lambda_4 = 0$.
6) For $u = 0$, $\lambda_0 = \lambda_1 = \lambda_2 = 0$.
7) For $u = -c$, $\lambda_3 = 0$.

Note

1) This result has been established only for a hard-sphere gas. For other intermolecular force laws, the equilibration at $x = \infty$ should be slower than exponential.

2) For the BGK model, this result had been derived earlier.

3) The nonlinear problem has not been analyzed. The main difficulty is that linearization needs to be performed about the correct Maxwellian distribution at infinity, which is not exactly M. M must be modified as in formula (10).

IV. Shock Waves

Shock waves are an important example of nonlinear and nonequilibrium behavior in kinetic theory. A steady planar shock wave is a smooth connection between two Maxwellian equilibrium states, which translates steadily and depends on only one space dimension. The solution of Eq. (1) is sought in the form $F = F(x - st, \xi)$, which satisfies

$$(\xi_x - s)\partial_x F = Q(F,F) \quad -\infty < x < \infty \quad (12)$$

with boundary conditions

$$F(x = \infty, \underline{\xi}) = M_+(\underline{\xi})$$

$$= \rho_+(2\pi T_+)^{-3/2} exp(-|\underline{\xi} - \underline{u}_+|^2/2T_+)$$

$$F(x = -\infty, \underline{\xi}) = M_-(\underline{\xi})$$

$$= \rho_-(2\pi T_-)^{-3/2} exp(-|\underline{\xi} - \underline{u}_-|^2/2T_-) \quad (13)$$

Conservation laws and the inequality for the H integral lead to consistency conditions on the choice of the asymptotic states M_+ and M_- and the shock speed s. These are the usual Rankine-Hugoniot jump conditions and the entropy inequality from fluid dynamics; i.e,

$$\rho_-(u_- - s) = \rho_+(u_+ - s)$$

$$\rho_-((u_- - s)^2 + T_-) = \rho_+((u_+ - s)^2 + T_+) \quad (14)$$

$$\rho_-(u_- - s)(\frac{1}{2}(u_- - s)^2 + \frac{5}{2}T_-) = \rho_+(u_+ - s)(\frac{1}{2}(u_+ - s)^2 + \frac{5}{2}T_+)$$

$$(u_- - s)(\rho_+ - \rho_-) < 0 \quad (15)$$

In order to be definite, assume that the shock is moving to the right relative to the fluid; i.e., $s > u_\pm$. The entropy inequality then implies that $\rho_- > \rho_+$ and also that $T_- > T_+$. The strength of the shock can be measured by the ratio $\varepsilon = (T_- - T_+)/T_+$, which is the Mach number minus 1. For $\varepsilon = 0$, $s = u_- + c$, in which $c = \sqrt{(5/3)T_-}$ is the sound speed, whereas, for $\varepsilon > 0$, $s < u_- + c$.

For the weak shocks ($\varepsilon \ll 1$) described here nonequilibrium effects are small. Thus, the width of the shock is large, i.e., of size ε^{-1}. The solution F is thus sought as

$$F(x,\xi) = M_-(\xi) + \varepsilon f(y = x/\varepsilon, \xi)$$

satisfying

$$(\xi_x - s)\partial_y f = \varepsilon^{-1} Lf + Q(f,f) \tag{16}$$

in which $Lf = 2Q(M_-, f)$. This equation is like the Boltzmann equation in the fluid dynamic limit.

For a given asymptotic state M_- at $-\infty$, the shock wave can be thought of as a bifurcation from the constant state $f \equiv 0$, i.e., $F \equiv M_-$. The bifurcation occurs at $\varepsilon = 0$, and nontrivial shock waves exist only for $\varepsilon > 0$, as described later. The principal part of the bifurcated, nontrivial solution is sought as the solution of a generalized eigenvalue problem for condition (16), i.e.,

$$L\phi = \varepsilon \gamma (\xi_x - s) \phi \tag{17}$$

Because of the factor $(\xi_x - s)$ on the right, this is a generalized eigenvalue problem with eigenvalue γ. This problem has a solution only for $s < u_- + c$, which is for $\varepsilon > 0$. Actually, for collision operators other than hard spheres, only an approximate solution, with error ε^n for some n, can be found for Eq. (17).

Once Eq. (17) is (approximately) solved, the solution f of Eq. (16) is found as $f = h(y)\phi(\xi) + O(\varepsilon)$. The spatial profile function h is found by solving the Navier-Stokes equations in which the transport coefficients are those derived by the Chapman-Enskog expansion. At the present order of accuracy, the Navier-Stokes solution leads to a hyperbolic tangent profile

$$h(y) = c(1 + \tanh \alpha y)$$

This result is summarized in the following theorem of Caflisch and Nicolaenko:[12]

<u>Theorem</u> <u>4.1</u> : Assume that $\rho_\pm, T_\pm$ and s satisfy the jump conditions (14) and (15) and that the shock strength ε is sufficiently small. Let $\rho_{ns}(x), u_{ns}(x), T_{ns}(x)$ be the solution of the Navier-Stokes shock profile with speed

s and asymptotic states $\rho_\pm$, $u_\pm$, $T_\pm$, and let $M_{ns}(x,\xi)$ be the Maxwellian distribution corresponding to ρ_{ns}, u_{ns}, T_{ns}. Then there is a solution F of the Boltzmann shock profile problem (12),(13) satisfying

$$|F - M_{ns}| < c\varepsilon^2 \qquad (18)$$

Moreover, F is unique (up to spatial translation) among all solutions satisfying Eq. (18).

Since the spatial variation in F and M_{ns} is of size ε, the agreement between them show in Eq. (18) is nontrivial. Positivity of the solution has not been established because it is constructed by a procedure like the Chapman-Enskog expansion, which does not yield positive solutions at any finite order. Analysis of strong shock waves has not been carried out, although there has been a partial analysis of the infinite-strength limit in Grand[13] and Caflisch.[14]

V. Convergence of a Monte Carlo Method

Monte Carlo has been a very successful method for computational solution of the Boltzmann equation. Nanbu's method,[15] which will be discussed here, use Monte Carlo to solve the Boltzmann equation directly. The more popular direct simulation Monte Carlo (DSMC) method of Bird[16] performs a simulation of the statistical behavior of a gas of particles. Convergence of a solution of the DSMC method to a solution of the Boltzmann equation may be more difficult to establish because it may require establishment of the validity of the Boltzmann equation, in particular, verification of propagation of chaos. Derivation of the Boltzmann equation has been carried out for only short times or for a state near a vacuum (see Illner and Pulvirenti[17]).

Nanbu's method can be summarized roughly as follows: Space and time discretization sizes Δx and Δt are chosen. N particle positions and velocities $(\tilde{x}_i, \tilde{\xi}_i)$ $(i=1,\ldots,N)$ are chosen initially. Over every time interval, the particle configuration is changed in three steps.

1) <u>Convection</u>: Particles are moved for time Δt according to their velocities.

2) <u>Spatial grouping</u>: The spatial domain is split up into subdomains of size Δx^3, over which the particle distribution function F is considered to be homogeneous for the next collision step.

3) <u>Collisions</u>: In order to solve the homogeneous Boltzmann equation in each subdomain, two copies are made of the particles in that subdomain. For the collisions, these two copies are treated as independent; this is the difference between Nanbu's and Bird's methods. Collisions are calculated between a parti-

cle randomly chosen from the first copy and a particle randomly chosen from the second copy. For the subsequent steps of the method, only the first copy is used.

Babovsky and Illner[18] have proved convergence of this numerical method. The distribution function F from the Boltzmann equation is compared to the empirical distribution function $G(\underset{\sim}{x},\underset{\sim}{\xi},t) = N^{-1} \sum_{1}^{N} \delta(\underset{\sim}{x} - \underset{\sim}{x}_i(t)) \delta(\underset{\sim}{\xi} - \underset{\sim}{\xi}_i(t))$. Their result is as follows:

<u>Theorem 5.1</u> : Assume that the Boltzmann equation (1), with initial data F_0, has a smooth solution F, with $F \geq 0$ and $F \in L^1(\underset{\sim}{x},\underset{\sim}{\xi})$. Then, the solution G of Nanbu's method converges weakly in L^1 to F, in the sense that, for any $\phi(\underset{\sim}{x},\underset{\sim}{\xi}) \in L^\infty(\underset{\sim}{x},\underset{\sim}{\xi})$

$$\int \phi G \, d\underset{\sim}{x} \, d\underset{\sim}{\xi} \to \int \phi F \, d\underset{\sim}{x} \, d\underset{\sim}{\xi} \qquad (19)$$

as $N \to \infty, \Delta x \to 0, \Delta t \to 0$.

The convergence proof uses a weak formulation of the Boltzmann equation and an illuminating interpretation of the collision process due to Babovsky and Illner.[18] Suppose that there are N sample particles representing the density function $F(\underset{\sim}{\xi})$. Then the choice of two colliding particles is a choice of a pair of particles from the N^2 possible pairs $(\underset{\sim}{\xi},\underset{\sim}{\eta})$ that represent the product density $F(\underset{\sim}{\xi})F(\underset{\sim}{\eta})$. A random choice of N such pairs gives a valid representation of the product.

This interpretation of Nanbu's method has led to development of the new low discrepancy method of Babovsky,[19] in which randomness is eliminated. This promising method seems to be amenable to the same mathematical analysis.

VI. Global Existence for the Initial Value Problem

Global (i.e., for all time) existence, uniqueness and regularity of solutions with arbitrary large initial data for the Boltzmann equation, is a basic goal of mathematical kinetic theory. Short time existence for arbitrary data and global existence for small data (close to either a vacuum or a Maxwellian state) have been proved using linearized theory, monotonicity arguments, or bounds on the rate of particle dispersion (see DiPerna and Lions[1] for a complete list of references). Recently, the global existence question has been solved by DiPerna and Lions[1]. Uniqueness and regularity for their solutions are still unknown.

The solutions constructed by DiPerna and Lions are called mild solutions because they lack regularity; i.e., they may not be bounded or smooth at some points in phase space. In order to define the idea of a mild solution, denote

$$F^{\#}(x,\xi,t) = F(x+t\xi,\xi,t) \tag{20}$$

which is the solution of the transport part of the Boltzmann equation. A function $F(x,\xi,t)$ is said to be a mild solution of the Boltzmann equation if

$$F^{\#}(x,\xi,t) - F^{\#}(x,\xi,s) = \int s' Q(F,F)^{\#}(x,\xi,\sigma)d\sigma \tag{21}$$

for all $0 < s < t < \infty$. This equation is required to be valid only for almost all (x,ξ) (i.e., except on a set of measure zero), corresponding to possibly lack of regularity of F. If F were regular, then Eq. (21) would be equivalent to the Boltzmann equation (1), upon application of the transport operator $\partial_t + \xi \cdot \partial_x$.

The global existence result of DiPerna and Lions[1] is the following.

<u>Theorem 6.1</u> : Suppose the initial data F_0 satisfies

$$F_0 \geq 0$$

$$\int F_0 (1 + |x|^2 + |\xi|^2 + |\log F_0|)\, dx\, d\xi < \infty.$$

Then, there is a mild solution F of Eq. (21) for all $0 < t < \infty$ with $F(t=0) = F_0$, $F \geq 0$ almost everywhere, $F \in C([0,\infty); L^1(x,\xi))$, and

$$\sup_{t \geq 0} \int F(1 + |x - \xi t|^2 + |\log F|)\, dx\, d\xi < \infty \tag{22}$$

Note

1) This result is valid under general conditions on the collision operator Q, which are satisfied by the collision operator for hard spheres and by the so-called hard collision operator with angular cutoff.

2) From the bound (22), it can be shown that mass and momentum are conserved for this mild solution, i.e., for all t,

$$\int F\, dx\, dx = \int F_0\, dx\, d\xi$$

$$\int \xi F\, dx\, d\xi = \int \xi F_0\, dx\, d\xi \tag{23}$$

Conservation of kinetic energy has not been established. This is because the existence proof involves a weak limit for which energy may decrease. The usual equation for flux of entropy has been proved to hold

3) Uniqueness of mild solutions for a given F_0 is not known.

We shall briefly describe the main ideas of the proof. There are two main difficulties in analysis of the Boltzmann equation: nonlinearity and spatial singularity.

The quadratic nonlinearity of the collision operator is a potential source of blowup for the solution. A simple model problem $F_t = F^2$ has blowup at time $t = F(t=0)^{-1}$. Boundedness of the mass, energy, and H integral, as in bound (22), is not sufficient in itself to prevent blowup in some parts of phase space. DiPerna and Lions handle this problem by considering what they call the "renormalized Boltzmann equation": Define $G = log(1 + F)$. The renormalized Boltzmann equation is

$$\partial_t G + \underset{\sim}{\xi} \cdot \partial_x G = (1+F)^{-1} Q(F,F) \qquad (24)$$

which is obtained by dividing Eq. (1) by $(1 + F)$. If F blows up, the renormalized equation may be significantly different from the original equation. However, they show that renormalized and the mild reformulations of the Boltzman equation are equivalent.

DiPerna and Lions also use the H inequality to show that the gain term Q_+, which contributes to blowup, is not much worse than the loss term Q_-, which helps prevent blowup. They show that, for any $k > 1$,

$$Q^+(F,F) < kQ^-(F,F) + (log\, k)^{-1} e(F) \qquad (25)$$

in which $e(F)$ is a nonlinear integral of F that can be controlled.

The second difficulty with the Boltzmann equation is the spatial singularity of the collision operator; collisions occur only between particles with equal spatial coordinates $\underset{\sim}{x}$. In fact, if the collision law is modified to include a nonzero collision distance, Povzner[20] gave an easy proof of global existence. Although the actual collision distance is nonzero, it is so small that dependence on it is unsatisfactory.

DiPerna and Lions[1] overcome this problem using the slight smoothness of the inverse of the transport operator, $(\partial_t + \underset{\sim}{\xi} \cdot \partial_x)^{-1}$ from the following lemma, which was proved by Golse et al.[21,22].

<u>Lemma 6.2</u> : Assume that g^n and G^n are sequences of functions in a weakly compact set of $L^1((0,T) \times \mathbf{R}^3 \times \mathbf{R}^3)$ and satisfy

$$(\partial_t + \underset{\sim}{\xi} \cdot \partial_x) g_n = G_n \qquad (26)$$

Then, for each $\Psi \in L^\infty((0,T) \times \mathbf{R}^3 \times \mathbf{R}^3)$ with compact support, the velocity average $\int g^n \Psi d\underset{\sim}{\xi}$ is in a compact set of $L^1((0,T) \times \mathbf{R}^3)$.

The result arrived at by DiPerna and Lions which was described earlier, settles the longstanding global existence problem for the Boltzmann equation. The remaining related problems of uniqueness and regularity are even more challenging. Their successful existence result should compel further study of these important mathematical problems.

This survey was prepared with partial support from the Air Force Office of Scientific Research under URI grant #AFOSR 86-0352.

References

[1] DiPerna, R. and Lions, P. L., "On the Cauchy Problem for Boltzmann Equations: Global Existence and Weak Stability," Annals of Mathematics, to be published.

[2] Arkeryd, L., "On the Boltzmann Equation, Part I: Existence," Archive for Rational Mechanics and Analysis, Vol. 45, 1972, pp. 1-16.

[3] Arkeryd, L., "On the Boltzmann Equation, Part II: The Full Initial-Value Problem," Archive for Rational Mechanics and Analysis, Vol. 45, 1972, pp. 17-34.

[4] Elmroth, T., "On the H-Function and Convergence Towards Equilibrium for a Space-Homogeneous Molecular Density," SIAM Journal of Applied Mathematics, Vol. 44, 1984, pp. 150-159.

[5] Arkeryd, L., "L^∞ Estimates for the Space Homogeneous Boltzmann Equation," Journal of Statistical Physics, Vol. 31, 1983, pp. 347-361.

[6] Gustaffson, T., "L^p − Estimates for Nonlinear Space Homogeneous Boltzmann Equation," Archive for Rational Mechanics and Analysis, Vol. 92, 1986, pp. 23-57.

[7] Ukai, S., and Asano, K., "Steady Solutions of the Boltzmann Equation for a Gas Flow Past an Obstacle, I. Existence," Archive for Rational Mechanics and Analysis, Vol. 84, 1983, pp. 249-291.

[8] Ukai, S., and Asano, K., "Steady Solutions of the Boltzmann Equation for a Gas Flow Past and Obstacle," II. Stability," Publications of Research Institute of Mathematical Sciences, Kyoto, Japan, Vol. 22, 1986, pp. 1035-1062.

[9] Coron, F., Golse, F., and Sulem, C., "A classification of Well-Posed Kinetic Layer Problems," Communications on Pure and Applied Mathematics, Vol. 41, 1988, pp. 409-435.

[10] Bardos, C., Caflisch, R. E., and Nicolaenko, B., "The Milne and Kramers Problems for the Boltzmann Equation of a Hard Sphere Gas." Communications on Pure and Applied Mathematics, Vol. 49, 1986, pp. 323-352.

[11] Cercignani, C., "Half-Space Problems in the Kinetic Theory of Gases," Trends in Applications of Pure Mathematics to Mechanics, edited by Kroener and Kirchgaessner, Lecture Notes in Physics, Springer-Verlag, New York, Vol. 249, 1986, pp. 35-50.

[12] Caflisch, R. E. and Nicolaenko, B., "Shock Profile Solutions of the Boltzmann Equation." Communications in Mathematical Physics, Vol. 86, 1982, pp. 161-194.

[13] Grad, H., "Singular and Nonuniform Limits of Solutions of the Boltzmann Equation," SIAM-AMS Proceedings I, Transport Theory, Providence, R.I., 1969, pp. 296-308.

[14] Caflisch, R. E., "The Half-Space Problem for the Boltzmann Equation at Zero Temperature," Communications on Pure and Applied Mathematics, Vol. 38, 1985, pp. 529-547.

[15] Nanbu, K., "Direct Simulation Scheme Derived From the Boltzmann Equation," Japanese Journal of Physics, Vol. 40, 1980, p. 2042.

[16] Bird, G.A., Molecular Gas Dynamics Clarendon Press, Oxford, UK, 1976.

[17] Illner, R., and Pulvirenti, M., "Global Validity of the Boltzmann Equation for a Two-Dimensional Rare Gas in Vacuum," Communications in Mathematical Physics, Vol. 105, 1986, pp. 189-203.

[18] Babovsky, H. and Illner, R., "A Convergence Proof for Nanbu's Simulation Method for the Full Boltzmann Equation." SIAM Journal Numerical Analysis, to be published.

[19] Babovsky, H., Groppengeisser, F., Neunzert, H., Struckmeier, J., and Wiesen, B., "Low Discrepancy Methods for Solving the Boltzmann Equation," Progress in Astronautics and Aeronautics: Rarefied Gas Dynamics, edited by E.P. Muntz, D.H. Campbell, and D.P. Weaver, AIAA, Washington, DC, 1989.

[20] Povzner, A. Ya., "The Boltzmann Equation in the Kinetic Theory of Gases," American Mathematical Society (Trans.), Vol. 47, 1962, pp. 193-216.

[21] Golse, F., Perthame, B., and Sentis, R., "Un Resultat pour les Equations de Transport et Application au Calcul de la Limite de la Valeur Propre Principale d'un Operateur de Transport," Comptes Rendus Hebdomadaires des Seances del'Academie des Sciences, Paris, Vol. 301, 1985, pp. 341-344.

[22] Golse, F., Lions, P. L., Perthame, B., and Sentis, R., "Regularity of the Moments of the Solution of a Transport Equation," Journal of Functional Analysis, Vol. 76, 1988, pp. 110-125.

Stationary Flows from a Model Boltzmann Equation

Y. Y. Azmy* and V. Protopopescu*
Oak Ridge National Laboratory, Oak Ridge, Tennessee

Abstract

The one-dimensional Boltzmann equation model of Ianiro and Lebowitz is generalized by allowing the collision frequency to be an arbitrary positive function of position and by including more general boundary conditions. Conservation laws and explicit solutions are derived, and macroscopic quantities obtained from the stationary Bhatnagar-Gross-Krook (BGK) variant of the model are compared with available experimental data.

I. Introduction

In a recent paper, Ianiro and Lebowitz (IL)[1] considered a model Boltzmann equation to describe the behavior of a gas between two infinite parallel plates maintained at different constant temperatures. The model is one-dimensional (translational symmetry is assumed along directions parallel to the plates), has constant collision frequency, and perfectly thermalizing boundary conditions at the walls (perfect diffusion). Despite its simplicity and unlikely adequacy to realistic situations, the model is interesting because it is one of the few models in nonequilibrium statistical mechanics that could be more or less explicitly solved in the steady case. It also shows how far we remain from a microscopic understanding of nonequilibrium phenomena.

The purpose of this paper is twofold:

1) We generalize the model to space-dependent collision frequencies, and we include more general boundary conditions allowing for both diffuse and specular reflections at the walls. Therefore, because of some special features of the model, these cases are also almost entirely tractable.

This paper is declared a work of the U.S. Government and is not subject to copyright protection in the United States.

*Research Staff, Engineering Physics and Mathematics Division.

2) For a simpler variant of the model (its Bhatnagar-Gross-Krook (BGK) version), we pursue the analysis somewhat further than IL by computing the thermal conductivity and the density as a function of temperature for this model, and compare it with available experimental data in order to assess better the model's plausibility.

II. Generalized IL Model

The model consists of point particles similar to "beads on a string" moving along the x axis on a segment of length $2L$. The particles undergo "model collisions" with each other and with the walls located at $x = -L$ and L. Two particles that meet at a point x, $-L < x < L$, can either exchange velocities:

$$(v_1, v_2) \rightarrow (v_2, v_1) \tag{1}$$

or have their velocities reversed:

$$(v_1, v_2) \rightarrow (-v_1, -v_2) \tag{2}$$

It is assumed that whenever the particles travel in the same direction they undergo the first type of collision; when the particles travel in opposite directions, the first type of collision occurs with a probability $1 - p$ and the second type with probability p, $0 \leq p \leq 1$. The collision (Eq. (1)) is what actually happens in one dimension, and it does not change the velocity distribution in the system. As we shall see, the model collisions (Eq. (2)) preserve mass and energy, but do not preserve momentum.

Denoting by $f(x, v, t)$ the one–particle distribution function in the gas as

$$\begin{aligned} f(x,v,t) &= f_+(x,v,t) & v > 0 \\ &= f_-(x,-v,t) & v > 0; \quad -L < x < L \end{aligned} \tag{3}$$

we write the kinetic equations of the model in the form[1]

$$\frac{\partial f_+}{\partial t}(x,v,t) + v \frac{\partial f_+}{\partial x}(x,v,t) = p(x) \int_0^\infty (v+v')[f_-(x,v,t)f_+(x,v',t)$$
$$- f_+(x,v,t)f_-(x,v',t)]dv' \tag{4}$$

$$\frac{\partial f_-}{\partial t}(x,v,t) - v \frac{\partial f_-}{\partial x}(x,v,t) = p(x) \int_0^\infty (v+v')[f_+(x,v,t)f_-(x,v',t)$$
$$- f_-(x,v,t)f_+(x,v',t)]dv' \tag{5}$$

At the boundaries $x = -L$ and L, the gas is in contact with heat baths at specified temperatures T_- and T_+, respectively. The interaction of a particle with the boundary is described by surface-scattering kernels $R_\pm(v' \to v)$ of the form

$$R_\pm(v' \to v) = \alpha_\pm \delta(v - v') + \beta_\pm v H_\pm(v) \qquad (6)$$

describing a combination of specular and diffuse reflection at the boundary. The case $\alpha = 1$, $\beta = 0$, corresponds to purely specular reflection, whereas $\alpha = 0$, $\beta = 1$ corresponds to purely diffuse reflection. In principle, the coefficients α, β can be different at $x = -L$ and L, but, for simplicity, we shall take them equal. Using Eq. (6), the boundary conditions (BC) read

$$v f_+(-L, v, t) = \alpha v f_-(-L, v, t) + \beta v H_-(v) \int_0^\infty v' f_-(-L, v', t) dv' \qquad (7)$$

$$v f_-(L, v, t) = \alpha v f_+(L, v, t) + \beta v H_+(v) \int_0^\infty v' f_+(L, v', t) dv' \qquad (8)$$

Current conservation at the boundaries requires that

$$\int_0^\infty R_\pm(v' \to v) dv = 1,$$

that is,

$$\int_0^\infty v H_\pm dv = 1 \qquad \text{and} \qquad \alpha + \beta = 1$$

(If particles are partially absorbed at the boundaries, i.e., $\alpha + \beta < 1$, then a steady nontrivial distribution cannot be achieved in the absence of external sources.) Since the walls are thermal reservoirs at temperatures $T_\pm$, we take

$$H_\pm(v) = \frac{m}{kT_\pm} \exp[-mv^2/2kT_\pm]$$

If the solutions are stationary, summing Eqs. (4) and (5) yields

$$f_-(x, v) - f_+(x, v) = c(v) \qquad (9)$$

Take $x/L \equiv s \in [-1, 1]$, define $q(s) = p(s)L$, and rewrite Eq. (3) as

$$v \frac{\partial f_+}{\partial s}(s, v) = q(s) \int_0^\infty (v + v')[c(v) f_+(s, v') - c(v') f_+(s, v)] dv' \qquad (10)$$

Since there is no net transport of particles in the stationary state, we have

$$\int_0^\infty vc(v)dv = \int_0^\infty v[f_+(s,v) - f_-(s,v)]dv = 0$$

and

$$\int_0^\infty vf_+(s,v)dv = \int_0^\infty vf_-(s,v)dv = d = \text{independent of } s \quad (11)$$

Then, we can rewrite Eq. (10) as

$$v\frac{\partial f_+}{\partial s}(s,v) = q(s)(vc(v)n_+(s) + dc(v) - vaf_+(s,v)) \quad (12)$$

where

$$a \equiv \int_0^\infty c(v)dv$$

3

$$n_+(s) \equiv \int_0^\infty f_+(s,v)dv \quad (13)$$

Integrating Eq. (12) over v, we get

$$\frac{dn_+(s)}{ds} = q(s)db \quad (14)$$

with

$$b \equiv \int_0^\infty \frac{c(v)}{v}dv$$

Defining the primitive of $q(s)$ by $Q(s) \equiv \int_{-1}^s q(s')ds'$, we obtain from Eq. (14):

$$n_+(s) = dbQ(s) + N_+$$

The constants N_+ and b can be determined from the BC for $n_+(s)$, obtained, in turn, from Eqs. (7) and (8):

$$n_+(-1) = \alpha n_-(-1) + (1-\alpha)dh_- \quad (15)$$

$$n_-(1) = \alpha n_+(1) + (1-\alpha)dh_+, \quad (16)$$

where

$$h_\pm = \int_0^\infty H_\pm(v)dv = \sqrt{\frac{\pi m}{2kT_\pm}}$$

Then, taking into account Eqs. (15) and (16) and the relationship $n_-(s) - n_+(s) = a$, we get

$$n_+(-1) = \frac{\alpha a + (1-\alpha)dh_-}{1-\alpha} = N_+ \qquad (17)$$

$$n_+(1) = \frac{(1-\alpha)dh_+ - a}{1-\alpha} = dbQ(1) + N_+$$

which determine N_+ (Eq. (17)) and db as

$$db = \frac{[d(h_+ - h_-)(1-\alpha) - a(1+\alpha)]}{Q(1)(1-\alpha)}$$

Finally,

$$n_+(s) = \frac{d(h_+ - h_-)(1-\alpha) - a(1+\alpha)}{Q(1)(1-\alpha)} Q(s) + \frac{\alpha a + (1-\alpha)dh_-}{1-\alpha} > 0 \qquad (18)$$

and the local particle density $n(s) = n_+(s) + n_-(s) = 2n_+(s) + a$, i.e.,

$$n(s) = 2\frac{d(h_+ - h_-)(1-\alpha) - a(1+\alpha)}{Q(1)(1-\alpha)} Q(s)$$

$$+ \frac{2(1-\alpha)dh_- + a(1+\alpha)}{1-\alpha} \qquad (19)$$

We notice that a stationary profile can be reached only if $\alpha < 1$, i.e., if there is a finite amount of stochastic scattering at the walls to couple the system with the heat baths. The profile of $n(s)$ is dependent on α, the temperatures of the walls, and on $Q(s)$. Since the cross section $q(s)$ cannot be negative, this means that the profile is a monotonous function of s. For $q(s) = q$, $\alpha = 0$, and $Q(1) = 2q$, we recover the result of Ianiro and Lebowitz.[1] From Eq. (19), one can compute the value of d by using the value of the average particle density:

$$\bar{\rho} = \frac{N}{2} = \frac{1}{2}\int_{-1}^{1} n(s)ds$$

With Eq. (18), we can solve Eq. (12) as

$$\frac{\partial f_+(s,v)}{\partial s} = q(s)c(v)n_+(s) + q(s)\frac{dc(v)}{v} - q(s)af_+(s,v)$$

$$f_+(s,v) = A_+(s,v)e^{-aQ(s)}$$

$$\frac{\partial A_+}{\partial s} = e^{aQ(s)}\left(q(s)c(v)n_+(s) + q(s)\frac{dc(v)}{v}\right) \quad (20)$$

Inserting $n_+(s)$ given by Eq. (18) into Eq. (20), performing the integration, and re-grouping terms, we finally get

$$f_+(s,v) = \frac{c(v)}{a}\left[n_+(s) - \frac{db}{a} + \frac{d}{v}\right] + B_+(v)e^{-aQ(s)}$$

The compatibility relations Eqs. (13) and (11) imply that

$$\int_0^\infty B_+(v)dv = 0, \qquad \int_0^\infty vB_+(v)dv = 0 \quad (21)$$

respectively. The BC at $s = -1$ yields

$$B_+(v) = dH_-(v) - \frac{c(v)}{a}dh_- + \frac{c(v)}{a}\frac{db}{a} - \frac{d}{a}\frac{c(v)}{v} \quad (22)$$

which is consistent with Eq. (21). The BC at $s = 1$ gives

$$f_+(1,v) = \frac{(1-\alpha)dH_+(v) - \alpha c(v)}{1-\alpha}$$

Using this expression in Eq. (21), with $B_+(v)$ given by Eq. (22), we get

$$\hat{c}(v) \equiv \frac{c(v)}{a} = \left\{(1-\alpha)d(H_+(v) - H_-(v)e^{-aQ(1)})v\right\}$$
$$\div \left\{(1-\alpha)d(1 - e^{-aQ(1)}) + v\left[\left[n_+(1) - n_+(-1)e^{-aQ(1)}\right.\right.\right.$$
$$\left.\left.\left. - \frac{db}{a}(1 - e^{-aQ(1)})\right](1-\alpha) + a(1 + \alpha e^{-aQ(1)})\right\}\right\}$$

From this point on, further analyses of the general IL model will have to rely on numerical computations such as those in Ref. 1.

III. Conservation Laws

In general, any quantity conserved in a binary collision is also conserved on the macroscopic scale yielding conservation laws for corresponding macroscopic quantities.[4] As a result of the type of collisions assumed in the IL model, this requires only that the quantity be an even function of the velocity as shown subsequently. Since collisions of particles moving in the same direction do not alter the distribution function, they have no influence on any quantity conserved in the collision. Suppose $\chi(v)$ is a quantity conserved in a binary collision between particles 1 and 2, i.e., $\chi(v_1) + \chi(v_2) = \chi(v_1') + \chi(v_2')$. If we consider only collisions for which $v_1 v_2 \leq 0$, then the IL model requires, without loss of generality, that

$$\chi(v_1) + \chi(-v_2) = \chi(-v_1) + \chi(v_2), \quad v_1 \text{ and } v_2 \geq 0$$

so that

$$\chi(v) - \chi(-v) = \text{const}, \quad v \geq 0 \qquad (23)$$

If χ is assumed continuous at $v = 0$, then the constant in Eq. (23) must vanish, proving that, in the IL model, conservation of $\chi(v)$ implies that χ is an even function of v

$$\chi(v) = \chi(-v), \quad v \geq 0 \qquad (24)$$

It is now easy to derive[4] a general conservation law for the model.

Multiplying Eqs. (4) and (5) by χ_+ and χ_-, respectively, and integrating over v, we obtain

$$\frac{\partial}{\partial t} <n\chi_+>_+ + \frac{\partial}{\partial x} <nv\chi_+>_+ =$$
$$+ p \int_0^\infty dv \int_0^\infty dv'(v+v')\chi_+ [f_+' f_- - f_+ f_-'] \qquad (25)$$

$$\frac{\partial}{\partial t} <n\chi_->_- - \frac{\partial}{\partial x} <nv\chi_->_- =$$
$$- p \int_0^\infty dv \int_0^\infty dv'(v+v')\chi_- [f_+' f_- - f_+ f_-'] \qquad (26)$$

where we have defined
$$< \cdot >_\pm \equiv \frac{1}{n} \int_0^\infty dv\, f_\pm(v).$$

$$n(t,x) \equiv \int_0^\infty dv[f_+ + f_-]$$

Adding Eqs. (25) and (26), we get

$$\frac{\partial}{\partial t}[<n\chi_+>_+ + <n\chi_->_-] + \frac{\partial}{\partial x}[<nv\chi_+>_+ - <nv\chi_->_-]$$

$$= p \int_0^\infty dv \int_0^\infty dv'(v+v')[\chi_+ - \chi_-][f'_+ f_- - f_+ f'_-] \quad (27)$$

The right-hand side of Eq. (27) vanishes because of Eq. (24), and we define the macroscopic quantities as

$$X_\chi(t,x) \equiv <\chi_+>_+ + <\chi_->_- \quad (28)$$

and
$$Y_\chi(t,x) \equiv <v\,\chi_+>_+ - <v\chi_->_-$$

Equation (27) becomes

$$\frac{\partial}{\partial t}(nX_\chi) + \frac{\partial}{\partial x}(nY_\chi) = 0$$

which is the general conservation law for the system.

Take $\chi(v) \equiv m$, the particle mass, and define the local mass density as
$$\rho(t,x) \equiv mn(t,x)$$

Equation (28) immediately yields
$$X_m(t,x) = m \quad (29)$$

For this choice of χ, Y_χ can be identified with the "fluid" flux, mu, where

$$Y_m = mu(t,x) \equiv \frac{m}{n} \int_{-\infty}^\infty dv\, v f(v) = \frac{m}{n} \int_0^\infty dv[vf_+ - vf_-] \quad (30)$$

Finally, we obtain the standard continuity equation for mass as

$$\frac{\partial \rho}{\partial t} + \frac{\partial}{\partial x}(\rho u) = 0$$

Take $\chi \equiv E = \frac{1}{2}mv^2$, the particle energy, and define the temperature and the heat flux

$$T(t,x) \equiv X_E = \frac{m}{2kn}\int_0^\infty dv\, v^2[f_+ + f_-]$$

$$\Phi(t,x) \equiv \rho Y_E = \frac{m\rho}{2n}\int_0^\infty dv\, v^3[f_+ - f_-] \qquad (31)$$

Equation (29) then becomes the continuity equation for energy

$$\frac{\partial}{\partial t}(\rho k T) + \frac{\partial \Phi}{\partial x} = 0 \qquad (32)$$

Absent from Eq. (32) is the convection term. This is due to the "no-flow" condition that IL imposed in their paper, which implies that $u(x) = 0$, so that the standard heat equation reduces to Eq. (32).

In steady state, the conservation equations become

$$\frac{d}{dx}(\rho u) = 0 \qquad (33)$$

$$\frac{d\Phi}{dx} = 0 \qquad (34)$$

We note that the preceding derivation is very general.[4] Alternatively, it is possible to derive Eqs. (33) and (34) directly from Eqs. (30) and (31), respectively, by using Eq. (9), which is specific to the IL model.

IV. The BGK Model

The BGK version of the IL model for constant collision frequency is[1]

$$\frac{\partial f_+}{\partial t} + v\frac{\partial f_+}{\partial x} = \frac{1}{\tau}(f_- - f_+) \qquad (35)$$

$$\frac{\partial f_-}{\partial t} - v\frac{\partial f_-}{\partial x} = \frac{1}{\tau}(f_+ - f_-) \qquad (36)$$

where we identified the collision frequency with the inverse of the mean free path, τ. Two situations seem worth considering: constant mean free time, τ, and constant mean free path, $\ell \equiv v\tau$. For the hydrodynamic limit, it is advantageous to take $\ell = $ const. Repeating the calculations in Sec. II for Eqs. (35) and (36), we get

$$f_+(x,v) = \frac{d}{2}[H_+(v) + H_-(v)] + c(v)\left[\frac{x}{\ell} - \frac{1}{2}\right]$$

$$f_-(x,v) = \frac{d}{2}[H_+(v) + H_-(v)] + c(v)\left[\frac{x}{\ell} + \frac{1}{2}\right]$$

and

$$c(v) = \frac{(H_+(v) - H_-(v))d\ell}{\ell + 2L}$$

$$d = \bar{\rho}/(h_+ + h_-)$$

with $\bar{\rho}$, the average particle density, given by

$$\bar{\rho} = \frac{1}{2L}\int_0^\infty \int_{-L}^L [f_+(x,v) + f_-(x,v)]dxdv$$

The density profile is given by

$$n(x) = \bar{\rho} + \frac{2\bar{\rho}}{\ell + 2L}\frac{h_+ - h_-}{h_+ + h_-}x \tag{37}$$

and the local temperature is

$$T(x) = \frac{d}{n(x)}\left\{\left(1 + \frac{2x}{\ell + 2L}\right)T_+h_+ + \left(1 - \frac{2x}{\ell + 2L}\right)T_-h_-\right\} \tag{38}$$

The temperatures of the gas at the walls are the following:

$$T(-L) = \frac{\pi m}{2k}\frac{\ell/h_+ + (\ell + 4L)/h_-}{\ell h_+ + (\ell + 4L)h_-}$$

$$T(L) = \frac{\pi m}{2k}\frac{\ell/h_- + (\ell + 4L)/h_+}{\ell h_- + (\ell + 4L)h_+}$$

and we notice that, in general, these temperatures are not equal to T_- and T_+, respectively. However, in the hydrodynamic limit $\ell \ll L$, the temperature jump goes to zero as it should[5] and $T(\pm L) = T_\pm$. Also, in the same limit,

$$T(x) \cong \frac{\pi m}{2(h_+ + h_-)k}\left[\left(\frac{1}{h_+} + \frac{1}{h_-}\right) + \frac{x}{2L}\frac{1/h_- - 1/h_+}{h_+h_-(h_+ + h_-)}\right]$$

One can eliminate x from Eqs. (37) and (38) and get a relationship between T and n in the following form:

$$T(n) = \frac{\pi m}{2k}\left(\frac{1/h_+ - 1/h_-}{h_+ - h_-}\right) + \frac{\bar{\rho}\pi m}{kn}\left(\frac{h_+/h_- - h_-/h_+}{h_+^2 - h_-^2}\right)$$

Conversely,

$$\frac{n(T)}{\bar{\rho}} = \frac{1}{\frac{1}{2} + \frac{h_+ h_-}{\pi m} kT} \qquad (39)$$

This is a somewhat strange result because it implies that the temperature coefficient of the density is not constant for a given material but rather depends on the wall temperatures that are specific to a given experimental setting. However, we may look at it as an empirical result in which the regularity of an inferred law is dependent on the previously collected experimental data, in particular on $T_\pm$.

Ianiro and Lebowitz[1] showed that for the BGK model the heat flux is given by

$$\Phi_h = \bar{\rho}\sqrt{\frac{2}{\pi}} \frac{\ell}{((1/\sqrt{T_+} + 1/\sqrt{T_-})} \frac{T_- - T_+}{2L + \ell} \qquad (40)$$

Identifying Eq. (40) with Fourier's law,

$$\Phi_h = -K(n(T), T)\frac{dT}{dx} \qquad (41)$$

we get

$$K(n(T(x)), T(x)) = \frac{\ell + 2L}{2} \frac{h_+ h_-}{h_+^2 + h_-^2} \frac{n^2}{\bar{\rho}^2} (h_+ + h_-)^2$$

Since Fourier's law contains the gradient of temperature, we cannot hope to get a scaleless thermodynamic limit of Eq. (41) such as in equilibrium. The preceding identification implicitly *assumed* that Fourier's law *is* valid at the scale $\ell \ll L$. Rigorous derivations of the validity for special models can be found in Refs. 2 and 3.

Using the relationship between n and T, we get

$$K(T) = A/\left(\frac{1}{2} + BT\right)^2 \qquad (42)$$

Given the value of K at two values of T, $K_j = K(T_j)$, $j = 1, 2$, one determines A and B as

$$B = \frac{1}{2}\left[\sqrt{K_1/K_2} - 1\right]\left[T_2 - T_1\sqrt{K_1/K_2}\right]^{-1}$$

$$A = K_1\left[\frac{1}{2} + BT_1\right]^2 = K_2\left[\frac{1}{2} + BT_2\right]^2$$

Similarly, using the value of the density at two different values of the temperature, $n_j \equiv n(T_j)$, one can determine the constant $\bar{\rho}$ and $C \equiv 2h_+h_-/\pi m$ from Eq. (39) as

$$C = \frac{1}{2}(n_1 - n_2)/(n_2 T_2 - n_1 T_1)$$

$$\bar{\rho} = n_1\left(\frac{1}{2} + CT_1\right) = n_2\left(\frac{1}{2} + CT_2\right)$$

In other words, we use formulas (39) and (42) as if they were empirical formulas for $n(T)$ and $K(T)$, respectively, with the constants appearing in these expressions determined from available experimental data. Indeed, the empirical nature of this part of our study is underscored by the observation that even though the model yields $B \equiv C$, we still calculate each one separately from the thermal conductivity and density data, respectively. In order to check the validity of these empirical formulas, we tried to find a fit with experimental data. The model *cannot* be used to represent heat conduction phenomena in gases and common liquids because, in these materials, the thermal conductivity increases with temperature, thus violating Eq. (42), since $B \geq 0$. Therefore, we compared the model with available experimental data for five different liquid metals: lead, lithium, potassium, sodium, and tin.[6] First, we define the reduced temperatures, density, and thermal conductivity by

$$\bar{T}^n \equiv CT \tag{43}$$

$$\bar{T}^K \equiv BT \tag{44}$$

$$\bar{n}^n \equiv n/\bar{\rho} \tag{45}$$

$$\bar{K}^K \equiv K/A \tag{46}$$

For each material, we calculate the empirical constants using the two points that have the largest temperature difference, then we use

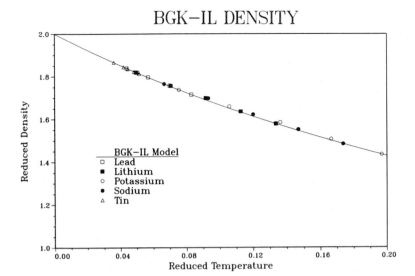

Fig. 1 Comparison of the experimentally determined[6] values to theoretically calculated values of the reduced mass density, Eq. (45), as a function of the reduced temperature, Eq. (43), for five liquid metals.

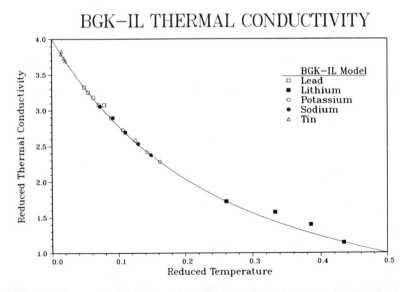

Fig. 2 Comparison of the experimentally determined[6] values to theoretically calculated values of the reduced thermal conductivity, Eq. (46), as a function of the reduced temperature, Eq. (44), for five liquid metals.

Eqs. (43-46) to calculate the reduced properties for this material. The resulting values for $\bar{n}^n(\bar{T}^n)$ and $\bar{K}^K(\bar{T}^K)$, together with the corresponding theoretical expressions, are plotted in Figs. 1 and 2, respectively. The good agreement between the BGK-IL model and the experimental results evident from these plots suggests that this model may be used to represent "some" phenomena involving liquid metals. However, since there is almost no information in the model that distinguishes different states of the materials used, the good agreement between the BGK-IL model and the experimental values for liquid metals has to be recognized at this stage more as a fortuitous coincidence rather than a systematically derivable fact. This is consistent with the empirical nature of the formulas alluded to above and should not be completely surprising for two reasons: 1) the range of validity of the Boltzmann equation – originally proposed for dilute gases – proved to go far beyond its assigned limits, and this is true even for simplified models like the IL model; 2) in the hydrodynamic limit $\ell \ll L$, the Boltzmann equation is supposed to describe well some phenomena occurring in liquids. Thus, although we do not attach too much importance to this agreement, we do not want to discard it until later studies will have brought out a better understanding of the model.

V. Summary and Conclusions

We have generalized the Ianiro-Lebowitz model of the Boltzmann equation to the case of space-dependent collision frequency and general boundary conditions. We have derived an analytical solution for our generalized system, which makes it interesting in spite of its simplicity.

In order to determine the domain of applicability of the Bhatnagar-Gross-Krook (BGK) version of this model on the macroscopic scale, we calculated the thermal conductivity and mass density and compared their behavior as a function of local temperature with existing experimental data.

References

1 Ianiro, N. and Lebowitz, J. L., "Stationary Nonequilibrium Solution of Model Boltzmann Equation," Foundations of Physics, Vol. 15, May 1985, pp. 531-544.

2 Lebowitz, J. L. and Spohn, H., "Microscopic Basis for Fick's Law of Self-Diffusion," Journal of Statistical Physics, Vol. 28, 1982, pp. 539-556.

3 Lebowitz, J. L. and Spohn, H., "Transport Properties of the Lorentz Gas: Fourier's Law," Journal of Statistical Physics, Vol. 19, 1978, pp. 633-654.

4 Huang, K., Statistical Mechanics, Wiley, New York, 1963.

5 Cercignani, C., The Boltzmann Equation and Its Applications, Springer-Verlag, New York, 1988.

6 Nuclear Engineering Handbook, edited by Harold Etherington, 1st Ed., McGraw-Hill, New York, 1958, pp. 9.18-9.22.

A Tensor Banach Algebra Approach to Abstract Kinetic Equations

W. Greenberg*
*Virginia Polytechnic Institute and State University,
Blacksburg, Virginia*
and
C. V. M. van der Mee†
University of Delaware, Newark, Delaware

Abstract

By proving the injective tensor product $L^1(\mathbb{R}) \otimes_\epsilon L(H)B$ to be a Banach algebra, we obtain the unique solvability of the Wiener-Hopf equation $\psi - k * B\psi = w$ in $L^\infty(\mathbb{R}, H)$ or $C(\mathbb{R}, H)$. This demonstrates the solvability of the equivalent abstract kinetic equation for non-self-adjoint collision operators $A = I - B$ without any regularity condition assumed, where B is a finite-rank operator on a complex Hilbert space H and k is an element in $L^1(\mathbb{R}) \otimes_\epsilon L(H)$. Conventional treatment of such kinetic equations fails because of the lack of integrability of the convolution kernel.

I. Introduction

During the past several years, the authors have attempted to develop a complete Hilbert space and Banach space theory of the abstract kinetic equation boundary-value problem

$$(T\psi(x))' = -A\psi(x), \quad 0 < x < \infty \tag{1a}$$

$$Q_+\psi(0) - RJQ_-\psi(0) = \varphi_+ \tag{1b}$$

$$\|\psi(x)\| = 0(1) \text{ or } o(1) \quad (x \to \infty) \tag{1c}$$

on the half-space $x \geq 0$, and Eq. (1c) replaced by

$$Q_-\psi(a) - RJQ_+\psi(a) = \varphi_- \tag{1ĉ}$$

in slab geometry $0 \leq x \leq a$. Here, $T(d/dx)$ and A are linear transformations that correspond to the streaming operator and the collision operator, repectively, $Q_\pm$ are projections onto the positive/negative parts of T (in physical language,

---

Copyright © 1989 by the American Institute of Aeronautics and Astronautics, Inc. All rights reserved.
* Professor, Department of Mathematics.
† Assistant Professor, Department of Mathematics.

projections onto incoming/outgoing fluxes), RJ is the reflection operator at the boundary, and $\varphi_\pm$ are the specified fluxes entering the medium at the boundaries. Such an abstract equation includes most stationary linear model kinetic equations with planar symmetry in neutron transport, radiative transfer, and rarefied gas dynamics which have been considered. In this paper, we will outline a concrete algebraic construction that provides the existence theory for boundary-value problems of the type of Eqs. (1a-c), when the collision operator A is an accretive finite-rank perturbation of the identity operator in a Hilbert space H.

The solution of stationary linear kinetic equations in one-dimensional geometries has been an active industry ever since the fundamental contribution of Case almost 30 years ago on half-range eigenfunction expansions.[1,2] (See also Ref. 3.) The essential element of proving completeness of the eigenfunction expansion is the Wiener-Hopf factorization of a dispersion function.

Traditional *Caseology* suffers from two weaknesses. First, the Wiener-Hopf factorization question, which generally was dealt with by direct complex analysis methods, is very difficult, if not impossible, except for the simplest models, at least in the case of conservative systems. Second, the analysis utilized fairly detailed information about the generalized eigenfunctions, information not readily available in difficult models and not, certainly, in the abstract case.

The modern theory of Wiener-Hopf factorizations has been developed in connection with matrix and operator convolution equations, and so it is natural to look in this direction for a more powerful treatment of abstract kinetic equations. Indeed, this is a return to the earliest approach to stationary equations: beginning with the work of Milne[4], Hopf[5] and others in the 1920s on the equations of radiative transfer, such kinetic equations had been attacked by studying the related (scalar) convolution equations.

The boundary-value problem [Eqs. (1a-c)] may be seen to be equivalent to the integral equation

$$\psi(x) - \int_0^\infty [\mathcal{H}(x-y)B\psi(y) + e^{-xT^{-1}}RJ\mathcal{H}(-y)](I-A)\psi(y)\,dy = e^{-xT^{-1}}\varphi_+ \quad (2)$$

for $\psi \in L^\infty(\mathbb{R}_+, H)$ with the propagator function $\mathcal{H}$ defined by

$$\mathcal{H}(x) = \pm T^{-1} e^{-xT^{-1}} Q_\pm, \quad 0 < \pm x < \infty \quad (3)$$

This is actually not quite a convolution equation unless the reflection operator RJ is zero (*incoming flux* boundary condition). For simplicity in this presentation, we will assume that there is no reflection at the boundary.

In the case where T and A are self-adjoint operators on a Hilbert space H, which is typical of simple Bhatnagar-Gross-Krook and nonpolarized radiative transfer equations, solvability of the convolution equation (with A positive and appropriate geometric constraints on the boundary value for conservative systems) follows by application of the Bochner-Phillips Theorem[6]. This generalization of Wiener's lemma on Fourier integrals provides for invertibility of the convolution operator whenever the *symbol* of the convolution equation is invertible on the extended imaginary axis. This hypothesis is generally straightforward to verify, inasmuch as the symbol $W(\lambda)$ is expressible as a quite simple operator-valued function:

$$W(\lambda) \equiv I - \int_{-\infty}^{\infty} e^{x/\lambda} \mathcal{H}(s)(I - A)\, ds = (\lambda - T^{-1})^{-1}(\lambda - T^{-1}A)$$

Indeed, the dispersion function a la Case is just the (dimensionally reduced) symbol of the related convolution equation.

All of the preceding assumes that the integrand in Eq. (2) makes sense for arbitrary $\psi \in L^\infty(\mathbb{R}_+, \mathcal{H})$, more precisely, that $\mathcal{H}(x)(I - A) \in L^1(\mathbb{R}, L(H))$. One may show that this is the case if T, A satisfy a regularity condition of the type $\operatorname{Ran}(I - A) \subset \operatorname{Ran}|T|^\alpha \cap \operatorname{Ran}|T|^{-\beta}$ for certain $\alpha, \beta > 0$. In general, however, one may conclude only that $\mathcal{H}(x)(I - A) \in \text{w*-}L^1(\mathbb{R}, L(H))$, the closure of $L^1(\mathbb{R}, L(H))$ with respect to the weak-star (predual) topology.

In the present paper, we will utilize an algebraic generalization of the Bochner-Phillips Theorem to study solvability of the abstract boundary-value problem without any regularity condition. The systems of most interest are those with conservation laws reflected by the collision operator A having a nontrivial kernel. Although such systems *can* be treated in the fashion to be outlined here by exploiting a perturbation technique (See Ref. 7), we will assume throughout that Ker $A = 0$ in order to concentrate on the particular difficulties engendered by the lack of regularity of the collision operator.

In Sec. II, we obtain explicitly a Banach algebra in which the convolution kernel acts. In Sec. III, this result is utilized to prove a perturbation theorem for bisemigroups, which then plays a vital role in solving Eqs. (1a-c).

II. Tensor Banach Algebra

Throughout we assume that $B \equiv I - A$ is a finite-rank operator, i.e.,

$$Bh = \sum_{i=1}^{m} \alpha_i (h, e_i) e_i' \tag{4}$$

where $\{e_i\}$ and $\{e_j'\}$ are orthonormal sets in the Hilbert space H. Without loss of generality, we may assume that $\alpha_i > 0$ for $i \leq i \leq m$. The injective tensor product[8] $L^1(\mathbb{R}) \otimes_\epsilon L(H)$ of Banach spaces $L^1(\mathbb{R})$ and $L(H)$ is, by definition, the completion of the algebraic tensor product $L^1(\mathbb{R}) \otimes L(H)$ with respect to the norm

$$\|u\|_\epsilon = \sup\left\{ \left|\sum_{i=1}^{n} g(f_i)g'(T_i)\right| : g \in (L^1(\mathbb{R}))^*, g' \in (L(H))^*, \|g\| \leq 1, \|g'\| \leq 1 \right\}$$

for $u = \sum_{i=1}^{n} f_i \otimes T_i$ with $f_i \in L^1(\mathbb{R})$ and $T_i \in L(H)$.

Let us denote by $K(H)$ and $K_1(H)$ the set of compact operators and trace-class operators on H, respectively. Let $\|\ldots\|_0$ be the trace norm on $K_1(H)$. Since $(K_1(H))^* = L(H)$, it is not difficult to see that $L^1(\mathbb{R}) \otimes_\epsilon L(H)$ is the Banach space of w*-integrable $L(H)$-valued function on $\mathbb{R}$, with norm

$$\|f\|_\epsilon = \sup\{\| <f(\cdot), C>\|_1 : C \in K_1(H), \|C\|_0 \leq 1\} \tag{5}$$

One may show by fairly straightforward arguments that the $\|\ldots\|_\epsilon$ closure of the linear manifold $L^1(\mathbb{R}) \otimes (L(H)B)$ is equal to $(L^1(\mathbb{R}) \otimes_\epsilon L(H))B$.

Define a multiplication on $L^1(\mathbb{R}) \otimes (L(H)B)$ in the natural way as

$$u_1 \cdot u_2 = \sum_{i=1}^{n_1}\sum_{j=1}^{n_2} (f_{1i} * f_{2j}) \otimes T_{1i}BT_{2j}B$$

where the asterisk indicates convolution in $L^1(\mathbb{R})$.

LEMMA 2.1. *There exists a norm $\|\ldots\|_m$ on the Banach space $\overline{L^1(\mathbb{R}) \otimes (L(H)B)}^\epsilon$, which is equivalent to the injective tensor norm $\|\ldots\|_\epsilon$ and such that the space $\overline{L^1(\mathbb{R}) \otimes_\epsilon (L(H)B)}^\epsilon$ is a Banach algebra with the preceding multiplication and this norm.*

PROOF: We recall that $m = \text{Rank}\{B\}$. Let $C \in K_1(H)$ with $C = \sum \nu_i(\cdot, \hat{e}_i')\hat{e}_i$ and $\|C\|_0 = 1$. For any $u = u_1 B$ and $v = v_2 B \in L^1(\mathbb{R}) \otimes (L(H)B)$, one has

$$\|\text{tr}(Cu \cdot v)\|_1 \leq \sum \nu_i \sum_{j=1}^m \alpha_j(u_1 e_j, \hat{e}_i')\|_1 \|v_2 B \hat{e}_i', e_j'\|_1$$

$$\leq \sum_{j=1}^m \alpha_j \sup\{\|(u_1 e_j, e)\|_1 : e \in H, \|e\| \leq 1\} \|v_2 B\|_\epsilon \qquad (6)$$

By considering the trace-class operators $C_j = (\cdot, e)e_j'$, one obtains ($j = 1, 2, \ldots, m$):

$$\|\text{tr}(C_j u_1 B)\|_1 = \|(u_1 B e_j', e)\|_1 = \|\alpha_j(u_1 e_j, e)\|_1$$

Hence, by Eq. (6),

$$\|u \cdot v\|_\epsilon \leq m \|u_1 B\|_\epsilon \|v_2 B\|_\epsilon = m \|u\|_\epsilon \|v\|_\epsilon \qquad (7)$$

It follows from Eq. (7) that the multiplication on $L^1(\mathbb{R}) \otimes (L(H)B)$ can be extended to the linear space $\overline{L^1(\mathbb{R}) \otimes (L(H)B)}^\epsilon$ and

$$\|u \cdot v\|_\epsilon \leq m \|u\|_\epsilon \|v\|_\epsilon$$

for any $u, v \in \overline{L^1(\mathbb{R}) \otimes (L(H)B)}^\epsilon$. Therefore, according to Gelfand's Theorem, there exists an equivalent norm $\|\ldots\|_m$ such that $\overline{L^1(\mathbb{R}) \otimes (L(H)B)}^\epsilon$ is a Banach algebra with respect to $\|\ldots\|_m$.

In order to prove the unique solvability of the convolution equation (2), we extend $L^1(\mathbb{R})$ and $L(H)B$ to Banach algebras $Z_I = \mathbb{C} \oplus L^1(\mathbb{R})$ and $\mathcal{F}_I = \mathbb{C} \oplus (L(H)B)$ with units. The Banach algebra norm $\|\ldots\|_m$ and the algebra multiplication can be extended in the usual manner and will be designated by the same notation. For $\hat{u}, \hat{v} \in Z_I \otimes \mathcal{F}_I$, one has $\|\hat{u} \cdot \hat{v}\|_m \leq \|\hat{u}\|_m \|\hat{v}\|_m$. Let $\mathcal{A}$ be the completion of $Z_I \otimes \mathcal{F}_I$ with respect to the norm $\|\ldots\|_m$. Therefore, the multiplication defined on $Z_I \otimes \mathcal{F}_I$ can be extended to $\mathcal{A}$, and $\mathcal{A}$ is a Banach algebra with unit.

Note that the commutative Banach algebra Z_I is in the center of $\mathcal{A}$ as a subalgebra of $\mathcal{A}$. The following lemma shows further that $\mathcal{A}$ is a so-called $Z_I \otimes \mathcal{F}_I$-algebra, i.e., for every multiplicative functional φ on Z_I, the induced homomorphism $\phi : Z_I \otimes \mathcal{F}_I \to \mathcal{F}_I$ defined by

$$\phi\left(\sum_{i=1}^{n}(\alpha_i I + f_i) \otimes (\beta_i I + T_i B)\right) = \sum_{i=1}^{n} \varphi(\alpha_i I + f_i)(\beta_i I + T_i B) \qquad (8)$$

is a bounded operator.

LEMMA 2.2. For each multiplicative functional φ on Z_I, the homomorphism ϕ induced by φ via Eq. (8) is a bounded operator from $Z_I \otimes \mathcal{F}_I$ to $\mathcal{F}_I$.

PROOF: Let

$$\hat{u} = \sum_{i=1}^{n}(\alpha_i I + f_i) \otimes (\beta_i I + T_i B) \in Z_I \otimes \mathcal{F}_I \qquad (9)$$

Then

$$\|\phi(\hat{u})\| \leq \|\sum_{i=1}^{n}\alpha_i\beta_i\varphi(I)\| + \|\sum_{i=1}^{n}\alpha_i\varphi(I)T_i B\| + \|\sum_{i=1}^{n}\beta_i\varphi(f_i)I\| + \|\sum_{i=1}^{n}\varphi(f_i)T_i B\|$$

We need show only that

$$\|\sum_{i=1}^{n}\varphi(f_i)T_i B\| \leq M\|\sum_{i=1}^{n} f_i \otimes T_i B\|_m \qquad (10)$$

for some constant $M > 0$.

In fact, if $\varphi(f) = 0$ for all $f \in L^1(\mathbb{R})$, then Eq. (10) is true for any $M > 0$. If φ is not zero on $L^1(\mathbb{R})$, then[9,10] there exists a $\lambda \in \mathbb{R}$ such that

$$\varphi(f) = \int_{-\infty}^{\infty} e^{-\lambda t} f(t)\, dt \qquad (11)$$

for $f \in L^1(\mathbb{R})$ and $|\varphi(f)| \leq \|f\|_1$. In this case, choose $C \in K_1(H)$ with $C = \sum \nu_i(\cdot, \hat{e}'_i)\hat{e}_i$ and $\|C\|_0 = 1$. Then we have

$$|\operatorname{tr}\{C\sum_{i=1}^{n}\varphi(f_i)T_i B\}| \leq \sum_{i=1}^{\infty}\nu_i\|\sum_{j=1}^{n}(T_j B\hat{e}_i, \hat{e}'_i)f_j\| \leq \|\sum_{j=1}^{n} f_j \otimes T_j B\|_\epsilon$$

Taking the supremum over the trace-class operators of norm 1, we obtain

$$\|\sum \varphi(f_i)T_i B\| \leq \|\sum f_i \otimes T_i B\|_\epsilon \qquad (12)$$

Hence, it follows from the equivalence of the norms $\|\ldots\|_\epsilon$ and $\|\ldots\|_m$ that the inequality Eq. (10) holds.

The following result, which is an immediate consequence of the fact that $\mathcal{A}$ is a $Z_I \otimes \mathcal{F}_I$-algebra, is the algebraic generalization of the Bochner-Phillips Theorem[6,11,12].

THEOREM 2.1. An element $\hat{u}$ of $\mathcal{A}$ is invertible if and only if $\phi(\hat{u})$ is invertible in $\mathcal{F}_I$ for every induced homomorphism ϕ.

Obviously, the Banach algebra $\mathcal{B} \equiv \mathbb{C} \oplus (L^1(\mathbb{R}) \otimes_\epsilon L(H))B$ can be identified as a Banach subalgebra of $\mathcal{A}$. Let $\hat{u} = \alpha I + uB$ be an element of the subalgebra. Define an operator on $L^\infty(\mathbb{R}, H)$ induced by $\hat{u}$ via the multiplication

$$\hat{u} \cdot \psi(t) = \alpha\psi(t) + uB * \psi(t) = \alpha\psi(t) + \int_{-\infty}^{\infty} u(t - t_1) B\psi(t_1)\, dt_1 \qquad (13)$$

THEOREM 2.2. Every element of the Banach subalgebra $\mathcal{B}$ induces a bounded operator from $L^\infty(\mathbb{R}, H)$ to itself by Eq. (13). If the symbol

$$W(\lambda) = I + \int_{-\infty}^{\infty} e^{-\lambda t} u(t) B\, dt$$

is invertible in $\mathcal{F}_I$ for all λ on the extended imaginary axis $i\mathbb{R}^*$, then for each $w \in L^\infty(\mathbb{R}, H)$ the Wiener-Hopf equation

$$\psi(t) + \int_{-\infty}^{\infty} u(t - t_1) B\psi(t_1)\, dt_1 = w(t) \qquad (14)$$

has a unique solution in $L^\infty(\mathbb{R}, H)$ given by

$$\psi(t) = w(t) - \int_{-\infty}^{\infty} v(t - t_1) B w(t_1)\, dt_1 \qquad (15)$$

where v is some element of $L^1(\mathbb{R}) \otimes_\epsilon L(H)$.

PROOF: Since $L^\infty(\mathbb{R}, H) = L^\infty(\mathbb{R}) \otimes_\epsilon H$, for each $h \in H$ with $\|h\| = 1$ we have

$$|(uB * \psi(t), h)| = |\sum_{i=1}^{m} \alpha_i(ue_i, h) * (\psi(t), e_i')| \le$$

$$\le \sum_{i=1}^{\infty} \alpha_i \|(ue_i, h)\|_1\, \|(\psi(t), e_i')\|_\infty \le \|B\|_0 \|u\|_\epsilon \|\psi\|_\infty$$

Taking the supremum over $t \in \mathbb{R}$ and $h \in H$ with $\|h\| = 1$, we obtain

$$\|uB * \psi\|_\infty \le \|B\|_0 \|u\|_\epsilon \|\psi\|_\infty \qquad (16)$$

Therefore, $\hat{u}$ acts via multiplication as a bounded operator on $L^\infty(\mathbb{R}, H)$.

By Lemma 2.2, one can write $W(\lambda) = I + u_1(\lambda)B$ for some $u_1(\lambda) \in L^1(\mathbb{R}) \otimes_\epsilon L(H)$. It is easy to show that $W(\lambda)$ is invertible in $\mathcal{F}_I$ for $\lambda \in i\mathbb{R}^*$. Since the set of multiplicative functionals on Z_I is the set of multiplicative functionals on $L^1(\mathbb{R})$ plus τ_∞, where τ_∞ is 1 on the unit of Z_I and zero on $L^1(\mathbb{R})$, it can be identified as the extended imaginary axis $i\mathbb{R}^*$ via Eq. (13). We conclude that $\phi(\hat{u})$ is invertible in $\mathcal{F}_I$ for each induced homomorphism ϕ. By Theorem 2.1, $\hat{u} = I + uB$ is invertible in $\mathcal{A}$. Let $\hat{v}$ be its inverse in $\mathcal{A}$. It is not difficult to show that $\hat{v} \in \mathbb{C} \oplus (L^1(\mathbb{R}) \otimes_\epsilon L(H))B$. Consequently, by virtue of the first part of the present theorem, we can apply $\hat{v}$ to both sides of Eq. (14) to obtain Eq. (15).

III. Unique Solvability of Abstract Kinetic Equations

Throughout this section, we assume that the operators T and A satisfy the following conditions:
1) T is a bounded, injective, self-adjoint operator on H.
2) $A = I - B$ is accretive and B has finite rank.
3) $\operatorname{Ker} A = \operatorname{Ker}(A + A^*) = 0$.

It is easy to verify that $-T^{-1}$ generates a strongly continuous bisemigroup $E(t)$ on H [13,14]; i.e., $E(t)$ for $t \in \mathbb{R} \setminus \{0\}$ is a family of bounded operators on H with the following properties:
a) $E(\cdot)$ is strongly continuous.
b) $E(t_1)E(t_2) = \pm E(t_1 + t_2)$ if $\operatorname{sign}(t_1) = \operatorname{sign}(t_2)$ and $E(t_1)E(t_2) = 0$ otherwise.
c) $\Pi_+ + \Pi_- = I$, where the separating projectors $\Pi_\pm \equiv \text{s-}\lim_{\pm t \to 0}(\pm B(t))$.

In fact, $E(t)$ has the form
$$E(t) = \begin{cases} e^{-tT^{-1}}Q_+, & x > 0 \\ e^{-tT^{-1}}Q_-, & x < 0 \end{cases}$$

where $Q_\pm$ are the projections corresponding to the positive / negative parts of T.

The results of the preceding section can be used to prove a perturbation theorem for bisemigroups, interesting in itself and critical for solving Eqs. (1).

THEOREM 3.1. The operator $-T^{-1}A$ generates a strongly continuous bisemigroup $E_A(t)$ with separating projectors $P_\pm$. For $t \in \mathbb{R} \setminus \{0\}$, $E_A(t) - E(t)$ and $P_\pm - Q_\pm$ are compact. Furthermore, $E_A(t)$ is strongly decaying.

PROOF: First we note that the propagator $\mathcal{H}(t) = T^{-1}E(t)$ is in $L^1(\mathbb{R}) \otimes_\epsilon L(H)$. In fact,
$$\mathcal{H}(T) = \begin{cases} T^{-1}e^{-tT^{-1}}Q_+ = \int_0^\infty \frac{1}{\mu}e^{-t/\mu}\,d\sigma(\mu) & t > 0 \\ -T^{-1}e^{-tT^{-1}}Q_- = -\int_{-\infty}^0 \frac{1}{\mu}e^{-t/\mu}\,d\sigma(\mu) & t < 0 \end{cases}$$

where $\sigma(t)$ is the resolution of the identity for T. For any $C \in K_1(H)$ with $C = \sum \nu_i(\cdot, \hat{e}_i')\hat{e}_i$ and $\|C\|_0 = 1$, we have

$$\|\operatorname{tr}(C\mathcal{H}(t))\|_1 = \left\| \sum_{i=1}^\infty \nu_i(\mathcal{H}(t)\hat{e}_i, \hat{e}_i') \right\|_1 \leq$$

$$\leq \sum_{i=1}^\infty \nu_i \left[\int_0^\infty dt \left| \int_0^\infty \frac{1}{\mu}e^{-t/\mu}\,d(\sigma(\mu)\hat{e}_i, \hat{e}_i') \right| + \int_{-\infty}^0 dt \left| \int_{-\infty}^0 \frac{1}{\mu}e^{-t/\mu}\,d(\sigma(\mu)\hat{e}_i, \hat{e}_i') \right| \right]$$

$$\leq 2 \sum_{i=1}^\infty \nu_i = 2\|C\|_0$$

The argument then yields $\|\mathcal{H}(t)\|_\epsilon \leq 2$.

Therefore, by Theorem 2.2, the operator $\mathcal{L}$, defined by

$$\mathcal{L}\psi(t) = \int_{-\infty}^{\infty} \mathcal{H}(t-t_1)B\psi(t_1)\,dt_1$$

is bounded on $L^\infty(\mathbb{R}, H)$. The symbol of this equation is

$$W(\lambda) = I + (\lambda T^{-1} - I)^{-1}B = (\lambda - T^{-1})^{-1}(\lambda - T^{-1}A), \quad \text{Re}\,\lambda = 0$$

We see that $W(\lambda)$ is a compact perturbation of the identity and does not have eigenvalues on the extended imaginary axis. Hence, $W(\lambda)$ is invertible for $\lambda \in i\mathbb{R}^*$. Theorem 2.2 shows that the operator $I - \mathcal{L}$ is invertible in $L(L^\infty(\mathbb{R}, H))$ and its inverse is given by $I + \mathcal{L}_1$, where

$$\mathcal{L}_1\psi(t) = \int_{-\infty}^{\infty} \mathcal{H}_1(t-t_1)B\psi(t_1)\,dt_1$$

for some $\mathcal{H}_1(t) \in L^1(\mathbb{R}) \otimes_\epsilon L(H)$.

Let us define $E_A(t)h = (I + \mathcal{L}_1)E(t)h$ for $h \in H$ and $t \in \mathbb{R} \setminus \{0\}$. It is easily verified that $E_A(t)$ is a strongly continuous bisemigroup generated by $-T^{-1}A$ and also is strongly decaying.[7,13]

It remains to show that $E_A(t) - E(t)$ and $P_\pm - Q_\pm$ are compact operators for $t \in \mathbb{R} \setminus \{0\}$. Indeed, let $\phi_n(t) = |t|^{1/n}X_{[n,n]}(t)$, where $X_{[-n,n]}$ is the characteristic function of the interval $[-n,n]$ for $n \in \mathbb{Z}$. By the functional calculus for T^{-1}, it is easy to check that $\|(\phi_n(T^{-1}) - I)h\| \to 0$ and therefore $\|(\phi_n(T^{-1}) - I)G\| \to 0$ as $n \to \infty$ for each $h \in H$ and compact operator G on H. From this we can show that $\|(\phi_n(T^{-1}) - I)B\|_0 \to 0$ as $n \to \infty$. Using Eq. (6) and the strong decay of $E_A(t)$, we have the estimate

$$\left\| \int_{-\infty}^{\infty} \mathcal{H}(t-t_1)(\phi_n(T^{-1}) - I)BE_A(t_1)h\,dt_1 \right\| \leq 2\|G(\phi_n(T^{-1}) - I)B\|_0 \cdot \|E_A(t)\|_\infty \cdot \|h\|$$

for any $h \in H$ and $t \in \mathbb{R}$. This shows that the operator $\mathcal{L}_n E_A(t)$, defined by

$$\mathcal{L}_n E_A(t) = \int_{-\infty}^{\infty} \mathcal{H}(t-t_1)\phi_n(T^{-1})BE_A(t_1)\,dt_1$$

approaches $\mathcal{L}E_A(t)$ in operator norm as $n \to \infty$. However, $\mathcal{H}(t)\phi_n(T^{-1})B$ is a Bochner-integrable $K(H)$-valued function. Then $\mathcal{L}_n E_A(t)$ and $\mathcal{L}_n E_A(0)$ are compact operators, and, consequently, $E_A(t) - E(t)$ and $P_\pm - Q_\pm$ are compact. This completes the proof.

Finally, we consider the unique solvability of Eqs. (1a–c). We say that a function $\psi(x): [0, \infty) \to H$ is a solution of Eqs. (1) if $\psi(x)$ is continuous on $[0, \infty)$ and $T\psi(x)$ is strongly differentiable for $0 < x < \infty$, such that Eqs. (1) are satisfied. Using the separating projectors $P_\pm$, we can prove that $\psi(x)$ is a solution of Eqs. (1a–c) if and only if $\psi(x) = E_A(x)P_+h$ for $0 \leq x < \infty$, where $Q_+P_+h = \varphi_+$ for some $h \in H$.[7]

Define the operator V on H by $V = Q_+P_+ + Q_-P_-$. If V is invertible, then Eqs. (1) have the unique solution $\psi(x) = E_A(x)V^{-1}\varphi_+$. Therefore, it suffices to show the invertibility of the operator V to prove the unique solvability of Eqs. (1). This, however, follows from the accretive assumption on A, as derived in Ref. 13. To summarize these results, we have the following theorem.

THEOREM 3.2. Under the preceding assumptions (1-3) on T and A, Eqs. (1a–c) have the unique solution

$$\psi(x) = E_A(x)V^{-1}\varphi_+$$

where $V = Q_+P_+ + Q_-P_-$.

Acknowledgment

This work was supported in part by DOE Grant DE-FG05 87ER25033 and NSF Grant DMS 8701050.

References

[1] Case, K. M., "Plasma Oscillations," Annals of Physics (New York), Vol. 7, 1959, pp. 349–364.

[2] Case, K. M., "Elementary Solutions of the Transport Equation," Annals of Physics (New York), Vol. 9, 1960, pp. 1–23.

[3] van Kampen, N. G., "On the Theory of Stationary Waves in Plasmas," Physica, Vol. 21, 1955, pp. 949–963.

[4] Milne, E. A., "Radiative Equilibrium in the Outer Layers of a Star: the Temperature Distribution and the Law of Darkening," Notices of the Royal Astronomical Society, Vol. 81, 1921, pp. 361–375.

[5] Hopf, E., *Mathematical Problems of Radiative Equilibrium*, Cambridge University Press, Cambridge, 1934.

[6] Bochner, S. and Phillips, R. S., "Absolutely Convergent Fourier Expansions for Noncommutative Normed Rings, Annals of Mathematics, Vol. 43, 1942, pp. 409–418.

[7] Greenberg, W., van der Mee, C. V. M., and Protopopescu, V., *Boundary Value Problems in Abstract Kinetic Theory*, Birkhäuser Verlag, Basel, 1987.

[8] Diestel, J, and Uhl, J. J., Jr., *Vector Measures*, American Mathematical Society, Providence, 1977.

[9] Larsen, R., *Banach Algebras: An Introduction*, Marcel Dekker, New York, 1973.

[10] Gelfand, L. M., Raikov, D. A., and Shilow, G. Z., *Commutative Normed Rings*, Chelsea, New York, 1964.

[11] Allan, G. R., "On One-sided Inverses in Banach Algebras of Holomorphic Vector-valued Functions," Journal of London Mathematical Society, Vol. 42, 1967, pp. 463–470.

[12] Gohberg, I. C. and Leiterer, J., "Factorization of Operator Functions with Respect to a Contour, I,II,IV," Mathematische Nachrichten, Vol. 52, 1972, pp. 259–282; Vol. 54, 1972, pp. 41–74; Vol. 55, 1973, pp. 33–61.

[13] Ganchev, A. H., Greenberg, W. and van der Mee, C. V. M., " Abstract Kinetic Equations with Accretive Collision Operators," Integral Equations and Operator Theory, Vol. 11, 1988, pp. 332–350.

[14] Bart, H., Gohberg, I. C., and Kaashoek, M. A., "Wiener-Hopf Factorization, Inverse Fourier Transforms and Exponentially Dichotomous Operators," Journal of Functional Analysis, Vol. 68, 1986, pp. 1–42.

Singular Solutions of the Nonlinear Boltzmann Equation

J. Polewczak*
*Virginia Polytechnic Institute and State University,
Blacksburg, Virginia*

Abstract

New estimations of the nonlinear Boltzmann collision operator $J(F)$ are given. For potentials with angular cutoff and scattering kernels $B(\theta, |V|)$ bounded by $\cos\theta |V|^\beta$, $\beta > -1$, it is shown that $|Q(F,G)| \leq c\|g\|_{L^p} g(|v|) m(|v|)(1+|v|)^{\alpha_\beta}$ and $|FR(G)| \leq c\|g\|_{L^p} g(|v|) m(|v|)(1+|v|)^\beta$ if $|F|, |G| \leq g(|v|) m(|v|)$. Here, $J(F) = Q(F,F) - FR(F)$, g is a nonincreasing L^p function with $1 \leq p \leq \infty$; and $m(|v|) = \exp(-\alpha v^2)$, $\alpha > 0$, or $m(|v|) = (1+v^2)^{-r/2}$, $r > 2$. In addition, the exponent $\alpha_\beta < 0$ for some $p < \infty$, even when $\beta > 0$. The preceding result includes all known estimations of $J(F)$ in supnorm. These new estimations are used to prove the existence of a mild solution to the nonlinear Boltzmann equation in R^3. The solution is global in time if the initial data decay fast enough at infinity and L^p norm of g is small. Furthermore, the existence theorem admits solutions that are unbounded both in the space variable and velocity. Finally, if g is bounded and the initial value is smooth, the existence theorem provides a classical solution together with the asymptotic behavior.

Introduction

We consider the initial value problem for the Boltzmann equation

$$\frac{\partial F}{\partial t} + v\frac{\partial F}{\partial x} = J(F), \quad t > 0, \quad F(0, v, x) = f_0(v, x) \tag{1}$$

where $F : [0, T] \times R^3 \times R^3 \to R$ is the one-particle distribution function that depends on the velocity $v \in R^3$, the position $x \in R^3$, and whose time evolution is governed by Eq. (1). The collision operator J with a cutoff, $J(F) = Q(F, F) - FR(F)$, is given in terms of the following operators:

$$Q(F,F)(v) = \int_{S_+^2 \times R^3} B(\theta, |w-v|) F(v') F(w') d\omega dw \tag{2a}$$

Copyright © 1989 by the American Institute of Aeronautics and Astronautics, Inc. All rights reserved.
* Department of Mathematics and Center for Transport Theory and Mathematical Physics.

$$FR(F)(v) = \int_{S^2_+ \times R^3} B(\theta, |w - v|) F(v) F(w) d\omega dw \qquad (2b)$$

where $F(v)$ denotes $F(t, v, x)$.

Because the kinetic energy and the linear momentum are conserved during the collision, we have the following relations among $v, w, v', $ and w':

$$v' = v - \omega \langle v - w, \omega \rangle, \quad w' = w + \omega \langle v - w, \omega \rangle \qquad (3)$$

where $\langle \cdot , \cdot \rangle$ is the inner product in R^3, $\omega \in S^2_+ = \{\omega \in S^2 : \langle v - w, \omega \rangle \geq 0\}$, and $S^2 = \{\omega \in R^3 : |\omega| = 1\}$. The relations in Eq. (3) are equivalent to

$$v + w = v' + w', \quad v^2 + w^2 = v'^2 + w'^2 \qquad (4)$$

The angle $\theta \in [0, \pi/2]$ is given by $\cos \theta = \langle v - w, \omega \rangle / |w - v|$. The function $B(\theta, |v|)$ is defined on $[0, \pi/2] \times R_+$ and is continuous. Throughout this paper, we will assume the following bound on $B(\theta, |v|)$:

$$\frac{|B(\theta, |v|)|}{|\cos \theta|} \leq c |v|^\beta \qquad (5)$$

where $c > 0$ and $\beta > -1$. For inverse power potentials, $\mathcal{F}(r) = r^{-s}$ and $B(\theta, |v|) = b(\theta)|v|^{(s-4)/s}$. The function $b(\theta)$ is nonnegative and has a singularity at $\pi/2$ of the type $(\cos \theta)^{-\lambda}$, where $\lambda = (s+2)/s$. Assuming the usual angular cutoff hypothesis[1], inequality (5) is satisfied for all inverse power potentials with $s > 2$. The rigid spheres model $(B(\theta, |v|) = |v| \cos \theta)$ corresponds to $\beta = 1$.

After integration along the characteristics, Eq. (1) can be written formally as

$$F(t) = U(t)f_0 + \int_0^t U(t - s) J(F(s)) \, ds \qquad (6)$$

where $(U(t)f_0)(v, x) = f_0(v, x - tv)$ for $t \in R$ and $(v, x) \in R^3 \times R^3$.

New estimations of $J(F)$

In this section we show that for potentials with angular cutoff and scattering kernels $B(\theta, |v|)$, as in the preceding section $|Q(F,G)| \leq C\|g\|_{L^p} g(|v|) m(|v|)(1 + |v|)^{\alpha_\beta}$ and $|FR(G)| \leq c\|g\|_{L^p} g(|v|) m(|v|)(1 + |v|)^\beta$ if $|F|, |G| \leq g(|v|)m(|v|)$.

This property holds for a large class of bounding functions g, which are not necessarily bounded, and for $m(|v|) = \exp(-\alpha v^2)$, $\alpha > 0$, or $m(|v|) = (1+v^2)^{-r/2}$, $r > 2$. Moreover, for $m(|v|) = \exp(-\alpha v^2)$, the gain term $Q(F, F)$ possesses certain mollifying properties as compared to the loss term $FR(F)$. This fact is used in a very simple local existence theorem to Eq. (1).

Let g be a nonincreasing, positive, and continuous function defined on $(0, \infty)$ with the property that there exists $K > 1$ such that $g(s/2) \leq Kg(s)$ for $s > 0$. Note that, since g is nonincreasing, the last property of g is equivalent to: for each $0 < d < 1$, there exists $K(d)$ such that $g(ds) \leq K(d)g(s)$ for $s > 0$. Furthermore, we require that $g(|\cdot|) \in L^p(R^3)$, and the norm of $g(|\cdot|)$ will be denoted by $\|g\|_{L^p(R^3)}$.

SOLUTIONS OF THE BOLTZMANN EQUATION 41

For $\alpha > 0$, we define $X(g,\alpha) = \{F : F$ is Lebesgue measurable on $R_+ \times R^3 \times R^3$ and $|F(t,v,x)| \leq cg(|v|)\exp(-\alpha v^2)$ for some $c > 0$ a.e. (almost everywhere) in $(t,v,x)\}$. Then $X(g,\alpha)$ with the norm given by

$$\|F\|_{g,\alpha} = \sup_{(t,v,x)} \{g^{-1}(|v|)\exp(\alpha v^2)|f(t,v,x)|\}$$

becomes a Banach space.

THEOREM 2.1 (Generalized Grad estimation) Let $F, G \in X(g,\alpha)$, $-1 < \beta \leq 1$ and $1 \leq p \leq \infty$. We have

i $g^{-1}(|v|)\exp(\alpha v^2)|Q(F,G)| \leq M_1(p,\alpha,K(1/\sqrt{2}))\|F\|_{g,\alpha}\|G\|_{g,\alpha}$

$\times \left\{\|g\|_{L^p(R^3)}(1+v^2)^{(\beta-2)/2} + g(|v|)(1+v^2)^{(1+\beta)/2}\right\}$ a.e. in (t,v,x) (7)

ii $g^{-1}(|v|)\exp(\alpha v^2)|FR(G)| \leq M_2(\beta,p,\alpha)\|F\|_{g,\alpha}\|G\|_{g,\alpha}\|g\|_{L^p(R^3)}(1+v^2)^{\beta/2}$

a.e. in (t,v,x) (8)

where $p > 3/(3-|\beta|)$ if $\beta < 0$.

PROOF: To prove part i we need the following integral representation of $Q(F,G)$, originally obtained by Carleman[3] for the special case of hard spheres (see also Arkeryd[4] for a treatment of the general case):

$$Q(F,G)(v) = \int_{R^3} F(v') \int_{E_{v,v'}} G(w')B(\theta,|v'-w'|)|w'-v'|^{-2}\cos^{-2}\theta dE'dv'$$

where $E_{v,\bar{v}} = \{z \in R^3 : \langle \bar{v}-v, z-v\rangle = 0\}$ and dE' denotes the Lebesgue measure on $E_{v,v'}$. Next, using Eq. (5) and the fact that $|v'-v| = \cos\theta|w'-v'|$, we have $Q(F,G)(v) \leq cQ_0(F,G)(v)$ a.e., where

$$Q_0(F,G)(v) = \int_{R^3} \frac{F(v')}{|v'-v|} \int_{E_{v,v'}} \frac{G(w')}{|w'-v'|^{1-\beta}} dE'dv'$$

Using the symmetry of Q_0 [$(Q_0(F,G)(v) = Q_0(G,F)(v)$ a.e.(see Ref. 3 p. 35)] and the conservation of energy (see Ref. 3 pp. 37-38, and Ref. 4 pp. 359-360), we obtain the following for the nonnegative functions $F(v)$:

$$Q_0(F,F)(v) \leq 2\int_{R^3} \frac{F(v)}{|v'-v|} \int_{|w'|\geq |v|/\sqrt{2}} \frac{F(w')}{|w'-v'|^{1-\beta}} dE'dv'$$

where $|w'| \geq |v|/\sqrt{2}$ is shorthand notation for the set $\{w' \in E_{v,v'} : |w'| \geq |v|/\sqrt{2}\}$.

Now we can write

$$|Q(F,G)(t,v,x)| \leq c\exp(-\alpha v^2)\int_{R^3} \frac{g(|v'|)\exp\{-\alpha[(v'-v)+v^{(1)}]^2\}}{|v'-v|}$$

$$\times \int_{|w'|\geq |v|/\sqrt{2}} \frac{g(|w'|)\exp\{-\alpha[(w'-v)+v^{(2)}]^2\}}{|w'-v'|^{1-\beta}} dE'dv'$$

where $v = v^{(1)} + v^{(2)}$, with $v^{(1)} \perp (w' - v)$ and $v^{(2)} \perp (v' - v)$. From the last inequality, part i follows easily. Indeed, since $\beta \leq 1$, we have

$$|Q(F,G)| \leq c\exp(-\alpha v^2)g(|v|)K(1/\sqrt{2})\int_{R^3} \frac{g(|v'|)\exp\{-\alpha[(v'-v)+v^{(1)}]^2\}}{|v'-v|^{2-\beta}}dv'$$

Since the preceding integral can be estimated by $\text{const}|v|^{\beta-2}\|g\|_{L^p(R^3)}$ for $|v'| \leq |v|/\sqrt{2}$ or $|v'| \geq 2|v|$, and for $\beta \geq -1$,

$$\int_{|v|/\sqrt{2}\leq|v'|\leq 2|v|} g(|v'|)|v'-v|^{\beta-2}\,dv' \leq \text{const}g(|v|)|v|^{1+\beta}$$

we obtain Eq. (7).

Part ii follows after an easy integration.

We remark that the Grad's estimation (see Ref. 5 pp. 180-182) is a special case of Eq. (7) and (8) with $g(|v|) = (1+v^2)^{-r/2}$, $r \geq 1$. Grad[5] did not obtain the decay in Eq. (7). We also point out that the decay in Eq. (7) is an archetype for the mollifying property of the integral part of the linear collision operator that is derived from Q.

Next, when $(1+v^2)^{(1+2\beta)/2}g(|v|) \in L^\infty(R^3)$ and $\beta \leq 1$, Eq. (7) implies that

$$g^{-1}(|v|)e^{\alpha v^2}|Q(F,G)(t,v,x)| \leq M(p,\alpha,k)$$
$$\times \|F\|_{g,\alpha}\|G\|_{g,\alpha}\|(1+v^2)^{(1+2\beta)/2}g(|\cdot|)\|_{L^\infty(R^3)}(1+v^2)^{-\beta/2} \text{ a.e. in } (t,v,x)$$
(7a)

Thus, the decay in (7a) can destroy the growth in Eq. (8). This fact is used in the following local existence theorem to Eq (1).

COROLLARY 2.2 Let $\alpha > 0$, $\beta \geq 0$, $(1+v^2)^{(1+2\beta)/2}g \in L^\infty(R^3)$ and $R > 0$. Suppose that $f_0 \geq 0$ is such that $(1+v^2)^{\beta/2}f_0 \in X_{g,\alpha}$ and $\|f_0\|_{g,\alpha} \leq R$. Then there exists $T > 0$ (which depends on f_0) such that the following problem has a nonnegative solution in $[0,T]$:

$$f(t,v,x,) = f_0(v,x) + \left(\int_0^t U(-s)J(U(s)f(s))ds\right)(v,x) \quad (9)$$

where the preceding integral is the Lebesgue integral computed for each $(v,x) \in R^3 \times R^3$.

PROOF: First, we note that Eq. (9) is obtained from Eq. (6) by acting with $U(-t)$ on both sides of Eq. (6) and introducing $f(t) = U(-t)F(t)$. Next, one can write Eq. (9) in the following integral form:

$$f(t) = Af(t) = f_0\exp\{-\int_0^t U(-s)R(U(s)f(s))ds\}$$
$$+ \int_0^t U(-s)Q(U(s)f(s),U(s)f(s))\exp\{-\int_s^t U(-\tau)R(U(\tau)f(\tau))d\tau\}ds$$
(10)

Now, $|\exp(-z) - \exp(-y)| \le |z-y|$ and $(1+v^2)^{\beta/2}f_0 \in X(g,\alpha)$ together with Eq. (7a) imply that A maps $X(g,\alpha)$ into itself and is a contraction on the set $\{f \in X(g,\alpha) : f(t,v,x) \ge 0 \text{ a.e. in } (t,v,x) \text{ and } \|f\|_{g,\alpha} \le 2R\}$ for small enough $T \ge 0$. Furthermore, each solution to Eq. (10) is solution to Eq. (9) since the function

$$(\int_0^t U(-s)R(U(s)f(s))\,ds)(v,x) \in L^1_{\text{loc}}(R_+ \times R^3 \times R^3)$$

We do not know whether one can prove a global in time existence theorem to Eq. (9) in $X(g,\alpha)$. Recently, DiPerna and Lions[2] have proved **a global existence theorem to the Boltzmann equation for large data in L^1**. In this paper, however, we study the Boltzmann equation in **Banach spaces with various supnorms**. We impose certain smallness conditions on initial data, and, as a result, we are able to obtain various types of explicit singularities in the solutions to the Boltzmann equation in both v and x variables.

Let $X(g,r)$ denote $X(g,\alpha)$, with $\exp(-\alpha v^2)$ replaced by $(1+v^2)^{-r/2}$ and $\|\cdot\|_{g,r}$ its norm. For $m(|v|) = (1+v^2)^{-r/2}$ we have the following version of Theorem 2.1:

THEOREM 2.3 Let $F, G \in X(g,r)$, $r > 3/2$, $-1 < \beta \le 1$. We have

$$(1+v^2)^{r/2}|Q(F,G)(t,v,x)| \le C(r, K(1/\sqrt{2}))\|F\|_{g,r}\|G\|_{q,r}\|g\|_{L^\infty}g(|v|)(1+v^2)^{\beta/2}$$

and

$$(1+v^2)^{r/2}|FR(G)(t,v,x)| \le C(r, K(1/\sqrt{2}))\|F\|_{g,r}\|G\|_{q,r}\|g\|_{L^\infty}g(|v|)(1+v^2)^{\beta/2}$$

a.e. in (t,v,x).

Next we want to prove existence theorems to Eq. (9) that are global in time. Such results have been obtained previously by many authors.[6-12] They showed that if $|f| \le h(|x|)\exp(-\alpha v^2)$, then the operator $H(f) = \int_0^t U(-s)J(U(s)f(s))ds$ satisfies $|(Hf)(t,v,x)| \le c\|h\|_{L^1(0,\infty)}\|f\|^2 h(|x|)\exp(-\alpha v^2)$ a.e. in (t,v,x), where $\|f\| = \sup_{(t,v,x)} h^{-1}(|v|)\exp(\alpha v^2)|f(t,v,x)|$. We will generalize this result to show that h can be a function of both the $|x|$ and $|v|$ variables. These new estimations allow us to prove the global existence to Eq. (9) in a class of unbounded functions.

We start with a simple generalization of the Bellomo-Toscani estimation. Let $H(z,s)$ be a positive, continuous function defined on $(0,\infty) \times (0,\infty)$. We also assume that $H(z,s)$ is nonincreasing with respect to each variable and has the property that there exists $K > 1$ such that $H((1/2)z,(1/2)s) \le KH(z,s)$ for $z, s > 0$. For $r > 0$, we define $X(H,r) = \{f : f$ is Lebesgue measurable on $R_+ \times R^3 \times R^3$ and $|f(t,v,x)| \le cH(|v|,|x|)(1+v^2)^{r/2}$ for some $c > 0$ and a.e. in $(t,v,x)\}$. Then $X(H,r)$ with the norm given by

$$\|f\| = \sup_{(t,v,x)} \{H^{-1}(|v|,|x|)|f(t,v,x)|\}$$

becomes a Banach space. Let

$$\tilde{Q}(f_1,f_2)(t,v,x) = \int_0^t U(-s)Q(U(s)f_1, U(s)f_2)ds$$

and
$$\tilde{R}(f_1, f_2)(t, v, x) = \int_0^t f_1(s) U(-s) R(U(s) f_2) ds$$

THEOREM 2.4 Let $f_1, f_2 \in X(H, r)$, $-1 < \beta \leq 1$, $r > 2 + \delta$. Furthermore, assume that
$$\sup_{z \geq 0} [\int_0^\infty H(z, s) ds] = M_H < \infty$$
Then
$$\|\tilde{Q}(f_1, f_2)\| \leq C(\beta, r, H) M_H \|f_1\| \|f_2\|$$
and
$$\|\tilde{R}(f_1, f_2)\| \leq C(\beta, r, H) M_H \|f_1\| \|f_2\|$$

PROOF: First, we notice that Lemma 2.1 of Bellomo and Toscani[9] is true with h replaced by H. Indeed, it is enough to observe that $v' = v + v' - v$ and $w' = v + w' - v$ with $<v'-v, w'-v> = 0$. Next, an application of Lemma 2.3 of Bellomo and Toscani[9] concludes our proof.

We remark that the Bellomo-Toscani result is a special case of Theorem 2.4 with $H(|v|, |x|) = h(|x|)$. Furthermore, $H(|v|, |x|)$ can be an unbounded function of $|x|$.

Our final estimation allows function H to be unbounded in both $|v|$ and $|x|$. This is done for $m(|v|) = \exp(-\alpha v^2)$. Let $Y(g, h, \alpha)$ denote $X(H, v)$, with $H(|v|, |x|) = g(|v|) h(|x|)$ and $(1 + v^2)^{-r/2}$ replaced by $\exp -\alpha v^2$. Here g is a function from Theorem 2.1, and h is a function defined on $(0, \infty)$ with the same properties as g and such that it belongs to $L^1(0, \infty)$. The norm at h is denoted by $\|h\|_{L^1(0,\infty)}$. We have

THEOREM 2.5 Let $f_1, f_2 \in Y(g, h, \alpha)$, $-1 < \beta \leq 1$, $p_1 > 3$ and $p_2 > 2$. We have

i $\|\tilde{Q}(f_1, f_2)\| \leq M_1(K_g, K_h) \|h\|_{L^1} (\|g\|_{L^{p_1}(R^3)} + \|g\|_{L^{p_2}(R^2)}) \|f_1\| \|f_2\|$ (11)

ii $\|\tilde{R}(f_1, f_2)\| \leq M_2(K_h) \|h\|_{L^1} \|g\|_{L^p(R^3)} \|f_1\| \|f_2\|$ (12)

where $K_g \geq 1$ is such that $g(s/\sqrt{2}) \leq K_g(s)$ for $s > 0$ and K_h is such that $h((1 - 1/\sqrt{2})s) \leq K_h h(s)$ for $s \geq 0$.

PROOF: To prove part i we notice that by using the proof of Lemma 2.1 in Bellomo-Toscani[9] and the integral representation of Q_0 we obtain

$$|\tilde{Q}(f_1, f_2)(t, v, x)| \leq c K_h h(|x|) \exp(-\alpha v^2) \|h\|_{L^1(0,\infty)} \max\{I_1, I_2\}$$

where
$$I_i = \int_{R^3} \frac{F_i(v') \exp\{-\alpha[(v'-v) + v^{(1)}]^2\}}{|v'-v|}$$
$$\times \int_{E_{v,v'}} \frac{G_i(w') \exp\{-\alpha[(w'-v) + v^{(2)}]^2\}}{|w'-v'|^{1-\beta}} dE' dv', \quad i = 1, 2$$

with $F_1(v') = g(v')/|v'-v|$, $G_1(w') = g(w')$, $F_2(v') = g(v')$, and $G_2(w') = g(w')/|w'-v|$, and $v^{(1)}, v^{(2)}$, as in the proof of Theorem 2.1.

Next, using the conservation of energy, we have

$$\begin{aligned}I_i = &\int_{R^3} \frac{F_i(v')\exp\{-\alpha[(v'-v)+v^{(1)}]^2\}}{|v'-v|}\\&\times \int_{|w'|\geq |v|/\sqrt{2}} \frac{G_i(w')\exp\{-\alpha[(w'-v)+v^{(2)}]^2\}}{}|w'-v'|^{1-\beta}\,dE'dv'\\&+ \int_{|v'|\geq |v|/\sqrt{2}} \frac{F_i(v')\exp\{-\alpha[(v'-v)+v^{(1)}]^2\}}{|v'-v|}\\&\times \int_{|w'|\leq |v|/\sqrt{2}} \frac{G_i(w')\exp\{-\alpha[(w'-v)+v^{(2)}]^2\}}{|w'-v'|^{1-\beta}}\,dE'dv',\ i=1,2.\end{aligned}$$

From the preceding representation the estimation for I_1 follows easily. Indeed, for $p_1 > 3$ and $p_2 > 2$, we have

$$I_1 \leq K_g g(|v|)\left[c\|g\|_{L^{p_1}}\left(\int_{R^3}\frac{\exp\{-\alpha q_1[(v'-v)+v^{(1)}]^2\}}{|v'-v|^{2q_1}}\right)^{\frac{1}{q_1}} + \frac{c}{|v|^{1-\beta}}\|g\|_{L^{p_2}}\right]$$

Since the estimation for I_2 is very similar to the one obtained earlier for I_1, we conclude the proof of part i.

Part ii follows easily after one uses Lemma 2.1 of Bellomo-Toscani[9] and integrates over $w \in R^3$.

Theorems 2.1 and 2.3-2.5 show that positive, nonincreasing functions Γ that satisfy the condition $\Gamma(s/2) \leq K\Gamma(s)$ for $s > 0$ with K independent of s play an important role in the estimations of $J(F)$. In addition, Γ can be unbounded as in the following example:

$$\begin{aligned}H(s) &= \exp(r-d)/s^r \quad 0 < s \leq e\\&= 1/(s^d \ln^\gamma s) \quad s \geq e\end{aligned}$$

where $0 < r < 1$, $d \geq 1$ and $\max(d, \gamma) > 1$, if $h(s) = \Gamma(s)$ or $0 < r < 2$ and $d > 2$, $\gamma > 0$ are such that $1/(s^{d-2}\ln^\gamma s) \in L^p(e, \infty)$ if $g(s) = \Gamma(s)$.

In the preceding example, $s^d \ln^\gamma s$ is an N function that satisfies the Δ_2 condition for large s (see Ref. 13 p. 23). In fact, one can take Γ defined by

$$\begin{aligned}\Gamma(s) &= (1/M(s_0))(s_0/s)^r \quad 0 < s \leq s_0\\&= 1/M(s) \quad s \geq s_0\end{aligned}$$

where $r < 1$ for h and $r < 2$ for g and $M(s)$ is any N function that satisfies the Δ_2 condition for $s \geq s_0$ and such that $1/M(s) \in L^1(s_0, \infty)$ or $s^2/M(s) \in L^p(s_0, \infty)$.

We point out that there exists a large class of functions $M(s)$. They are used in the definition of various Orlicz spaces.[13]

COROLLARY 2.6 Suppose that the conditions of Theorem 2.5 are satisfied and $f_0 \geq 0$ is such that $\|f_0\| \leq 1$. Then, for $\|h\|_{L^1(0,\infty)}(\|g\|_{L^p(R^3)}) + \|g\|_{L^p(R^2)})$ sufficiently small, there exists a nonnegative solution to Eq. (9) in $Y(g, h, \alpha)$ with $\|f\| \leq 2$.

PROOF: As in the proof of Corollary 2.2, we use Eq. (10). Now Theorem 2.5 implies that the operator A is a contraction on the set $\{f \in Y(g,h,\alpha) : f(t,v,x) \geq 0$ a.e. in (t, v, x) and $\|f\| \leq 2\}$ if $\|f_0\| \leq 1$ and the number $\|h\|_{L^1(0,\infty)}(\|g\|_{L^p(R^3)} + \|g\|_{L^p(R^2)})$ is small enough.

A similar global existence theorem can be obtained from Theorem 2.4 if M_H is sufficiently small.

So far, the solution $f(t, v, x)$ of Eq. (9) could be unbounded only at $x = 0$ and $v = 0$. A simple modification of the earlier analysis can give any finite number of points $x_i \in R^3$, $i = 1, \ldots, n$, at which the solution is unbounded. To do this let, $n \geq 1$ and h_i satisfy the same conditions as the function g in Theorem 2.1. For $x_i \in R^3$, $i = 1, \ldots, n$, $v_0 \in R^3$, and $\alpha > 0$, we define $Y(g, h, \alpha) = \{f : f$ is measurable on $R^+ \times R^3 \times R^3$ and $|f(t,v,x)| \leq cg(|v-v_0|)\exp\{-\alpha(v-v_0)^2\}h_1(|x-x_1|)\ldots h_n(|x-x_n|)$ for some $c > 0$ and a.e. in $(t, v, x)\}$ with the supnorm as for the space $Y(g, h, \alpha)$. For $r_i \geq 1$, such that

$$\sum_{i=1}^{n} \frac{1}{r_i} = 1$$

assume that $h_i \in L^{r_i}(0,\infty)$. By using Lemma 2.1 of Ref. 9 for each h_i, $i = 1, \ldots, n$, and the Hölder inequality, we can obtain the same estimations for Q and R as in Theorem 2.5 with $\|h\|_{L^1(0,\infty)}$ replaced by $\|h_1\|_{L^{r_1}(0,\infty)} \cdots \|h_n\|_{L^{r_n}(0,\infty)}$.

We also note that, as opposed to many authors,[6-12] we did not use the iterative scheme of Kaniel and Shinbrot[14] to obtain a nonnegative solution to Eq. (9). We used, instead, the integral form [Eq. (10)] that preserves nonnegativity.

We note that the solution to Eq. (9) obtained in Corollary 2.6 becomes a classical solution if h and g are bounded.[12] Polewczak[12] also gives the asymptotic behavior of the solution. As $t \to \infty$, the solution to the Boltzmann equation can be approximated by the solution to the free motion problem. This is equivalent to the existence of the corresponding wave operator associated with the Boltzmann collision operator. Furthermore, by using the explicit bound on the collision operator proven in Ref. 12 (Corollary 4.1), one can show that the collisions become less important as compared to the translational motion of the molecules of the gas. This behavior is due to escape of the gas as $t \to \infty$ from any bounded set of R^3.

Finally, we want to point out that all results given in this paper are also valid for the Enskog equation with the Y function locally Lipschitz (see Résibois,[15] Toscani and Bellomo,[9,16] and Polewczak[17] for definitions). In the case of the Enskog equation, h must be bounded but g can be an unbounded function if Y is bounded on $(0, \infty)$.

Acknowledgment

This work was supported in part by DOE Grant DEF60587ER25033 and NSF Grant DMS87015050.

References

[1] Grad, H., "Asymptotic Theory of the Boltzmann II, "Rarefied Gas Dynamics I, edited by J. A. Laurmann, Academic Press, 1963, pp. 26-59.

[2] DiPerna, R. L. and Lions, P. L., "On the Cauchy Problem for the Boltzmann Equations: Global Existence and Weak Stability," Annals of Mathematics (to be published).

[3] Carleman, T., Problèmes Mathématiques dans la Théorie Cinétique des Gaz, Almiqvist, Uppsala, Sweden, 1957.

[4] Arkeryd, L., "L^∞ Estimates for the Space Homogenous Boltzmann Equation," Journal of Statistical Physics,Vol. 31, 1983, pp. 347-361.

[5] Grad, H.,"Asymptotic Equivalence of the Navier-Stokes and Nonlinear Boltzmann Equations, in Applications of Nonlinear Partial Differential Equations in Mathematical Physics," Proceedings of the Symposium on Applied Mathematics, Vol.17, AMS, Providence, RI, 1965, pp. 154-184.

[6] Illner R. and Shinbrot, M., "Global Existence for a Rare Gas in Infinite Vacuum," Communications in Mathematical Physics, Vol. 95, 1984, pp.117-126.

[7] Bellomo, N. and Toscani, G., "On the Cauchy Problem for the Non-linear Boltzmann Equation, Global Existence, Uniqueness and Asymptotic Stability," Journal of Mathematical Physics, Vol. 26, 1985, pp. 334-338.

[8] Toscani, G., "On the Non-linear Boltzmann Equation in Unbounded Domains," Archive for Rational Mechanics and Analysis, Vol. 95, 1986, pp. 37-49.

[9] Bellomo, N. and Toscani, G., "Lecture Notes on the Cauchy Problem for the Nonlinear Boltzmann Equations," Internal Rept. 16, Dipartimento di Matematica, Politecnico di Torino, Italy, Sept. 1986.

[10] Bellomo, N. and Toscani, G., "On the Cauchy Problem for the Nonlinear Boltzmann Equation: Global Existence Uniqueness and Asymptotic Stability," in Proceedings of the Workshop on Mathematical Aspects of Fluid and Plasma Dynamics, edited by C. Cercignani, S. Rionero, and M. Tessarotto, Trieste, Italy, May 1987, pp. 45-60.

[11] Toscani, G. and Bellomo, N., "The Nonlinear Boltzmann Equation: Analysis of the Influence of the Cutoff on the Solution of the Cauchy Problem," in Rarefied gas dynamics XV, Vol. 1, edited by V. Boffi and C. Cercignani, Teubner, Stuttgart, FRG, 1986, pp. 167-174.

[12] Polewczak, J., "Classical Solutions of the Nonlinear Boltzmann Equation in all R^3. Asymptotic Behaviour of Solutions," Journal of Statistical Physics, Vol. 50, 1988, pp. 611-632.

[13] Krasnoselskii, M. A. and Rutickii, Ya, B., Convex Functions and Orlicz Spaces, Noordhoff, The Netherlands, Groningen, 1961.

[14] Kaniel, S. and Shinbrot, M., "The Boltzmann Equations I. Uniqueness and Local Existence," Communications in Mathematical Physics, Vol. 58, 1978, pp. 65-84.

[15] Résibois, P. and DeLeener, M., Classical Kinetic Theory of Fluids, John Wiley & Sons, New York, 1977.

[16] Toscani, G. and Bellomo, N., "The Enskog-Boltzmann Equation in the Whole Space R^3: Some Global Existence, Uniqueness and Stability Results," Computational Mathematics and Applications, Vol. 13, No. 9-11, 1987, pp. 851-859.

[17] Polewczak, J., "Global Existence and Asymptotic Behavior for the Nonlinear Enskog Equation," SIAM Journal of Applied Mathematics (to be published).

Spatially Inhomogeneous Nonlinear Dynamics of a Gas Mixture

V. C. Boffi*
University of Bologna, Bologna, Italy
and
G. Spiga†
University of Bari, Bari, Italy

Abstract

A 2x2 semilinear hyperbolic system governing the number densities $\rho_1(\underline{x},t)$ and $\rho_2(\underline{x},t)$ of a mixture of two gases of interacting particles, having masses m_1 and m_2 and diffusing, upon the action of external constant forces $\underline{F}_1$ and $\underline{F}_2$, respectively, in a host medium consisting of certain other particles of fixed number density ρ_3, is studied in detail. In connection with an appropriate specialization of the physical inputs, a number of explicit analytical solutions, which seem to exhaust the class of all such solutions, are constructed for the case in which the two drift velocities $\underline{v}_{10}+(\underline{F}_1/m_1)t$ and $\underline{v}_{20}+(\underline{F}_2/m_2)t$ coincide; that is, when $\underline{v}_{10}=\underline{v}_{20}$ and $\underline{a}_1\equiv(\underline{F}_1/m_1)=(\underline{F}_2/m_2)\equiv\underline{a}_2$, $\underline{v}_{j0}$ being the velocity at which the particles of species j are injected into the mixture at time t=0, and $\underline{a}_j$ denoting the corresponding acceleration (j=1,2). The general case $\underline{v}_{10}\neq\underline{v}_{20}$; $\underline{a}_1\neq\underline{a}_2$ is finally illustrated, and its main features commented upon.

Introduction

A physical model has been recently proposed[1,2], according to which the system of the N nonlinear integro-partial differential Boltzmann equations governing the time evolution of the distribution functions $f_1, f_2, \ldots, f_N$ of the N gases of a spatially inhomogeneous, unbounded mixture of

---

Copyright © 1989 by the American Institute of Aeronautics and Astronautics, Inc. All rights reserved.
*Professor, Nuclear Engineering Laboratory.
†Professor, Department of Mathematics.

interacting particles has been solved for the general $f_j = f_j(\underline{x},\underline{v},t)$ in terms of the number densities $\rho_1, \rho_2, \ldots, \rho_N$ of the N gases considered, $\rho_j = \rho_j(\underline{x},t) = \int_{R_3} f_j(\underline{x},\underline{v},t) d\underline{v}$. (For the case N=1, see also Ref. 3 and 4. For the spatially homogeneous case, see, instead, Ref. 5). In such a model, not only scattering (as usual in classical kinetic theory), but also removal, creation, and host medium effects are accounted for [the host medium acting as the (N+1)th component of the mixture], and the so-called scattering kernel formulation has been adopted in establishing the Boltzmann system to be dealt with. The remarkable feature that the distribution functions are determined by their zeroth-order velocity moments is, of course, a consequence of the particular mathematical models used for the description of the physical processes. The set of assumptions can be summarized as follows: 1) All the microscopic collision frequencies are taken to be constant. 2) The scattering and creation probability distributions are taken to depend only on the velocity after collision (synthetic kernels) and in a discrete velocity kind of approach. 3) The initial data are separable in space and velocity and are monokinetic (that is, of the form $f_j(\underline{x},\underline{v},0) = Q_j(\underline{x}) \delta(\underline{v} - \underline{v}_{j0})$, with $Q_j(\underline{x})$ bounded and positive for $\underline{x} \in R_3$, and $\underline{v}_{j0} \neq 0$ a certain assigned velocity at which the particles of species j are injected at time t=0 into the host medium).

Once the original Boltzmann system is thus solved for f_j in terms of the initial datum and of all the ρ_j, we can integrate such a solution over the velocity domain to get a system for the ρ_j, which can be classified as a semilinear functional hyperbolic system, reading as

$$\frac{\partial \rho_j}{\partial t} + (\underline{v}_{j0} + \underline{a}_j t) \cdot \frac{\partial \rho_j}{\partial \underline{x}} + \frac{\partial}{\partial \underline{x}} \cdot \underline{J}_j^{sc}(\underline{x},t)$$

$$+ \sum_{l=1}^{N+1} g_{jl}^A \rho_l(\underline{x},t) \rho_j(\underline{x},t) \tag{1}$$

$$- \sum_{k=1}^{N} \sum_{l=k}^{N+1} g_{kl,j}^C \eta_{kl}^j \rho_k(\underline{x},t) \rho_l(\underline{x},t) = 0$$

to be integrated upon the initial condition

$$\rho_j(\underline{x},0) = Q_j(\underline{x}) \tag{1a}$$

In Eq. (1), $\underline{a}_j = \underline{F}_j/m_j$, and

$$g_{j1}^A = g_{j1}^R + \sum_{k=1}^{N} g_{j1,k}^C,$$

$g_{k1,j}^C$, and η_{k1}^j are physical constant parameters including removal and creation effects (at this stage, in fact, the scattering effects have already been integrated out). More precisely, g_{j1}^R is the microscopic frequency of collisions resulting in removal of particles j by particle 1; $g_{k1,j}^C = g_{1k,j}^C$ is, instead, the microscopic frequency of events, in which a particle j is created by a collision between a particle k and a particle 1; and η_{k1}^j, possibly greater than unity, is the mean number of such created particles j. Finally, and this is the most important feature of the system, Eq. (1), $\underline{J}_j^{sc}(\underline{x},t)$ is a cumbersome functional depending on all the ρ_j and their previous history. Here, we shall restrict our attention to the case in which $\underline{J}_j^{sc}(\underline{x},t)$ can be neglected. There are, indeed, two limiting cases in which $\underline{J}_j^{sc}$ vanishes exactly: the first one occurs when only removal takes place between the particles of the mixture; and the other is the quite idealistic case in which, for any j, particles j are given the same velocity $\underline{v}_{j0}$ after any interaction (either scattering or creation), and the external force $\underline{F}_j$ vanishes. We shall consider henceforth the semilinear hyperbolic system

$$\frac{\partial \rho_j}{\partial t} + (\underline{v}_{j0} + \underline{a}_j t) \cdot \frac{\partial \rho_j}{\partial \underline{x}} + \sum_{l=1}^{N} C_{jl} \rho_l(\underline{x},t) \rho_j(\underline{x},t) = 0 \qquad (2)$$

where the constant coefficients C_{jl} are to be specialized according to the removal and creation properties of the mixture considered. We refer thus to the case in which creation reduces to self-generation only and no interactions of the chemical reaction type occur. Therefore, the C_{ii} are always non-negative and, if C_{ij} is negative, then C_{ji} must be positive ($i \neq j$), where $C_{ij} < 0$ means that self-generation prevails over removal in the overall balance. For Eq. (2) to be physically consistent, it will be also implicitly understood that $\underline{a}_j$ is equal to zero whenever particles j can be generated by a creation mechanism and, in particular, when one of the C_{jl} is negative.

The Case N=2

When N=2, we rewrite Eq. (2) as

$$\frac{\partial \rho_1}{\partial t} + (\underline{v}_{10}+\underline{a}_1 t)\cdot\frac{\partial \rho_1}{\partial \underline{x}} + \sum_{l=1}^{3} C_{1l}\rho_l(\underline{x},t)\rho_1(\underline{x},t)=0, \quad \rho_1(\underline{x},0)=Q_1(\underline{x}) \quad (3)$$

$$\frac{\partial \rho_2}{\partial t} + (\underline{v}_{20}+\underline{a}_2 t)\cdot\frac{\partial \rho_2}{\partial \underline{x}} + \sum_{l=1}^{3} C_{2l}\rho_l(\underline{x},t)\rho_2(\underline{x},t)=0, \quad \rho_2(\underline{x},0)=Q_2(\underline{x})$$

where ρ_3 will be taken as a given constant, and $C_{ij3}\rho_3$ will be denoted below by ν_j. Each of the ρ_j can be expressed in terms of the other, either by solving the relevant equation by the characteristics method (i,j=1,2)

$$\rho_i(\underline{x},t) = Q_i(\underline{x} - \underline{v}_{i0}t - \frac{1}{2}\underline{a}_i t^2)$$

$$\times \left\{\exp\{C_{ij}\int_0^t \rho_j[\underline{x} - \underline{v}_{i0}(t-u) - \frac{1}{2}\underline{a}_i(t^2-u^2),u]du + \nu_i t\}\right.$$

$$+ C_{ii}Q_i(\underline{x} - \underline{v}_{i0}t - \frac{1}{2}\underline{a}_i t^2)\int_0^t \exp\{C_{ij}\int_u^t \rho_j[\underline{x}$$

$$\left. - \underline{v}_{i0}(t-u') - \frac{1}{2}\underline{a}_i(t^2-u'^2),u']du' + \nu_i(t-u)\}du\right\}^{-1} \quad (4)$$

or extracting it in differential terms from the other equation, namely,

$$\rho_i(\underline{x},t) = -\frac{1}{C_{ij}}\left[\frac{1}{\rho_j}\frac{\partial \rho_j}{\partial t} + \frac{1}{\rho_j}(\underline{v}_{j0}+\underline{a}_j t)\cdot\frac{\partial \rho_j}{\partial \underline{x}} + C_{jj}\rho_j+\nu_j\right] \quad (5)$$

Equation (5) becomes meaningless if one of the C_{ij} (i≠j) vanishes but, in such a case, Eq. (4) provides the full solution of the problem as we obtain first one of the two ρ_j explicitly and then the other number density as a function of the previous one.

The case in which both C_{12} and C_{21} are different from zero, as it will be assumed in the sequel, is thus of interest.

Let us now face the problem of constructing explicit analytical solutions to the system, Eq. (3), with the twofold aim of achieving a better understanding for the non-linear structure of the considered transport problem and of making available, at least for some specializations of the physical parameters, exact benchmark solutions for testing numerical codes. We consider first the simple case of equal injection velocities and forces per unit mass (as it occurs, for instance, for the gravity field),

$$\underline{v}_{10} = \underline{v}_{20} = \underline{v}, \quad \underline{a}_1 = \underline{a}_2 = \underline{a} \tag{6}$$

and introduce correspondingly the new independent variables

$$\xi = \underline{x} - \underline{v}t - \frac{1}{2}\underline{a}t^2, \quad \tau = t \tag{7}$$

associated with the characteristic common to each of the two equations separately, obtaining the system

$$\frac{\partial n_1}{\partial \tau} + C_{11}n_1^2 + C_{12}n_1n_2 + \nu_1 n_1 = 0$$

$$\frac{\partial n_2}{\partial \tau} + C_{21}n_1n_2 + C_{22}n_2^2 + \nu_2 n_2 = 0 \tag{8}$$

where $n_i(\xi,\tau) \equiv \rho_i(\underline{x},t)$, and ξ simply plays the role of a parameter.

<u>Solutions to the System, Eq. (8)</u>

From the first of Eq. (8), we get

$$n_2 = -\frac{1}{C_{12}} + (\frac{1}{n_1}\frac{\partial n_1}{\partial \tau} + C_{11}n_1^2 + \nu_1) \tag{9}$$

which is introduced in the second of Eq. (8) to yield

$$C_{12}n_1\ddot{n}_1 - (C_{12}+C_{22})\dot{n}_1^2 + [(C_{11}C_{12}+C_{21}C_{12}-2C_{11}C_{22})n_1^2$$
$$+ (C_{12}\nu_2 - 2C_{22}\nu_1)n_1]\dot{n}_1 + C_{11}(C_{12}C_{21}-C_{11}C_{22})n_1^4$$
$$+ (C_{12}C_{21}\nu_1 + C_{11}C_{12}\nu_2 - 2C_{11}C_{22}\nu_1)n_1^3$$
$$+ \nu_1(C_{12}\nu_2 - C_{22}\nu_1)n_1^2 = 0 \tag{10}$$

the dot indicating differentiation with respect to τ. Methods from the theory of ordinary differential equations can in fact be exploited since the hypothesis, Eq. (6), reduces the number of independent variables essentially to one. Equation (10) can be regarded as a semilinear ordinary differential equation for n_1 and, as such, is treatable on the basis of comprehensive, well-known results (see, for instance, the classical monographs of Refs. 6-8, where the corresponding spatially homogeneous problem has been investigated in detail). Some attention must be paid, of course, to the presence of the parameter ξ, appearing in the arbitrary functions contained in the general solution and playing its role when the initial conditions are applied to ρ_1 and ρ_2. Some classes of explicit analytical solutions to Eq. (10) have been determined for suitable specialization of the six coefficients C_{ij} and ν_j and are presented later, omitting, for simplicity, the quite cumbersome, but otherwise standard, details of their derivation.

First,

$$\nu_1 = \nu_2 = 0, \quad C_{21} = \frac{2C_{22} - C_{12}}{C_{22}} C_{11} \quad (C_{22} > 0, \; C_{12} \neq 2C_{22}) \quad (11)$$

Accounting for Eq. (6), the solution (ρ_1, ρ_2) to the system, Eq. (3), satisfying the assigned initial conditions $Q_1(\underline{x})$ and $Q_2(\underline{x})$, is then verified, via Eqs. (7)-(10), to be

$$\rho_1(\underline{x},t) = \widehat{Q}_1(C_{11}\widehat{Q}_1 + C_{22}\widehat{Q}_2)\left\{C_{11}\widehat{Q}_1\left[1+(C_{11}\widehat{Q}_1+C_{22}\widehat{Q}_2)t\right]\right.$$
$$\left. + C_{22}\widehat{Q}_2\left[1+(C_{11}\widehat{Q}_1+C_{22}\widehat{Q}_2)t\right]^{C_{12}/C_{22}}\right\}^{-1} \quad (12a)$$

$$\rho_2(\underline{x},t) = \widehat{Q}_2(C_{11}\widehat{Q}_1 + C_{22}\widehat{Q}_2)\left\{C_{11}\widehat{Q}_1\right.$$
$$\left. \times \left[1+(C_{11}\widehat{Q}_1+C_{22}\widehat{Q}_2)t\right]^{(2C_{22}-C_{12})/C_{22}}\right.$$
$$\left. + C_{22}\widehat{Q}_2\left[1+(C_{11}\widehat{Q}_1+C_{22}\widehat{Q}_2)t\right]\right\}^{-1} \quad (12b)$$

where $\hat{Q}_i$ is a shorthand notation for $Q_i(\underline{x}-\underline{v}t-\tfrac{1}{2}\underline{a}t^2)$ and corresponds to the free motion of particles of species i when no collisions take place in the mixture.

A particular case of the solution, Eqs. (12), can be obtained when $C_{12}=-C_{22}$, that is,

$$\rho_1(\underline{x},t) = \frac{\hat{Q}_1[1+(C_{11}\hat{Q}_1+C_{22}\hat{Q}_2)t]}{1+2C_{11}\hat{Q}_1 t+C_{11}\hat{Q}_1(C_{11}\hat{Q}_1+C_{22}\hat{Q}_2)t^2} \qquad (13a)$$

$$\rho_2(\underline{x},t) = \frac{\hat{Q}_2}{[1+(C_{11}\hat{Q}_1+C_{22}\hat{Q}_2)t][1+2C_{11}\hat{Q}_1 t+C_{11}\hat{Q}_1(C_{11}\hat{Q}_1+C_{22}\hat{Q}_2)t^2]} \qquad (13b)$$

Second,

$$\nu_1 = \nu_2 = 0, \qquad C_{11} = C_{22} = 0 \qquad (14)$$

Now we get immediately $C_{21}\dot{n}_1 - C_{12}\dot{n}_2 = 0$ leading to a linear relationship between n_1 and n_2, with an additive arbitrary function of $\underline{\xi}$, and then to a homogeneous Riccati equation for n_1, which can be solved and generates a second arbitrary function of $\underline{\xi}$. Going back to original variables and applying initial conditions finally yield

$$\rho_i(\underline{x},t) = \hat{Q}_i \frac{C_{ij}\hat{Q}_j - C_{ji}\hat{Q}_i}{C_{ij}\hat{Q}_j \exp[(C_{ij}\hat{Q}_j - C_{ji}\hat{Q}_i)t] - C_{ji}\hat{Q}_i} \qquad (15)$$

which, in the degenerate case $\hat{Q}_2(\underline{x})/\hat{Q}_1(\underline{x}) = C_{21}/C_{12}$, reduces to

$$\rho_i(\underline{x},t) = \frac{\hat{Q}_i}{1 + C_{ji}\hat{Q}_i t} \qquad (16)$$

Third,

$$C_{12} = -C_{22} \;\; (C_{22}>0), \quad C_{11} = 0, \quad \nu_1 = \nu_2 = \nu \qquad (17)$$

Here the procedure involves first a quadrature, leading to a inhomogeneous Riccati equation, which is solved by a proper substitution. After some algebra, the solution reads as

$$\rho_1(\underline{x},t) = \widehat{Q}_1\widehat{k}\exp(-\nu t)\left[(\widehat{k}+1)\exp\left(\widehat{k}C_{21}\widehat{Q}_1 \frac{1-\exp(-\nu t)}{\nu}\right)\right.$$

$$\left. - (\widehat{k}-1)\right]\left\{(\widehat{k}+1)\exp\left[\widehat{k}C_{21}\widehat{Q}_1 \frac{1-\exp(-\nu t)}{\nu}\right] + (\widehat{k}-1)\right\}^{-1} \quad (18a)$$

$$\rho_2(\underline{x},t) = \widehat{Q}_2 4\widehat{k}\exp(-\nu t)\left\{(\widehat{k}+1)^2\exp\left[\widehat{k}C_{21}\widehat{Q}_1 \frac{1-\exp(-\nu t)}{\nu}\right]\right.$$

$$\left. - (\widehat{k}-1)^2\exp\left[-\widehat{k}C_{21}\widehat{Q}_1 \frac{1-\exp(-\nu t)}{\nu}\right]\right\}^{-1} \quad (18b)$$

in which

$$k^2(\underline{x}) = 1 + \frac{2C_{22}Q_2(\underline{x})}{C_{21}Q_1(\underline{x})} \quad (19)$$

and $\widehat{k}$ stands one more for $k(\underline{x}-\underline{v}t-\tfrac{1}{2}\underline{a}t^2)$. In the limit of $\nu \to 0$, Eqs. (18) reduce to

$$\rho_1(\underline{x},t) = \widehat{Q}_1\widehat{k} \frac{(\widehat{k}+1)\exp(\widehat{k}C_{21}\widehat{Q}_1 t) - (\widehat{k}-1)}{(\widehat{k}+1)\exp(\widehat{k}C_{21}\widehat{Q}_1 t) + (\widehat{k}-1)} \quad (20a)$$

and

$$\rho_2(\underline{x},t) = \widehat{Q}_2 \frac{4\widehat{k}}{(\widehat{k}+1)^2\exp(\widehat{k}C_{21}\widehat{Q}_1 t) - (\widehat{k}-1)^2\exp(-\widehat{k}C_{21}\widehat{Q}_1 t)} \quad (20b)$$

respectively.
Fourth,

$$C_{12} = C_{21} = C \quad (C>0) \quad (21)$$

Here we distinguish two subcases. In the first one, under the restriction

$$C_{11} = C_{22} = C \quad (22)$$

we find, after some manipulations, for $i=j=1,2$; $i \neq j$,

$$\rho_i(\underline{x},t) = \widehat{Q}_i \nu_1 \nu_2 \left\{ (\nu_1 \nu_2 + \nu_1 c\widehat{Q}_2 + \nu_2 c\widehat{Q}_1) \exp(\nu_i t) \right.$$

$$\left. - \nu_i c\widehat{Q}_j \exp[(\nu_i - \nu_j)t] - \nu_j c\widehat{Q}_i \right\}^{-1} \qquad (23)$$

Particular cases are

$$\rho_1(\underline{x},t) = \frac{\nu_2}{\nu_2(1+c\widehat{Q}_1 t) + c\widehat{Q}_2[1-\exp(-\nu_2 t)]} \qquad (24a)$$

$$\rho_2(\underline{x},t) = \frac{\widehat{Q}_2}{\widehat{Q}_1} \rho_1(\underline{x},t) \exp(-\nu_2 t) \qquad (24b)$$

for $\nu_1 = 0$, and

$$\rho_i(\underline{x},t) = \frac{\widehat{Q}_i}{1+c(\widehat{Q}_1 + \widehat{Q}_2)t} \qquad (24c)$$

for $\nu_1 = \nu_2 = 0$.

The second subcase refers instead to the additional assumption

$$c_{11} = c_{22} = 0, \qquad \nu_1 = \nu_2 = \nu \qquad (25)$$

in which the solution (ρ_1, ρ_2) reads as ($i=j=1,2$; $i \neq j$)

$$\rho_i(\underline{x},t) = \widehat{Q}_i(\widehat{Q}_j - \widehat{Q}_i) \exp(-\nu t)$$

$$\times \left\{ \widehat{Q}_j \exp\left[c(\widehat{Q}_j - \widehat{Q}_i) \frac{1-\exp(-\nu t)}{\nu} \right] - \widehat{Q}_i \right\}^{-1} \qquad (26)$$

If, furthermore, ν vanishes, the solution becomes

$$\rho_i(\underline{x},t) = \frac{\widehat{Q}_i(\widehat{Q}_j - \widehat{Q}_i)}{\widehat{Q}_j \exp[c(\widehat{Q}_j - \widehat{Q}_i)t] - \widehat{Q}_i} \qquad (27)$$

with the obvious limiting value for $Q_1(\underline{x}) = Q_2(\underline{x})$.

Other Classes of Solutions

Of course, the solutions obtained so far do not exhaust the class of all analytical solutions to the considered problems, and new insight into the problem itself, as well as new solutions, can be achieved by different approaches. So, for instance, it is sufficient to set $n_1=1/y$ in Eq. (10) in order to get the new differential equation

$$y\ddot{y} + (\nu_2 - 2\frac{C_{22}\nu_1}{C_{12}})y\dot{y} + (\frac{C_{22}}{C_{12}} - 1)\dot{y}^2 + (C_{11}+C_{21}$$

$$- 2\frac{C_{11}C_{22}}{C_{12}})\dot{y} + (\frac{C_{22}}{C_{12}}\nu_1 - \nu_2)\nu_1 y^2 + (2\frac{C_{11}C_{22}\nu_1}{C_{12}}$$

$$- C_{21}\nu_1 - C_{11}\nu_2)y + (\frac{C_{11}C_{22}}{C_{12}} - C_{21})C_{11} = 0 \qquad (28)$$

while the other density is related to y by

$$n_2 = \frac{1}{C_{12}}(\frac{\dot{y}}{y} - \frac{C_{11}}{y} - \nu_1) \qquad (29)$$

Clearly, Eq. (28) has a different structure with respect to Eq. (10), and it is possible to solve it explicitly in some particular cases not covered by the previous results. Some examples are listed below.

First,

$$C_{11} = C_{21}, \quad C_{22} = C_{12} \qquad (30)$$

In this case, Eq. (28) becomes linear and reads as

$$\ddot{y} + (\nu_2 - 2\nu_1)\dot{y} + (\nu_1 - \nu_2)\nu_1 y - C_{21}(\nu_1 - \nu_2) = 0 \qquad (31)$$

with general solution

$$y(\xi,\tau) = -(C_{21}/\nu_1) + A(\xi)\exp(\nu_1\tau) + B(\xi)\exp[(\nu_1 - \nu_2)\tau] \qquad (32)$$

where A and B are arbitrary functions. Going back to original variables and applying initial conditions, we end up with the sought solution (i=1,2)

$$\rho_i(x,t) = \widehat{Q}_i \left\{ (1 + \frac{c_{21}\widehat{Q}_1}{\nu_1} + \frac{c_{12}\widehat{Q}_2}{\nu_2}) \exp(\nu_1 t) \right.$$

$$\left. - \frac{c_{12}\widehat{Q}_2}{\nu_2} \exp[(\nu_1-\nu_2)\tau] - \frac{c_{21}\widehat{Q}_1}{\nu_1} \right\}^{-1} \qquad (33)$$

Second,

$$\nu_1 = \nu_2 = 0, \quad c_{11} = 0, \quad c_{22} = c_{12} \qquad (34)$$

In this case, we are left with

$$y\ddot{y} + c_{21}\dot{y} = 0 \qquad (35)$$

which, after a quadrature, is amenable to the implicit form solution

$$\int \frac{dy}{A(\xi) - c_{21}\ln y} = \tau + B(\xi) \qquad (36)$$

where again A and B denote arbitrary functions and where the indefinite integral can be expressed in terms of the logarithmic integral function.

Discussion of the General Case

If we relax the condition, Eq. (6), the problem becomes much more complicated. In the monodimensional case, say, $\rho_i = \rho_i(x,t)$, with $v_{i0x} = v_i$ and $a_{ix} = a_i$, Eq. (3) reads as

$$\frac{\partial \rho_1}{\partial t} + (v_1 + a_1 t) \frac{\partial \rho_1}{\partial x} + \sum_{l=1}^{2} c_{1l}\rho_1\rho_l + \nu_1\rho_1 = 0, \quad \rho_1(x,0) = Q_1(x)$$

$$\frac{\partial \rho_2}{\partial t} + (v_2 + a_2 t) \frac{\partial \rho_2}{\partial x} + \sum_{l=1}^{2} c_{2l}\rho_2\rho_l + \nu_2\rho_2 = 0, \quad \rho_2(x,0) = Q_2(x)$$

(37)

Introducing the new independent variables

$$\xi_i = x - v_i t - \frac{1}{2} a_i t^2 \qquad (38)$$

NONLINEAR DYNAMICS OF A GAS MIXTURE

and setting, as before, $n_i(\xi_1,\xi_2) \equiv q_i(x,t)$, the system, Eq. (37), becomes

$$\left[(v_1-v_2)^2 - 2(a_1-a_2)(\xi_1-\xi_2)\right]^{1/2} \frac{\partial n_1}{\partial \xi_2} + \sum_{l=1}^{2} C_{1l} n_1 n_l + \nu_1 n_1 = 0 \qquad (39)$$

$$\left[(v_1-v_2)^2 - 2(a_1-a_2)(\xi_1-\xi_2)\right]^{1/2} \frac{\partial n_2}{\partial \xi_1} + \sum_{l=1}^{2} C_{2l} n_2 n_l + \nu_2 n_2 = 0$$

which appears hardly treatable by purely analytical means, even under the simplifying assumption $a_1 = a_2$. Because it is not restrictive to assume $v_1 > v_2$, we may further define

$$a_{il} = \frac{C_{il}}{v_1 - v_2} \quad , \quad b_i = \frac{\nu_i}{v_1 - v_2} \qquad (40)$$

extract the second unknown as

$$n_2 = -\frac{1}{a_{12}}\left(\frac{1}{n_1}\frac{\partial n_1}{\partial \xi_2} + a_{11}n_1 + b_1\right) \qquad (41)$$

and get for n_1 the self-contained semilinear hyperbolic partial differential equation

$$\frac{1}{n_1}\frac{\partial^2 n_1}{\partial \xi_1 \partial \xi_2} - \frac{1}{n_1^2}\frac{\partial n_1}{\partial \xi_1}\frac{\partial n_2}{\partial \xi_2} + \frac{a_{22}}{a_{12}}\frac{1}{n_1^2}\left(\frac{\partial n_1}{\partial \xi_2}\right)^2 + a_{11}\frac{\partial n_1}{\partial \xi_1}$$

$$+ \left[\left(\frac{2a_{11}a_{22}}{a_{12}} - a_{21}\right) + \left(\frac{2a_{22}b_1}{a_{12}} - b_2\right)\frac{1}{n_1}\right]\frac{\partial n_1}{\partial \xi_2}$$

$$+ \left(\frac{a_{11}^2 a_{22}}{a_{12}} - a_{11}a_{21}\right)n_1^2 + \left(\frac{2a_{11}a_{22}b_1}{a_{12}} - a_{21}b_1 - a_{11}b_2\right)n_1$$

$$+ \frac{a_{22}b_1^2}{a_{12}} - b_1 b_2 = 0 \qquad (42)$$

with characteristics ξ_1=const, ξ_2=const. Cases of interest are those with both a_{12} and a_{21} different from zero, since otherwise the original set of equations can be solved as indicated by Eq. (4). In any case, in order to solve Eq. (42), or any other formulation of the same problem, we must plan to resort to iterative, or perturbative, or similarity methods. Work is in progress along these lines, and results will be the object of a future paper.

Acknowledgments

The authors acknowledge the financial support of both Ministry of Public Education and C.N.R.'s National Group for Mathematical Physics in the frame of the 40% Project Program and of the Applied Mathematics Project, respectively.

References

[1] Boffi, V.C., and Aoki, K., "A System of Conservation Equations Arising in Nonlinear Dynamics of a Gas Mixture," Il Nuovo Cimento, Vol. 10D, 1988, p. 145.

[2] Boffi, V.C., "Systems of Conservation Equations in Nonlinear Particle Transport Theory," in Proceedings of Wing Conference on Invariant Imbedding Transport Theory and Integral Equations, edited by P.Nelson et al., Marcel Dekker Inc., (Santa Fe, NM, 20-22 January 1988), to be published.

[3] Boffi, V.C., and Spiga, G., "An Analytical Study of Space Dependent Evolution Problems in Particle Transport Theory," Il Nuovo Cimento, Vol. 7D, 1986, p. 327.

[4] Boffi, V.C., and Spiga, G., "Spatial Effects in the Study of Nonlinear Evolution Problems of Particle Transport Theory," Transport Theory and Statistical Physics, Vol. 17, 1988, p. 241.

[5] Boffi, V.C., Franceschini, V., and Spiga, G., "Dynamics of a gas Mixture in an Extended Kinetic Theory," Physics of Fluids, Vol. 28, 1985, p. 3232.

[6] Ince, E.L., Ordinary Differential Equations, Dover, New York, 1956.

[7] Murphy, G.M., Ordinary Differential Equations and Their Solutions, Van Nostrand, Princeton, NJ, 1960.

[8] Boffi, V.C., and Spiga, G., "Extended Kinetic Theory for Gas Mixtures in the Presence of Removal and Regeneration Effects," Zeitschrift für Angewandte Mathematik und Physik, Vol. 37, 1986, p. 27.

Diffusion of a Particle in a Very Rarefied Gas

Bernard Gaveau*
Université Pierre et Marie Curie, Paris, France
and
Marc-André Gaveau†
Centre d'Etudes Nucléaires de Saclay, Gif-sur-Yvette, France

Abstract

An exact stochastic equation is derived for the momentum transfer per collision for a test particle interacting with a very rarefied gas of hard spheres in equilibrium at temperature T. This result is obtained without any assumption on the mass ratio of particles and is a Langevin-type equation with a friction term and a stochastic force. The covariance matrix of this noise force is different from the covariance of the white noise force in Langevin equation, since in our case its depends on the test particle momentum before collision. Finally, when the number of collisions undergone by the test particle increases, average and variance of its momentum tend to their values at équilibrium at temperature T.

I. Introduction

The evolution of a particle in a gas is usually described by a linearized Boltzmann equation or by a Fokker-Planck equation. This last equation is also equivalent to a stochastic differential equation called the Langevin equation, in which the increment of the momentum of a particle in time dt is of the type

$$d\vec{p}(t) = -\alpha \vec{p}(t)\, dt + d\vec{F}(t)$$

Copyright © 1989 by the American Institute of Aeronautics and Astronautics, Inc. All rights reserved.
*Professor, Département de Mathématiques.
†Dr. ès Sciences, Département de Physico-Chimie.

where α is a friction coefficient and $d\vec{F}(t)$ is a white noise force that is Gaussian, of average 0 satisfying

$$\langle d\vec{F}(t)\ d\vec{F}(s)\rangle = 2D\delta(t-s)$$

where D is the diffusion coefficient and is equal to $kT\alpha$ by the Einstein relation.

These equations are derived under two kinds of hypotheses: one assumes that the particle is very near from equilibrium[1-4] or that there are many collisions per unit of time, or that the ratio M/m of the mass of the test particle to the mass of the bath particle is very large.[1-5] If one of these hypotheses is not fulfilled, in particular, if the particle is far from equilibrium, or if the gas of bath particles is very rarefied (i.e., the test particle can remain out of equilibrium for a very long time), the arguments leading to the Fokker-Planck equation or to the Langevin equation cannot be applied.

Several attempts have been made to justify the Langevin equation far from equilibrium. Ford et al.[6] derived the Langevin equation for the diffusion of a particle in a crystal under very special hypothesis. Fox and Kac[7] derived a Langevin-type equation in one dimension and a Fokker-Planck equation with time-dependent coefficients without universality and showed that the fluctuation dissipation relation did not hold far from equilibrium (see also Ref. 5 and 8).

In this work, we study the diffusion of a test particle of mass M in a rarefied gas of particles of mass m. The test particle is out of equilibrium and the gas particles are in thermal equilibrium at temperature T. We assume that the gas is sufficiently rarefied so that we can neglect all recollisions and that the thermal equilibrium of the bath is not perturbed by the presence of the test particle. We do not suppose that the ratio M/m is large. Under these hypotheses, our results are exact.

Our point of view is to analyze a single collision event in order to derive the probability distribution of the momentum transfer per collision and a stochastic equation per collision for the evolution of the momentum of the particle; the stochastic equation is of the Langevin-type except that the diffusion coefficient per collision depends on the state of the particle. Throughout all of this work, we have only considered hard sphere collisions. More general potentials and situations with internal degrees of freedom will be considered in subsequent publications.

II. Analysis of a collision

We consider a test particle of mass M, momentum $\vec{P}$ in a bath of molecules of mass m, momentum $\vec{p}$. After a collision, the respective momenta are $\vec{P}'$ and $\vec{p}'$. We want to compute $\vec{P}' - \vec{P}$. Let $(\vec{V}, \vec{v})$ and $(\vec{V}', \vec{v}')$ represent the velocities of the molecules before and after collision. The velocity of the center of mass is

$$\vec{v}_G = \frac{m\vec{v} + M\vec{V}}{m + M}$$

The velocities of the particles in the center-of-mass frame are denoted by $(\vec{C}, \vec{c})$ and $(\vec{C}', \vec{c}')$ before and after collision, respectively. They are given by

$$\vec{C} = \frac{m(\vec{V} - \vec{v})}{m + M} \quad , \quad \vec{c} = -\frac{M(\vec{V} - \vec{v})}{m + M}$$

The conservation of kinetic energy and momentum gives

$$|\vec{C}'| = |\vec{C}| = \frac{m}{m + M} |\vec{v} - \vec{V}|$$

The direction of $\vec{C}'$ is given by a unit vector $\vec{n}$, which is uniformly distributed on the unit sphere in the case of hard sphere collisions.

Coming back to the laboratory frame, we obtain the transfer of momentum per collision

$$\vec{P}' - \vec{P} = \frac{mM}{m + M} (|\vec{v} - \vec{V}| \vec{n} + \vec{v} - \vec{V}) \tag{1}$$

III. The Probability Distribution of Momentum Transfer per Collision

Let us choose the axis Oxyz so that $\vec{V}$ has the direction of the positive z axis. We shall call

$$\vec{q} = \frac{\vec{v} - \vec{V}}{|\vec{v} - \vec{V}|} \tag{2}$$

so that formula (1) can be rewritten as

$$\vec{P}' - \vec{P} = \frac{mM}{m + M} |\vec{v} - \vec{V}| (\vec{n} + \vec{q}) \tag{3}$$

We shall use θ to denote the angle of $\vec{q}$ with the z axis and φ the angle of the projection of $\vec{q}$ on the x,y plane with the x axis, so that

$$\vec{q} = (\sin\theta \cos\varphi, \sin\theta \sin\varphi, \cos\theta) \qquad (4)$$

We use χ to denote the angle $(\vec{q},\vec{n})$ and ψ to denote the rotation angle of $\vec{n}$ around the $\vec{q}$ vector; then,

$$|\vec{n} + \vec{q}| = 2 \cos \chi/2 \qquad (0 \leq \chi < \pi) \qquad (5)$$

Let us finally denote by (ρ,λ) the polar angles of $\vec{n} + \vec{q}$ in the xyz frame (Fig. 1).

By writing $\cos\chi/2$ in terms of ρ, λ, θ, φ, we can prove that, for fixed (ρ,λ), the angles (θ,φ) must be in the domain

$$D_{\rho,\lambda} = \{(\theta,\varphi) \in [0,\pi[\times [0,2\pi] / \sin\rho \cos(\lambda-\varphi) \sin\theta + \cos\rho \cos\theta \geq 0 \} \qquad (6)$$

because $0 \leq \chi/2 \leq \pi/2$.

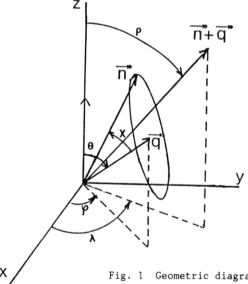

Fig. 1 Geometric diagram of vectors $\vec{n}$, $\vec{q}$, $\vec{n}+\vec{q}$ in the collision.

DIFFUSION OF A PARTICLE IN RAREFIED GAS

In hard sphere collisions, $\vec{n}$ is distributed according to the law

$$d\sigma = \frac{1}{4\pi} \sin\chi \, d\chi \, d\psi = 4 \cos \chi/2 \, \frac{\sin\rho \, d\rho \, d\lambda}{4\pi} \qquad (7)$$

Equation (3) tells us that the transfer of momentum in a single collision event is given by

$$\frac{m\,M}{m+M} |\vec{v} - \vec{V}| (\vec{n} + \vec{q}), \qquad \vec{q} = \frac{\vec{v} - \vec{V}}{|\vec{v} - \vec{V}|}$$

We now want to study the probability distribution of this vector for a given $\vec{V}$. We state that

$$\vec{\Delta} = |\vec{v} - \vec{V}| (\vec{n} + \vec{q})$$

The polar angles of $\vec{\Delta}$ with respect to the z axis are ρ and λ by definition.

We compute the joint probability law of $|\vec{\Delta}|$ and $\vec{\Delta}/|\vec{\Delta}|$ by successive conditioning: first, we take conditional average with respect to $\vec{v} - \vec{V}$, i.e., $\vec{v}$. Then, we average a test function $f(|\vec{\Delta}|)$ on $\vec{v} - \vec{V}$. Finally, we obtain

$$\text{Prob}(|\vec{\Delta}| \in dx, \frac{\vec{\Delta}}{|\vec{\Delta}|} \in d\rho \, d\lambda \, |\vec{V}) = \frac{x^2 \, dx \, \sin\rho \, d\rho \, d\lambda}{8\pi \, (2\pi\frac{kT}{m})^{3/2}} \Phi(\rho, x, |\vec{V}|) \qquad (8)$$

$$\Phi(\rho, x, |\vec{V}|) = \int_0^{2\pi} d\varphi \int_0^{\pi} \sin\theta \, d\theta \, \frac{1_{D_{\rho,0}}(\theta,\varphi)}{(\cos\varphi\sin\theta \, \sin\rho + \cos\rho \, \cos\theta)^2}$$

$$\times \exp\left\{ -\frac{m}{2kT} \left[\frac{x^2}{4(\cos\varphi \sin\theta \, \sin\rho + \cos\rho \, \cos\theta)^2} \right.\right.$$

$$\left.\left. + \frac{2x \, |\vec{V}| \, \cos\theta}{2(\cos\varphi \sin\theta \, \sin\rho + \cos\rho \, \cos\theta)} + |\vec{V}|^2 \right] \right\} \qquad (9)$$

and we recall that the domain $D_{\rho,0}$ is defined by Eq. (6) with $\lambda = 0$.

Moreover, we recall that this has been derived under the assumption that ρ, λ are the angles of the transfer of momentum with respect to the $\vec{V}$ axis.

IV. Statistical Analysis of a Collision Event

A. Probability Distribution

The probability distribution of the transfer of momentum in a collision is rather complicated. On the other hand, we can directly use formula (1) giving the momentum transfer to study some simple properties of the statistics of a collision event.

In this section, we shall use Oxyz to denote the laboratory system. We do not assume that $\vec{V}$ has the z direction.

B. Average and Friction Term

It is clear that the conditional average of $\vec{P}' - \vec{P}$ knowing $\vec{V}$ is given by

$$\langle \vec{P}' - \vec{P} | \vec{V} \rangle = -\frac{m}{m+M} \vec{P} \tag{10}$$

which is a friction law of the usual type.

C. Covariance Matrix of Eq. (1)

The variance of $|\vec{w}|$ $(\vec{n} + \vec{q})$ is a 3 × 3 matrix. Here $\vec{w} = \vec{v} - \vec{V}$. We have to compute $\langle |\vec{w}|^2 (n_x + q_x)^2 | \vec{V} \rangle$. First, we fix $\vec{v}$, to obtain

$$\langle |\vec{w}|^2 (n_x + q_x)^2 | \vec{v}, \vec{V} \rangle = |\vec{w}|^2 \langle n_x^2 | \vec{v}, \vec{V} \rangle + w_x^2 + 2|\vec{w}| q_x \langle n_x | \vec{v}, \vec{V} \rangle$$

But $\vec{n}$ is, conditionally with respect to $\vec{v} - \vec{V}$, uniformly distributed; thus, the last term is 0. Then

$$|\vec{w}|^2 \langle n_x^2 | \vec{v}, \vec{V} \rangle = \frac{1}{3} |\vec{w}|^2$$

and finally

$$\langle |\vec{w}|^2 (n_x^2 + q_x^2) | \vec{V} \rangle = \frac{4}{3} \left(\frac{kT}{m} + V_x^2\right) + \frac{1}{3}\left(\frac{2kT}{m} + V_y^2 + V_z^2\right) \tag{11}$$

Also,

$$\langle |\vec{w}|^2 (n_x + q_x)(n_y + q_y) | \vec{V} \rangle = V_x V_y \tag{12}$$

The covariance matrix is then the matrix

$$\Gamma_{ij} = \langle |\vec{w}|^2 (n_i+q_i)(n_j+q_j)|\vec{V}\rangle - \langle |\vec{w}|(n_i+q_i)|\vec{V}\rangle \langle |\vec{w}|(n_j+q_j)|\vec{V}\rangle$$

where $i,j = x, y, z$. The nondiagonal terms are 0 and the diagonal terms are

$$\Gamma_{ii} = \frac{2kT}{m} + \frac{1}{3}|\vec{V}|^2 \qquad (13)$$

V. Exact Stochastic Equation per Collision

We see now that we can rewrite the equation for the momentum transfer in a single collision as follows:

$$\Delta \vec{P} = -\frac{m}{m+M}\vec{P} + \Delta\vec{F} \qquad (14)$$

where $\Delta\vec{F}$ is a pure stochastic impulsive force :

$$\Delta\vec{F} = \frac{mM}{m+M}(|\vec{v}-\vec{V}|\vec{n}) + \frac{M}{m+M}\vec{P}$$

of average 0 and of conditional variance given by Eq. (13) for $i,j = x,y,z$. The $\vec{P}$ is the momentum of the bath particle:

$$\langle (\Delta\vec{F})_j (\Delta\vec{F})_k | \vec{P}\rangle = [(\frac{mM}{m+M})^2 \frac{2kT}{m} + \frac{1}{3}(\frac{m}{m+M})^2 |P|^2]\delta_{jk} \qquad (15)$$

Equation (14) is exact. It decomposes the transfer of momentum per collision into a pure friction term $-[m/(m+M)]\vec{P}$, which is of the Langevin-type and a pure stochastic force with variance given by Eq. (15), and which depends on the state of the particle just before the collision event. This feature is the main difference with respect to the usual Langevin equation, where the covariance of the white noise force is independent of the state of the particle.

VI. Convergence to Equilibrium

A molecule of mass M and momentum $\vec{P}_o$ has a momentum $\vec{P}_n$ after n collisions, which is

$$\vec{P}_n = \vec{P}_o + \Delta_1\vec{P} + \ldots + \Delta_n\vec{P}$$

where $\Delta_j\vec{P}$ is the momentum transfer in collision j.

It is possible to prove that the conditional average of $P^2_{n,j}$ knowing $\vec{P}_o$ has the limiting thermal behavior

$$\langle P^2_{n,j} | \vec{P} \rangle \to kTM \quad \text{for} \quad j = x, y, z$$

We can also prove that, after n collisions, we have still a memory of the initial state $\vec{P}_o$ of the particle. In fact, the rate of convergence to thermal equilibrium is exponentially fast and is given by the equations

$$\left\langle \begin{pmatrix} P^2_{n,x} \\ P^2_{n,y} \\ P^2_{n,z} \end{pmatrix} \middle| \vec{P}_o \right\rangle - kTM \begin{pmatrix} 1 \\ 1 \\ 1 \end{pmatrix} = -kTM \left[\frac{m^2 + M^2}{(m+M)^2} \right]^n \begin{pmatrix} 1 \\ 1 \\ 1 \end{pmatrix}$$

$$+ \frac{P^2_{o,x} + P^2_{o,y} + P^2_{o,z}}{3} \left[\frac{m^2 + M^2}{(m+M)^2} \right]^n \begin{pmatrix} 1 \\ 1 \\ 1 \end{pmatrix}$$

$$+ \left[\frac{M}{m+M} \right]^{2n} \begin{pmatrix} \frac{2}{3} P^2_{o,x} - \frac{1}{3}(P^2_{o,y} + P^2_{o,z}) \\ \frac{2}{3} P^2_{o,y} - \frac{1}{3}(P^2_{o,x} + P^2_{o,z}) \\ \frac{2}{3} P^2_{o,z} - \frac{1}{3}(P^2_{o,y} + P^2_{o,z}) \end{pmatrix} \quad (16)$$

VII. Conclusion

We have derived an exact Langevin equation per collision for a particle far from equilibrium in a bath at thermal equilibrium. This Langevin equation is exact because it is a reinterpretation of the collision equation for the momentum transfer. We have seen that this equation has a friction term that is of the Langevin type and a diffusion term that depends explicitly on the state of the test particle; thus there is no universal relation of the type "fluctuation dissipation," confirming the analysis[7] done by Fox and Kac using a Fokker-Planck-type equation.

References

[1] Landau, L. D. and Lifshitz, E. M., "Statistical Physics, Part I," Course of Theoretical Physics, Vol. 5, Pergamon, Oxford, UK, 1980.

[2] Landau, L. D. and Lifshitz, E. M., "Physical Kinetics," Course of Theoretical Physics, Vol. 10, Pergamon, Oxford, UK, 1980.

[3] Onsager, L. and Machlup, S., "Fluctuations and Irreversible Processes," Physical Review, Vol. 91, N° 6, Sept. 1953, pp. 1505-1515.

[4] Fox, R. F. and Uhlenbeck, G. E., "Contributions to Non-equilibrium Thermodynamics I and II," Physics of Fluids, Vol. 13, 1970, pp. 1893-1903, 2881-2890.

[5] Wang-Chang, C. S. and Uhlenbeck, G. E., "The Kinetic Theory of Gases," Studies in Statistical Mechanics, Vol. 5, North-Holland, Amsterdam, 1970.

[6] Ford, G. W., Kac, M., and Mazur, P., "Statistical Mechanics of Assemblies of Coupled Oscillators," Journal of Mathematical Physics, Vol. 6, N° 4, April 1965, pp. 504-515.

[7] Fox, R. F. and Kac, M., "A Remark on the Theory of Fluctuations Far from Equilibrium," Biosystems, Vol. 8, 1977, pp. 187-191.

[8] Logan, J. and Kac, M., "Fluctuations and the Boltzmann Equations. I.," Physical Review A, Vol. 13, N° 1, Jan. 1976, pp. 458-470.

Heat Transfer and Temperature Distribution in a Rarefied Gas between Two Parallel Plates with Different Temperatures: Numerical Analysis of the Boltzmann Equation for a Hard Sphere Molecule

Taku Ohwada,* Kazuo Aoki,† and Yoshio Sone‡
Kyoto University, Kyoto, Japan

Abstract

The behavior of a rarefied gas between two parallel plates with different temperatures is studied numerically for the whole range of the Knudsen number on the basis of the standard Boltzmann equation for a hard sphere molecule and the diffuse reflection. The velocity distribution function of the gas molecules, the temperature and density distributions in the gas, and the heat transfer between the plates are presented. The results of various approximate analyses are also shown for comparison.

I. Introduction

Analysis of the heat transfer and temperature distribution in a rarefied gas between two parallel infinite plates at rest with different temperatures is one of the most fundamental problems in rarefied gas dynamics and has been tried by various authors (e.g., Refs. 1-8). The accurate temperature distribution and heat transfer in the gas are obtained on the basis of the Boltzmann-Krook-Welander equation (BGK model of the Boltzmann equation).[6,8] As for the original Boltzmann equation, only the results based on arbitrary assumption on the form of the velocity distribution function (moment method,[1-3,5] variational method,[7] etc.) have been reported, but they are not good

---

Copyright © 1988 by Yoshio Sone, Kazuo Aoki, and Taku Ohwada. Published by the American Institute of Aeronautics and Astronautics, Inc. with permission.
  * Graduate Student, Department of Aeronautical Engineering.
  † Associate Professor, Department of Aeronautical Engineering.
  ‡ Professor, Department of Aeronautical Engineering.

enough in describing the local behavior of the gas, such as the temperature distribution, the velocity distribution function, especially for moderate and large Knudsen numbers, though they give fairly or very good results for the heat flow.

In our previous paper[9], we developed an accurate numerical method (finite-difference method) to analyze the standard Boltzmann equation for a hard sphere molecule and analyzed the temperature jump problem, giving the temperature jump coefficient and its associated Knudsen layer. In this paper, we apply the method to the above problem of heat transfer and temperature distribution between two parallel plates with different temperatures. The success of the present analysis promises the further development of the numerical study of rarefied gas flows at arbitrary Knudsen numbers by the standard Boltzmann equation.

II. Problem and Assumptions

Problem: Consider a rarefied gas between two parallel infinite plates at rest with different uniform temperatures. Let the temperature of one plate [at $X_1 = -D/2$, $(D > 0)$, X_i: rectangular space coordinates] be $T_0(1 - \Delta\tau)$ and that of the other (at $X_1 = D/2$) be $T_0(1 + \Delta\tau)$. Investigate the steady behavior of the gas (the velocity distribution function, temperature distribution, and heat transfer in the gas) for the entire range of the Knudsen number (the mean free path of the gas molecules divided by the distance between the plates D) on the basis of the standard Boltzmann equation.

We analyze the problem under the following assumptions:

1) The gas molecules are hard spheres of a uniform size and undergo complete elastic collisions between themselves.

2) The gas molecules are reflected diffusely on the plates.

3) The difference of the temperatures of the plates is so small ($|\Delta\tau| \ll 1$) that the governing equation and the boundary condition can be linearized around a uniform equilibrium state at rest.

III. Basic Equation

We first summarize the notations used in this paper: ρ_0 is the reference density (the density at $X_1 = 0$); $p_0 = R\rho_0 T_0$; R is the specific gas constant; ℓ_0 is the mean free path of the gas molecules at the equilibrium state at rest with density ρ_0 and temperature T_0; $Kn = \ell_0/D$ (the Knudsen number); $k = (\sqrt{\pi}/2)Kn$; $x_i = X_i/D$; $(2RT_0)^{1/2}\zeta_i$ is the velocity of the gas molecules; $E(\zeta_i) = \pi^{-3/2}\exp(-\zeta_i^2)$; $\rho_0(2RT_0)^{-3/2}E(\zeta_i)[1 + \phi(x_1, \zeta_i)]$ is the velocity distribution function of the gas molecules; $\rho_0[1 + \omega(x_1)]$ is the density of the gas; $T_0[1 + \tau(x_1)]$ is the temperature; and $\rho_0(2RT_0)^{1/2}Q_i$ is the heat flow vector, where we assume that all the quantities are independent of x_2, x_3.

The linearized Boltzmann equation for a steady state in the present spatially one-dimensional case is[10,11]

$$\zeta_1 \frac{\partial \phi}{\partial x_1} = \frac{1}{k}\{L_1[\phi] - L_2[\phi] - \nu(|\zeta_i|)\phi\} \tag{1}$$

$$L_1[\phi] = \frac{1}{\sqrt{2\pi}}\int \frac{1}{|\zeta_i - \xi_i|}\exp\left(-\xi_i^2 + \frac{|\zeta_i \times \xi_i|^2}{|\zeta_i - \xi_i|^2}\right)\phi(x_1, \xi_i)d\xi \tag{2a}$$

$$L_2[\phi] = \frac{1}{2\sqrt{2\pi}}\int |\zeta_i - \xi_i|\exp(-\xi_i^2)\phi(x_1, \xi_i)d\xi \tag{2b}$$

$$\nu(s) = \frac{1}{2\sqrt{2}}[\exp(-s^2) + (2s + \frac{1}{s})\int_0^s \exp(-c^2)dc] \tag{2c}$$

where $\zeta_i \times \xi_i$ is the vector product of ζ_i and ξ_i, and $d\xi = d\xi_1 d\xi_2 d\xi_3$.

The boundary condition for Eq. (1) at the plates ($x_1 = \pm 1/2$) is given as

$$\phi(x_1, \zeta_i) = \pm[2\sqrt{\pi}\int_{\xi_1 \gtrless 0}\xi_1\phi(x_1, \xi_i)E(\xi_i)d\xi + (\zeta_i^2 - 2)\Delta\tau]$$

at $x_1 = \pm 1/2$, for $\zeta_1 \lessgtr 0$

(The upper or lower signs go together.) (3)

The nondimensional density ω, temperature τ, and heat flow Q_1 are given as the moments of ϕ [the gas velocity is zero for Eqs. (1) and (3)]:

$$\omega = \int \phi E d\zeta \tag{4}$$

$$\tau = \frac{2}{3}\int (\zeta_j^2 - \frac{3}{2})\phi E d\zeta \tag{5}$$

$$Q_1 = \int \zeta_1 \zeta_j^2 \phi E d\zeta \tag{6}$$

where $d\boldsymbol{\zeta} = d\zeta_1 d\zeta_2 d\zeta_3$. In the present problem, Q_1 is independent of x_1, and $Q_2 = Q_3 = 0$.

IV. Numerical Analysis

In the same way as in Refs. 9, 15, a solution of the form

$$\phi = \phi(x_1, \zeta_1, \zeta_r), \quad \zeta_r = (\zeta_2^2 + \zeta_3^2)^{1/2} \qquad (7)$$

can be shown to be compatible with Eqs. (1) and (3). This reduces the number of the independent variables from four to three and saves a great amount of computation. Furthermore, the problem can be reduced to that for the half-interval $0 \le x_1 \le 1/2$ by imposing the reflection boundary condition at $x_1 = 0$:

$$\phi(0, \zeta_1, \zeta_r) = -\phi(0, -\zeta_1, \zeta_r), \quad (\zeta_1 > 0) \qquad (8)$$

which is derived from the symmetry property of Eqs. (1) and (3). The solution for $(-1/2 \le x_1 < 0)$ is given by

$$\phi(x_1, \zeta_1, \zeta_r) = -\phi(-x_1, -\zeta_1, \zeta_r) \qquad (9)$$

We determine the solution of Eq. (1) for $0 \le x_1 \le 1/2$ subject to the boundary conditions [Eq. (3) at $x_1 = 1/2$ and Eq. (8) at $x_1 = 0$] by pursuing the long-time behavior of the solution of the initial boundary value problem of the time-dependent Boltzmann equation [Eq. (1) with $\partial\phi/\partial t$ term added on the left-hand side] subject to Eq. (8) and Eq. (3) at $x_1 = 1/2$ and an initial condition (e.g., $\phi = 0$). The time-dependent problem is solved numerically by the finite-difference method developed by the authors.[9]

The scheme for numerical computation is the same as that in our previous paper.[9] The finite-difference scheme for the left-hand side of Eq. (1) with the additional $\partial\phi/\partial t$ term is a standard one. In the collision integral [the right-hand side of Eq. (1)], the distribution function $E\phi$ is expanded in terms of a system of basis functions such as used in the finite-element method. The basis functions are so chosen that $E\phi$ is approximated by a sectionally quadratic function of ζ_1 and ζ_r that takes the exact value at the lattice points in (ζ_1, ζ_r). In this system, the collision integral is in the form of the product of the matrix of the universal collision integrals of the basis functions and the column vector of the values of $E\phi$ at the lattice points. This method of collision integral is not only universal (in

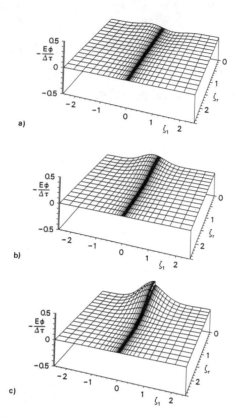

Fig. 1 The velocity distribution function $E\phi$ ($k = 0.1$).
a) $x_1 = 0$; b) $x_1 = 0.2515$; c) $x_1 = 0.5$.
In Figs. 1-4, each ζ_r = const. and each two ζ_1 = const. lattices in our computation are shown on the surface $E\phi/\Delta\tau$.

the sense that the data can be applied to other problems) but also very convenient to vectorize the program of computation.

In our finite-difference computation, the space interval $0 \leq x_1 \leq 1/2$ is divided into 100 sections, uniform for $k \geq 1$, and nonuniform for $k < 1$ with the minimum width $0.005k$ around $x_1 = 1/2$ and the maximum width $0.010 \sim 0.019$ around $x_1 = 0$. The velocity space (ζ_1, ζ_r) is limited to the finite region $-4 \leq \zeta_1 \leq 4$, $0 \leq \zeta_r \leq 4$. This region is divided into 24 uniform sections for ζ_r and 80 nonuniform sections for ζ_1 with the minimum width 0.005 around $\zeta_1 = 0$ and the maximum width 0.283 around $\zeta_1 = \pm 4$. The limitation of ζ_1 and ζ_r to the finite region is legitimate in view of the form of Eqs. (2), (3), and (4-6), where ϕ is multiplied by functions decaying rapidly as $|\zeta_1|$, $\zeta_r \to \infty$. The variation of Q_1 with x_1, which should be theoretically zero, is 1.29% for $k = 0.033$ and at most 0.145% for the other cases of our computation. This is a measure of accuracy of compuation. We have also made the preceding computation for $k = 0.1$ with 50 divisions of the x_1 interval. The maximum differences from the original computation in $\omega/\Delta\tau$ and $\tau/\Delta\tau$, which occur near the wall, are 0.187×10^{-3} and 0.257×10^{-3}, respectively. As another test of accuracy of the computation, we have analyzed the BKW equation by the present finite-difference scheme and compared the result with the very accurate result in Ref. 8. The difference is less than 10^{-5} for $\omega/\Delta\tau$ and $\tau/\Delta\tau$ at $x_1 = 0.5$, $k = 1/7$ and 10, and 2×10^{-5} for $Q_1/\Delta\tau$ at $k = 1/7$ and 10.

V. Result

The velocity distribution function $E\phi$ at three points in the gas are shown for $k = 0.1, 1, 10,$ and 40 in Figs. 1 - 4. The density and temperature distributions in the gas are

Table 1   Heat flow Q_1 vs k.

k	$-Q_1/\Delta\tau$		
	Present result	Ref. 7	Asymptotic theory
0.033	0.1377	0.1383	0.1381
0.1	0.3246	0.3264	0.3247
0.1585	0.4329	0.4360	0.4326
0.2512	0.5492	0.5537	0.5474
0.3981	0.6637	0.6686	0.6574
0.5	0.7170	0.7214	0.7068
0.6310	0.7682	0.7715	0.7528
1	0.8577	0.8576	0.8287
1.585	0.9304	0.9269	
2	0.9609	0.9561	
2.512	0.9871	0.9813	
3.981	1.030	1.023	
5	1.046	1.040	
6.310	1.061	1.055	
8	1.073	1.068	
10	1.083	1.079	
15.85	1.098	1.095	
20	1.104	1.102	
25.12	1.109	1.107	
40	1.116	1.114	
50	1.118	1.117	
63.10	1.120	1.119	
100	1.123	1.123	

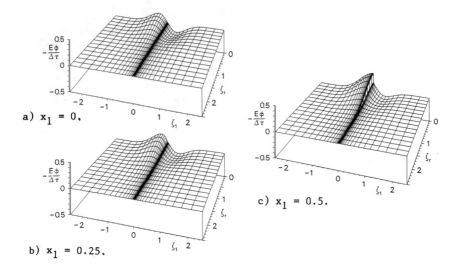

Fig. 2 The velocity distribution function $E\phi$ ($k = 1$).

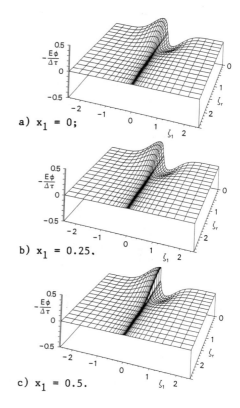

Fig. 3 The velocity distribution function E ($k = 10$).

HEAT TRANSFER BETWEEN TWO PARALLEL PLATES

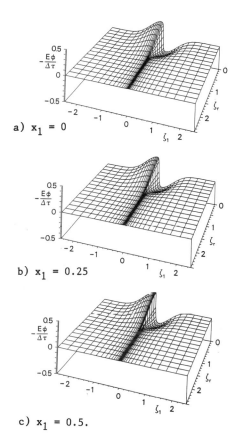

a) $x_1 = 0$

b) $x_1 = 0.25$

c) $x_1 = 0.5$.

Fig. 4 The velocity distribution function E (k = 40).

shown in Figs. 5 and 6, respectively, for various values of k [= $(\sqrt{\pi}/2)$Kn]. In these figures, the results by the moment method (full-range 4- and half-range 8-moment methods)[2] are also shown for comparison. For intermediate Knudsen numbers, the accuracy of the moment method, especially the four-moment method, is poor, although the method is still used without error estimate and further improvements in various problems. The analytical result for small k based on the asymptotic theory[12-14] (with the data of thermal conductivity[9,16] temperature jump coefficient[9] and Knudsen layer[9] for a hard sphere molecule) agrees very well with the present result. No difference can be found in Figs. 5 and 6 for k = 0.1.

The variation of the heat flow with k is shown in Fig. 7 and Table 1, together with other results for comparison. In Fig. 7, the square □ denotes the present numerical result; ① the curve connecting our numerical data smoothly; ② the result by the four-moment method[2]; ③ the result by

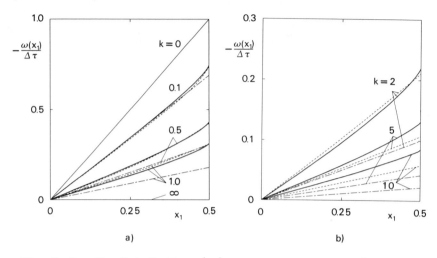

Fig. 5 Density distribution $\omega(x_1)$. ———, present result; —·—, result by the four-moment method[2]; -----, result by the eight-moment method[2]; ———, analytical results for the Navier-Stokes gas without temperature jump (k = 0) or the free molecular gas (k = ∞). a) k = 0, 0.1, 0.5, 1.0, and ∞; b) k = 2, 5, and 10.

the eight-moment method[2]; ④ the analytical result by the asymptotic theory[9,12-14]; ⑤ the analytical result for the free molecular gas (k = ∞), i.e., $-Q_1/\Delta\tau = 2/\sqrt{\pi}$; ⑥ the result for the Navier-Stokes gas without temperature jump (with the thermal conductivity for a hard sphere molecule[16]); the plus sign + the result for the BKW equation[8], where the mean free path of the BKW equation is converted to that of the hard sphere molecule through thermal conductivity or by the relation obtained by eliminating the thermal conductivity from the two relations between mean free path and thermal conductivity for the two cases $[\ell_0(\text{BKW}) = 1.922284\ell_0 \text{ (hard sphere)}]$. The BKW result is very close to the present one over the whole range of the Knudsen number. Our numerical data of the heat flow are shown in Table 1, where the result by variational method in Ref. 7 and that by the asymptotic analysis are also shown for comparison. The variational analysis in Ref. 7, where a very simple test function, linear in x_1, for the velocity distribution function is used and therefore the accurate local behavior such as the accurate temperature distribution cannot be expected, gives very accurate results for the heat flow over the whole range of the Knudsen number.

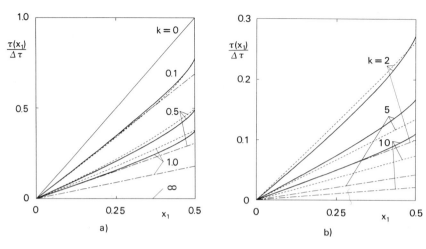

Fig. 6 Temperature distribution $\tau(x_1)$. ———, present result; —·—·—, result by the four-moment method[2]; - - - - -, result by the eight-moment method[2]; ———, analytical results for the Navier-Stokes gas without temperature jump ($k = 0$) or the free molecular gas ($k = \infty$). a) $k = 0, 0.1, 0.5, 1.0,$ and ∞; b) $k = 2, 5,$ and 10.

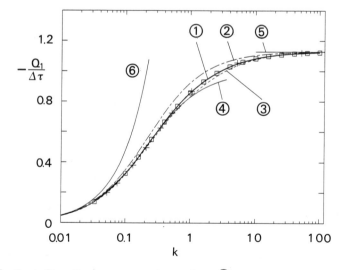

Fig. 7 Heat flow Q_1. □, present result; ①, curve connecting the present numerical data smoothly; ②, result by the four-moment method[2]; ③, result by the eight-moment method[2]; ④, analytical result by the asymptotic theory[9,12-14]; ⑤, result for the free molecular gas; ⑥, result for the Navier-Stokes gas without temperature jump; +, result for the BKW equation[8].

The numerical computation of the present study was carried out by FACOM VP-400 at the Data Processing Center, Kyoto University.

References

[1] Wang Chang, C. S. and Uhlenbeck, G. E., "The heat transport between two parallel plates as functions of the Knudsen number," Engineering Research Institute, University of Michigan, Michigan, Project M999, 1953.

[2] Gross, E. P. and Ziering, S., "Heat flow between parallel plates," The Physics of Fluids, Vol. 2, Nov. Dec. 1959, pp. 701-712.

[3] Liu, C. Y. and Lees, L., "Kinetic theory description of plane compressible Couette flow," Rarefied Gas Dynamics, edited by L. Talbot, Academic, New York, 1961, pp. 391-428.

[4] Willis, D. R., "Heat transfer in a rarefied gas between parallel plates at large temperature ratios," Rarefied Gas Dynamics, edited by J. A. Laurmann, Academic, New York, 1963, Vol. 1, pp. 209-225.

[5] Frankowski, F., Alterman, Z., and Pekeris, C. L., "Heat transport between parallel plates in a rarefied gas of rigid sphere molecules," The Physics of Fluids, Vol. 8, Feb. 1965, pp. 245-258.

[6] Bassanini, P., Cercignani, C., and Pagani, C. D., "Comparison of kinetic theory analyses of linearized heat transfer between parallel plates," International Journal of Heat and Mass Transfer, Vol. 10, Apr. 1967, pp. 447-460.

[7] Cipolla, J. W., Jr. and Cercignani, C., "Effect of molecular model and boundary conditions on linearized heat transfer," Rarefied Gas Dynamics, edited by D. Dini, Editrice Tecnico Scientifica, Pisa, Italy, 1971, Vol. 2, pp. 767-777.

[8] Thomas, J. R., Jr., Chang, T. S., and Siewert, C. E., "Heat transfer between parallel plates with arbitrary surface accommodation," The Physics of Fluids, Vol. 16, Dec. 1973, pp. 2116-2120.

[9] Sone, Y., Ohwada, T., and Aoki, K., "Temperature jump and Knudsen layer in a rarefied gas over a plane wall: Numerical analysis of the linearized Boltzmann equation for hard-sphere molecules," The Physics of Fluids A, Vol. 1, Mar. 1989.

[10] Grad, H., "Asymptotic theory of the Boltzmann equation, II," Rarefied Gas Dynamics, edited by J. A. Laurmann, Academic, New York, 1963, Vol. 1, pp. 26-59.

[11] Cercignani, C., Theory and Application of the Boltzmann Equation, Scottish Academic, Edinburgh, Scotland, U.K., 1975, Chap. IV.

[12] Sone, Y., "Asymptotic theory of flow of rarefied gas over a smooth boundary I," Rarefied Gas Dynamics, edited by L. Trilling and H. Y. Wachman, Academic, New York, 1969, Vol. 1, pp. 243-253.

[13] Sone, Y., "Asymptotic theory of flow of rarefied gas over a smooth boundary II," *Rarefied Gas Dynamics*, edited by D. Dini, Editrice Tecnico Scientifica, Pisa, Italy, 1971, Vol. 2, pp. 737-749.

[14] Sone, Y. and Aoki, K., "Steady gas flows past bodies at small Knudsen numbers — Boltzmann and hydrodynamic systems —," *Transport Theory and Statistical Physics*, Vol. 16, No. 2&3, 1987, pp. 189-199; "Asymptotic theory of slightly rarefied gas flow and force on a closed body," *Memoirs of the Faculty of Engineering, Kyoto University*, Vol. 49, July 1987, pp. 237-248.

[15] Sone, Y. and Aoki, K., "A similarity solution of the linearized Boltzmann equation with application to thermophoresis of a spherical particle," *Journal de Mécanique Théorique et Appliquée*, Vol. 2, No. 1, 1983, pp. 3-12.

[16] Pekeris, C. L. and Alterman, Z., "Solution of the Boltzmann-Hilbert integral equation II. The coefficients of viscosity and heat conduction," *Proceedings of the National Academy of Sciences of the U.S.A.*, Vol. 43, Nov. 1957, pp. 998-1007.

Chapter 2. Discrete Kinetic Theory

Low-Discrepancy Method for the Boltzmann Equation

H. Babovsky,* F. Gropengiesser,* H. Neunzert,† J. Struckmeier,*
and B. Wiesen*
*University of Kaiserslautern, Kaiserslautern,
Federal Republic of Germany*

Abstract

As an alternative to the commonly used Monte Carlo simulation methods for solving the Boltzmann equation, we have developed a new code with certain important improvements. We present results of calculations on the re-entry phase of a Space Shuttle. One aim was to test physical models of internal energies and of gas/surface interactions.

I. Introduction

Simulation methods are the most important tool for solving the Boltzmann equation in realistic settings. In the past, a number of so-called direct simulation Monte Carlo schemes have been developed (see Nanbu's review[1]). The most popular attitude for their derivation was: Imitate the behavior of real gas molecules but in a reduced particle system. Nanbu has gone one step beyond this interpretation when deriving his scheme from the Boltzmann equation. This scheme is now quite well understood (from a physical as well as a rigorous mathematical point of view[1]), and it has been proven to yield approximations of solutions of the Boltzmann equation, given a sufficiently large number of test particles.[2,3] However, this method can also be interpreted as an imitation of a physical

---

Copyright © 1989 by the American Institute of Aeronautics and Astronautics, Inc. All rights reserved.
  *Assistant, Department of Mathematics.
  † Professor, Department of Mathematics.

situation: the motion of particles in a fixed background gas.[4]

For the derivation of a powerful simulation code, we propose to forget about the physical situation and instead to search for a mathematical model yielding results as close to the Boltzmann equation as possible. Such a code (called the Low Discrepancy (LD) code) has been developed by our group. The philosophy behind it is completely different from that of Monte Carlo schemes, in that it replaces the purely random "microscopic" behavior by a behavior that is as regular as possible in order to cut down fluctuations. Our code, as far as we know at present, shows these essential improvements compared with all Monte-Carlo schemes in use (compare Nanbu[1]):

1) It imitates two particle collisions and thus satisfies strictly the conservation laws (in contrast to Nanbu's);

2) It is highly vectorizable (in contrast to Bird's), even in the treatment of internal energies and boundary conditions;

3) It has reduced fluctuations and thus allows reduction of particle numbers.

The main applications for our scheme have been calculations on the re-entry phase of the European space shuttle Hermes. To obtain results for two- and three-dimensional test cases, we had to develop an efficient adaptive grid, applicable to all geometries of interest, which allows for the reconstruction of even high gradients (shocks) within reasonable calculation times. Details are described in sec. 3.
We have carefully studied problems of modeling physical effects such as internal degrees of freedom and gas/surface interactions. The usual way of treating these effects is to apply robust models that are easy to implement and that produce plausible results. The commonly used models are the Larsen-Borgnakke model for internal energies and diffuse reflection with accommodation coefficient for the gas/surface interaction. To test their physical relevance, we have also implemented alternative models that seem better motivated from a physical point of view. The results are shown in the sec. 4 and 5. (Calculations concerning gas mixtures and chemical reactions are in progress and cannot be presented here. All calculations have been performed on the vector calculator Fujitsu VP 100.)

II. The Idea of Low Discrepancy

In order to explain the main idea of Low Discrepancy, we choose as simple a situation as possible. Therefore, in this section, we consider only the space homogeneous Boltzmann equation

$$\frac{\partial}{\partial t} f(v) = J[f,f](v) = \iint k \cdot \{f(v')f(w')-f(v)f(w)\} d\eta dw$$

(where η is a unit vector, $v' = v-\eta\langle v-w,\eta\rangle$, and $w' = w+\eta\langle v-w,\eta\rangle$), and its time discretization

$$f_{j+1}(v) = f_j(v) + \Delta t \cdot J[f_j,f_j](v)$$

$$= (1-\Delta t \cdot \iint k f_j(w) d\eta dw) f_j(v) + \Delta t \cdot \iint k f_j(v') f_j(w') d\eta dw$$

Multiplying a test function ϕ and integrating with respect to v, one can compress this formula to the following weak version:[2]

$$\int \phi(v) f_{j+1}(v) dv = \int \phi(\psi(v,w,b)) d^2 b f_j(v) dv f_j(w) dw$$

with impact parameter b and

$$\psi(v,w,b) = \begin{cases} v' & \text{if b indicates "collision"} \\ v & \text{if b indicates "no collision"} \end{cases}$$

This version is appropriate for our aims because it can be interpreted as follows: If f_j is the velocity distribution at the jth time step, then with probability

$$d^2 b f_j(v) dv f_j(w) dw$$

the velocity at the (j+1)st time step is $\psi(v,w,b)$.

This interpretation motivates the following general simulation scheme:

General Scheme (One Time Step):

Step 1: Start with an N point approximation

$$(v_i(0))_{i \leq N} = (v_1(0),\ldots,v_N(0))$$

of $f_0(0)dv$.

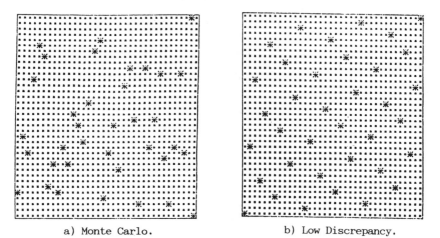

a) Monte Carlo.   b) Low Discrepancy.

Fig. 1   Selection of N points from N^2 points.

Step 2: Select for each $v_i(0)$ a "collision partner"
$w_i(0) = v_{n(i)}(0)$, and an "impact parameter" b_i
such that

$$(b_i, v_i(0), w_i(0))_{i \leq N}$$

is a good approximation of $d^2b\, f_0(v)dv f_0(w)dw$ ("factorization property" of selection mechanism).

Step 3: Define new velocities

$$v_i(1) = \psi(v_i(0), w_i(0), b_i)$$

(If the collection of pairs $(i, n(i))$ is "symmetric", i.e.,

$$n(i) = j \iff n(j) = i$$

and if $b_i = b_{n(i)}$, then the scheme satisfies the conservation laws.)

Taking the Monte Carlo version of this scheme (this is equivalent to Nanbu's), one has to choose b_i and $n(i)$ as independent random numbers. In this case, one can show that the simulation result is a good approximation of the solution of the Boltzmann equation if N is large enough.[2] (A similar statement is true in the space-dependent case.[3])

To construct an LD version, one has to find a selection algorithm with an optimal factorization property. The following simple example should clarify this somewhat: Suppose f_j depends only on $|v|$, and the velocities are arranged as follows:

$$|v_1| \leq \ldots \leq |v_N|$$

Then the best approximation of $f_j(v)dvf_j(w)dw$ by pairs (v_i,w_i) is that for which the pairs $(i,n(i))$ are spread over $\{1,\ldots,N\}\times\{1,\ldots,N\}$ as uniformly as possible. Fig. 1 shows a Monte Carlo and an LD choice for $(i,n(i))$. In this simple one-dimensional case it is possible to find a practicable strategy that is almost optimal. In higher-dimensional cases, a practicable alternative is to find sequences of pseudorandom numbers with a good factorization property (such as Hammersley sequences).

The idea of LD is not restricted to the approximation of $d^2bf_j(v)dvf_j(w)dw$. It may also be used (in an obvious manner) to treat initial and boundary conditions, internal energies, etc.

III. Generation of Adaptive Grid Structures

Recall that the collision operator of the Boltzmann equation is local in the space coordinates. In simulations, it is therefore necessary to homogenize the density function with respect to the space coordinates. This homogenization procedure is done according to a cell structure that is influenced by the properties of the flowfield. Therefore, the cell structure may vary from time to time.

To be acceptable in a simulation, a given grid must fulfill three conditions:

1) The approximate homogeneity of the density function over each cell must be guaranteed.

2) It must be easy to refind the particles in the cells after the free flow.

3) The number of cells must yield a reasonable computer storage.

Until now, several criteria have been given to refine or to coarsen an existing grid. Widely used are physically motivated criteria based, for example, on the particle density or on gradients of the macroscopic quantities. Instead of those, we have chosen a mathematical criterion based on a requirement of the proof of convergence of

Babovsky and Illner:
    We have to ensure that

$$\operatorname*{ess\,sup}_{t,x,v} |f(t,x+\Delta x,v) - f(t,x,v)| \exp(\alpha v^2) \leq B\Delta x \qquad (1)$$

for some $\alpha > 0$, $B > 0$ and all spatial displacements Δx.
    To perform this requirement, we use the following algorithm:
    For each time step,

1) Divide the domain of computation into rectangular cells two-dimensional or cubes three-dimensional of fixed shape.

2) Divide each rectangle (cube) into smaller rectangles (cubes) until requirement (1) is satisfied.

It is clear that this algorithm allows the indexing of the particles in a straightforward way (the only things you have to do are modulo operations and reorderings!).
    As a first test case for the performance of the algorithm, we have chosen the problem of the calculation of the flowfield around a two-dimensional ellipse. The input data were: 1) Flow velocity: Mach 20, 2) Wall temperature: 1000 K, 3) Gas temperature: 194 K, 4) Mean free path: 0,13 m, 5) Ellipse axes: 6,85 m; 2,055 m, 6) Angle of attack: 40 deg.
Figures 2a and 2b show the initial coarse grid and the refined grid in the stationary state at the end of the simulation. We have plotted the midpoints of the cells.

IV. Gas/Surface Interaction Laws

Usually the structure of a solid boundary is by far too complicated to compute the interaction potential between the surface and the incoming particles. Even if this were possible the incomplete knowledge of the state of the surface (roughness, chemical reactions, etc.) made such a calculation impossible.
    Therefore the description of gas/surface interaction phenomena is done by simple models that are motivated by phenomenological reasoning and that have some disposable parameters. These parameters have to be fitted on experimental results. The usual mathematical description of these models can be done in the frame of scattering kernels yielding an integral equation of the form:

$$|v \cdot n| f(x,v,t) = \int_{v' \cdot n < 0} R(v' \longmapsto v, \cdot x, t) f(x, v', t) |v' \cdot n| dv$$

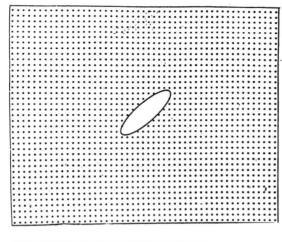

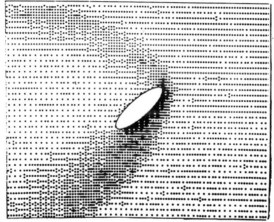

Fig. 2 Midpoints of a) the coarse and b) the refined grid for the flow around a two-dimensional ellipse.

Here n is the inner normal at the boundary point x and $R(v' \mapsto v, \cdot x, t)$ is the scattering kernel. The probabilistic interpretation of R is: Rdv is the probability that a particle that hits the wall at x with velocity v' leaves the wall with a velocity in the volume element around v.

In addition to the simple models 1) specular reflection (no parameters), 2) diffuse reflection (the wall temperature can be considered as parameter) and 3) Maxwell boundary (parameters: wall temperature and accomodation c oefficient), we have implemented the Cercignani-Lampis model.[5] This model treats the normal component v_n and the tangent component v_t of the scattered velocity in different ways.

The features of the Cercignani-Lampis model are:

1) Scattering in the tangent space and in the normal direction are independent.
2) Specular reflection and diffuse reflection are special cases of this model.
3) The scattering kernel satisfies the reciprocity condition.[5]
4) Good agreement with scattering experiments can be achieved by suitable choice of the accommodation coefficients.

Another advantage of the model is the easy implementation in the simulation procedure. The algorithm is as follows:

A) Scattering in tangent space:

Choose random numbers r_1, r_2

$$s \leftarrow \sqrt{-\alpha_t(2-\alpha_t)\log(1-r_1)}$$

$$v_t^{(1)} \leftarrow (1-\alpha_t)v_t'^{(1)} - s*\cos(2\pi r_2)$$

$$v_t^{(2)} \leftarrow (1-\alpha_t)v_t'^{(2)} - s*\sin(2\pi r_2)$$

Here $v_t'^{(1)}$, $v_t'^{(2)}$ are the two components of the incoming velocity in the tangent space.

B) Scattering in normal direction, Polya-Aeppli-distribution algorithm:

1) Generate Poisson $(\frac{1-\alpha_n}{\alpha_n} \cdot v_n'^2)$ random variable z.

2) Generate gamma (1+z) random variable G

$$v_n' \leftarrow \sqrt{\alpha_n G}$$

It should be noticed that (in this way of implementing) the gas/surface interaction procedure is completely vectorized and therefore not time-consuming.

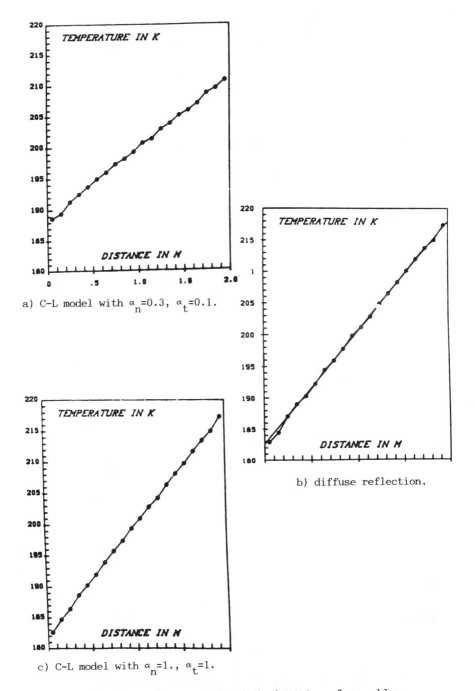

a) C-L model with $\alpha_n=0.3$, $\alpha_t=0.1$.

b) diffuse reflection.

c) C-L model with $\alpha_n=1.$, $\alpha_t=1.$

Fig. 3 Temperature profile of the heat-transfer problem (temperature, K).

As test case for gas/surface models, we have selected the heat-transfer problem in one dimension. Figure 4a shows the temperature profile of a monoatomic hard-sphere gas between two infinite walls. At the left boundary, we have a temperature of 180°K and, at the right boundary, a temperature of 220°K.

As can be seen by comparison of Figs. 3b and 3c, the results are the same for the diffuse reflection and Cercignani-Lampis models. Figure 3a shows that the profile becomes flatter if the accommodation coefficients are lower than 1.0. Thus we are able to adjust our results to measurements by fitting the parameters, but to make a good guess about the right values, we need measurements of the temperature profile in this simple case.

V. Treatment of Classical Internal Degrees of Freedom

Because of the temperatures that arise in the simulation of realistic gas flow problems, the internal states of the gas molecules have to be considered. In many cases, we deal with linear molecules (e.g., N_2, O_2), for which the rotations of the molecules are of particular importance.

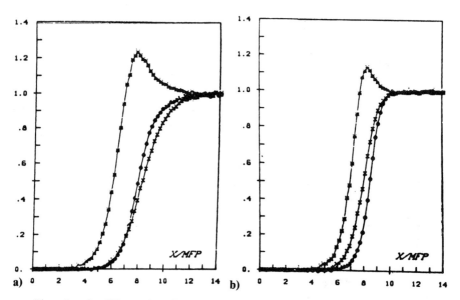

Fig. 4   Profiles of reduced density and temperatures for a shock wave of molecules with 2 internal degrees of freedom at upstream temperature of 200°K. Hard-sphere model: a) Z = 0.9. b) Z=0.6; curves 0: density profile, X: internal temperature, *: translational temperature.

Whereas there is little doubt about the right kinetic equation for monoatomic gases, a generalization to polyatomic molecules is not quite straightforward (one has to decide, for example, whether to treat the internal degree of freedom by means of quantum mechanics).

In this paper, we report on classical internal degrees of freedom. To save computer storage and time, we calculate the distribution of the internal energy only. The kinetic equation we use was described by Pullin:[6]

$$\left(\frac{\partial}{\partial t} + v \cdot \nabla_x\right) f(t,x,v,\varepsilon)$$

$$= \int_{R^3} \int_0^\omega \int_{\Delta_E} \int_{S^2} \|v-w\| \sigma(E; \varepsilon, \varepsilon_1; \varepsilon', \varepsilon_1'; \eta \cdot \eta')$$

$$\times \{f' f_*' - f \cdot f_*\} dw(\eta) d^2\varepsilon' d\varepsilon_1 d^3w \qquad (1)$$

with

$$\eta = \frac{v-w}{\|v-w\|}, \quad \eta' = \frac{v'-w'}{\|v'-w'\|}$$

$$E = \frac{m}{4} \|v-w\|^2 + \varepsilon + \varepsilon_1 = \frac{m}{4} \|v'-w'\|^2 + \varepsilon' + \varepsilon_1'$$

$$f = f(t,x,v,\varepsilon), \quad f_* = f(t,x,w,\varepsilon_1)$$

$$f' = f(t,x,v',\varepsilon') \quad f_*' = f(t,x,w',\varepsilon_1')$$

$$\Delta_E = \{(\varepsilon', \varepsilon_1') : 0 \leq \varepsilon', \ 0 \leq \varepsilon_1', \ \varepsilon' + \varepsilon_1' \leq E\}$$

It is clear that different models of the exchange of internal and translational energy are characterized by the particular form of the scattering cross section $\sigma(E; \varepsilon, \varepsilon_1; \varepsilon', \varepsilon_1'; \eta \cdot \eta')$.

A widely used model is that proposed by Larsen and Borgnakke.[7] For this model, the scattering cross section for rotating linear molecules reads

$$\sigma(E; \varepsilon, \varepsilon_1; \varepsilon', \varepsilon_1'; \eta \cdot \eta')$$

$$= Z(E) \sigma_0(\|v-w\|) h(\eta \cdot \eta') \delta(\varepsilon - \varepsilon') \delta(\varepsilon_1 - \varepsilon_1')$$

$$+ (1-Z(E)) \sigma_0(\|v-w\|) \|v'-w'\|^2 \sigma_0(\|v'-w'\|) h(\eta \cdot \eta') N(E) \qquad (2)$$

where N(E) is a normalization function and

$$\int_{S^2} h(\eta \cdot \eta') dw(\eta') = 1 .$$

The features of this model are:

1) The total cross section depends on $\|v-w\|$ only.
2) A part of the collisions is elastic. The ratio of elastic to inelastic collisions is controlled by the total collisional energy.
3) The "energy-scattering kernel" does not depend on ε and ε_1; it is determined by the total cross section σ_0.

According to this model, we have the following simulation algorithm to perform the collision process:

1) Define the collision partners in such a way that the pairs $[(v_i, \varepsilon_i), (w_i, \varepsilon_i)]$ are a good approximation of the product density function.

2) For each pair $[(v, \varepsilon), (w, \tilde{\varepsilon})]$,

$P_{coll} \leftarrow | \; n \cdot \sigma_0(\|v-w\|) \cdot \|v-w\| \Delta t'$

If $(1-r) \leq P_{coll}$

$E \leftarrow | \; \frac{m}{4} \|v-w\| + \varepsilon + \tilde{\varepsilon}; \quad \varepsilon_t \leftarrow | \; \frac{m}{4} \|v-w\|^2$

If $(r \geq Z(E))$,

generate r·v. μ_t, μ_1 according to P_t, P_1

generate r·v. η' according to $h(\eta \cdot \eta')$

$\varepsilon_t \leftarrow | \; \mu_t E$

$\varepsilon \leftarrow | \; (1-\mu_1)(1-\mu_t)E$

$\tilde{\varepsilon} \leftarrow | \; \mu_1(1-\mu_t)E$

And if

$\begin{pmatrix} v \\ w \end{pmatrix} \leftarrow | \; \frac{1}{2} [(v+w) \; \substack{+ \\ -} \; \eta' \sqrt{\frac{4}{m} \varepsilon_t} \;]$

Here we have

$P_t(\mu_t) = N(E)\sigma_0(\sqrt{\frac{4}{m} \mu_t E})\mu_t(1-\mu_t)$

$P_1(\mu_1) = 1 .$

It should be noticed that this simulation procedure is completely vectorizable because there is no need for the use of a time counter (remember that the introduction of time counters causes recurrence, which avoids the possibility of vectorization).
To study the influence of the parameters, we have calculated a one-dimensional shock wave of gases with two internal degrees of freedom. In all our calculations, the initial distribution was given by a Maxwellian with different parameters upstream, respectively downstream with a jump at zero. The downstream values are determined from the upstream values by the Rankine-Hugoniot relations. Equilibrium at infinity is assumed. In Figure 4 we show the results obtained by using a hard-sphere total cross section and, in Figure 5, those obtained with the help of the VHS scattering cross section of I. Kuscer:[8]

$$\sigma_o(\|v-w\|) = \sigma_\infty \cdot \left(1 + \frac{12KT_s}{m\|v-w\|^2}\right)$$

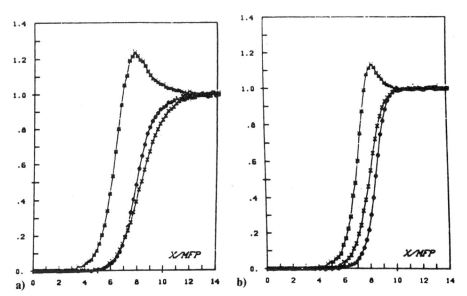

Fig. 5 Profiles of reduced density and temperatures for a shock wave of molecules with 2 internal degrees of freedom at upstream temperature of 200°K. Kuscer model:
($\sigma_\infty = 3 \cdot 10^{-19}$ m^2. $T_s = 107$ K).
a) $Z = 0.9$, b) $Z = 0.6$; curves 0: density profile, X: internal temperature, *: translational temperature.

Here T_S is the Sutherland temperature of the gas molecules that are to be simulated and σ_∞ is a constant that has to be adapted on the measured viscosity (notice that the viscosity calculated with the help of this scattering cross section obeys the Sutherland formula). In the results shown, the function $Z(E)$ has been kept constant at the values 0.9 and 0.6, respectively. As can be seen by comparison of the various results, both the temperature and the density profiles are influenced by the choice of the scattering cross section and the ratio of elastic collisions. Therefore, by comparison of calculated with measured shock profiles, it should be possible to make a decision about the interaction law that is suitable for a given gas type.

VI. Summary and Conclusions

We have developed a new code for the simulation of the Boltzmann's equation that is based on the LD method. Because of the structure of this algorithm, the vectorization of the code is straightforward. This vectorization property yields reasonably short computation times: to calculate the flowfield around a two-dimensional ellipse, we needed about 18 CPU min on the Fujitsu VP 100.

Also, the LD ideas are very appropriate for simulating boundary conditions. In this field, we implemented the Cercignani-Lampis model, which has, in our opinion, enough parameters to fit on experimental results.

The consideration of internal energies is also straightforward because the LD method is based on binary collisions. This property of the method has been demonstrated by the calculation of a one-dimensional shock wave of gases with 2 internal degrees of freedom.

The most important property of our algorithm is, in our opinion, the proof of convergence. This proof shows that the LD method is based on a good mathematical ground and does not rely on heuristics (as, e.g., Bird's scheme). This groundwork allows further consideration of the scheme as, for example, its behavior when the number of simulation particles is very small. The study of this behavior will be one of our main research topics in the future.

References

[1]Nanbu, K., "Theoretical Basis of the Direct Simulation Monte Carlo Method," <u>Proceedings of the Fifteenth International Symposium on Rarefied Gas Dynamics</u>, Vol. 1, edited by V. Boffi and C. Cercignani, Teubner, Stuttgart, FRG, 1986, pp. 369-383.

[2] Babovsky, H., "A Convergence Proof for Nanbu's Boltzmann Simulation Scheme," to appear in Eurepean Journal of Mechanics B.

[3] Babovsky, H., Illner, R., "A Convergence Proof for Nanbu's Simulation Method for the Full Boltzmann Equation," to appear in SIAM Journal of Numerical Analysis.

[4] Babovsky, H., "On a Simulation Scheme for the Boltzmann Equation, " Mathematical Methods in the Applied Sciences, Vol. 8, 1986, pp. 223-233.

[5] Cercignani, C., "The Boltzmann Equation and Its Applications," Springer, New York, 1988.

[6] Pullin, D.I., "Kinetic Models for Polyatomic Molecules with Phenomenological Energy Exchange," Physics of Fluids, Vol. 21.2, Feb. 1978, pp. 209-216.

[7] Borgnakke, C., Larsen, P., "Statistical Collision Model for Monte Carlo Simulation of Polyatomic Gas Mixture," Journal of Computational Physics, Vol. 18, 1975, pp. 405-420.

[8] Kuščer, I., "A Model for Monte Carlo Simulation of Rarefied Gas Flows," Talk presented at "Mathematical Problems in the Kinetic Theory of Gases," Oberwolfach, FRG, 1988.

Investigations of the Motion of Discrete-Velocity Gases

D. Goldstein,[*] B. Sturtevant,[†] and J. E. Broadwell[‡]
California Institute of Technology, Pasadena, California

Abstract

A model of molecular gasdynamics with discrete components of molecular velocity has been implemented for parallel computation, and two test problems have been calculated. When the molecular velocity components have integer values, and time is discretized for digital computation, the particles move on a regular array of points and the gas is called a "lattice gas." Calculations of molecular motions are thus simplified. The outcome of binary collisions between identical particles with discrete velocity components is determined by simple reflections about axes of symmetry in the center-of-mass system; thus, calculations of collisions are sped up. It is shown that fewer than 10 values of each component of molecular velocity are necessary to produce accurate results in calculations by direct simulation Monte Carlo methods of rarefied gas flows involving moderately strong shock waves. Thus, significant savings in memory required to store the molecular velocities are realized.

I. Introduction

With the design of lifting vehicles powered by air-breathing engines, which take off at sea level and fly at hypersonic speed in the earth's upper atmosphere, the demands on methods for numerical simulation of flows over complex bodies have increased significantly. When conditions are such that the governing partial differential equations change type, mapping out the flowfield around the vehicle requires that finite-difference solutions from different codes be overlaid, even in the simplest case of two-dimensional flow of a perfect gas. When complex chemical effects with widely differing relaxation times become important, new computational difficulties often arise. Under such circumstances simulation methods that are "exact" at the molecular level become attractive, since,

---

Copyright © 1989 by the American Institute of Aeronautics and Astronautics, Inc. All rights reserved.
[*]Research Assistant, Graduate Aeronautical Laboratories.
[†]Senior Scientist, Graduate Aeronautical Laboratories.
[‡]Professor, Graduate Aeronautical Laboratories.

presumably, once the correct treatment of the physics and chemistry has been ensured at the microscopic level, macroscopic fields will be calculated correctly, even in cases for which the Navier-Stokes equations are invalid. The major disadvantage of using molecular methods to calculate continuum flowfields is their extreme inefficiency, due to the level of detail at which the calculations are made. Thus, molecular simulation methods are most often used to calculate flows where their application is necessary, e.g., in rarefied gas flows. Nevertheless, for the reasons stated earlier, and also because often some parts of a flowfield are rarefied whereas others are not, there has always been interest in extending the application of molecular simulation methods well into the continuum flow regime. The objective of the present work is to study methods for simplifying the molecular approach in order to make it more amenable to application to continuum flows. In particular, we examine the consequences of modeling flows with molecules that move with only a few, rather than a continuum, of different velocities. Such discrete-velocity models have in the past stimulated many fundamental studies in kinetic theory (see, for example Ref. 1), and recently their computer implementation as cellular automata (CA) has generated a great deal of interest.[2] The implementation in the present work of one discrete-velocity model, using methods for concurrent computation, has provided further insight into the physics of nonequilibrium gas flow.

In this work we are concerned with methods that *directly* simulate molecular motions, and in particular, with the direct simulation Monte Carlo method[3] (DSMC). Such methods do not solve systems of partial differential equations; hence, the mesh can be independent of the coordinate system, and the calculation of flow over complex bodies is inherently simple. The present research is an investigation of simplifications of the DSMC model. Emphasis is placed on the influence of the simplification on the speed and accuracy of the calculation of supersonic flowfields. The models treated in this work follow the simplifications of the Boltzmann equation introduced by Carleman[4] and Broadwell[5,6] in which the molecular velocity components are discretized. In the early work, the molecular velocities were prescribed a priori, whereas in our treatment of the DSMC the discretized velocities emerge as needed in the course of the computation.

II. Integer Direct Simulation Monte Carlo Method (IDSMC)

The research version of the parallel DSMC that we use for comparison with the methods developed in this work treats the molecules as elastic hard spheres. Phenomenological models of real-gas effects, such as vibrational relaxation and dissociation, which have been developed at other laboratories, could easily be incorporated. In all of our calculations we use an adaptive grid of cells that at every time step automatically remeshes itself in the vertical and horizontal directions to keep the average number of particles in the rows and columns of cells constant. By this means we aid load balancing among the processors of the computing machine, the most important consideration for the efficient use of parallel computation.

In the conventional DSMC a relatively small sample of molecules (typically tens of thousands to hundreds of thousands) is taken to represent a flowing gas. Space is divided into cells whose size $(\Delta x, \Delta y, \Delta z)$ is small compared to the mean free path λ, and time is discretized into steps Δt, which are smaller than the mean molecular collision time τ. The calculation of molecular collisions is decoupled from the motion of the particles, and only the molecules within a given cell are considered as candidates for collisions. Only binary collisions are treated; thus, the gas is, by definition, dilute. During a given time step, collisions within a cell are calculated until the (known) collision frequency has been achieved. After collisions in all cells have been so calculated, the particles are moved in free flight to locations appropriate for the beginning of the next time step.

The goal of the modifications to the DSMC discussed here is to limit the amount of information concerning the molecular velocities that is developed, thus speeding up the calculations and freeing memory space for the treatment of more particles. The most direct way to ensure that the method for limiting the information kept on velocities does not result in spurious generation or loss of momentum or energy is to carry out the simplification in a way that ensures the conservation of momentum and energy in *every* collision, as in the conventional floating-point calculations. It is easy to do this for identical particles if the *components* of the molecular velocities are discretized. In the IDSMC the velocity components are integers. Though, in principle, the number of values that the integer velocity components can assume is infinite, in practice, the number depends on the integer size provided by the digital computer used for the numerical calculations. For 32-bit integers the number of velocities can be as large as 4×10^9, for 16 bits 65,536, and for 8 bits 256. As will be seen later in the paper, the latter provides sufficient resolution for flows even up to hypersonic Mach numbers; hence, we usually declare the velocities to be one byte long.

2.1 Discretization

An immediate consequence of the discretization of the components of molecular velocity is that velocity is quantized with the unit of velocity, say q. In any given problem, whether q is small or large depends on the characteristic thermal speed of the molecules, say, the rms thermal velocity,

$$c_s \equiv (\overline{c'^2})^{1/2} = \sqrt{3RT} \qquad (1)$$

where R is the gas constant and T is the local temperature. For high-temperature gases the velocity distribution function is "wide," and many molecular speeds occur; hence, q is small compared to c_s. For cold gases the velocity distribution function is "narrow"; thus, only a few molecular speeds occur, and q may be of order c_s. In the IDSMC the number of molecular speeds found at any point in space and time adjusts to the temperature there. Furthermore, as in the DSMC, the number of speeds (the width of the

MOTION OF DISCRETE-VELOCITY GASES 103

distribution function) does not affect the computational cost. This is not the case when the gas is treated as a cellular automaton,[2] where the computational cost increases rapidly with complexity.

Initially the velocity resolution is, in effect, set by the choice of the cell size, say Δx, compared to the distance traveled by a particle of speed q in time Δt, which we shall call the lattice spacing δ. For, as already stated, in the DSMC $\Delta t / \tau \equiv$ m and $\Delta x / \lambda \equiv$ l must both be somewhat smaller than unity. Furthermore,

$$\frac{\Delta x}{\delta} = \frac{\Delta x}{q \, \Delta t} = \frac{1}{m} \frac{\lambda}{q\tau} = \frac{1}{m} \frac{\overline{c'}}{q} \qquad (2)$$

where $\overline{c'} = \sqrt{8/3\pi} \, c_s$ is the mean thermal speed. Then for l ≈ m, if δ is small compared to Δx, q must be small compared to $\overline{c'}$. Thus, the velocity resolution improves as the cell contains more lattice sites. In practice, Δx and Δt are chosen on the basis of λ and τ in the region of highest expected density and temperature in the flow under consideration. (When, as in the present work, the mesh is coarsened during the calculation in low-density regions to keep l roughly constant, the right-hand side of Eq. (2) increases because 1/m, not $\overline{c'}/q$, increases. Therefore, in this case the velocity resolution does not increase.) In a typical example, we take $\delta = 0.1\lambda$, $\Delta x = 0.5\lambda$, $\Delta t = 0.2\tau$ and, from (2), $\overline{c'} = 2q$; thus, $RT = (\pi/2) \, q^2$. On the other hand, if we halve δ (to = 0.05 λ) and increase Δt so that l = m = 0.5, then $\overline{c'} = 10 \, q$ and the temperature is 25 times higher. In any application the dimensional value of q can be chosen to obtain the desired dimensional value of the reference temperature T, e.g., 300°K.

A further consequence of the discretization of the velocity components is that, if particles are initially distributed in space on a regular array coincident with the axes of the coordinate system at points with spacing δ, then the particles remain on the array for all time, and we have a lattice gas. By initially positioning the particles on a lattice, the spatial resolution is coarsened to a level consistent with the velocity resolution, and the calculation of particle motion is simplified and sped up; particle translations during the motion phase are obtained simply by counting lattice sites. Since the particle locations are integer numbers, storage requirements are also reduced.

In the IDSMC a (coarse) mesh of cells is superimposed on the (fine) lattice, and particles are drawn as candidates for collisions from all of the lattice sites within a cell. There may be as few as one lattice site in a cell, provided there are enough particles at the site that a sufficient number of collisions can be calculated during a time step to provide the necessary collision rate. This is in contrast to the procedure followed when the lattice gas is treated as a CA, in which case the number of particles at a site is limited by an exclusion principle, and every particle is treated as a candidate for a collision, consistent with a set of specified collision rules. For the CA the collision rate is an outcome of the calculation. In the IDSMC, as in the DSMC, a record is kept of the cell in which every particle resides, at the expense of additional storage. Since the adaptive grid of cells is superimposed on the fixed lattice, the cell sizes may not have any simple relationship to the lattice spacing, and two cells of the same size may contain a different number of lattice points. Thus, it is necessary to take the cell volume to be proportional to the number of lattice sites in the cell for the time increment for each collision to be properly computed. If there are many lattice sites per cell, the discretization of space is no longer significant. Thus, there is a continuum of IDSMC models of variable resolution.

2.2 Collisions

To date we have considered only the interaction of identical hardsphere molecules, and the following description of the method is limited to that case. As in the conventional DSMC, particle pairs are chosen for collision from among all the particles in a cell with a probability proportional to each pair's relative velocity. The mechanics of collisions of integer-velocity molecules can be understood by the following considerations. For simplicity we present examples for a two-dimensional gas. The generalization to a three-dimensional gas is straightforward, and, except where otherwise noted, all of the calculations we present were performed with a three-dimensional gas. Figure 1 shows a simple quantitative example of a collision in which one of the colliding particles initially has velocity $(u_i, v_i) = (4q, q)$ and the other $(u_j, v_j) = (2q, 5q)$. The relative velocity is the vector difference between these two velocities, and the center-of-mass velocity lies at the center of the relative velocity vector. The consequence of momentum and energy conservation is that after the

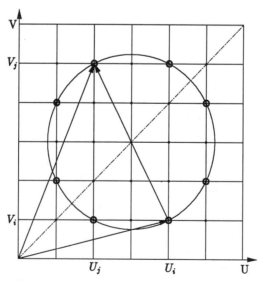

Fig. 1 Reflections about 0/1, 1/1, and 1/0 are possible. Parity: EE; relative velocity: (-2,4).

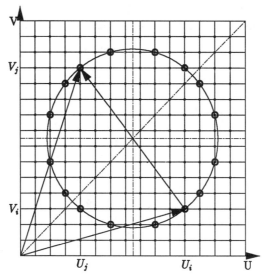

Fig. 2 Reflections about 0/1, 1/1, 1/0, 2/1, and 1/2 are possible. Parity: OO; relative velocity: (-7,9).

collision the relative velocity vector has the same magnitude and is simply rotated by the collision angle χ about the center-of-mass velocity. Candidates for outcome discrete velocities in the example of Fig. 1 are indicated by open circles on the perimeter of the large circle. It is clear that, if one colliding particle is fast and the other is slow, as might occur in a high-temperature gas, the relative velocity will be large; hence, the diameter of the circle defined by the relative velocity vector will be large, and there will be many possible outcome velocities. Thus, in this method new integer velocities are generated or canceled automatically, as required by the local temperature and collision dynamics.

In general, four different cases must be considered depending on the parity of the components of the relative velocity vector. In Fig. 1 both components are even [EE, e.g., $(2q, 4q)$, with a squared relative velocity of $20q^2$]. In this case the center-of-mass velocity falls on a lattice point in velocity space, and, in general, given a pair of initial velocities as shown, there are four possible pairs of output velocities (including that in which the initial velocity vectors are simply interchanged with no apparent change on the figure). It can be seen that the three pairs that are different from the input pair can be constructed by sequential reflections about the vertical, the 45 deg, and the horizontal axes. We designate the slope of these three axes by 1/0, 1/1, and 0/1, respectively. If both components of the relative velocity are odd (OO), the center of mass falls at the center of a unit cell of the lattice in velocity space, and, again, four outcome pairs obtained by the same reflections as described earlier are possible. On the other hand, when the relative-velocity components are of opposite parity

(EO), the center of mass falls on the edge of a unit cell in velocity space; thus, no reflection about the 1/1 axis occurs, and only two outcome pairs are possible.

At higher relative velocities, i.e., at higher temperatures, reflections about other axes can be made; as a result, more possibilities for outcomes arise. For example, the circle defined by a relative energy e_r of $130q^2$ intersects components $(7q, 9q)$ and $(3q, 11q)$, which cannot be obtained from each other by reflections about 1/0, 1/1/ or 0/1 (see Fig. 2); thus, there are eight possible outcome pairs for either of these OO input configurations. These outcomes can be constructed by reflection about axes of slope 1/2 and 2/1. In general, as the length of the relative velocity vector increases, symmetries about lines of slopes given by ratios of increasing values of whole numbers (e.g., 1/3, 2/3, etc.) arise. In Fig. 3 we indicate on the relative velocity lattice the number of points in the quadrant that are intersected by the circle about the origin that passes through that point. The boxed-in points are those that participate in symmetries more complex than 0/1, 1/0, and 1/1. If the relative velocity falls on an axis of

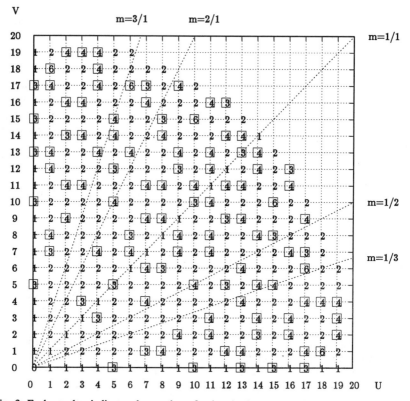

Fig. 3 Each number indicates the number of points in the quarter of the plane which lie on a circle centered at (0,0) and which pass through the given point. Lines of slope 1/3, 1/2, 1/1, 2/1, and 3/1 are also shown.

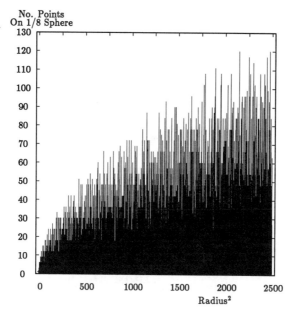

Fig. 4 The number of discrete points in the first octant on all spheres whose radius is less than 50.

symmetry, the possible outcomes are correspondingly reduced, and the case is degenerate. The algorithm for finding all the possible outcomes for a given relative velocity is relatively complex; hence, it is more efficient to do the calculation once and store the results in a look-up table for use during the Monte Carlo calculations.

As the preceding discussion suggests, the number of possible collisions increases rapidly as the velocity resolution increases. In three dimensions there are $2^3 = 8$ parity cases, each possessing slightly different symmetry possibilities and degeneracies. Figure 4 is a histogram of the number of integer points intersected by spheres with radius from 1 to 50. Table 1 gives the first few entries of the look-up table used for the collisions. This table contains the coordinates of the integer points on spheres centered at $(x, y, z) = (0, 0, 0)$ with different values of squared radius. Before using the table the velocities of the collision partners are transformed into the relative velocity frame. After choosing with uniform probability the outcome point from the table, under the condition that the parity of the components of the pre- and postcollision relative velocity vectors be the same, the final velocities are retransformed into the lab frame. Note that some spheres (e.g., squared radius = 7) are not intersected at all by the lattice. The complete table presently used has entries for 1143 spheres. A look-up table containing entries for spheres with diameters up to, say, $40q$, can be several kilobytes in size. Since memory requirements of this

Table 1 Sample from look-up table in three dimensions

Radius^2	Number of points on sphere	Coordinates of points on sphere
0	1	(0,0,0)
1	6	(-1,0,0) (0,-1,0) (0,0,-1) (0,0,1) (0,1,0) (1,0,0)
2	12	(-1,-1,0) (-1,0,-1) (-1,0,1) (-1,1,0) (0,-1,-1) (1,0,-1) (0,-1,1) (0,1,-1) (0,1,1) (1,-1,0) (1,0,1) (1,1,0)
3	8	(-1,-1,-1) (-1,-1,1) (-1,1,-1) (-1,1,1) (1,-1,-1) (1,-1,1) (1,1,-1) (1,1,1)
4	6	(-2,0,0) (0,-2,0) (0,0,-2) (0,0,2) (0,2,0) (2,0,0)
5	24	(-2,-1,0) (-2,0,-1) (-2,0,1) (-2,1,0) (-1,-2,0) (2,-1,0) (-1,0,-2) (-1,0,2) (-1,2,0) (0,-2,-1) (0,-2,1) (2,0,-1) (0,-1,-2) (0,-1,2) (0,1,-2) (0,1,2) (0,2,-1) (2,0,1) (0,2,1) (1,-2,0) (1,0,-2) (1,0,2) (1,2,0) (2,1,0)
6	24	(-2,-1,-1) (-2,-1,1) (-2,1,-1) (-2,1,1) (-1,-2,-1) (-1,-2,1) (-1,-1,-2) (-1,-1,2) (-1,1,-2) (2,-1,1) (-1,1,2) (-1,2,-1) (-1,2,1) (1,-2,-1) (2,1,-1) (1,-2,1) (1,-1,-2) (1,-1,2) (1,1,-2) (2,1,1) (1,1,2) (1,2,-1) (1,2,1) (2,-1,-1)
8	12	(-2,-2,0) (-2,0,-2) (-2,0,2) (-2,2,0) (0,-2,-2) (2,0,-2) (0,-2,2) (0,2,-2) (0,2,2) (2,-2,0) (2,0,2) (2,2,0)

magnitude are a matter of concern on many present-day parallel-computing machines in which the processors do not share memory, we choose to store the entries for only the octant $(x, y, z) > 0$ of each sphere and reflect the selected outcome configuration of each collision across the planes $x = 0$, $y = 0$ and $z = 0$, each with 50% probability. This reduces the size of the look-up table to 1/8 of that needed for the full sphere, at the expense of a small increase in computational time.

It is clear from Fig. 4 that, for collisions with moderately large relative velocities (typical of moderate temperatures), effectively a continuum of dynamic interactions is possible. On the other hand, for relatively low temperatures, where only a few velocities are used, the possible outcomes are limited, and it is easy to see that the resulting discrete distribution of, say, collision angle might be quite different from the expected uniform distribution of a continuous-velocity gas. Figure 5 is a histogram of the deflection angle of the relative-velocity vector in an equilibrium three-dimensional IDSMC gas at two different temperatures. The figure was constructed by allowing 32,000 particles in a box to collide 64,000 times. The box contained just one computational cell, and the calculations were performed on a small sequential computer. The solid continuous line is a sine distribution appropriate for a hardsphere gas. It can be seen that at the lower temperatures the angles 0, $\pi/2$, etc., occur very often and that the resulting distribution does not resemble that of a continuous-velocity gas, whereas at high temperatures the continuous

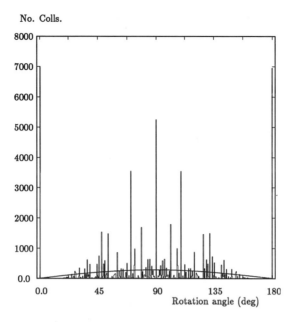

Fig. 5a Histogram distribution function of the rotation angle of the relative velocity vector in each collision. Solid line, theoretical distribution. The particle temperature was $RT = (\pi/2)q^2$.

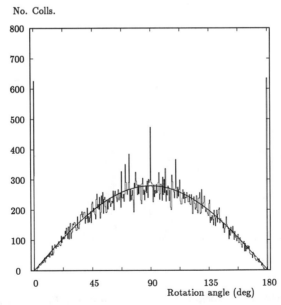

Fig. 5b Histogram distribution function where the particle temperature was $RT = 201.3q^2$.

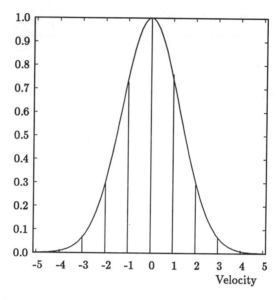

Fig. 6a Velocity component distribution function. Solid line, theoretical distribution normalized to cover the same area as the discrete distribution. The particle temperature was $RT = (\pi/2)q^2$.

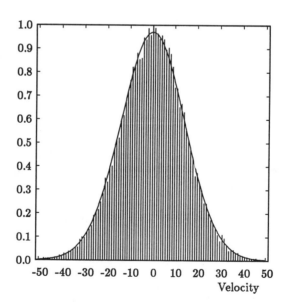

Fig. 6b Velocity component distribution function for $RT = 201.3q^2$.

distribution is well modeled. To a certain extent, the frequent occurrence of certain discrete collision angles compensates for the absence of neighboring values. However, in the integer-velocity gas there seem to be more occurrences of zero deflection angle than would account for the lack of small nonzero deflections. As will be seen below, at low temperatures (i.e., low velocity resolution) the excess of zero-angle collisions has the effect of artificially increasing the diffusivity of the gas, whereas at moderate temperatures the effect on macroscopic quantities is not noticeable. The sampled distribution of particle velocity components at two different temperatures in an equilibrium gas shows no such anomalies and agrees extremely well with the Maxwellian velocity distribution function (Fig. 6). Even with the low-temperature case of Fig. 6a, the nine available discrete-velocity components reproduce the Maxwellian distribution with excellent accuracy. We find, however, that the equilibrium distribution is not as sensitive a measure of the correctness of the collision process as the distribution function in a highly nonequilibrium flow.

2.3 Boundary Conditions

In the IDSMC, the boundaries, which in the calculations presented here are parallel to the lattice axes, are taken to lie midway between lattice points. Specular wall collisions are treated in the same way as in the conventional DSMC, by reflecting the particle trajectory across all wall segments necessary to ensure that the particle at the end of the time step is inside the flowfield, and by reversing its normal velocity after each reflection. For diffuse wall collisions the velocity components of the emitted particles are chosen from the integer Maxwellian distribution (cf. Fig. 6) corresponding to the wall temperature. The trajectories of colliding particles approaching and departing from walls are calculated with floating-point precision, and at the end of the time step the particle positions are rounded to the nearest lattice sites.

III. Results

In this section we present the results of calculations of two test problems that exhibit certain features of the IDSMC method.

3.1 Relaxation to Equilibrium

In this calculation, 16,000 particles in a box consisting of one computational cell are initially distributed bimodally with integer-valued velocity components, as indicated in the top rows of Figs. 7 and 8. The x-component molecular velocities are distributed in two narrow bands (each with $RT \approx 1.5q^2$) about $u = \pm 10q$, whereas the y- and z-component velocities have only one peak of the same width. The fluid is uniform in these calculations, and all particles in the box are candidates for collisions. Only the collisions are calculated; the particle motions are not. Figure 7 shows the results for the conventional DSMC method, and Fig. 8 gives the results for the IDSMC at the same times. Though in the DSMC calculation the initial distribution contains only integer-valued components, after the first collision the velocities become decimal numbers. In Fig. 7 the distributions are plotted as histograms with bin size q, and in Fig. 8 the plotted spikes represent the accumulated data at the corresponding discrete values of velocity. The molecular velocities are sampled after 1 (second row), 2 (third row), and 10 collisions per particle (fourth row), respectively. It can be seen that in both calculations the initial bimodal distribution evolves into Maxwellian distributions with temperature $RT = 35.02\,q^2$ (indicated by the solid curves in the bottom row), and that the IDSMC is essentially the same as the DSMC result. Using the Bhatanagar-Gross-Krook model of a discrete-velocity gas, Broadwell[7] showed that the equilibrium distribution has the form of a Maxwellian. Thus, the distributions shown in the bottom row of Fig. 8 are

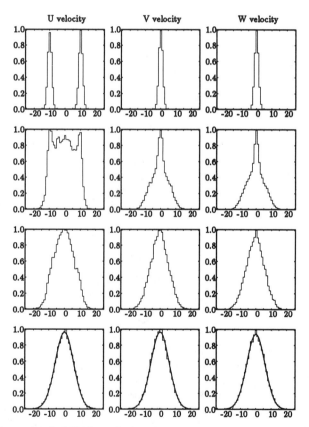

Fig. 7 Conventional DSMC method. Time development of velocity distribution functions. The initial bimodal distribution spreads to form a Maxwellian. Equilibrium Maxwellians are also drawn in the last row for $RT = 35.02q^2$. The first row is before any collisions, the second after 1 collision/particle, the third after 2 collisions/particle, and the last after 10 collisions/particle.

expected. The present calculations show that in addition the discrete and continuous distributions are very similar in nonequilibrium flows.

3.2 Normal Shock Wave

One way to rigorously test the new discrete-velocity method is to compare results for the structure of strong shock waves with DSMC calculations carried out on the same machine using comparable code with the same time step size, cell size, etc. Figure 9 shows the normalized density and temperature profiles obtained by the DSMC (solid line) and the IDSMC (points) for a normal shock wave of strength $M_s = 6.11$ in a perfect hardsphere gas. The space coordinate is normalized with the upstream mean free path. Also shown are discrete Maxwellian distributions of molecular thermal velocities corresponding to the measured uniform states upstream and downstream of the shock together with

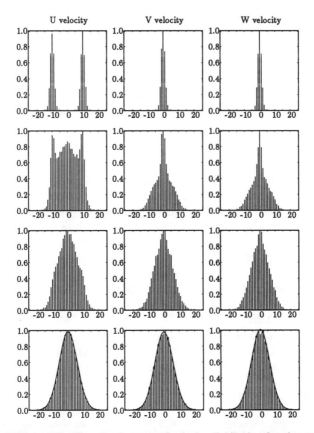

Fig. 8 IDSMC method. Time development of velocity distribution functions. The initial bimodal distribution spreads to form a Maxwellian. Equilibrium Maxwellians are also drawn in the last row for $RT = 35.02q^2$. Rows as in Fig. 7.

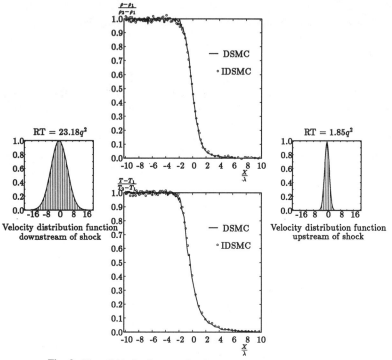

Fig. 9 $M_s = 6.11$ shock-wave density and temperature profiles.

the continuous Maxwellian for comparison. The calculation is carried out in a nonsteady frame; the left wall of a box is impulsively accelerated to a constant speed of $8q$ at time $t = 0$, and the shock profile is sampled after 180 time steps of size $\Delta t = 0.1085 \tau_1$, where τ_1 is the upstream mean free time. The cell size is about $\Delta x = 0.5 \lambda$. There are about 60 particles in each cell and 20 lattice sites per mean free path. To achieve a smooth profile using this approach, a total of about 140 million collisions between 6.1 million particles are calculated, using 64 processors of an Intel iPSC message-passing multicomputer. Physical space is assigned to the computer nodes in accordance with the load-balancing algorithm already described. When, during the move phase of the calculation, a particle moves from one node's domain to that of another, it is sent there as a message. It can be seen that excellent agreement with the continuous-velocity DSMC model is achieved starting with only seven values of each velocity component. In the uniform gas behind the shock 29 values of each component are found.

Figure 10 shows results from a similar problem, but with the upstream temperature in the IDSMC calculation four times smaller than in Fig. 9. In this case the velocity resolution is poor; thus, the discrete calculation does not agree as well with the continuous-velocity result. As discussed earlier, with just a

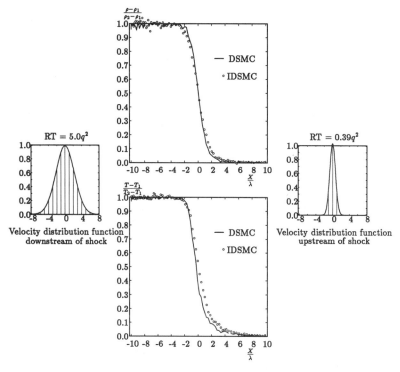

Fig. 10 $M_s = 6.16$ shock-wave density and temperature profiles.

few possible velocities, zero deflection angle occurs too often in the discrete calculations; hence, the particle diffusivity is too large. Thus, the hot downstream particles diffuse toward the front of the shock, and the shock becomes too thick.

A sensitive test of the absolute accuracy of shock structure calculations is obtained by plotting the normal component of the pressure tensor, $p_{xx} = \overline{\rho u'^2}$, vs the specific volume. According to the x-momentum equation, this should be a straight Rayleigh line. In Fig. 11 p_{xx}, normalized by its upstream value, which by definition is the upstream pressure, is plotted vs ρ_1/ρ for the same shock calculations as presented in Fig. 9. The cluster of points at (1, 1) are from samples near the upstream end of the shock and the cluster near (46.4, 0.27) are from the downstream end. In the calculations reported here the cell size and time step were selected to achieve the performance indicated in the figure; larger values would have resulted in an S-shaped curve that deviated more from the straight Rayleigh line. The figure shows that, with the same time step and cell size, the IDSMC and the DSMC perform about the same.

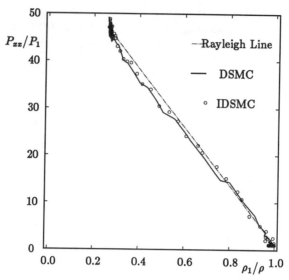

Fig. 11 $M_s = 6.11$ P_{xx} variation through shock.

IV. Summary and Discussion

It has been shown that, for rarefied gasdynamics problems in which there are fairly strong shock waves, accurate results in calculations by DSMC methods can be obtained when the upstream distributions of each component of molecular velocity contain fewer than 10 discrete values. When the molecular velocity components have integer values, the particles move on a lattice; thus, calculations of their motions are simplified. The outcome of binary collisions between identical particles with discrete-velocity components is determined by simple reflections about axes of symmetry in the center-of-mass system, and the results can be stored in look-up tables for rapid use during calculations of collisions.

Calculations of relaxation to equilibrium and of shock-wave structure have been performed to test the accuracy of the discrete-velocity method. Comparison has been made with known equilibrium results, and, for nonequilibrium, with calculations under identical conditions by the conventional DSMC. A test of absolute accuracy by plotting the Rayleigh line in normal shocks has been introduced. From these tests the velocity resolution cited earlier, necessary for achieving high accuracy, was deduced.

In the multicomputer used for the present calculations, provision for 256 different values of velocity, by declaring the molecular velocity components to be 1 byte long, allows 2.4 times more particles to be treated in a two-dimensional flow than when the particle velocities are stored as real numbers. With real numbers the storage per particle is 22 bytes (12 for the 3 components of velocity, 8 for 2 positions, and 2 for the particle index), whereas with the present method it is 9 bytes (3, 4, and 2). Thus, flows with Reynolds numbers 2.4 times greater can be calculated. Clearly, since many fewer than 256 velocity values are necessary, this result could be improved if the velocities were stored more compactly. For example, if only 5 bits were used for velocities and 10 bits for positions, the improvement could, in principle, be 3.5 fold.

By design, the IDSMC computes primarily with integer arithmetic and the DSMC with floating-point arithmetic. Therefore, comparison of the relative *speed* of the two methods is bound to be machine-dependent. For example, in the calculation of relaxation to equilibrium reported earlier, in which only collisions in one large cell are calculated, and the particles are not moved, the IDSMC ran about 3 times faster on a sequential SUN 3/60 microcomputer, which is apparently relatively efficient at integer calculations, than did the DSMC. On the other hand, in the strong shock-wave problem on the Intel iPSC multicomputer, which is evidently a rather efficient floating-point machine, the IDSMC ran only about 10% faster. In view of the fact that our codes have not yet been optimized for speed, it is clear that more work needs to be done to define definitive benchmarks.

V. Acknowledgments

This research was carried out with the support of Contract AFOSR-87-0155, Office of Scientific Research, United States Air Force and of the Lockheed-California Company. We express our gratitude to the Submicron Systems Architecture Project, Department of Computer Science, Caltech, for providing computer facilities and expert advice.

References

[1] Gatignol, R., "Théorie Cinétique des Gaz à Répartition Discrète de Vitesses," *Lecture Notes in Physics,* Vol. 36, Springer-Verlag, Berlin, FRG, 1975.

[2] Frisch, U., Hasslacher, B., and Pomeau, Y., "Lattice-gas Automata for the Navier-Stokes Equation," *Physical Review Letters,* Vol. 56, 1986, pp. 1505-1508.

[3] Bird, G. A., *Molecular Gasdynamics,* Clarendon, Oxford, UK, 1976.

[4] Carleman, T., *Problèmes mathématiques dans la théorie cinétique des gaz,* Almqvist and Wiksells, Uppsala, Sweden, 1957.

[5] Broadwell, J. E., "Study of Rarefied Shear Flow by the Discrete Velocity Method," *Journal of Fluid Mechanics,* Vol. 19, part 3, 1964, pp. 401-414.

[6] Broadwell, J. E., "Shock Structure in a Simple Discrete Velocity Gas," *The Physics of Fluids,* Vol. 7, Aug., 1964, pp. 1243-1247.

[7] Broadwell, J. E., "Study of Rarefied Shear Flows by the Discrete Velocity Method," Space Technology Laboratories, Inc., Rept. 9813-6001-RU000, March, 1963.

On Discrete Kinetic Theory with Multiple Collisions: Plane Six-Velocity Model and Unsteady Couette Flow

E. Longo* and R. Monaco*
Politecnico di Torino, Torino, Italy

Abstract

In the present paper we study a discrete plane six-velocity model of the Boltzmann equation with binary and triple collisions. We derive the kinetic evolution equation and discuss the thermodynamic equilibrium state showing that thanks to triple collision the Maxwellian density is uniquely defined in terms of the gas macroscopical quantities; this fact, allows in particular, the formulation of boundary-value problems with purely diffuse re-emission law at the walls. As an application we study the unsteady Couette flow problem.

I. Introduction

Discrete kinetic theory deals with the analysis of systems of gas particles with a discrete set of selected velocities and provides a useful substitute to the Boltzmann equation in terms of a system of nonlinear partial differential equations defining the space-time evolution of the number densities joined to each selected velocity.

Discrete kinetic theory, which leads to the so-called discrete Boltzmann equation, generally takes into account only binary collisions.[1-3] On the other hand, for moderately dense gases, multiple collisions effects can become relevant. Dealing with multiple collisions effects is interesting for

---

Copyright © American Institute of Aeronautics and Astronautics, 1988.
*Associate Professor, Dipartimento di Matematica.

MULTIPLE COLLISIONS DISCRETE KINETIC THEORY   119

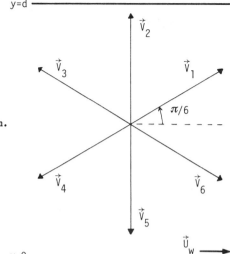

Fig. 1 Geometry of the system.

two main reasons: the thermodynamics of the system is, as we shall see, more accurately described and the evolution equations are modified so that the solution of fluid dynamic problems can show some substantial differences with respect to the solution obtained by models with binary collisions only. This kind of mathematical modeling may also be a useful tool for "cellular automata" theory,[4] when multiple collisions have to be dealt with.

Keeping this in mind, let us briefly outline the contents of this paper; in Sec.II we deal with the kinetics and thermodynamics of a plane six-velocity model with both binary and triple collisions. This analysis allows us to state suitable boundary conditions for problems with interaction between gas particles and solid walls. Accordingly, Sec.III is devoted to the analysis of the unsteady Couette flow problem with a diffuse re-emission law on the walls; a general discussion on the results obtained for the macroscopic flow variables is then performed.

II. Plane Six-velocity Model with Triple Collisions

Consider a plane regular discrete model with six selected velocities

$$j=1,..,6: \quad \vec{V}_j = c\{\cos(\pi(2j-1)/6)\vec{e}_1 + \sin(\pi(2j-1)/6)\vec{e}_2\} \quad (1)$$

where c is the velocity modulus; velocities $\vec{V}_j$ are shown in Fig. 1.

Let us introduce all of the collisions that preserve momentum and energy: that is, binary head-on collisions

$$(\vec{V}_1, \vec{V}_4) \leftrightarrow (\vec{V}_2, \vec{V}_5) \leftrightarrow (\vec{V}_3, \vec{V}_6) \tag{2}$$

and triple collision

$$(\vec{V}_1, \vec{V}_3, \vec{V}_5) \leftrightarrow (\vec{V}_2, \vec{V}_4, \vec{V}_6) \tag{3}$$

If $N_j = N_j(t,x,y)$ is the number density (i.e., the number of particles per unit volume) referred to velocity $\vec{V}_j$, then the six kinetic equations describing the space-time evolution of the N_j densities can be written as $\underline{N} = (N_1, \ldots, N_6)$:

$$\partial N_j / \partial t + (\vec{V}_j \cdot \vec{e}_1) \partial N_j / \partial x + (\vec{V}_j \cdot \vec{e}_2) \partial N_j / \partial y = \\ = J_j^b(\underline{N},\underline{N}) + J_j^t(\underline{N},\underline{N},\underline{N}) \tag{4}$$

where J_j^b and J_j^t are the collision operators due to binary and triple collisions, respectively. Note that J_j^b is a quadratic functional on $\underline{N}$, whereas J_j^t is a cubic functional. Details will be now given on the evaluation of these two functionals.

Binary Collisions

Binary collisions have already been studied in discrete kinetic theory.[1] Nevertheless, let us recall that the $\vec{V}_i$ particles which can collide in the unit time and volume with a given $\vec{V}_j$ particle are those contained in the right cylinder with base S (cross-collisional section) and height $|\vec{V}_i - \vec{V}_j|$; the number of such binary encounters is then

$$S|\vec{V}_i - \vec{V}_j|N_i \tag{5}$$

and consequently the related binary collisional frequency δ_j^b of each $\vec{V}_j$ particle is

$$\delta_j^b = S|\vec{V}_i - \vec{V}_j| a^b N_{j+3} \tag{6}$$

where a^b are the transition probability densities of the i-j collisions. According to the actual model with the usual hypothesis of equally probable collisional products ($a^b=1/3$), we have

$$\delta_j^b = (2Sc/3)N_{j+3} \tag{7}$$

and therefore

$$J_j^b = (2cS/3)\{N_{j+1}N_{j+4}+N_{j+2}N_{j+5}-2N_jN_{j+3}\} \tag{8}$$

<u>Triple Collisions</u>

For a given $\vec{V}_j$ particle the admissible triple collision involves $\vec{V}_{j+2}$ and $\vec{V}_{j+4}$ particles {see Eq.(3)}. According to Eq.(5) the number of binary encounters j+2, j+4 occurring per unit time and volume is $\forall j=1,\ldots,6$

$$\Gamma_{j+2,j+4} = S|\vec{V}_{j+2}-\vec{V}_{j+4}|N_{j+2}N_{j+4} = \sqrt{3}cSN_{j+2}N_{j+4} \tag{9}$$

To perform the triple collisional frequancy of each $\vec{V}_j$ particle, let us remark that the particles, traveling with velocity $\vec{V}_{j+2}$ (or $\vec{V}_{j+4}$), which in unit time can collide with a given $\vec{V}_j$ particle, are contained in a right cylinder with axis along the vector $\vec{V}_{j+2}-\vec{V}_j$ ($\vec{V}_{j+4}-\vec{V}_j$, respectively), height $|\vec{V}_{j+2}-\vec{V}_j| = |\vec{V}_{j+4}-\vec{V}_j| = \sqrt{3}c$, and base $S = \pi r^2$, where r is the molecular interaction radius of the particles. The volume Ω of the intersection between the above cylinders can be expressed by

$$\Omega = 5\sqrt{3}\,\pi r^3/6 = 5\sqrt{3/\pi}\,S^{3/2}/6 \tag{10}$$

Consequently, we assume that the triple collisional frequency of each $\vec{V}_j$ particle is

$$\delta_j^t = \Gamma_{j+2,j+4}\Omega a^t = 5cS^{5/2}/(4\sqrt{\pi})\,N_{j+2}N_{j+4} \tag{11}$$

where $a^t=1/2$ is the transition probability density of the triple collision. In the same way as in the binary collision case, we obtain

$$J_j^t = 5cS^{5/2}/(4\sqrt{\pi})\{N_{j+1}N_{j+3}N_{j+5}-N_jN_{j+2}N_{j+4}\} \tag{12}$$

By introducing expressions (8) and (12) into Eq.(4) we obtain the kinetic equations of the actual model.

The knowledge of the number densities N_j leads to evaluate the macroscopical variables of the system, i.e.,
Numerical density:
$$n(t,x,y) = \sum_1^6 {}_j N_j(t,x,y) \tag{13}$$

Mass velocity:
$$\vec{v}(t,x,y) = (1/n) \sum_1^6 {}_j \vec{V}_j N_j(t,x,y) \tag{14a}$$

$$v_x = \vec{v} \cdot \vec{e}_1 = \sqrt{3}c/(2n)\{N_1 + N_6 - N_3 - N_4\} \tag{14b}$$

$$v_y = \vec{v} \cdot \vec{e}_2 = c/(2n)\{N_1 + 2N_2 + N_3 - N_4 - 2N_5 - N_6\} \tag{14c}$$

Consider now a thermodynamic equilibrium state of the system; according to H theorem the Maxwellian densities[5] $M_j = M_j(t,x,y)$ are such that the vector $\log \vec{M} = (\log M_1, \ldots, \log M_6)$ is a collisional invariant. From Eqs.(2) and (3) it follows that

$$M_1 M_4 = M_2 M_5 = M_3 M_6 \quad , \quad M_1 M_3 M_5 = M_2 M_4 M_6 \tag{15}$$

or, equivalently,

$$M_1 = (\tfrac{1}{2}n/\mu)\exp(\alpha) \;,\; M_2 = (\tfrac{1}{2}n/\mu)\exp(\alpha+\beta) \;,\; M_3 = (\tfrac{1}{2}n/\mu)\exp(\beta)$$

$$M_4 = (\tfrac{1}{2}n/\mu)\exp(-\alpha) \;,\; M_5 = (\tfrac{1}{2}n/\mu)\exp(-\alpha-\beta) \;,\; M_6 = (\tfrac{1}{2}n/\mu)\exp(-\beta)$$

with
$$\mu = ch(\alpha) + ch(\beta) + ch(\alpha+\beta) \tag{16}$$

where n is given by Eq.(13) and the variables $\alpha = \alpha(t,x,y)$ and $\beta = \beta(t,x,y)$ are uniquely related to the components v_x and v_y of the mass velocity {Eqs(14b) and (14c)}. In fact, it results in

$$\{sh(\alpha) - sh(\beta)\}/\{ch(\alpha) + ch(\beta) + ch(\alpha+\beta)\} = 2v_x/(\sqrt{3}c) \tag{17a}$$

$$\{sh(\alpha) + sh(\beta) + 2sh(\alpha+\beta)\}/\{ch(\alpha) + ch(\beta) + ch(\alpha+\beta)\} = 2v_y/c \tag{17b}$$

The set of Eqs.(17), although not generally invertible in elementary form, defines a one-to-one map $(\alpha,\beta) \leftrightarrow (v_x, v_y)$;

in fact, it can be easily verified that the Jacobian of the previous map is everywhere different from zero.

For the particular case $v_y=0$ (which will be used in the application of Sec. III), one has

$$\alpha=-\beta=\log\{(v_x/(\sqrt{3}c)+(1-v_x^2/c^2)^{\frac{1}{2}})/(1-2v_x/(\sqrt{3}c))\}$$

$$0 \leq v_x < \sqrt{3}c/2 \qquad (18)$$

III. Unsteady Couette Flow Problem

Consider the unsteady Couette flow between parallel plates located at y=0 and y=d. The upper plate is at rest, whereas the lower one is moving with velocity $\vec{U}_w(t)=cu_w(t)\vec{e}_1$ ($0 \leq u_w < \sqrt{3}/2$), with $u_w(t=0)=0$ (see Fig. 1).

This problem will be studied, using the kinetic equations deduced in Sec. II, under the usual hypothesis of x invariance (i.e., $\partial N_j/\partial x=0$). Moreover, the following initial and boundary conditions will be introduced:

1) At t=0 the gas is in absolute Maxwellian equilibrium, that is,

$$\forall j=1,\ldots,6: \quad N_j(t=0,y) = n_0/6 \qquad (19)$$

where n_0 is the numerical density at t=0.

2) As in classical kinetic theory,[5] the boundary conditions can be written in the following general form:

$$h=1,2,3: \quad (\vec{V}_h \cdot \vec{e}_2)N_h(t,y=0)=\sum_{4}^{6} B_{kh}^{\dagger}|\vec{V}_k \cdot \vec{e}_2|N_k(t,y=0) \qquad (20a)$$

$$k=4,5,6: \quad |\vec{V}_k \cdot \vec{e}_2|N_k(t,y=d)=\sum_{1}^{3} B_{hk}(\vec{V}_h \cdot \vec{e}_2)N_h(t,y=d) \qquad (20b)$$

where $B_{kh}^{\dagger}$ (B_{hk}) are the transition probability rates that a particle hitting the plate y=0 (y=d) with velocity $\vec{V}_k$ ($\vec{V}_h$) will be re-emitted with velocity $\vec{V}_h$ ($\vec{V}_k$, respectively).

Assuming the incident particles flux equal to the re-emitted one, the following condition must be satisfied:

$$\sum_{1}^{3} B_{kh}^{\dagger} = \sum_{4}^{6} B_{hk} = 1 \qquad (21)$$

To evaluate the quantities $B_{kh}^{\dagger}$ and B_{hk}, we will assume here that the re-emission law on the two plates is the purely diffusion one; in other words, the re-emitted gas particles are in Maxwellian equilibrium with the walls. Therefore, the transition probability rates $B_{kh}^{\dagger}$ and B_{hk} do not depend on the first subscript; we will indicate thereinafter $B_{kh}^{\dagger}=B_h$ and $B_{hk}=B_k$. Accordingly, on the upper plate we have

$$y=d: \quad B_4 = 1/4, \quad B_5 = 1/2, \quad B_6 = 1/4 \quad (22a)$$

On the lower plate, where the x component of the mass velocity of the re-emitted particles is equal to the wall velocity, it is easy to obtain from Eq. (18)

$$y=0: \quad B_1 = \exp(\alpha)/\{\exp(\alpha)+\exp(-\alpha)+2\} \quad (22b)$$

$$B_2 = 2/\{\exp(\alpha)+\exp(-\alpha)+2\} \quad (22c)$$

$$B_3 = \exp(-\alpha)/\{\exp(\alpha)+\exp(-\alpha)+2\} \quad (22d)$$

where

$$\alpha = \alpha(u_w) = \log\{(u_w/\sqrt{3} + (1-u_w^2)^{\frac{1}{2}})/(1-2u_w/\sqrt{3})\}$$

We can now formulate the initial boundary-value problem corresponding to the unsteady Couette flow problem.

According to Eqs. (3), (8), and (12), the evolution equations of the actual problem (where $j=1,\ldots,6$)

$$\partial N_j/\partial t + c\sin(\pi(2j-1)/6)\,\partial N_j/\partial y$$
$$= (2cS/3)\{N_{j+1}N_{j+4}+N_{j+2}N_{j+5}-2N_jN_{j+3}\}$$
$$+ 5cS^{5/2}/(4\sqrt{\pi})\{N_{j+1}N_{j+3}N_{j+5}-N_jN_{j+2}N_{j+4}\} \quad (23)$$

Let us introduce the following rescaling:

$$n_j = N_j/n_o, \quad \eta = y/d, \quad \tau = ct/d$$

together with the following definition of the Knudsen number $Kn = 1/(dSn_o)$. Accordingly, Eq. (23) becomes

$$\partial n_j/\partial \tau + \sin(\pi(2j-1)/6)\,\partial n_j/\partial \eta$$
$$= 2/(3Kn)\{n_{j+1}n_{j+4}+n_{j+2}n_{j+5}-2n_jn_{j+3}\}$$
$$+ 5S^{\frac{1}{2}}/(4\sqrt{\pi}Kn^2 d)\{n_{j+1}n_{j+3}n_{j+5}-n_jn_{j+2}n_{j+4}\} \quad (24)$$

MULTIPLE COLLISIONS DISCRETE KINETIC THEORY   125

Moreover, the initial and boundary conditions {Eqs.(19) and (22)} to be joined to Eq.(24) are given by

$$\tau=0, \quad \eta\in[0,1], \quad j=1,\ldots,6 : \quad n_j = 1/6 \quad (25a)$$

$\forall \tau \geq 0, \quad \eta=0, \quad h=1,2,3 :$

$$n_h(\tau) = B_h(u_w)/\{\sin(\pi(2h-1)/6)\} \sum_{k=4}^{6} |\sin(\pi(2k-1)/6)| n_k(\tau)$$

$\forall \tau \geq 0, \quad \eta=1, \quad k=4,5,6 :$

$$n_k(\tau) = B_k/|\sin(\pi(2k-1)/6)| \sum_{h=1}^{3} \sin(\pi(2h-1)/6) n_h(\tau) \quad (25b)$$

where B_h and B_k have been deduced in Eqs. (22).

The initial boundary-value problem {Eqs.(24) and (25)} will now be numerically solved following the "Differential Quadrature Method" suggested in Ref.6.

Let us discretize the space interval $[0,1]$ into N-1 equal subintervals $[\eta_m, \eta_{m+1}]$, $m=1,\ldots,N$. In addition, assume that the function $n_j(\tau,\eta)$ are interpolated, using N nodal points as follows:

$$n_j(\tau,\eta) \simeq \sum_{r=1}^{N} L_r(\eta) n_{jr}(\tau), \quad n_{jr}(\tau) := n_j(\tau; \eta=\eta_r) \quad (26)$$

where $L_r(\eta)$ are the Lagrangian polynomials defined by

$$L_r(\eta) := \prod_{k=1}^{N} (\eta-\eta_k)/(\eta_r-\eta_k) \quad k \neq r$$

From hypothesis (26) one has

$$\partial n_{jm}(\tau)/\partial \eta := \partial n_j/\partial \eta \big|_{\eta=\eta_m} \simeq \sum_{r=1}^{N} a_{mr} n_{jr}(\tau), \quad m=1,\ldots,N \quad (27)$$

where

$$a_{mr} = dL_r/d\eta \big|_{\eta=\eta_m}$$

Details regarding the computation of the coefficients a_{mr} can be found in Ref.7. Introducing expression (27) into the Eq. (24) and into the boundary conditions {Eqs. (25b)}, one obtains the following set of ordinary differential equations and limit conditions:

$$m=1; \quad h=1,2,3, \quad k=4,5,6$$

$$n_{h1} = B_h(u_w)/\{\sin(\pi(2h-1)/6)\} \sum_{4}^{6} k n_{k1} |\sin(\pi(2k-1)/6)| \quad (28)$$

$$dn_{k1}/d\tau = -\sin(\pi(2k-1)/6) \sum_{1}^{N} {}_r a_{1r} n_{kr}$$

$$+2/(3Kn)\{n_{(k+1)1} n_{(k+4)1} + n_{(k+2)1} n_{(k+5)1} - 2n_{k1} n_{(k+3)1}\}$$

$$+5S^{\frac{1}{2}}/(4\pi^{\frac{1}{2}}Kn^2 d)\{n_{(k+1)1} n_{(k+3)1} n_{(k+5)1} - n_{k1} n_{(k+2)1} n_{(k+4)1}\} \quad (29)$$

$$m=2,\ldots,N-1; \quad j=1,\ldots,6$$

$$dn_{jm}/d\tau = -\sin(\pi(2j-1)/6) \sum_{1}^{N} {}_r a_{mr} n_{jr}$$

$$+2/(3Kn)\{n_{(j+1)m} n_{(j+4)m} + n_{(j+2)m} n_{(j+5)m} - 2n_{jm} n_{(j+3)m}\}$$

$$+5S^{\frac{1}{2}}/(4\pi^{\frac{1}{2}}Kn^2 d)\{n_{(j+1)m} n_{(j+3)m} n_{(j+5)m} - n_{jm} n_{(j+2)m} n_{(j+4)m}\} \quad (30)$$

$$m=N; \quad h=1,2,3, \quad k=4,5,6$$

$$dn_{hN}/d\tau = -\sin(\pi(2h-1)/6) \sum_{1}^{N} {}_r a_{Nr} n_{hr}$$

$$+2/(3Kn)\{n_{(h+1)N} n_{(h+4)N} + n_{(h+2)N} n_{(h+5)N} - 2n_{hN} n_{(h+3)N}\}$$

$$+5S^{\frac{1}{2}}/(4\pi^{\frac{1}{2}}Kn^2 d)\{n_{(h+1)N} n_{(h+3)N} n_{(h+5)N} - n_{hN} n_{(h+2)N} n_{(h+4)N}\} \quad (31)$$

$$n_{kN} = B_k/|\sin(\pi(2k-1)/6)| \sum_{1}^{3} {}_h n_{hN} \sin(\pi(2h-1)/6) \quad (32)$$

$$m=1,\ldots,N; \quad j=1,\ldots,6$$

$$n_{jm}(\tau=0) = 1/6 \quad (33)$$

Let us note that Eq. (30) represents the density evolution in the internal nodal points selected by the spatial

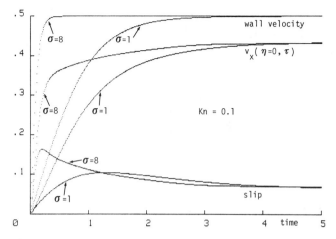

Fig. 2 Wall velocity, tangential component of the mass velocity near the wall, and slip coefficient vs time.

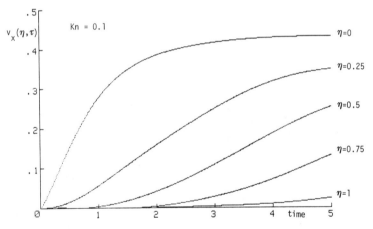

Fig. 3 Distribution of the tangential component of the mass velocity vs time.

discretization. Conversely, Eqs.(28) and (32) reproduce, in the "differential quadrature" scheme, the boundary conditions (25b). The remaining Eqs. (29) and (31) are the evolution equations of the incoming particles densities on the lower and upper plate, respectively. Equation (33) states then, the initial conditions for each density at each nodal point.

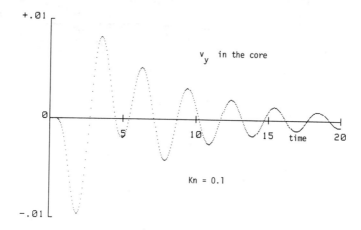

Fig. 4 Normal component of the mass velocity in the core of the flow vs time.

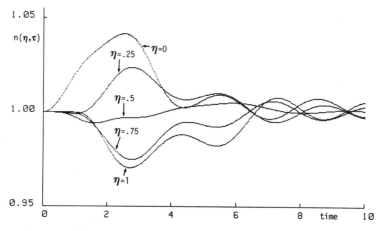

Fig. 5 Distribution of the numerical density vs time for the Knudsen number Kn = 0.1.

The set of Eqs. (28-33) can be now solved using a standard fourth-order Runge-Kutta-Gill routine.

Some numerical experiments were developed using the wall velocity law $u_w = \frac{1}{2}\text{th}(\sigma\tau)$ in order to examine the transient behavior and the trend to steady-state flow.

In Fig. 2 the wall velocity, the x component $v_x(\eta=0,\tau)$ of the mass velocity on the moving wall, and the slip coefficient $(u_w(\tau)-v_x(\eta=0,\tau))$ are plotted vs τ, for $\sigma=1$ and $\sigma=8$.

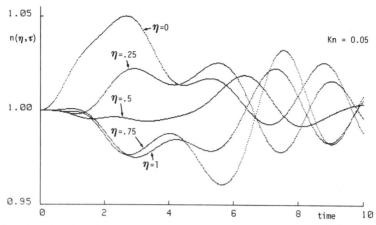

Fig. 6 Distribution of the numerical density vs time for the Knudsen number Kn = 0.05.

Let us note in particular the presence of the peculiar overshoot of the slip coefficient and the fast trend of the plotted quantities to steady-state.

In Fig. 3 we show the behavior of the component $v_x(\eta,\tau)$ for different values of $\eta \in [0,1]$, vs time.

In Fig. 4 the y component of the mass velocity in the flow core, $v_y(\eta=0.5,\tau)$, is plotted in a wider range of time. The picture shows the behavior of a damped wave which propagates backward and forward between the plates; in fact, it follows immediately from the kinetic equations of the model that the steady-state conditions prescribe $v_y=0$ everywhere.

In Figs. 5 and 6 the time-space behavior of the total numerical density $n(\eta,\tau)$ is shown; the two pictures correspond to Kn=0.1 and Kn=0.05, respectively. Let us note that the trend to steady-state is lower in the case of the denser gas, where the effect of the triple collisions plays a major role.

The preceding results qualitatively agree both with the ones obtained in classical kinetic theory (see, for instance Kogan's book[5]) and with those performed by Gatignol,[8] who used a discrete four-velocity model. A more accurate comparison with Ref.8 cannot be shown because the aforementioned discrete model with only binary collisions does not provide the dependence of the results on the Knudsen number, which, as we have seen, plays a relevant role in the proposed multiple collisions scheme.

Acknowledgment

This work has been partially supported by the National Council for the Research, project Applied Mathematics in Industry and Technology of Gruppo Nazionale della Fisica Matematica, and by the Ministery of Education.

References

[1] Gatignol, R.,"Théorie cinétique des gaz à répartition discrète de vitesses,"Springer-Verlag, New York, 1975, Vol. 36 (Lecture Notes in Physics).

[2] Cabannes, H., "The Discrete Boltzmann Equation: Theory and Applications", lectures at Univ. California, Berkeley, 1980.

[3] Platkowski, T. and Illner, R.,"Discrete Velocity Models of the Boltzmann Equation: a Survey on the Mathematical Aspects of the Theory,"Society of Industrial and Applied Mathematics Review, Vol. 30, June 1988, pp. 213-255.

[4] Wolfram, S.,"Cellular Automaton Fluids: Basic Theory,"Journal of Statistical Physics, Vol. 45, June 1986, pp. 471-525.

[5] Kogan, M.,"Rarefied Gas Dynamics,"Plenum, New York, 1969.

[6] Bellman, R., Kashef, B. and Casti, J.,"Differential Quadrature: a Technique for the Rapid Solution of Nonlinear Partial Differential Equations,"Journal of Computational Physics, Vol. 10, January 1972, pp. 40-52

[7] Lachowicz, M. and Monaco, R., "Existence and Quantitative Analysis of the Solutions to the Initial Value Problem for the Discrete Boltzmann Equation in all Space," Journal of Society of Industrial and Applied Mathematics, 1989 (to be published)

[8] Gatignol, R., "Unsteady Couette Flow for a Discrete Velocity Gas," Proceedings of the XI Symposium on Rarefied Gas Dynamics, Vol. 1, Edited by R. Campargue, Commissariat à l'Energie Atomique, Paris, France, 1978, pp. 195-204.

Exact Positive (2+1)-Dimensional Solutions to the Discrete Boltzmann Models

H. Cornille*
Physique Théorique CEN-CEA Saclay, Gif-sur Yvette, France

I. Abstract

The exact solutions for the discrete Boltzmann models are sums of similarity shock waves: two for the (1+1)-dimensional solutions and three for the (2+1)-dimensional ones. We review the previously found (2+1)-dimensional solutions and present new ones. We discuss the four velocity planar model, the Broadwell model and an hypercubic p-dimensional model. The difficult problem is the positivity one. Three classes of solutions are known: 1) solutions relaxing toward nonuniform Maxwellians (here we determine such solutions for the hypercubic p-dimensional model) 2) shock waves and semiperiodic waves.

II. Introduction

For the discrete Boltzmann models, the velocity can only take discrete values $\vec{v}_i$, $|\vec{v}_i|=1$, $i=1,\ldots,2p$ with p couples of opposite velocities $\vec{v}_{2i-1}+\vec{v}_{2i}=0$ $i=1,\ldots,p$. To each velocity $\vec{v}_i$ a density $N_i(\vec{x},t)$ with for spatial coordinate $x_1, x_2, x_3,\ldots$ is associated.

The simplest solutions are the exact similarity shock wave solutions

$$N_i = n_{0i} + n_i/(1+d\ \exp(\rho t+\vec{\gamma}\cdot\vec{x})), \quad d>0 \quad \vec{\gamma}\cdot\vec{x} = \sum_q \gamma_q x_q \qquad (1)$$

Copyright © 1989 by the American Institute of Aeronautics and Astronautics, Inc. All rights reserved.
*Director of Research.

while the exact multidimensional solutions are simply the sums of such solutions:

$$N_i = n_{0i} + \sum_{j=1}^{J} n_{ji}/D_j, \qquad D_j = 1 + d_j \exp(\rho_j t + \vec{\gamma}_j \cdot \vec{x}) \quad j=1,\ldots,J \qquad (2)$$

The (1+1)-dimensional solutions have two components or $J=2$ and a great number of classes have been found[1].

Here we review the (2+1)-dimensional results for $J=3$ and present new solutions. We consider three models[2-3] the $4v_i$ planar model, the Broadwell model and the $2pv_i$ hypercubic model with $\vec{v}_{2q-1}$ along the $0\vec{x}_q$ axis $q=1;\ldots;p$. For $p=2,3$ we recover the $4v_i, 6v_i$ previous models respectively.

Three classes of (2+1)-dimensional solutions[4-8] are known: 1) solutions relaxing toward nonuniform Maxwellians, 2) shock waves, and 3) semiperiodic solutions. For the hypercubic $p>3$ models these solutions have not been obtained. Here we present new results: nonuniform Maxwellians for the hypercubic model.

In section II we construct, for the p-dimensional model, the solutions relaxing toward nonuniform Maxwellians. The Maxwellians $N_i^M(\vec{x})$, with two independent spatial coordinates, are sums of two time-independent similarity waves, and for the $N_i = N_i^M(\vec{x}) + p_i/D_3(t)$ we add a third, only time-dependent, similarity wave. We construct the linearized eigenfunctions $N_i = N_i^M + \delta_i \exp(\mu t)$ and prove that the eigenvalues μ cannot be positive.

In section III we discuss the shock waves. Let us choose situations for which the two first $\vec{\gamma}_i \cdot \vec{x}$ are proportional. Then in the spatial coordinate plane exist four asymptotic shock limits which are plateaus and which must be positive. For the $4v_i$ model we introduce two new coordinates, $x=x_1$ and y, with $\vec{\gamma}_i \cdot \vec{x} = y\gamma_{i2}$, $j=1,2$. We define the total mass $M = \Sigma N_i = m_0 + \Sigma m_i/D_j^j$ and, as illustred in Figs 1 and 2 we present examples for both N_i and M. In the x,y plane we draw the equidensity lines $M=$const at initial time $t=0$ and observe the shock strip and the shock limits in both upstream and downstream domains. We study the motion of the shock when the time is growing. We observe, at intermediate time, the appearance of bumps not present at initial time or at equilibrium.

In section IV, for both the $4v_i$ and the Broadwell models we study the semiperiodic solutions which are sums of periodic waves N_i^p, M^p plus a third similarity shock wave component: $\tilde{n}_{3i}/D_3$, m_3/D_3. They represent periodic waves, submitted to a strong perturbation and rebuilt beyond the shock. In Figs. 3 and 4 we present, in the x,y plane, the equidensity lines M=const at t=0 and t large where the solutions become similarity shock waves $M \simeq m_0 + m_3/D_3$ with two shock limits.

II. Nonuniform Maxwellians for a p-Dimensional Hypercubic Model (p>1)

The nonlinear equations for the densities N_i are

$$N_{2i-1t} + N_{2i-1x_i} = N_{2it} - N_{2ix_i} = Col_i, \quad \sum_i Col_i = 0, \quad i=1,2,\ldots,p$$

$$Col_i = -(p-1)a_{2i-1,2i} N_{2i-1} N_{2i} + \sum_{\substack{k=1 \\ k \neq i}}^{p} a_{2k-1,2k} N_{2k-1} N_{2k}, \quad a_{i,j} = a_{ji} \quad (3)$$

with the conservation laws

$$\sum_{i=1}^{p} N_{it} + N_{i+pt} + N_{2i-1x_i} - N_{2ix_i} = 0, \quad N_{2i-1t} - N_{2it} + N_{2ix_i} - N_{2i-1x_i} = 0$$

If $a_{i,j} \neq 1$, with microreversibility violated, we define as usual the relative entropy

$H = \Sigma N_i \log(N_i/b_i)$, $b_i > 0$, $b_{2i} b_{2i-1} a_{2i-1,2i}$ = const (i-independent)

and find $H_t + \Sigma \partial_{x_i} \ldots < 0$.

We define the nonuniform Maxwellians N_i^M as the sums of two time-independent similarity components and add a third time-dependent component for N_i

$$N_i^M = n_{0i} + \sum_{j=1}^{2} n_{ji}/(1+d_j \exp(\vec{\gamma}_j \cdot \vec{x})), \quad N_i = N_i^M + p_i/D_3, \quad \rho_3 \neq 0 \quad (4)$$

and it will be shown that $\vec{\gamma}_3 = 0$ so that the third component is only t-dependent.

A. Relations for the parameters of the N_i^M

$$n_{j2i-1} + n_{j2i} = 0, \quad \sum_{k=1}^{p} n_{j2k-1} \gamma_{jk} = 0, \quad a_{2i-1,2i} n_{j2i-1} n_{m2i} = \text{const}, \quad j \neq m$$

$$a_{2i-1,2i} n_{02i-1}^0 n_{02i}^0 = \text{const}, \quad i=1,\ldots,p, \quad j=1,2 \tag{5}$$

with constants i-independent and

$$\gamma_{ji} n_{j2i-1} = (p-1) a_{2i-1,2i} n_{j2i-1}^2 - \sum_{k \neq i} a_{2k-1,2k} n_{j2k-1}^2, \quad i=1,\ldots,p$$

$$= (p-1) a_{2i,2i-1} n_{j2i-1} n_{2i,2i-1}^0 - \Sigma'$$

$$\Sigma' = \sum_{k \neq i} a_{2k-1,2k} n_{j2k-1} n_{2k,2k-1}^0, \quad \text{with } n_{p,q}^0 = n_{0p}^0 n_{0q}^0 \tag{6}$$

We note that the last i=p relation in Eq. (6) is equivalent to the second Eq.(5) relation. We remark that once the n_{0i} are known, then the wave vector $\vec{\gamma}$ or γ_{ji} can be deduced. Furthermore if n_{ii} for i odd is known, then we obtain n_{ji} for i even, so that we are faced with the determination of the 4p parameters n_{0i}, n_{ji} (i odd) submitted to $4(p-1)$ relations (p-1 relations for the n_{0i} alone in Eq.(5), p-1 relations for the n_{ii} alone in Eq.(5) and $2p-2$ relations mixing n_{0i} and n_{ii} in Eq.(6)). In principle we only have four independent parameters for the n_{0i}, n_{ii}, but in fact identities will occur and the solutions will have p+2 arbitrary parameters. With substractions Eq.(6) relations can be rewritten:

$$a_{i,i+1} n_{ji}^2 - a_{2,1} n_{j,1}^2 = a_{i,i+1} n_{ji} n_{i+1,i}^0 - a_{2,1} n_{j1} n_{2,1}^0, \quad \text{i odd} \tag{6'}$$

and conversely by linear superposition of Eq.(6') we recover the Eq.(6). We divide both sides by n_{ji}, substract the j=1,2 relations and taking into account the n_{ji} alone relations Eq.(5) we find

$$(n_{1i} n_{21} - n_{2i} n_{11})(n_{1i} + n_{2i} + n_{i+1,i}^0) = 0, \quad \text{i odd} \tag{6''}$$

We seek solutions such that the first factor is different of zero or more generally $n_{1m} n_{2m'} - n_{1m'} n_{2m} \neq 0$, and the n_{0i}, n_{ji}

relations become

$$n_{0i}+n_{1i}+n_{2i}=n_{0i+1}, \quad i \text{ odd} =1,3,\ldots,2p-1 \tag{7}$$

B. Construction of the N_i^M

We start with p+2 arbitrary parameters (n_{0i} i odd, n_{02}, n_{11}), construct all other parameters, and prove that all relations in Eqs.5-6') are satisfied. First, we get n_{0i} i even and n_{21}:

$$n_{02i}=n_{01}n_{02}a_{2,1}/n_{02i-1}a_{2i-1,2i}, \quad n_{21}=n_{02}-n_{01}-n_{11} \tag{8}$$

Second, from the Eq.(5) relation $a_{1,2}n_{11}n_{21}=a_{i,i+1}n_{ji}n_{j'i}$, i odd, $j \neq j'$, if we substitute relation Eq.(7), then the n_{ji} satisfy quadratic relations with known coefficients

$$n_{ji}^2-n_{ji}n_{i+1,i}^0+n_{11}n_{21}a_{2,1}/a_{i,i+1}, \quad i \text{ odd } j=1,2$$

$$2z_i^{\pm} = n_{0i+1}-n_{0i}^{\pm}\sqrt{\Delta_{i+1,i}}, \quad n_{1i}=z_i^{\pm}, \quad n_{2i}=z_i^{\mp}$$

$$\Delta_{i+1,i}=(n_{0i+1}-n_{0i})^2-4n_{11}n_{21}a_{2,1}/a_{i,i+1} \tag{9}$$

Third, we find n_{ji} for i even and at this stage all n_{0i}, n_{ji} being known we deduce the wave vector components γ_{jq}. From the first Eq.(5) relation we notice:

$$N_i^M+N_{i+1}^M=n_{0i}+n_{0i+1}, \quad i \text{ odd} \tag{4'}$$

We must verify that the n_{ji} constructed from the free parameters satisfy Eqs. (5-6'^{ji}). From the quadratic n_{ji} relation (9) we remark that the product of the two roots is just the first i-independent relation (5). Finally the relation (2.6') is obtained by substractions of the quadratic relations (9) for both i≠1 and i=1.

C. Positivity of the N_i^M solutions

Because $D_j>1$, the N_i^M must have four asymptotic positive limits:

$$n_{0i}>0, \quad n_{0i}+n_{1i}+n_{2i}>0, \quad n_{0i}+n_{ji}>0, \quad j=1,2, \; j=1,2,\ldots,2p \tag{10}$$

Theorem: Sufficient conditions on the arbitrary parameters leading to the eight positivity constraints Eq.(10) are

$$n_{0i} > 0 \quad i \text{ odd and } i=2, \quad 0 < n_{11} < n_{02} < n_{01} \qquad (11)$$

From Eqs.(8-11) all n_{0i} are positive and all the 2p first conditions Eq.(10) are satisfied. The 2p following are deduced both from Eq.(7) and the first Eq.(5) relation. The 4p conditions $n_{0i} + n_{ji} > 0$ remain, and because of the first Eq.(5) relations they are equivalent to

$$-n_{0i} < n_{ji} < n_{0i+1}, \quad j=1,2, \quad i \text{ odd} \qquad (12)$$

In fact, if Eq.(12) holds, then for i odd $n_{ji} + n_{0i} > 0$ and $n_{ji+1} + n_{0i+1} - n_{ji} > 0$. The inequality i=1 for n_{11}^{ji} in Eq.(12) is a consequence of the assumption Eq.(11) and from Eq.(7) we deduce the corresponding one for n_{21}. For other odd i values we rewrite Eq.(9):

$$n_{0i+1} - z_i^{\mp} = n_{0i} + z_i^{\pm} = (n_{0i} + n_{0i+1} \mp (\Delta_{i+1,i})^{1/2})/2 \qquad (9')$$

and all the last 2p inequalities for $n_{ji} + n_{0i}$ are satisfied if the last Eq.(9') terms are positive, or, equivalently, if $(\Delta_{i+1,i})^{1/2} < n_{0i} + n_{0i+1}$. For this ultimate result, which guarantees the positivity conditions Eq.(10), we first notice that $n_{21} < 0$ (see Eqs.(7-9)). Consequently, $\Delta_{i+1,i} > 0$ and real z^+, z^- exist. Second, we write down a set of inequalities: $-n_{11} n_{21} = n_{11}(n_{11} + n_{01} - n_{02}) < n_{11} n_{01} < n_{02} n_{01}$ or

$$-n_{11} n_{21} a_{2,1}/a_{i:1,i} < n_{01} n_{02} a_{2,1}/a_{i+1,i} = n_{ci} n_{0i+1} \text{ or } \Delta_{i+1,i} < (n_{0i} + n_{0i+1})^2.$$

When the limits in Eq.(10) are positive, we can always manage the arbitrary parameters $d_j > 0$ parameters in D_j so that the N_i^M are positive. We notice that the a_{ij} parameters do not enter into the positivity conditions Eq.(10) or in the existence of the algebraic determination of the parameters. In particular all of these results hold for all $a_{ij} = 1$. These Maxwellians exist whether the microreversibility is violated or not.

D. Time-dependent N_i solutions

Let us add to N_i^M a third $\vec{x}$ and t-dependent component $p_i/(1+d \exp(\rho_3 t+\vec{\gamma}_3 \cdot \vec{x}^i))$. The terms $1/(D_3 D_j)$ j=1,2 only present in Col$_i$ lead to the relations

$$a_{2i-1,2i} n_{j2i-1}(p_{2i}-p_{2i-1}) = \text{const (i and j -independent)}.$$

Consequently, the ratios $n_{ji}/n_{ji'}$, j=1,2, i and i' odd are j-independent or $n_{1i}n_{2i'}-n_{2i}^j n_{1i'}^{ji}=0$, and contrary to our previous assumptions, the first Eq.(6") factors are zero. The only escape is $p_{2i}=p_{2i-1}$ or $\gamma_{3i}=0$ from the linear parts of Eq.(3). In conclusion, we can only add a time-dependent component, and we write

$$N_i(\vec{x},t) = N_i^M(\vec{x}) + p_i/(1+d \exp(\rho t)) \qquad (13)$$

The N_i^M being the Maxwellians, the formalism must necessarily require $\rho>0$ or the third component must vanish when $t \to \infty$. If this is true, choosing d large enough, the third component is small for any $t \geqslant 0$. The relations for the third component lead to p-2 relations between the arbitrary parameters that built up the N_i^M. We must verify that these new relations are compatible with the positivity conditions Eq.(10). We notice that finally the N_i depend on four parameters. We define $n_{k,m}^+ = n_{0k}+n_{0m}$ and write down the relations for the third component:

$$p_i = p_{i+1}, \text{ i odd}, \qquad \sum_1^p p_{2i-1}=0 \qquad (14)$$

$$\rho p_{2i-1} = -(p-1) p_{2i-1}^2 a_{2i-1,2i} + \Sigma a_{2k-1,2k} p_{2k-1}^2, \quad k \neq i, \; i=1,\ldots,p$$

$$= (p-1) a_{2i-1,2i} p_{2i-1} n_{2i,2i-1}^+ - \Sigma a_{2k-1,2k} p_{2k-1} n_{2k,2k-1}^+ \quad (15)$$

with the last Eq.(15) relation equivalent to the second Eq.(14) relation. We have 2p+1 new parameters p_i, ρ submitted to 3p-1 relations, leaving p-2 new relations for the arbitrary N_i^M parameters. We begin with p=2 and find

$$\rho = p_1(a_{3,4}-a_{1,2}) = a_{1,2} n_{2,1}^+ + a_{3,4} n_{4,3}^+ > 0, \quad p_1 = p_2 = -p_3 = -p_4 \qquad (16)$$

without any supplementary constraints on the four arbitrary parameters which have only to satisfy Eq.(10). However, $a_{1,2}=a_{3,4}$ equal or not to 1 leads to $\rho=0$. The $4\vec{v}_i$ model is the only one for which a violation of microreversibility is necessary for the existence of time-dependent N_i. For the other p>2 models, solutions with either $a_{m,m'}=1$ or $\neq 1$ exist, and both classes of solutions were previously given for the Broadwell p=3 model.

Here we present a general solution for p>3, valid when the microreversibility is satisfied or $a_{m,m'}=1$ in Eq.(3).

$$p_i=p_1, \quad i=2,3,\ldots,2p-2, \quad p_{2p}=p_{2p-1}=-(p-1)p_1$$

$$\rho=p_1 p(p-2)=n_{2,1}^+ + (p-1)n_{2p,2p-1}^+, \quad n_{0i}=n_{01}, \quad i=3,5,\ldots,2p-3 \quad (17)$$

The solutions depend on four arbitrary parameters (n_{01}, n_{02}, n_{2p-1}, n_{11}) satisfying the positivity constraints Eq.(11). We notice that p=2 is the only p model for which the solution Eq.(17) does not exist.

E. Linearized eigenfunctions $N_i = N_i^M + \delta_i e^{\mu t}$

Linearizing around N_i^M and assuming t-dependent perturbations $N_i = N_i^M + \delta N_i(t)$, $\delta N_i = \delta_i \exp(\mu t)$, δ_i and $\mu \neq 0$ being constants, we only want to verify that $\mu > 0$ eigenvalues do not exist. We substitute into Eq.(3),, take into account $N_i^M + N_{i+1}^M = n_{i+1,i}^+$, i odd, and at the linear level (excluding $\delta N_k \delta N_m$ terms) we obtain

$$\delta_{2i-1}=\delta_{2i}, \quad \sum_{k=1}^{p} \delta_{2k-1}=0, \quad i=1,\ldots,p \quad (18)$$

$$\mu \delta_{2i-1} = -(p-1)a_{2i-1,2i} n_{2i,2i-1}^+ \delta_{2i-1} - \sum_{k \neq i} a_{2k-1,2k} n_{2k,2k-1}^+ \delta_{2k-1} \quad (19)$$

with still the last i=p Eq.(19) relation equivalent to the second Eq.(18) one. For simplicity in the writing let us define $s_i = a_{2i-1,2i} n_{2i,2i-1}^+$ and the symmetric functions $S_1 = \Sigma s_i$, $S_2 = \Sigma s_k s_m$, $S_k = \Sigma s_{i_1} s_{i_2} \ldots s_{i_k}$. In Eq.(19) we eliminate

δ_{2p-1}, obtain a linear system for $\delta_1, \delta_3, \ldots, \delta_{2p-3}$, and from the vanishing of the determinant obtain the eigenvalue polynomial

$$\mu^{p-1} + \sum_{k=1}^{p-1} S_k(p-k)p^{k-1}\mu^{p-1-k} = 0, \quad S_k > 0 \qquad (20)$$

This formula, verified up to p=6, is conjectured for higher p values. The essential point is that $\mu > 0$ roots cannot exist. For p=2 we find $\mu + S_1 = 0$, and for p=3
$\mu^2 + 2\mu S_1 + S_2 = 0$ with $2(S_1^2 - S_2) = (s_1-s_2)^2 + (s_1-s_3)^2 + (s_2-s_3)^2$
leads to two negative eigenvalues μ.

III. (2+1)-Dimensional Shock Waves for the $4v_i$ Model

The $4v_i$ model is the p=2 model of sec.II and the four N_i satisfy

$$N_{1t} + N_{1x_1} = N_{2t} - N_{2x_1} = -N_{3t} - N_{3x_2} = -N_{3t} + N_{3x_2} = aN_3 N_4 - N_1 N_2 \qquad (21)$$

The (2+1)-dimensional shock waves are sums of three similarity waves:

$$N_i = n_{0i} + \sum_{j=1}^{3} n_{ji}/D_j, \quad D_j = 1 + d_j \exp(\vec{\gamma}_j \cdot \vec{x} + \rho_j t), \quad d_j > 0, \quad \vec{\gamma}_j = (\gamma_{j1}, \gamma_{j2}) \qquad (22)$$

What are the positivity conditions? In one spatial coordinate x_1, we only have two asymptotic shock limits when $|x_1| \to \infty$. If these limits are positive, we can manage the positive d_j in D_j so that the N_i are positive for all x_1, t values. For a superposition of three waves $\vec{\gamma}_j \cdot \vec{x}$ spaning a two-dimensional coordinate space, then $\vec{\gamma}_3 \cdot \vec{x}$ is a linear combination of the two other waves. Let us assume, for instance, that $\vec{\gamma}_3 \cdot \vec{x}$ is in the first quadrant of the $\vec{\gamma}_1 \cdot \vec{x}, \vec{\gamma}_2 \cdot \vec{x}$ plane. Then for each N_i six asymptotic shock limits exist that are plateaus in this plane: $n_{0i}, \Sigma n_{ji}, n_{0i} + n_{ji}, n_{0i} + n_{ji} + n_{3i}$, j=1,2, and i=1,...,2p. This is quite complicated, and we must try to find simpler situations. Let us choose situations for which the two first $\vec{\gamma} \cdot \vec{x}$ are proportional with a constant ratio called λ. Then, in the same coordinate plane as that given earlier we find only four asymptotic limits depending on the sign of λ:

$$\lambda>0, \quad n_{0i}, \quad n_{0i}+n_{3i}, \quad \sum_{j=0}^{2} n_{ji}, \quad \sum_{j=0}^{3} n_{ji} \tag{23a}$$

$$\lambda<0, \quad n_{0i}+n_{ji}, \quad n_{0i}+n_{ji}+n_{3i}, \quad j=1,2 \tag{23b}$$

For the $4v_i$ model we choose the shock limits conditions in Eq.(23a). We substitute Eq.(22) into Eq.(21) and get

$$n_{ji}(\rho_j+\gamma_{j1})=n_{j2}(\rho_j-\gamma_{j1})=-n_{j3}(\rho_j+\gamma_{j2})=-n_{j4}(\rho_j-\gamma_{j2})$$

$$=an_{j3}n_{j4}-n_{j1}n_{j2}=-a(n_{03}n_{j4}+n_{04}n_{j3})+n_{j1}n_{02}+n_{j2}n_{01}$$

$$an_{03}n_{04}=n_{01}n_{02}, \quad a(n_{m4}n_{q3}+n_{m3}n_{q4})=n_{m1}n_{q2}+n_{m2}n_{q1}, \quad m \neq q \tag{24}$$

Our assumptions are 1) $\vec{\gamma}_j \cdot \vec{x}$, $j=1,2$ are proportional, or the two first components depend on the same spatial coordinates $y=x_2+x_1\gamma_{11}/\gamma_{12}$; and 2) $\vec{\gamma}_j \cdot \vec{x}=\gamma_{j2}y$ $j=1,2$ with $\gamma_{12}\gamma_{22}>0$ and the asymptotic shock limits that guarantee the positivity of the N_i are

$$n_{0i}>0, \quad \Sigma_i^{03}=n_{0i}+n_{3i}>0, \quad \Sigma_i^2=\sum_{j=0}^{2} n_{ji}>0, \quad \Sigma_i^3=\sum_{j=0}^{3} n_{ji}>0 \tag{25}$$

A. Solutions

We have found solutions for two classes of models: 1) $a \neq 1$, $a<1/3$, with $\gamma_{j1}=0$ $j=1,2$ or the two first components depend only on $y=x_2$. We find 24 parameters and 19 relations leaving five arbitrary parameters. 2) $a=1$ with $\gamma_{j1} \neq 0$. For these models exist 20 relations, 25 parameters, and still 5 arbitrary parameters. In both cases, the five parameters that reconstruct all other are

$$\text{def. } z_j=n_{j4}/n_{j3} \rightarrow (P=z_1z_2, \; S=z_1+z_2, \; n_{0i}>0 \; i=1,2,3) \tag{26}$$

For both models we find that the shock limits Σ_i can be written as

$$\Sigma_i n_{03}=\Omega_i(n_{03}-n_{01}A_{ki})(n_{03}-n_{02}A_{k',i}) \tag{27}$$

with $\Omega_i, A_{k,i}, A_{k',i}$ well-defined (P,S)-dependent functions. For each $n_{03}\Sigma_i$, second-degree n_{03} polynomial, we determine the n_{03} interval (depending on n_{01}, n_{02}, and P,S) in which Σ_i is positive. Then we must find the 12 Σ_i positivity intervals having a nonempty intersection. We find theorems of the following type: 1) if (P,S) belongs to a well-defined domain of the P,S plane; 2) if n_{01}/n_{02} has a well-defined (P,S)-dependent upper bound; and 3) if n_{03}/n_{01} has well-defined, (P,S)-dependent upper and lower bounds, then all the Σ_i are positive. With these results and appropriate choices of the d_j parameters in D_j, then we can construct positive N_i.

B. Physical Discussion and Numerical Examples

We introduce the total mass $M = \Sigma N_i$ that we rewrite as

$$M = m_0 + \sum_j m_j/D_j, \quad m_0 = \Sigma n_{0i}, \quad m_j = \sum_i n_{ji} \quad (28)$$

and study examples in the y, $x = x_1$ plane at fixed t.

1. Models with $a \neq 1$. In Fig.1 we present a model with $P = 0.5$ $S = -10$ $n_{01} = 10^{-3}$ $n_{02} = 1$ $n_{03} = 6.7 \ 10^{-3}$ studied in Ref.6. In Fig. 1a we present the equidensity lines $M(y, x, t=0) =$ const. We observe the shock in a strip around the x axis and the four asymptotic plateaus (two in the upstream domain and two in the downstream one) corresponding to the shock limits of Eq.(25). We write $D_3 = 1 + d_3 \exp(\rho_3 t + \gamma_3 x + \tau_3 y)$ and we see that asymptotically the equidensity lines are parallel either to x or to $\gamma_3 x + \tau_3 y = 0$. In this example $\rho_j > 0$ j=1,3 while $\rho_2 < 0$. Then, in the finite x,y plane for large t, we can approximate M by the second component similarity shock wave: $M = m_0 + m_2/D_2$. The shock strip that at initial time was at y=0 has shifted toward $\gamma_{22} y = -\rho_2 t > 0$. However, contrary to the semi periodic waves of sec. IV, the approximation by a similarity component cannot hold for asymptotic x,y values because, for instance, in $D_1 = 1 + d_1 \exp(\rho_1 t + \gamma_{12} y)$, even if the t contribution is very large, for finite t and $\gamma_{12} y \to -\infty$, $D_1 \to 1$, and the first component contributes to the sum in M. For this example $m_i = 8.5, -8.55, 16.\hat{a}', 4.2$ j=0, 1,2,3. Consequently, if t is growing the negative m_1/D_1 term becomes less important and we can observe bumps. In the space, populations of particles larger than at initial

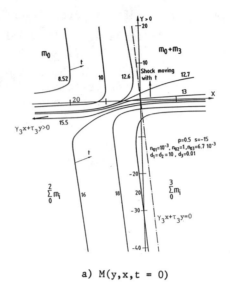

a) $M(y,x,t=0)$

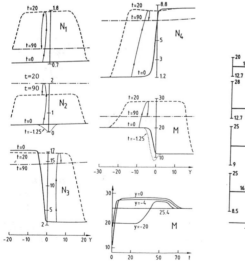

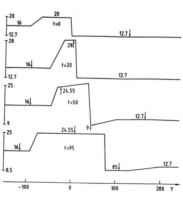

b) $N_i(y, x = 10, t)$

c) $a = 19.7 \; 10^{-3}$, $M(y, x = 10$, $t = 0, 20, 50, 95)$.

Fig. 1 Shock Waves for $4v_i$ model ($a = 19.7 \; 10^{-3}$).

time or at equilibrium can appear. In Fig. 1b we observe these bumps both for the N_i and M.

2. **Model with a=1.** In Fig. 2 we present a solution with $P=0.1$, $S=-2.2$, $n_{01}=10^{-9}$, $n_{02}=1$, $n_{03}=1.23\ 10^{-9}$ leading to $\rho_j=-4.3, 0.72, 0.13$ $j=1,2,3$ $m_j=1.56, 8.6, -1.44, -0.26$ $j=0,1,2,3$. For the $M(y,x,t=0)=\text{const}^j$ equidensity lines we observe the shock strip around the x axis and the four asymptotic Eq.(25) plateaus (two in the upstream and two in the downstream). A bump is present at $t=0$ in the shock domain so that upstream and downstream domains are isolated (contrary to the example in 1). As seen earlier, one ρ_j being negative, for t large and y,x finite we can write $M=m_0+m_1/D_1$ and the shock strip moves from y=0 toward $y\gamma_{12}=-\rho_1 t>0$.

IV. Semiperiodic Solutions for the $4v_i$ and the Broadwell models

These solutions represent periodic waves, submitted to a strong perturbation, and rebuilt after the shock. We build up (1+1)-dimensional periodic solutions and add a third similarity shock wave component.

A. Broadwell model (Eq.(3) for p=3 with $a_{2k-1,2k}=1$)

$N_i=P_i+n_{3i}/D_3$, $N_5=N_3$, $N_6=N_4$, $P_i=n_{0i}+2\text{Re}(n_i/D)$,

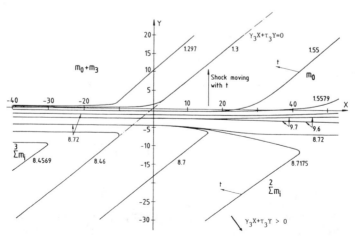

Fig. 2 The $4v_i$ model: a = 1.

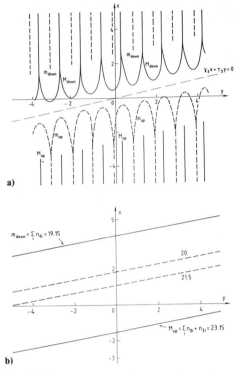

Fig. 3 Semiperiodic waves, total mass $M(y,x,t)$ at $t=0$ and 4: a) Broadwell model, $t=0$; b) Broadwell model, $t=4$.

$$P_{i+1}=n_{0i+1}+2\text{Re}(n_i^*/D), D=1+d\ \exp(\rho t+i(\gamma_1 x_1+\gamma_2(x_2+x_3))),$$
$$D_3=1+d_3\ \exp(\rho_3 t+\vec{\gamma}_3\cdot\vec{x}),\ \vec{\gamma}_3\cdot\vec{x}=\gamma_{31}x_1+\gamma_{32}(x_2+x_3) \tag{29}$$

These solutions depend on two spatial coordinates x_1 and x_2+x_3, and we choose the time dependence of the P_i such that $\rho>0$ with the asymptotic positivity conditions for the N_i alone:

$$n_{0i}>0,\ \Sigma_i^2=n_{0i}+n_{3i} \tag{30}$$

Three arbitrary parameters ($s=n_{1I}/n_{1R}, n_{01}=1, n_{03}>0$) exist, and we prove a theorem saying that, if n_{03} is chosen inside well-defined, positive, s^2-dependent upper and lower bounds, then the asymptotic shock limits Eq. (30) are posi-

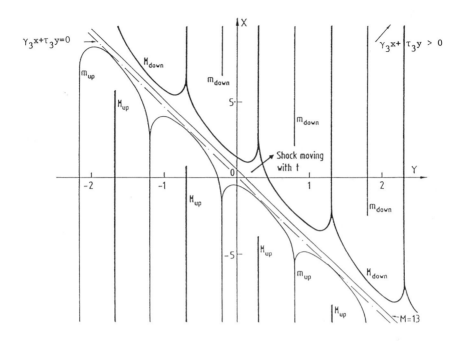

Fig. 4 The $4v_i$ model; semiperiodic waves: total mass $M(y,x,t=0)$.

tive. For the physical discussion, it is convenient to introduce two new spatial coordinates: $\vec{\gamma}\cdot\vec{x}=2\pi y$ or $D=1+d\exp(\rho t+i2\pi y)$, $x=x_1$ and $\vec{\gamma}_3\cdot\vec{x}=\gamma_3 x+\tau_3 y$. We write the total mass $M=M^P+m_3/D_3$, $M^P=m_0+2\mathrm{Re}(m/D)$, and in Fig. 3a we present the equidensity lines parallel to the x axis. Around $\vec{\gamma}_3\cdot\vec{x}=0$ we observe the shock strip. The periodic waves are perturbed by a shock and reconstructed beyond the shock. At fixed t, the shock strip has moved toward a shock around $\vec{\gamma}_3\cdot\vec{x}=-\rho_3 t$ and for t large, M becomes equal to a similarity shock wave $M=m_0+m_3/D_3$ with two shock limits (Fig. 3b)

B. The $4v_i$ Model

The equations for the four N_i are

$$N_{1t}+N_{1x_1}=N_{2t}-N_{2x_1}=-N_{3t}-N_{3x_2}=-N_{4t}+N_{4x_2}=N_3N_4-N_1N_2$$

$$N_i=P_i+n_{3i}/D_3, \quad P_i=n_{0i}+2\mathrm{Re}(n_i/D), \quad D=1+d\exp(\rho_R t+i(\rho_I t+\vec{\gamma}\cdot\vec{x}))$$
(31)

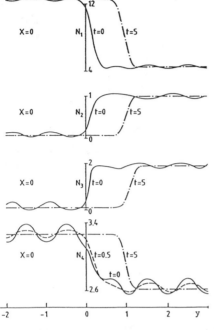

a) the $4v_i$ model.

Fig. 5 Semiperiodic waves, densities N_i.

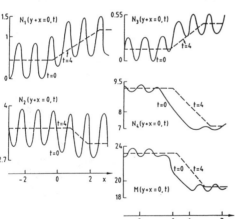

b) Broadwell model.

The solutions depend on four arbitrary parameters: $(S=2\mathrm{Re}(n_4/n_3)$, $P=|n_4/n_3|^2$, $n_{02}>0$, $n_{03}>0)$. We choose $\rho_R>0$, and the asymptotic positivity conditions to be satisfied are still Eq. (30). A theorem is proved showing that, if (P,S) belong to a well-defined domain of the (P,S) plane, $n_{02}>0$ and n_{03}/n_{02} remain between two well-defined, positive (P,S)-dependent upper and lower bounds, then conditions in

Eq. (30) are satisfied. For a physical discussion, like in the previous case, we define $x=x_1$, $2\pi y=\vec{\gamma}\cdot\vec{x}$ and in Fig. 4 present the equidensity lines for the total mass M. We observe essentially the same structures at t=0 and the same evolution at t≠0.

For both models N_i and M one-dimensional profiles are drawn in Fig. 5. With small changes in d that can violate positivity for N_i and not for M. This recalls the fact that macroscopic positivity does not guarantee microscopic positivity.

References

[1] Cornille, H., "Exact Solutions for the Discrete Boltzmann Models with Specular Reflection," Letters in Mathematical Physics Vol. 16, 1988, pp. 245-250 with all previous references included.

[2] Gatignol, R., "Théorie Cinétique des Gaz à Répartition Discrète de Vitesses," Lecture Notes in Physics, Vol. 36, Springer-Verlag, Berlin, Heidelberg, New York, 1975.

[3] Broadwell, J.E., "Shock Structure in a simple Discrete Velocity Gas," Physics of Fluids Vol. 7, 1964, pp. 1243-1248.

[4] Cornille, H., "Construction of Positive Exact (2+1)-Dimensional Shock Wave Solutions for two Discrete Boltzmann Models," Journal of Statistical Physics, Vol. 52, 1988, pp. 897-949.

[5] Cornille, H., "Exact (2+1)-Dimensional Solutions for two Discrete Boltzmann Models with four Independent Densities," Journal of Physics A, Vol. 20, 1987, pp. 1063-1067 L.

[6] Cornille, H., "Construction of Positive (2+1)-Dimensional Solutions for Three Discrete Boltzmann Models," published in "Some Topics on Inverse Problems," edited by P.C. Sabatier, World Scientific, Singapore, New Jersey, Hong Kong, 1988, pp. 104-133.

[7] Cornille, H., "Positive (2+1)-Dimensional Exact Shock Waves Solutions to the Broadwell Model," to be published in Journal of Mathematical Physics (New York) 1989.

[8] Cornille, H., "Construction of Positive (2+1)-Dimensional Exact Solutions for the Broadwell Model," to be published in Transport Theory and Statistical Physics 1989.

Initial-Value Problem in Discrete Kinetic Theory

Shuichi Kawashima*
Kyushu University, Fukuoka, Japan
and
Henri Cabannes†
Université Pierre et Marie Curie, Paris, France

Abstract

From former results obtained by Cabannes ("Comportement asymptotique des solutions de l'équation de Boltzmann discrète" Comptes Rendus de l'Académie des Sciences de Paris, Vol. 202, série 1, 1986, pp. 249-253), Beale ("Large-time behavior of discrete velocity Boltzmann equation" Communications in Mathematical Physics, Vol. 106, 1986, pp. 659-678) and Crandall and Tartar("Existence globale pour un système hyperbolique de le théorie cinétique des gaz" Séminaire Goulaouic-Schwartz, Ecole Polytechnique, 1975), we prove that the initial-value problem in discrete kinetic theory has, in one-dimensional space, a global solution in time, when the initial densities are nonnegative and bounded. This result is valid for all models satisfying the microreversibility condition. By another method the same result has been obtained by Bony ("Solutions globales bornées pour les modèles discrets de l'équation de Boltzmann en dimension 1 d'espace" Journées Equations aux Dérivées Partielles, Centre de Mathématique de l'Ecole Polytechnique, 1987).

Introduction

The discretisation of space velocities in kinetic theory allows one to replace the Boltzmann equation by a system of semi-linear hyperbolic partial differential equations:

$$\frac{\partial N_i}{\partial t} + \vec{u}_i \cdot \vec{\nabla} N_i = \sum_{kl} \left(A^{kl}_{ij} N_k N_l - A^{ij}_{kl} N_i N_l \right) \qquad (1)$$

Copyright © 1989 by the American Institute of Aeronautics and Astronautics, Inc. All rights reserved.
*Faculty of Engineering, Department of Applied Sciences.
†Laboratoire de Modelisation en Mecanique.

where vectors $\vec{u}_i$ (discrete velocities) are constant vectors whose number is p; the indices i,j,k and l vary from 1 to p; the unknown functions N_i, in fact $N_i(x,y,z,t)$ represent the densities of molecules with velocity $\vec{u}_i$; coefficients A_{ij}^{kl} are positive or zero; and

$$A_{ij}^{kl} \neq 0 \rightarrow \vec{u}_i + \vec{u}_j = \vec{u}_k + \vec{u}_l \ , \ \vec{u}_i^{\,2} + \vec{u}_k^{\,2} = \vec{u}_k^{\,2} + \vec{u}_l^{\,2} \qquad (2)$$

$$A_{ij}^{kl} = A_{kl}^{ij} \qquad (3)$$

where the couples (i,j) and (k,l) are not ordered.

Equations (1) have been derived for the first time in a simple case by Broadwell [4] and in the general case by Gatignol in 1970 [5,6]. The initial value problem (1)-(4)

$$N_i(x,y,z,0) = N_{i0}(x,y,z) \qquad (4)$$

always possesses a local solution. The first global existence theorem was obtained by Nishida and Mimura [7] in 1974; it was applicable to the Broadwell model, in the one-dimensional case, with nonnegative and "small" initial densities. Numerous extensions have been made since that date, by Illner [8], Kawashima [9], Bellomo and Toscani [10], Hamdache [11] and Tartar [12]; in all those extensions solutions are not always one-dimensional, but the initial data are always nonnegative and "small". One case concerning initial data partially negative has been studied by Balabane [13]. For the one-dimensional solutions, the case of initial data non negative but not necessarily small, was studied for the first time in 1975 by Crandall and Tartar [3] for a two-dimensional model with four velocities and has been extended then to more complex models [14,15]. Recently the existence of uniform bounds has been proved by Beale [2] and Alvès [16] for diverse models, and explicit bounds have been obtained by Bony [17]. Bony has also proved the global existence in time of the solutions, when the initial densities are positive and bounded. The method of Crandall and Tartar is based on the decrease of the Boltzmann H(t) function, whereas the method of Bony is based on the decrease of the function

$$L(t) = \sum_i \sum_j \int_{\mathbb{R}\times\mathbb{R}} (u_i - u_j) \, \text{Sgn}(\alpha-\beta) \, N_i(\alpha,t) \, N_j(\beta,t) \, d\alpha \, d\beta$$

The densities depend on only one space variable $N_i = N_i(x,t)$ and u_i denotes the component of the vector $\vec{u}_i$ on the axis of abscissae.

In this paper we prove that the Crandall and Tartar method can be applied not only to some particular models, but also to the general case. Of course the coefficients A_{ij}^{kl} must satisfy the relation (3), which represents the microreversibility assumption. This assumption is not necessary with the proof given by Bony.

Case of Small Initial Data

When the initial densities are nonnegative and depend only on the space variable x, then the local solution depends only on x and is defined for $x \in \mathbb{R}$ and $0 \leq t \leq T$; as in Ref. 1 we put:

$$M_r(x,t) = \sum_{u_i = u_r} N_i(x,t) \tag{5}$$

$$E_0 = \max_i \sup_{x \in \mathbb{R}} N_{i0}(x)$$
$$E(t) = \max_i \sup_{\substack{0 \leq s \leq t \\ x \in \mathbb{R}}} N_i(x,s) \tag{6}$$

By addition of eq. (1) for which $u_i = u_r$, one obtains

$$\frac{\partial M_r}{\partial t} + u_r \frac{\partial M_r}{\partial x} = \sum_{u_i = u_r} \sum_{kl} A_{ij}^{kl} (N_k N_l - N_i N_j) \tag{7}$$

Then by integration over the characteristic associated with the velocity u_r

$$M_r(x,t) \leq p E_0 + C_1 E(t) \int_0^t \sum_{u_k \neq u_r} N_k(x - u_r s, t-s)\, ds \tag{8}$$

where C_1 is a positive constant; and then $N_r(x,t) \leq M_r(x,t)$.

Let $\tilde{u}$ be a constant such that $\tilde{u} > \max_i |u_i|$. We can integrate the equation of conservation of mass over the triangle $x - \tilde{u}(t-\tau) \leq \xi \leq x + \tilde{u}(t-\tau)$, $0 \leq \tau \leq t$, in the ξ, τ

plane. This gives

$$\int_0^t \sum_i (\tilde{u}-u_i) N_i(x-\tilde{u}s, t-s) \, ds + \int_0^t \sum_i (\tilde{u}+u_i) N_i(x+\tilde{u}s, t-s) \, ds$$

$$= \int_{x-\tilde{u}t}^{x+\tilde{u}t} \sum_i N_{i0}(\xi) \, d\xi \qquad (9)$$

Therefore denoting by C_2 a new constant, we obtain:

$$\int_0^t \sum_i N_i(x-\tilde{u}s, t-s) \, ds + \int_0^t \sum_i N_i(x+\tilde{u}t, t-s) \, ds$$

$$\leq C_2 \int_{x-\tilde{u}t}^{x+\tilde{u}t} \sum_i N_{i0}(\xi) \, d\xi \qquad (10)$$

Furthermore, the conservation equations of mass and momentum allow one to write:

$$\frac{\partial}{\partial t} \sum_i (u_i - u_r) N_i + \frac{\partial}{\partial x} \sum_i u_i (u_i - u_r) N_i = 0 \qquad (11)$$

Next over the triangle $x-\tilde{u}(t-\tau) \leq \xi \leq x-u_r(t-\tau)$, $0 \leq \tau \leq t$, in the ξ, τ plane, we integrate the relation (11). This gives

$$\int_0^t \sum_i (\tilde{u}-u_i)(u_i-u_r) N_i(x-\tilde{u}s, t-s) \, ds$$

$$+ \int_0^t \sum_i (u_i-u_r)^2 N_i(x-u_r s, t-s) \, ds$$

$$= \int_{x-\tilde{u}t}^{x-u_r t} \sum_i (u_i - u_r) N_{i0}(\xi) \, d\xi \qquad (12)$$

This equality combined with eq.(10) yields

$$\sum_{u_k \neq u_r} \int_0^t N_k(x-u_r s, t-s) ds \leq C_3 \int_{x-\tilde{u}t}^{x+\tilde{u}t} \sum_i N_{i0}(\xi) d\xi \leq C_3 m \quad (13)$$

$$E(t) \leq p\, E_0 + E(t)\, C_3 m \quad (14)$$

C_3 is again a constant, and m is the total mass. From the last inequality, for $2mC_3 < 1$, we deduce $E(t) \leq 2\, p\, E_0$, which proves the global existence in time when the mass is sufficiently small; the (classical) proof of this result is given, for example, in ref. 15.

Case of Bounded Initial Data

By using the Boltzmann H theorem, consequence of the kinetic eq. (1) and of the microreversibilty assumption, Crandall and Tartar have established new inequalities that allows one to drop the conditions concerning the initial mass m. We recall the main points of the method, explained in ref. 3 and 15.

In a first step, one assumes that the initial densities $N_{i0}(x)$ are nonnegative, bounded by E_0, and are periodic with period P. The microreversibility assumption allows one to write the following inequality

$$\sum_{i=1}^p \left(\frac{\partial}{\partial t} + u_i \frac{\partial}{\partial x}\right)(N_i \log N_i) = \frac{1}{4} \sum_{ikl} A_{ij}^{kl}(N_k N_l - N_i N_j) \log \frac{N_i N_j}{N_k N_l} \leq 0 \quad (15)$$

$$I(t) = \sum_{i=1}^p \int_0^P N_i(x,t) \log \frac{N_i(x,t)}{E_0}\, dx \leq 0 \quad (16)$$

If x is positive, the inequality $x(|\log x| - \log x) < 1$ may be used to obtain, for $2T < P$

$$\sum_{i=1}^p \int_{x-T}^{x+T} N_i(x,t)\, dx < m_1 = \frac{4pPE_0}{1 - \log\frac{2T}{P}} \quad (17)$$

One then defines functions $R_i(x,t)$ which satisfy, for all $t_1 > 0$, the conditions

$$R_i(x,t_1) = N_1(x,t_1) \quad \text{for } |X-x| \leq T$$
$$R_i(x,t_1) = 0 \quad \text{for } |X-x| > T \quad (18)$$

Therefore, for all t_1 one has

$$\int_{-\infty}^{\infty} \sum_{i=1}^{p} R_i(\xi,t_1) \, d\xi < m_1 = \frac{4pPE_0}{1 - \log\frac{2T}{P}} \quad (19)$$

As m_1 goes to 0 with T, the functions $R_i(x,t)$ exist for all $x \in \mathbb{R}$ and $t \geq t_1$. The functions $N_i(x,t)$ coincide with the functions $R_i(x,t)$ in the band $t_1 \leq t \leq t_1 + T_1$, with $T_1 = T/\max|u_i|$; therefore they exist also globally. One goes finally to the case of bounded but nonperiodic data by considering a period that increases indefinitely.

Conclusion

Theorem: *If the initial densities are continuous, integrable, nonnegative and bounded functions, if furthermore they depend on only one space variable, then the initial value problem for the kinetic equations (1) possesses a global solution in time.*

Acknowledgments

This work was supported by grants from DRET 86/1577 (Direction des Recherches et Etudes Techniques), during the visit of one of the authors (S.K.) at the Laboratoire de Modélisation en Mécanique, associé au CNRS (Centre National de la Recherche Scientifique), of the University Pierre et Marie Curie.

References

[1] Cabannes, H., "Comportement asymptotique des solutions de l'équation de Boltzmann discrete," Comptes Rendus de l'Académie des Sciences de Paris Vol. 302, série 1, 1986, pp. 249-253

[2] Beale, T., "Large-time behavior of discrete velocity Boltzmann equation," Communications in Mathematical Physics, Vol. 106, 1986, pp. 659-678.

[3] Tartar, L., "Existence globale pour un système hyperbolique de la théorie cinétique des gaz," Séminaire Goulaouic-Schwartz, 1975, Ecole Polytechnique, Palaiseau, France..

[4] Broadwell, J., " Study of rarefied shear flow by discrete velocity method, " Journal of Fluids Mechanics, Vol. 19, 1964, pp. 401-414.

[5] Gatignol, R., "Théorie cinétique d'un gaz à répartition discrète de vitesses," Zeitschrift für Fugwissenschaften, Vol. 18, 1970, pp. 93-97.

[6] Gatignol, R.,"Théorie cinétique des gaz à répartition discrète de vitesses," Lecture Notes in Physics, Vol. 36, 1975, Springer-Verlag, Berlin, Heidelberg.

[7] Nishida, T. and Mimura, M, " On the Broadwell model for a simple discrete velocity gas," Proceedings of the Japan Academy, Vol. 50, 1974, pp. 812-817.

[8] Illner, R., "Global existence results for discrete velocity models of the Boltzmann equation in several dimlensions," Journal de Mécanique Théorique et Appliquée, Vol. 1 , 1982 , pp. 611-622.

[9] Kawashima, S., " Global Existence and Stability of Solutions for Discrete Velocity Models," Lecture Notes in Numerical and Applied Analysis, Vol. 6, 1983, pp. 59-96.

[10] Bellomo, N. and Toscani, G., " On the Cauchy problem for nonlinear Boltzmann equation, global existence, uniqueness and asymptotic stability,' Journal of Mathematical Physics, New York Vol. 26, 1985, pp. 334-338.

[11] Hamdache, K., "Existence globale et comportement asymptotique pour l'équation de Boltzmann à répartition discrète de vitesses," Journal de Mécanique Théorique et Appliquée, Vol. 3, 1984, pp. 761-785.

[12] Tartar, L. "Some Existence Theorems for Semilinear Hyperbolic Systems in one Space Variable," Univ. of Wisconsin-Madison, MRC Technical Summary Report, 1980.

[13] Balabane, M. "Un résultat d'existence globale pour le système de Carleman," Comptes Rendus de l'Académis des Sciences de Paris, Vol. 303, série 1, 1986, pp. 919-922.

[14] Cabannes H. "Solution globale en théorie cinétique discrète: modèle plan," Comptes Rendus de l'Académie des Sciences de Paris, Vol. 284, série A, 1977, pp. 269-272.

[15] Cabannes, H., " Solution globale du problème de Cauchy en théorie cinétique discrète," Journal de Mécanique, Vol. 17, 1978, pp. 1-22.

[16] Alvès, A., " Comportement asymptotique des solutions de l'équation de Boltzmann pour un gaz à 14 vitesses," Comptes Rendus de l'Académie des Sciences de Paris, Vol. 302, série 1, 1986, pp. 367-370.

[17] Bony, J.M., "Solutions globales bornées pour les modèles discrets de l'équation de Boltzmann, en dimension 1 d'espace," Journées Equations aux Dérivées Partielles, 1987, Ecole Polytechnique, Palaiseau, France.

Study of a Multispeed Cellular Automaton

B. T. Nadiga,* J. E. Broadwell,† and B. Sturtevant‡
California Institute of Technology, Pasadena, California

Abstract

Most cellular automata intended to describe fluid motion simulate single-speed particles moving on square or hexagonal lattices. In the latter case, two-dimensional low-Mach-number flows have been shown by Frisch et al. ("Lattice Gas Automata for the Navier-Stokes Equation", *Physics Review Letters*, Vol. 56, 1986, pp. 1505-1508) to obey the Navier-Stokes equations for incompressible flow. These authors also discuss the various difficulties associated with the models, in particular, the restriction to low speeds. Furthermore, it is clear that with only one allowed molecular speed, temperature or energy cannot be specified independently of the velocity. d'Humières et al. ("Lattice Gas Models for 3d Hydrodynamics", *Europhysics Letters*, Vol. 2, 1986, pp. 291-297) describe what appears to be the simplest multispeed model for flows in both two and three dimensions. The present paper describes the results of an exploratory investigation of heat conduction and shock-wave formation with the two-dimensional model. The irreversible macroscopic behavior of this microscopically reversible system is also examined.

I. Cellular Automata

The macroscopic behavior of a fluid near equilibrium is expected to be nearly independent of the details of the motion of the molecules that constitute it. For example, low-Mach-number flow of a gas and of a liquid are described by the same equations. This idea forms the basis for the cellular automaton (CA) simulation of fluids.[1] The aim of this approach is to maximally simplify the molecular dynamics while retaining the essential physics. This simplification of the molecular dynamics involves a full discretization of phase space i.e., of both velocities and positions. This is in contrast with the discrete-velocity models,[2-5] where only the velocity space is discretized, and

---

Copyright © 1989 by the American Institute of Aeronautics and Astronautics, Inc. All rights reserved.
*Research Assistant, Graduate Aeronautical Laboratories.
†Senior Scientist, Graduate Aeronautical Laboratories.
‡Professor, Graduate Aeronautical Laboratories.

with various finite-difference formulations, where only position space is discretized. Discretization of both velocity and position gives rise to the notion of discrete time, the unit of time being that taken by the slowest moving particle to travel the smallest unit of distance in the direction of its velocity. All other particles move an integer number of link lengths in the direction of their velocities in the same time. The evolution of the system is then reduced to a set of discrete move and collide phases. At each instant, each lattice site collects the relevant information from its nearest neighbors (i.e., the particles convect) and performs a simple transformation on it (i.e., the particles collide). Such a limiting simplification is a cellular automaton and is implemented by mapping onto a digital computer.

To study the simplest of these models, only a few velocities are considered. Particles are then identified by their velocities, so that we have a small number of distinguishable particle types. For no reason other than to keep computation per time step small, an exclusion principle is adopted, namely, no lattice site is allowed to have more than one particle of a particular type. In the present work the rules for collisions conserve mass, momentum, and energy. The choice of candidates for collision is arbitrary; thus, the collision rate of CA, and therefore the mean free path, is model-dependent. This is in contrast to the procedure used in Monte-Carlo methods used for directly simulating molecular flows (Bird 1976), in which candidates are chosen to ensure that the collision rate is correct.

1.1. *Implementation*

The implementation of CA on a digital computer is simple, elegant and highly efficient. In the present work a lattice site is represented by a computer word. The computational domain is then an array of words. A particle of a particular type (i.e., a particle with a certain velocity) is identified with a particular bit in the word. A word therefore has to have at least as many bits as there are velocities in the model. The presence or absence of a type of particle at a lattice site is indicated by the presence or absence (on or off) of the corresponding bit in the word representing the lattice site. When only a few velocities are present in the model, the move phase is accomplished by a small number of simple binary operations on the array of words representing the computational domain, while the collide phase is reduced to a table look-up. With more velocities, the move phase requires more binary operations, and the look-up table becomes bulky, necessitating a functional implementation of the collisions. In a variant of the preceding implementation, not used in the present work, the presence or absence of a particle of a particular type at a group of lattice sites is compacted into a word. A set of words, as many as the number of different velocities, then represent several lattice sites, as many as the size of the word. In this scheme, the move phase amounts to shifting words bitwise in the appropriate direction and the collision phase to the evaluation of Boolean functions representing the collisions.

In either case, the simplicity of the move and collide steps makes it possible to simulate huge numbers of particles, in comparison to other direct simulation methods. Furthermore, since only nearest neighbors interact, the evolution itself is highly localized and hence is ideally suited for concurrent computation. The communication overhead between the nodes of a parallel processor is proportional to the ratio of the perimeter of the physical space represented by a node to its area, i.e., to the inverse of the aspect ratio of the computational domain. The complete synchrony between the various parts of the computational domain obviate the need for balancing load between the various processors dynamically.

The present study has concentrated on the simulation of two-dimensional fluids because most essential ideas can be described with them. An extension to three dimensions is straightforward.

II. Single- and Multiple-Speed Models

In the classical cellular automaton, known as the HPP model,[7] and in most subsequent models, the particles move with a single speed. Figure 1 shows the lattice for the HPP model, in which the particles can move in four different directions. The particles move along the horizontal or vertical links to cover one link length in one unit of time. The only nontrivial collisions are the horizontal and vertical head-on collisions. Then, if all particles are initially on the lattice, they remain so. Figure 1 also shows a second single-speed model with a better symmetry, the FHP model.[1] Particles now have one of six different velocities, all of the same speed. There are now a number of nontrivial collisions but spurious conservations by two-body collisions alone necessitate the implementation of three-body collisions as well. It has been

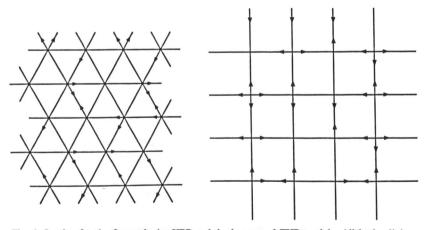

Fig. 1 Lattice for the four-velocity HPP and the hexagonal FHP models. All lattice links are of equal length, and all particles have the same speed. A particle traverses one link length in one unit of time. The FHP model has better symmetry, but has spurious conservations with binary collisions alone.

shown by Frisch et al.[1,8] for two dimensions and d'Humières et al.[9] for three dimensions that for the single-speed model on the hexagonal lattice, the fluid mechanical velocities, obtained by averaging over many sites, satisfy the incompressible Navier-Stokes equations under certain conditions. [On the square lattice (Fig 1), the cross-derivative convective terms in the averaged momentum equation are missing, leading to a spurious conservation of momentum.[1]] However, the problem with single-speed models is that the temperature cannot be represented independently of the velocity. In a two-speed model, even though differing proportions of the particles with the two allowable speeds can represent differing amounts of energy, there is no mechanism by which, in collisions conserving momentum and energy, particles can change speed. This implies that there is no dynamic balancing between the particles of differing speeds or that equilibration is only partial. Thus, the simplest model with temperature as an independent degree of freedom is a three-speed model. Preliminary investigation of one such model, a three-speed, nine-velocity model, is presented here.

A major shortcoming of these lattice gases in modeling real fluids is that the state variables of such a gas depend on the frame of reference; hence, they are not Galilean-invariant. Addition of more velocities will, of course, mitigate this problem. Even with just a few velocities, however, at very low speeds compared to the particle speeds (i.e., in the low Mach number limit), the effect of Galilean noninvariance is negligibe.

2.1. The Nine-Velocity Model

Figure 2 shows the allowable velocities in the three-speed model, and the two-dimensional lattice on which the particles move. The slow particles, which

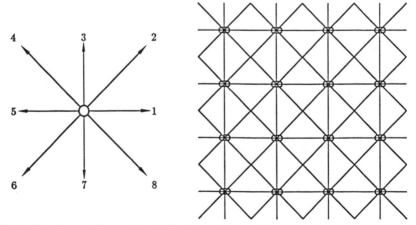

Fig. 2 The nine-velocity model and the lattice. Four particle types have unit speed, four other particle types have $\sqrt{2}$ units of speed, and one particle type has zero velocity. Slow-moving particles move along the horizontal and vertical links, and the fast-moving particles move along the diagonals.

have unit speed, say q, are restricted to move on the horizontal and vertical links, whereas the fast particles, which have a speed of $\sqrt{2}\,q$, move on the diagonal links. The zero-speed particles exist only to take part in collisions, to allow interaction between the other two speeds. Each lattice site has eight nearest neighbors, four at a distance δ away and four others a distance $\sqrt{2}\,\delta$ away where δ is the distance traveled at speed q in unit time Δt. To model a dilute gas with short-range intermolecular interactions, collision rules are defined in which only the nearest neighbors influence a lattice site. Then all possible collisions, each conserving mass, momentum, and energy, of the types shown in Fig 3, take place, subject to the exclusion condition. Since only one particle of a given velocity is allowable at a site, the site may not be able to accommodate some of the particles resulting from some collisions; hence, those collisions are excluded.

As stated earlier, the transport coefficients of a lattice gas depend on the details of the microscopic dynamics. To examine this behavior the experimental simulations have been carried out with two different sets of rules (collisions). In the first (rule set 1), only binary collisions are implemented; i.e., a collision occurs at a site if and only if there are exactly two particles at that site and if at least one component of their momenta is oppositely directed. This set of rules implements just 10 direct collisions and their inverses; thus, a total of 20 states of the $2^9 = 512$ possible at any lattice site are changeable. In the second (rule set 2), a collision occurs at a site, irrespective of the number of particles present there, provided only one collision is possible. Thus, in some of these collisions two particles collide while others at the same site are not affected. The same two-body collisions are implemented in this rule set as in

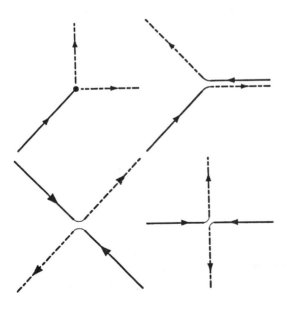

Fig. 3 Binary collisions in the nine-velocity model.

rule set 1, but this rule set transforms a total of 156 of the possible 512 states at a lattice site to a different configuration. Hence, the collision rate is substantially increased. Other rules are possible, including that implementing all possible conservative collisions at a site in a certain specific order, or randomly, but the two rule sets previously described are sufficient for our arguments.

2.2. Boltzmann Equations

An approximate description of the evolution of spatial averages of the nine populations in the automaton is given by Boltzmann equations for the corresponding discrete-velocity gas. They are formulated by assuming that 1) the gas is dilute, so that only binary collisions are important; 2) there is no exclusion; and 3) molecular chaos prevails; i.e. the joint two-particle distribution function can be replaced by the product of two one-particle distribution functions. The Boltzmann equations for the nine-velocity model are given schematically in Table 1. The gain and loss terms that appear on the right-hand sides of the equations for the 10 different collisions are indicated in the second row, and the relative velocity c_r for each of the collisions is indicated in the first row. The left-hand sides of each of the nine equations for the nine classes of particles are indicated in the first column. The entries in the matrix give the sign with which the gain/loss terms appear in the corresponding equation.

III. Observations and Discussion

3.1. Equilibrium State

A first step toward understanding the nature of the lattice gas is to study its equilibrium. If there are no spatial gradients, the above Boltzmann equations have solutions that approach a steady equilibrium state. This behavior is conveniently described by defining the H function

$$H = \sum_i n_i \log n_i \ , \qquad (1)$$

and showing that it must decline to a constant. According to the H theorem, at thermodynamic equilibrium detailed balancing prevails; thus, the gain/loss combination from each collision goes to zero individually. These are the equilibrium equations for the model, and they make up the following set of five independent equations.

$$n_1 n_3 = n_0 n_2 \ , \ n_3 n_5 = n_0 n_6 \ , \ n_5 n_7 = n_0 n_6 \ , \ n_7 n_1 = n_0 n_8 \ , \ n_3 n_7 = n_1 n_5 \qquad (2)$$

Customarily, equilibrium conditions are expressed in terms of thermodynamic variables of state, in this case the density n, the energy of the system e_t and the mean velocities $\bar{u}$ and $\bar{v}$. Then, the definitions of these variables provide four

Table 1. Boltzmann Equations: 9 Velocity Model

c_r =	q	$\sqrt{2}q$			$2q$	$2\sqrt{2}q$		$\sqrt{5}q$		
	$n_1n_3-n_0n_2$	$n_3n_5-n_0n_4$	$n_5n_7-n_0n_6$	$n_7n_1-n_0n_8$	$n_3n_7-n_1n_5$	$n_4n_8-n_2n_6$	$n_5n_2-n_1n_4$	$n_5n_8-n_1n_6$	$n_3n_8-n_7n_2$	$n_3n_6-n_7n_4$
$\dfrac{\partial n_0}{\partial t} =$	+	+	+	+						
$\dfrac{\partial n_2}{\partial t}+q\dfrac{\partial n_2}{\partial x}+q\dfrac{\partial n_2}{\partial y} =$	+					+	−		+	
$\dfrac{\partial n_4}{\partial t}-q\dfrac{\partial n_4}{\partial x}+q\dfrac{\partial n_4}{\partial y} =$		+				−	+			+
$\dfrac{\partial n_6}{\partial t}-q\dfrac{\partial n_6}{\partial x}-q\dfrac{\partial n_6}{\partial y} =$			+			+		+		−
$\dfrac{\partial n_8}{\partial t}+q\dfrac{\partial n_8}{\partial x}-q\dfrac{\partial n_8}{\partial y} =$				+		−		−	−	
$\dfrac{\partial n_1}{\partial t}+q\dfrac{\partial n_1}{\partial x} =$	−			−	+		+	+		
$\dfrac{\partial n_5}{\partial t}-q\dfrac{\partial n_5}{\partial x} =$		−	−		+		−	−		
$\dfrac{\partial n_3}{\partial t}+q\dfrac{\partial n_3}{\partial y} =$	−	−			−				−	−
$\dfrac{\partial n_7}{\partial t}-q\dfrac{\partial n_7}{\partial y} =$			−	−	−				+	+

other equations to be satisfied,

$$n = n_0 + n_1 + n_2 + n_3 + n_4 + n_5 + n_6 + n_7 + n_8 \tag{3a}$$

$$\bar{u} = (n_1 + n_2 + n_8 - n_4 - n_5 - n_6)/n \tag{3b}$$

$$\bar{v} = (n_2 + n_3 + n_4 - n_6 - n_7 - n_8)/n \tag{3c}$$

$$e_t = \frac{q^2}{2n}\left[n_1 + n_3 + n_5 + n_7 + 2\left(n_2 + n_4 + n_6 + n_8\right)\right] . \tag{3d}$$

Thus, there are nine equations for the nine variables n_i, and the system is uniquely determined. The equilibrium equations are solved for some simple cases. In particular, for $u = v = 0$,

$$n_0 = n\left[1 - \frac{kT}{q^2}\right]^2 \tag{4a}$$

$$n_2 = n_4 = n_6 = n_8 = \frac{n}{4}\frac{kT^2}{q^2} \tag{4b}$$

$$n_1 = n_3 = n_5 = n_7 = \frac{n}{2}\frac{kT}{q^2}\left[1 - \frac{kT}{q^2}\right] \tag{4c}$$

The solutions for small velocity in one of the directions x or y are obtained by a linearization about the above stationary states. The solution is also obtained for the case of a small velocity in an arbitrary direction.

Figure 4 shows the evolution of the H function for lattice-gas particles in a box initially distributed randomly with the state variables set to the desired final

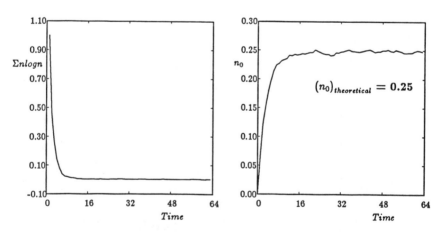

Fig. 4 Equilibration. Evolution of $H = \sum n \log n$ and n_o, the stagnant particle ratio with time for a box of lattice gas. ($kT = 0.5q^2$; $u = v = 0$; $\rho = 0.3$) Hydrodynamic quantities are constant with time, and the initial state was chosen randomly.

state. It is seen that, though the initial condition is in the proper macroscopic state, the system is initially out of equilibrium, but equilibrates in a few collisions. Experimental simulations of the equilibrium conditions are in accord with the solution of the preceding set of equations, for example, as shown in Fig 4.

3.2. Characterization of the Mean Free Path

The mean free path is of the order of the lattice spacing in the range of densities of interest. Above a certain density, the effects of exclusion are large, so the mean free path between effective collisions becomes large. The mean free path is determined empirically in the following way. The initial condition in a box with doubly periodic boundary condition is set to correspond to prescribed values of density, temperature, and velocity. The boundary conditions were made doubly periodic, because otherwise, shocks and rarefactions, in the cases of nonzero macroscopic velocity, dissipate directed kinetic energy and change the thermodynamic state of the system. The initial particle distributions are calculated numerically under the simplifying assumptions of binary collisions and absence of exclusion. In the simulation, exclusion may cause the actual particle distribution to be slightly different from that calculated initially. After the system is allowed to relax to equilibrium, the number of collisions in the box is counted for a number of time steps. Also monitored is the ratio of particle population of different speeds. Then, under the assumption of ergodicity of the behavior of particles, the typical collision frequency ν and the mean particle speed $\bar{c}$ are calculated and the mean free path obtained from the relation $\lambda = \bar{c}/\nu$.

The variation of the mean free path with density, temperature, and velocity for the two different rule sets was studied. The behavior of the mean free path with the two rule sets differs appreciably only in the regions of high density. Recall that the two rule sets implement the same collisions. The only difference between them is in the number of collisions that are effective.

A power regression between the mean free path and the density at low densities gives $\lambda \propto 1/\rho$ as in real gases (see Fig. 5). The anomalous behavior at higher densities comes from the exclusion of some collisions. This study gives bounds on the density that can be used in these cellular automatons to study gasdynamics. The variation of the mean free path with temperature comes from the collision rules and probably cannot be given physical significance. The Galilean noninvariance of the model, is illustrated by the changing mean free path with velocity (see Fig. 5). Note, however, that changes are small up to velocities of about $0.3q$.

3.3. Heat Conduction

To examine the nature of the model and, in particular, its ability to deal with heat conduction, we have studied heat transfer between two walls at rest (Fig. 6). The walls are maintained at different temperatures, and the

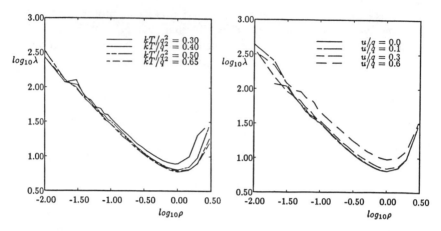

Fig. 5 Variation of λ, the mean-free-path with ρ, the density at different temperatures and velocities. Note the unphysical increase of mean free path with density above a certain density. Plots are shown for rule set 1. Behavior is similar for rule set 2.

temperature distribution in the gas is computed as a function of time. Temperatures are maintained by imposing, at each instant of time at the row of wall lattice sites, the equilibrium particle distribution corresponding to the wall conditions. The parameter of interest in the problem is the Knudsen number, $Kn = \lambda/L$, where L is the wall separation. The primary aim is to determine the nature of the steady-state profiles and the effect of a variation of Kn on these profiles. Both rule sets were used for some runs. The left cold wall was maintained at temperature of $0.6q^2/k$ and the right at $0.7q^2/k$. The prescribed density was 0.3 particles per site. From uniform initial conditions the system was allowed to relax to a steady state, after which the temperature profile was computed by averaging over about 1600 stations along the wall and 2048 time steps. Computations were made for three Knudsen numbers, 0.44, 0.22, and 0.10; the profiles for rule set 1 are shown in Fig. 6. As in the case for real gases, the profiles are linear with temperature jumps at the walls. Furthermore, as expected, the magnitude of the jumps decline with Kn, approaching zero in the continuum limit. The temperature profiles for the first two cases was computed with rule set 2. The results for the two rule sets were seen to be similar.

It should be noted that, in the free molecule flow limit, $Kn \to \infty$, the model does not work. In the absence of collisions, the zero-speed particles populations cannot adjust properly to those emitted from the walls to yield the correct mean temperature. This appears to be an inherent flaw in the model at this condition.

The time to reach steady state increases roughly as the square of the wall spacing, a result that suggests a similarly parameter of the form $y/\sqrt{vt}$, where y

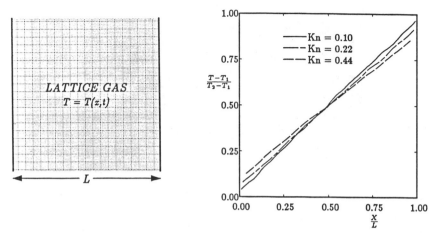

Fig. 6 The heat conduction problem: The two walls are maintained at different temperatures, and the temperature profile in the lattice gas is sought. The left wall is maintained at $kT/q^2 = 0.60$, and the right wall is maintained at $kT/q^2 = 0.70$. Also, the effect of the interwall spacing L is studied. The steady-state temperature profiles at three different Knudsen numbers is shown. The average density is 0.3 particles in all the three cases. Plots are shown for rule set 1. Behavior is similar for rule set 2.

is the distance normal to the walls and ν an appropriate diffusion coefficient. (The transient behavior of the temperature profile also indicates the same.) In summary, similarities between the automaton results and those in real gases include 1) linearity of the steady state profile, 2) increasing temperature jump at the wall with increasing Knudsen number, and 3) existence of a similarity parameter of the form $y/\sqrt{\nu t}$.

3.4. Normal Shock Wave

Another classic test of molecular models is the normal shock wave. The particle population ratios in a box (Fig. 7) are set to correspond to a certain temperature, density, and momentum in one direction, the momentum in the other direction being zero. The particles are allowed to relax to equilibrium by making the box doubly periodic. After that, at time zero, the boundaries perpendicular to the direction of mass motion are made specularly reflective (the other walls may be either periodic or specular). Subsequently, a shock forms at one end and propagates away from the wall at a speed that depends on the initial Mach number. At the other, a rarefaction wave forms. By varying the initial temperature and velocity, shocks of different strengths can be generated and their characteristics computed.

Temperature and density profiles for several Mach numbers computed with the two rule sets are given in Figs. 7 and 8. Here, in the stronger shocks, where the flow is further from equilibrium, results for the two rule sets may be expected to differ considerably. Again, as in the preceding example, the

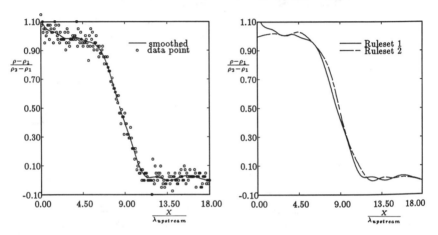

Fig. 7 Shock phenomena: Density profile across a 1.13-Mach shock at an average density of 0.3 particles per site. Rule set 1 was used. The density ratio $\rho_2/\rho_1 = 1.12$. The smoothing was done by using a low pass filter. The density profiles got by the two rule sets are compared for the 1.13-Mach shock.

qualitative behavior is similar to that in real gases. In particular, 1) the density ratio increases and the thickness declines with initial Mach number, and 2) the temperature rise precedes the density rise.

Quantitative comparisons with the behavior of an ideal gas with a specific heat ratio of two, the proper value for a two-dimensional gas, have not yet been made because of several uncertainties. First the thermodynamic properties of the lattice gas, i.e., the specific heats, are difficult to work out. Next, the Galilean noninvariance of the model precludes a comparison of the jump conditions in the present unsteady flow to those of the steady shock. Finally, there is a degree of arbitrariness in the definition of the mean free path. In the case of molecules having a continuum of velocities, all collisions resulting in a deflection of more than a certain threshold are counted in calculating the mean free path, but in the present model, a particle is deflected, if at all, by either 45

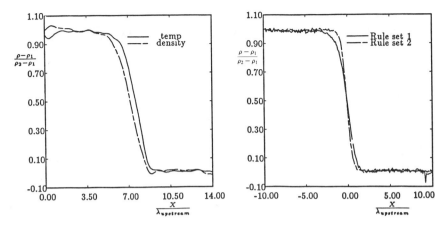

Fig. 8 The (smoothed) density and temperature profiles across a 1.48-Mach shock show that the temperature shock propagates ahead of the density shock ($\rho_2/\rho_1 = 1.47$); and a comparison of the density profiles got by the two rule sets at a higher density of 1 particle a site shows that the two rule sets simulate different viscosities at high densities (the density ratio $\rho_2/\rho_1 = 1.71$ is the same).

or 90 deg. Thus, whereas counting only the nonzero deflection collisions gives a larger than correct mean free path, counting all collisions i.e., even those collisions that leave the particle velocities unchanged, errs in the other direction.

3.5. The Arrow of Time

It is interesting to note that since each operation on the automaton is reversible, so are the macroscopic processes it models. More precisely, this is true when there is no external forcing, as in the shock and expansions wave flows discussed earlier. (In the heat conduction example, since the temperature at the walls is imposed at each instant, the procedure of reversal is less clear.) Therefore, if at any time during the unsteady wave computation all molecular velocities are reversed, the system returns to its initial state. How is this behavior in accord with the second law of thermodynamics? This section describes a brief investigation of this question.

The one-dimensional flow computed in Sect. III, Pt. 4, was repeated for a shorter box and longer lengths of time, so that with reasonable computer time the approach to a stationary mean state could be examined. Figure 9, between times $t = 0$ and $t = 600$, shows the decay of the mean velocity in the box as a function of time. Between the same times the entropy of the system increases as the gas comes to rest. If, as already noted, at any time, say $t = 600$, all molecular velocities are reversed, the system reverses and exactly retraces the forward path as indicated in Fig. 9 between $t = 600$ and $t = 1200$.

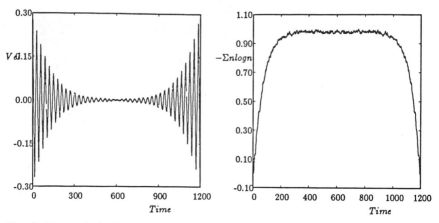

Fig. 9 Mean velocity in a box decays with time, and the entropy increases. On full reversal at $t = 600$, the system retraces its path and the initial state is fully recovered, going through a decrease in entropy, demonstrating micro-reversibility.

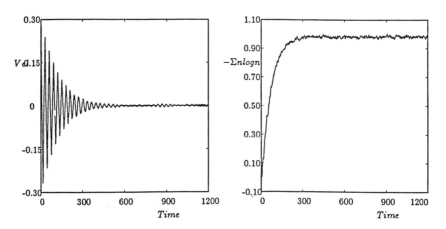

Fig. 10 A small error in the reversal at $t = 600$ effectively prevents any recovery of the macroscopic velocity, and the entropy behaves more physically. This shows the instability of the reversed path.

Nevertheless, to see that the system has a preferred direction in time, we consider the effect of a small "error" on the system overall mean velocity when it moves forward and backward in time. First, at $t = 600$, molecules at 0.1% of the sites in the box are reversed. The subsequent history is virtually the same as the original, and there are only microscopic differences. However, when the same error is then introduced at $t = 600$ as the system runs backward in time, i.e., 99.9% of the particles are reversed instead of 100%, the effects are drastically different, as illustrated in Fig. 10. There is no resemblance to the macroscopic history of the system running backward with no error.

This behavior, and presumably that of real systems also microscopically reversible, may be summarized by saying that the solution describing the state history is stable in the forward direction but highly unstable in the reverse. A different way of looking at this behavior is that the CA evolution is an iterated Boolean map possessing a macroscopic equilibrium state. From yet another point of view, the automaton is a nonlinear system whose mean state is extremely sensitive to initial conditions in one direction in time but not in the other.

This short and incomplete study suggests that the CA could be a new tool for examining the irreversible behavior of reversible systems and the further resolution of the Loschmidt and Zermelo paradoxes. There is, of course, an enormous amount of literature on these subjects, but it has not yet been examined for relationships to the present work. A recent review by Coveney[10] describes many current ideas.

IV. Conclusion

This work should be considered to be an exploration of the usefulness of CA in the study of gasdynamics. It is premature to discuss the accuracy of the model (as the identification of several important difficulties makes clear) or of relative computation speeds. The primary usefulness of the model may, in fact, turn out to be as a mechanism for investigating concepts.

V. Appendix: Sound Propagation

The equations describing propagation of weak disturbances can be found by linearizing the Boltzmann equations for the model. The result is a set of first order linear partial differential equations that can be combined into a single equation for any variable, ϕ (a population density, for instance). The equation is

$$\left[\frac{\partial^2}{\partial t^2}\left(\frac{\partial^2}{\partial t^2} - q^2\frac{\partial^2}{\partial x^2}\right)^2 + \frac{\partial}{\partial t}\left(f_{11}\frac{\partial^4}{\partial t^4} - q^2 f_{12}\frac{\partial^4}{\partial t^2 \partial x^2} + q^4 f_{13}\frac{\partial^4}{\partial x^4}\right) \right.$$

$$\left. + \left(f_{21}\frac{\partial^4}{\partial t^4} - q^2 f_{22}\frac{\partial^4}{\partial t^2 \partial x^2} + q^4 f_{23}\frac{\partial^4}{\partial x^4}\right) + \frac{\partial}{\partial t}\left(f_{31}\frac{\partial^2}{\partial t^2} - q^2 f_{32}\frac{\partial^2}{\partial x^2}\right)\right]\phi = 0 ,$$

where

$$f_{11} = 2\left[1 + \frac{kT}{q^2} - \left(\frac{kT}{q^2}\right)^2\right] \quad f_{12} = \left[2 + 4\frac{kT}{q^2} - 3\left(\frac{kT}{q^2}\right)^2\right]$$

$$f_{13} = \frac{kT}{q^2}\left[2 - \frac{kT}{q^2}\right] \quad f_{21} = \left[1 + 4\frac{kT}{q^2} - 4\left(\frac{kT}{q^2}\right)^2\right]$$

$$f_{22} = \left[1 + 2\frac{kT}{q^2} + \left(\frac{kT}{q^2}\right)^2 - 3\left(\frac{kT}{q^2}\right)^3\right] \quad f_{23} = \left(\frac{kT}{q^2}\right)^3 \left(1 - \frac{kT}{q^2}\right)$$

$$f_{31} = 2\frac{kT}{q^2}\left(1 - \frac{kT}{q^2}\right) \quad f_{32} = \frac{kT}{q^2}\left[1 - \left(\frac{kT}{q^2}\right)^2\right]$$

This equation represents a hierarchy of waves. Those of the higher order are important at early times and the disturbances propagate along characteristics of the highest order term, whereas at long times the lowest- order waves prevail. For a discussion of equations of this type see Whitham.[11] The speed of propagation of the disturbances at long times is seen to be $q\sqrt{(1+kT/q^2)/2}$.

VI. Acknowledgments

This research was carried out with the support of Contract AFOSR-87-0155, Office of Scientific Research, United States Air Force and of the Lockheed-California Company. We express our gratitude to the Submicron Systems Architecture Project, Department of Computer Science, Caltech, for providing computer facilities and expert advice.

VII. References

[1] Frisch, U., Hasslacher, B. and Pomeau, Y., "Lattice-gas automata for the Navier-Stokes equation", *Physics Review Letters*, Vol. 56, 1986, pp. 1505-1508.

[2] Carleman, T., *Problèmes mathématiques dans la théorie cinétique des gaz*, Almqvist and Wiksells, Uppsala, Sweden, 1957.

[3] Broadwell, J. E., "Study of Rarefied Shear Flow by the Discrete Velocity Method." *Journal of Fluid Mechanics*, Vol. 19, 1964, pp. 401-414.

[4] Broadwell, J. E., "Shock Structure in a Simple Discrete Velocity Gas", *Physics of Fluids*, Vol. 7, 1964, pp. 1243-1247.

[5] Gatignol, R., "Théorie Cinétique des Gaz à Répartition Discrète de Vitesses," *Lecture Notes in Physics*, Vol. 36, 1975, Springer-Verlag, Berlin.

[6] Bird, G. A., *Molecular Gasdynamics*, Clarendon, Oxford, UK, 1976.

[7] Hardy, J., de Pazzis, O., and Pomeau, Y., *Physical Review A*, Vol. 13, 1976, pp. 1949-1961.

[8] Frisch, U., d'Humières, D., Hasslacher, B., Lallemand, P., Pomeau, Y., and Rivet, J. -P., "Lattice Gas Hydrodynamics in Two and Three Dimensions", *Complex Systems*, Vol. 1, 1987, pp. 649-707.

[9] d'Humières, D., Lallemand, P., and Frisch, U., Lattice Gas Models for 3d Hydrodynamics, *Europhysics Letters*, Vol. 2, 1986, pp. 291-297.

[10] Coveney, P. V., "The Second Law of Thermodynamics: Entropy, Irreversibility and Dynamics", *Nature*, London, Vol. 333, 1988, pp. 409-415.

[11] Whitham, G. B., "Some Comments on Wave Propagation and Shock Wave Structure with Application to Magnetohydrodynamics", *Communications of Pure and Applied Mathematics*, Vol. XII, 1959, pp. 113 -158.

Direct Statistical Simulation Method and Master Kinetic Equation

M. S. Ivanov,* S. V. Rogasinsky,† and V. Ya. Rudyak*
USSR Academy of Sciences, Novosibirsk, USSR

Abstract

In this work the direct statistical simulation method for spatially uniform relaxation in rarefied gas is developed on the basis of a probabilistic interpretation for the integral representation of the master kinetic equation (Kats equation). It is shown that under certain conditions the conventional schemes for the relaxation simulation follow directly from this equation. A new accurate simulation scheme is proposed, the majorant frequency scheme, which requires a computing capacity that scales directly with the number of sampling particles. The relation between the solutions of the master kinetic equation of rarefied gas and the Boltzmann equation is studied for the uniform case. It is shown that the correlations occurring in the finite-number particle system affect significantly the statistical simulation results. The criterion for estimating the influence of such correlations during the computation process is suggested.

Introduction

At present, the direct statistical simulation method,[1] based on splitting up the evolution of a gas system in two stages, is widely used in the dynamics of rarefied gas. The method is realized in the following way. The flowfield computed is divided into cells of a finite size Δx and, according to the initial distribution function, N sampling particles are placed into each cell. Then the spatially-uniform relaxation stage and the stage of a free-molecular transition are successively carried out in all the cells. The free-molecular transition simulation can be performed without difficulties. In this case, the realization of the spatially-uniform relaxation stage is of primary importance.

Copyright © 1989 by the American Institute of Aeronautics and Astronautics, Inc. All rights reserved.
*Institute of Theoretical and Applied Mechanics.
†Computing Center.

For the simulation of collisional relaxation, the numerical schemes suggested in Refs. 1-4 are used. All these schemes are derived from heuristic considerations on the basis of physical understanding of the relaxation process in an actual gas; as a result there is no direct relation with the Boltzmann kinetic equation. The heuristic character of these schemes allow comparative analysis only qualitatively, with the use of the "Boltzmannian" collision frequency as the main criterion.[5-7] The stochastic process for an approximate solution of the Boltzmann equation is constructed in Ref. 8 using the Euler scheme for the Boltzmann spatially-uniform equation with its further randomization. In such an approach, for colliding particles the conservation laws are not valid, and this is the basic difference from the schemes given in Refs. 1-4.

It seems reasonable to consider the known numerical schemes for the statistical simulation of rarefied gas flows in light of a general theory of Monte Carlo techniques. Such a unified consideration enables one to carry out a comparative analysis to show the inner relation between these schemes and also justifies the use of various Monte Carlo weight techniques.[9]

Derivation of Simulation Technique from Master Kinetic Equation

In the construction of the Monte Carlo numerical technique we shall directly proceed from the master kinetic equation for the N-particle distribution function, which in the spatially-uniform case has the following form[10,11]

$$\frac{\partial}{\partial t} f_N(t,\bar{c}) = \frac{n}{N} \sum_{1 \leq i < j \leq N} \int_0^{2\pi} \int_0^{\infty} \left\{ f_N(t,\bar{c}'_{ij}) - f_N(t,\bar{c}) \right\} |\bar{v}_i - \bar{v}_j| \, b_{ij} \, db_{ij} \, d\varepsilon_{ij} \tag{1}$$

where $\bar{c} = (\bar{v}_1, \ldots \bar{v}_N)$ is a 3N-dimensional vector, $\bar{c}_{ij} = (\bar{v}_1, \ldots \bar{v}'_i, \ldots \bar{v}'_j, \ldots \bar{v}_N)$, $(\bar{v}_i, \bar{v}_j)$ and $(\bar{v}'_i, \bar{v}'_j)$ the velocities of a pair of particles (i, j) before and after the collision connected by the laws of impulse and energy conservation, b_{ij} and ε_{ij} the collisional parameters of a pair (i,j), and $\int f_N(t, \bar{c}) \, d\bar{c} = 1$ the numerical density.

With the initial condition taken into account Eq. (1) in its integral representation has the following form:[12]

$$\varphi(t,\bar{c}) = \int_0^t \int K_1(\bar{c}' \to \bar{c}) K_2(t' \to t \mid \bar{c}) \, \varphi(t',\bar{c}') \, d\bar{c}' \, dt' + \varphi_0(t,\bar{c}) \tag{2}$$

where $\varphi(t,\bar{c}) = \upsilon(\bar{c}) f_N(t,\bar{c})$ is the collision density.

$$\varphi_0(t,\bar{c}) = f_N(o,\bar{c})\upsilon(\bar{c})\exp\{-\upsilon(\bar{c})t\}$$

$$K_1(\bar{c}' \to \bar{c}) = \frac{n}{N}\sum_{i<j} \upsilon^{-1}(\bar{c})W(\bar{v}_i',\bar{v}_j'|\bar{v}_i,\bar{v}_j) \prod_{\substack{m \neq ij \\ m=1}}^{N} \delta(\bar{v}_m - \bar{v}_m') \quad (3)$$

$$W(\bar{v}_i,\bar{v}_j \to \bar{v}_i',\bar{v}_j')\, d\bar{v}_i'\, d\bar{v}_j' = g_{ij}\sigma(g_{ij},x_{ij})\sin x_{ij}\, dx_{ij}\, d\varepsilon_{ij}$$

$$K_2(t' \to t/\bar{c}) = \upsilon(\bar{c})\exp\{-\upsilon(\bar{c})(t-t')\} \quad (4)$$

$$\upsilon(\bar{c}) = \frac{n}{N}\sum_{i<j}\int W(\bar{v}_i,\bar{v}_j|\bar{v}_i',\bar{v}_j')\, d\bar{v}_i'\, d\bar{v}_j' = \frac{n}{N}\sum_{i<j} g_{ij}\sigma_t(g_{ij}) < \infty \quad (5)$$

$$\sigma(g_{ij}) = \int \sigma(g_{ij},x_{ij})\sin x_{ij}\, dx_{ij}\, d\varepsilon_{ij}\,, \quad g_{ij} = |\bar{v}_i - \bar{v}_j|$$

The probabilistic interpretation for the kernel and free term of Eq. (2) enables one to formulate the algorithm of the direct statistical simulation [Markov master process (MMP)] for evaluating the functionals of the master kinetic equation solution.[12]

Omitting the detailed description of the given algorithm note the following. In the N-particle system the transition from the $(t',\bar{c}')$ to $(t,\bar{c})$ state depends on the whole system state [see Eqs. (3 - 5)] at an instant t'. The collision itself is only realized by a single pair of selected particles (i, j). The computation of υ (c) requires about N^2 operations. For simplifying the simulation it is reasonable to consider such an approximate gas model where the time interval between collisions is only determined by the colliding pair. For such an approximate gas model an integral representation of the equation has the form[12]

$$\psi_B(t,\bar{c}) = \int_0^t \int K_B(t',\bar{c}' \to t,\bar{c})\,\psi_B(t',\bar{c}')\, d\bar{c}'\, dt' + \delta(t)f_N(o,\bar{c})$$

$$K_B(t',\bar{c}' \to t,\bar{c}) = \frac{n}{N}\sum_{i<j} W(\bar{v}_i',\bar{v}_j'|\bar{v}_i,\bar{v}_j)\upsilon^{-1}(\bar{c}')\delta(t-t'-\Delta t_{ij}(\bar{c}')) \cdot \prod_{\substack{m=1 \\ m \neq i,j}}^{N} \delta(\bar{v}_m - \bar{v}_m')$$

In an approximate and accurate model of a gas the value $\Delta t_{ij}(\bar{c}')$ is derived from the equality condition for the average number of collisions within the time interval (t',T). When the condition $\upsilon(\bar{c}')(T-t') \gg 1$ is satisfied, it follows that

$$\Delta t_{ij}(\bar{c}') = \frac{2}{N-1}\left\{ n g'_{ij}\ \sigma_t(g'_{ij}) \right\}^{-1}$$

Then the direct simulation process for the Eq. (1) coincides with the well known approximate Bird's scheme, the so-called "time-counter scheme." If Eq. (2) is considered for discrete times $t_N = n\Delta t$ then under the condition $\upsilon(\bar{c}')\Delta t \ll 1$ one can get the known approximate schemes given in Refs. 3-4.[12]

Thus, the approach given above enables one to obtain an accurate algorithm of the Markov master process (MMP) for the master kinetic equation and approximate algorithms of the Bird (time counter), Koura (collision frequency), and Belotserkovsky-Yanitsky (Bernoulli) schemes. It is well known that the approximate Bird scheme is the most efficient numerical algorithm. However, the scheme accuracy is determined by the inequality $\upsilon(\bar{c}')T \gg 1$. In the case of Maxwellian molecules an accurate algorithm of MMP compares favorably with the efficiency of Bird's algorithm, since the MMP does not require the recalculation of $\upsilon(\bar{c})$. Proceeding directly from the master kinetic equation one can construct a similar efficient and accurate algorithm.[13,14] To this end, for the total frequency of collisions $\upsilon(\bar{c})$ [see Eq. (5)] the υ_m-majorant is introduced and the master kinetic equation is then transformed to its integral form [Eq. (2)] with the subkernels

$$K_2^m(t' \to t/\bar{c}) = \upsilon_m \exp\{-\upsilon_m(t-t')\}$$

$$K_1^m(\bar{c}' \to \bar{c}) = \sum_{i<j} \frac{2}{N(N-1)} \left\{ \begin{array}{l} (1 - \dfrac{g'_{ij}\ \sigma_t(g'_{ij})}{[g\sigma_t(g)]_{max}})\ \delta(\bar{v}_i - \bar{v}'_i)\delta(\bar{v}_j - \bar{v}'_j) \\ + \dfrac{g'_{ij}\ \sigma_t(g'_{ij})}{[g\sigma_t(g)]_{max}} \cdot \dfrac{W(\bar{v}'_i, \bar{v}'_j / \bar{v}_i, \bar{v}_j)}{g_{ij}\ \sigma_t(g_{ij})} \end{array} \right\} \prod_{\substack{K=1 \\ K \neq i,j}}^{N} \delta(\bar{v}_K - \bar{v}'_K)$$

The simulation numerical scheme of this algorithm [majorant frequency scheme (MFS)] coincides with that for the MMP scheme but the colliding particle velocities at each transition $(\bar{c}' \to \bar{c})$ are varied with a probability $g'\sigma_t(g')/[g\ \sigma_t(g)]_{max}$ [see Eq. (6)].

It is worthwhile to note that the well-known method of maximum cross section[9] was used in Ref. 13 for constructing the "null-collision" numerical scheme. This scheme was substantiated on the heuristic level and only for the particular case of the hard sphere molecular model. The basis of the MFS method suggested in Ref. 12 is the generalization of the maximum cross section principle with its further randomization. In the particular case of the hard-sphere model these two approaches coincide, but the MFS technique is derived directly from the kinetic equation.

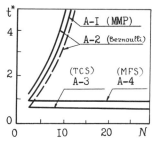

Fig. 1 Relative computation time as a function of the number of particles in the system for various schemes.

For the numerical evaluation of efficiencies for various numerical schemes of the direct statistical simulation method, the uniform relaxation for a number of initial data were computed. Figure 1 shows the relative computation time for an individual collision as a function of the number of particles N in the system $\bar{t}^* = t_{comp}/t_{comp}^{MFS}$ (t_{comp}^{MFS} is the time of computation according to the scheme given in Ref. 12). One can see that the MMP scheme[2] and Bernoulli testing scheme[4] are proportional to N^2, and the time counter scheme (TCS) and the MFS are linear with respect to N. The MFS is somewhat inferior to the TCS, but unlike the latter, it is an accurate scheme for the realization of the collisional relaxation process in a rarefied gas. This circumstance becomes fundamental in the case where the number of particles in a cell is small, i.e. $N_0 \lesssim 10$.

Comparisons to Boltzmann Equation Solution

In practical computations using the direct statistical simulation method a finite number of particles N is always used. Therefore, the correspondence of statistical simulation results to the Boltzmann equation solution is especially important.

To evaluate the dependence of the solution using Eq. (1) on N let us first derive the kinetic equation for a single-particle distribution function. To this end, let us introduce the S-particle distribution functions

$$f_s(\bar{v}_1,\ldots,\bar{v}_s,t) = \left(\frac{N}{V}\right)^s \int d\bar{v}_{s+1}\ldots d\bar{v}_N \, f_N$$

and integrate Eq. (1) over the phase variables of (N-1) particles. As a result, we get

$$\frac{\partial f_1}{\partial t} = \frac{N-1}{N}\int\int d\bar{\Omega}\, d\bar{v}_2 \, v_{12}(\tilde{f}_2\tilde{f}_1 - f_2 f_1) + \frac{N-1}{N}\int\int d\bar{\Omega}\, d\bar{v}_2 \, v_{12}(\tilde{g}_2 - g) \qquad (7)$$

At N→∞, if the molecular chaos condition is satisfied, this equation is reduced to the Boltzmann equation. In the general case, Eq. (7) differs substantially from the Boltzmann equation, since it depends on the variation of the two-particle pair correlation function $g = f_2 - f_1 f_1$. To get the equation for this function, let us integrate Eq. (1) over the phase variables of (N-2) particles and take into account only pair collisions. As a result, we get

$$\frac{\partial g}{\partial t} = \frac{1}{V} \int d\overline{\Omega} \, v_{12}(\tilde{g}-g + \tilde{f}_1 \tilde{f}_1 - f_1 f_1) -$$

$$\frac{1}{N} \cdot \sum_{i \neq j} \int d\overline{\Omega} \, d\overline{v}_3 \, v_{i3}[f_1(\tilde{v}_i) f_2(\tilde{v}_j) f_1(v_3) - f_1 f_1 f_1]$$

or

$$\frac{\partial}{\partial t} \cdot g = \frac{1}{V} L_2(g) - \frac{1}{N} Q(f_1) \tag{8}$$

The operator $L_2(g)$ can be represented in the form where

$$v = \frac{N}{V} \int d\overline{\Omega} \, v_{12} \quad ; \quad K_g = \frac{N}{V} \int d\overline{\Omega} \, v_{12} \tilde{g}$$

Then, the Eq. (8) takes the form

$$\frac{dg}{dt} = -\frac{v}{N} g + \frac{1}{N} K_g - \frac{1}{N} Q(f_1) \tag{9}$$

If the latter equation is integrated over time, for g one can get the second kind of Volterra integral equation, the solution for which can be written

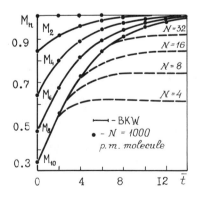

Fig. 2 Behavior of momenta.

with the use of the resolvent operator $\Gamma(t,t_1)$:

$$g(t) = g_0(t) + \frac{1}{N}\int_{t_0}^{t} \Gamma(t,t_1)g_0(t_1)\, dt_1$$

$$\Gamma(t,t_1) = \sum_{n=0}^{\infty} K_{n+1}(t,t_1)$$

$$g_0(t) = e^{-\frac{\nu}{N}(t-t_0)}g(t_0) - \frac{1}{N}\int_{t_0}^{t} e^{-\frac{\nu}{N}(t-t_1)}Q(f_1)\, dt_1 \tag{10}$$

where the iterated kernels K_{n+1} are determined by the recurrent formula

$$K_{n+1}(t,t_1) = \int_{t_1}^{t} K(t,t_2)\, K_n(t_2,t_1)\, dt_2$$

$$K(t,t_i) \equiv K_1(t,t_i) = \frac{1}{N}\exp\{-\frac{\nu}{N}(t-t_i)\}K$$

From the solution of Eq. (10) one can see that at any finite N and $t>t_0$ a two-particle function is nonmultiplicative for any initial condition $g(t_0)$ except for those that are locally-Maxwellian ($Q(f) = 0$). In those cases where there are no initial correlations, the $g(t)$ function evolution is determined by the terms that include the Q-operator. Whence, in particular, in an infinite system the molecular chaos state is conserved in time.

As a major approximation for all N particles, one can set $g(t) = g_0(t)$. Now, if one makes use of the mean-value theorem, for $g(t)$ we get the following estimate

$$g(t) = e^{-\frac{\nu}{N}(t-t_0)}\cdot g(t_0) + L[1-e^{-\frac{\nu}{N}(t-t_0)}] \tag{11}$$

where L is some function of the two-particle momenta.

According to Eq. (11), for a fixed instant $\tau = \upsilon(t-t_0)$, the larger the number of prarticles N, the less are the correlations. In a system of a finite number of particles, the correlations progress with the time growth and the correlations are larger for smaller number of partlcles.

The fact that in a system with a finite number of particles the correlation increases fast enough [even if $g(t_0) = 0$], results in the solutions of Eq. (1) being substantially different from the corresponding solutions of the Boltzmann equation. The level of this difference is shown by the behavior of momenta for a single-particle distribution function.

The numerical studies of the influence of the statistical dependence of particle velocities were carried out for a number of spatially-uniform problems. The majorant frequency numerical scheme was used in all the computations.

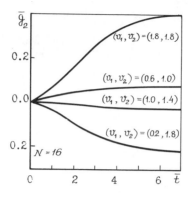

Fig. 3 Time behavior of normalized correlation function for various particle velocities.

The comparison with the exact BKW solution of the Boltzmann equation has shown that at N = 1000 the simulation results differed from the distribution function exact values by no more than 3% up to three thermal velocities. The higher-order momenta $M_{2\alpha}$ ($\alpha = 1,2,3,4,5$) at N=1000 are practically the same as the exact solution values. For small N-values and the sampling total amount being the same $L \cdot N = 10^6$ (L is the number of N-particle trajectories), the momenta behavior is given in Fig. 2. One can see that, in fact, for each value of N there is a time interval at which the simulation results coincide with the Boltzmann equation solution.

With the time growth the difference between the numerical data and exact values of momenta increases. Note that the time interval where these momenta coincide is larger for larger N. At large times the numerical momenta reach their asymptotic values that differ from the exact values.

Such behavior of the simulation results is directly related to the rise of a statistical dependence between the sample particles. The time behavior of the normalized value of the correlation function for various values of particle velocities

$$g_2(\tilde{v}_1,\tilde{v}_2)/f_1(\tilde{v}_1) \cdot f_2(\tilde{v}_2) \; ; \; \tilde{v}_i = v_i/v_0, \; v_0 = \sqrt{2RT}$$

is given in Fig. 3. Note that the computations of such correlation functions were never performed before. It is seen that correlations for groups of various velocities have a different behavior, but with the time growth all of them achieve their stationary values. In this case, the values depend essentially on the values of N (see Fig. 4) and decrease proportionally to 1/N, which is in agreement with theoretical predictions given earlier. Thus, the theoretical and numerical analyses carried out for velocity correlations in the N-particle model of a gas enable one to conclude that the they have a finite contribution to the macroparameters of a gas.

Figure 5 shows the correlation function behavior for the problem of a delta-peak [$f(0,\tilde{v}) = \delta(\sqrt{1.5})$] relaxation for the hard-sphere model and

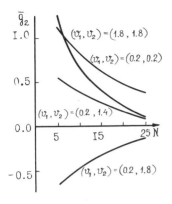

Fig. 4 Correlation function dependence on number of particles N.

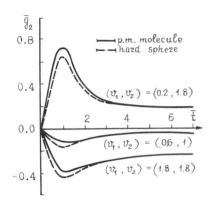

Fig. 5 Correlation function behavior for two molecular models.

for pseudo-Maxwellian molecules. The influence of the molecular model is observed only at small time values.

The computation of correlation functions is quite cumbersome and their values also depend considerably on the initial distribution form (compare Figs. 3 and 5). Therefore, it is difficult to use them in mass computations as a criterion for evaluating the relation with the Boltzmann equation. The correlation factor $\rho = <\bar{v}_i \bar{v}_j>$, used as such a criterion in Ref. 8, is conserved in time for our conservative numerical scheme.

The autocorrelation function $\mu(t) = <\bar{v}_i(0)\bar{v}_i(t)>$ is proposed as a criterion for the relation between the results of simulation and the Boltzmann equation solution. For the Boltzmannian gas ($N \to \infty$), $\mu(t)$ is approximately equal to $\exp(-t/A)$, where A depends on the model of intermolecular interaction. Such a behavior of $\mu(f)$ in a spatially-uniform gas is a consequence of the fluctuational-dissipative theorem.[15] At finite N for the pseudo-Maxwellian molecules the expression $\mu(t) = (1-N^{-1})\exp(-t/2)+N^{-1}$ is given in Ref. 16. The time behavior of $\ln \mu(t)$ for various N is given in Fig. 6. With the time growth, the deviation of $\mu_N(t)$ from $\mu(t)$ is observed and the asymptotical value N^{-1} is achieved.

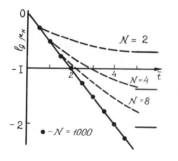

Fig. 6 Time behavior of ln μ(t) for various N.

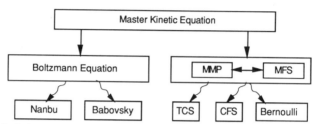

Fig. 7 Relations of the various numerical schemes with the master kinetic equation (→ = accurate scheme, ⤳ = approximate scheme).

It is worthwhile to mention that such a behavior of $\mu_N(t)$ does not depend on the initial data and is an intrinsic feature of the N-particle gas model. This fact was proved for a wide class of spatially-uniform problems, both with isotropic and nonisotropic distribution functions. If during the simulation process one computes $\mu_N(t)$, its deviation from the exponential attenuation indicates that the contribution of correlations is substantial. Thus, for the spatially uniform problems, $\mu_N(t)$ proves to be a universal criterion for evaluating the relation with the Boltzmann equation solution.

Conclusion

The construction of the direct statistical simulation technique directly from the master kinetic equation for the N-particle distribution function enabled us to establish the inner relation of the well known numerical schemes Markov master process (MMP),[2] time counter (TCS),[1] Bernoulli testing (Bernoulli scheme),[4] and majorant frequency (MFS).[12] The comparative analysis of the schemes mentioned, performed in Ref. 12, allows the presentation of their relations with the master kinetic equation as shown in Fig. 7.

The results of theoretical analysis of the influence of correlations in the finite particle-system and computational study of the statistical dependence of particles enable one to draw the conclusion that there is a finite contribution of this effect to the gas macroparameters.

References

[1] Bird, G. A., Molecular Gas Dynamics, Clarendon, Oxford, 1976.

[2] Denisik, S. A., Malama, J. G. O Lebedeve, S. N., and Polak, L. S., "Solution of Problems in Physical and Chemical Kinetics by Monte-Carlo Methods," In: Primenenie Vychisl. Matem. v Khim. i Fiz. Kinetike, Nauka, Moscow, 1969, pp. 179-231 (in Russian).

[3] Koura, K.O, "Transient Couette Flow of Rarefied Binary Gas Mixture," Physics of Fluids, Vol. 13, 1970, pp. 1457-1466.

[4] Belotserkovsky, O. M. and Yanitsky, V. E., "Statistical Method of Particles-in-Cells for solving Problems in Rarefied Gas Dynamics," Zh. Vychisl. Mat. Mat. Fiz., Vol. 15, No. 5, 1975, pp.1195-1208, and Vol. 15, No. 6, pp. 1553-1567 (in Russian).

[5] Bird, G. A., "Direct Simulation and the Boltzmann Equation," Physics of Fluids, Vol. 13, 1970, pp. 2677-2681.

[6] Belotserkovsky, O. M., Erofeev, A. I., and Yanitsky, V.E., "On a Non-Stationary Method of Direct Statistical Simulation of Rarefied Gas Flows," Zh. Vychisl. Mat. Mat. Fiz. , Vol. 20, 1980, pp. 1174-1204 (in Russian).

[7] Nanbu, K., "Theoretical Basis of the Direct Simulation Monte Carlo Method," In: Rarefied Gas Dynamics, V. Boffi and C. Cercignani, eds.,Vol. 1, B. G. Teubner, 1986, pp. 363-382.

[8] Nanbu, K., "Direct Simulation Scheme Derived from the Boltzmann Equation," Journal of the Physical Society of Japan, Vol. 49, 1980, pp. 2042-2049.

[9] Ermakov, S. M. and Mikhailov, G. A., Statistical Modelling, Nauka, Moscow, 1982.

[10] Kac, M., Probability and Related Topics in Physical Sciences, Interscience, London, 1959.

[11] Prigogin, I., Non-Equilibrium Statistical Mechanics, Interscience, London, 1966.

[12] Ivanov, M. S., Rogasinsky, S. V., "Comparative Analysis of Algorithms of Direct Statistical Simulation Method in Rarefied Gasdynamics," Zh. Vychisl. Mat. Mat. Fiz., Vol. 23, No. 7, 1988, pp. 1058-1070.

[13] Koura, K., "Null-Collision Technique in the Direct Simulation Monte Carlo Method," Physics of Fluids, Vol. 29, 1986.

[14] Rudyak, V. Ya., "On the Interrelation of the Master Kinetic Kac's Equation with Liouville and Boltzmann Equations," In: Abstracts of the All-Union Conference on Rarefied Gas Dynamics, Sverdlovsk, 1987, p. 29.

[15] Klimontovich, Yu. L., Statistical Physics, Nauka, Moscow, 1982 (in Russian).

[16] Nanbu, K., "Brownian Motion of Molecules in Pure Gases and Gas Mixtures," Journal of Chemical Physics, Vol. 82, No. 7, 1985, pp. 3329-3334.

Fractal Dimension of Particle Trajectories in Ehrenfest's Wind-Tree Model

Peter Mausbach*
Rheinisch Westfälische Technische Hochschule Aachen, Aachen, Federal Republic of Germany
and
Eckard Design GmbH, Cologne, Federal Republic of Germany

Abstract

Computer simulations of Ehrenfest's wind-tree model have been performed, and the scaling behavior of the particle trajectories has been studied via fractal analysis. The fractal dimension of $d_f > 2$ in the regime of "abnormal diffusion" represents a new class of scaling behavior as opposed to the fractal dimension of $d_f = 2$ for Brownian motion and the fractal dimensions of $d_f < 2$ that were observed in previous computer simulations of other fluid systems. Rather than giving support for the existence of a universal scaling exponent, the results imply that the fractal dimension is highly characteristic of the particular dynamics of a given system. In our simulations we also study for the first time the transition between the regimes of normal and abnormal diffusion.

Introduction

Over the past few years, the concept of fractal dimension has been shown to be a powerful tool for the classification of statistical phenomena with self-similar character. One intriguing example is the morphological analysis of particle trajectories in fluids. The total length of such trajectories is measured by using dividers

---

Copyright © 1989 by the American Institute of Aeronautics and Astronautics, Inc. All rights reserved.
*Research Engineer, Institute for Physical Chemistry.

of length ϵ. Stepping along the curves, the total length is

$$L(\epsilon) = C\epsilon^{-\alpha}. \qquad (1)$$

Here, the positive and finite Richardson coefficient α is the difference between fractal (d_f) and topological (d_t) dimension:

$$\alpha = d_f - d_t. \qquad (2)$$

For particle trajectories the topological dimension is $d_t=1$; hence, the fractal dimension is $d_f=1+\alpha$.

The first such studies of particle trajectories were carried out by Powles and Quirke for a Lennard-Jones liquid.[1] A fractal dimension of $d_f<2$ was found, in disagreement with the value of $d_f=2$ that is to be expected for strictly Brownian motion.[2] This observation initiated a vigorous debate.[3-7] By solving the generalized Langevin equation, it could finally be shown that Markovian trajectories do indeed yield a fractal dimension of $d_f=2$ if the stochastic forces are uncorrelated.[8] The empirical result $d_f<2$ therefore reflects a characteristic property of classical mechanical trajectories in real systems: because of memory effects the trajectories are not entirely stochastic even on relatively long time scales.

Particle motions that correspond to a fractal dimension of $d_f=2$ are governed by Einstein's classical diffusion equation. If the particle paths are very specifically mapped out in space, however, the average motion may well not obey Einstein's relation; then fractal dimensions of $d_f>2$ are possible. Ehrenfest's wind-tree model is an interesting example for this physical situation.

This model is a special case of the Lorentz gas which has been intensely investigated in the context of generalizations of the Boltzmann equation for dense fluids. In particular, the wind-tree model has figured as a testing ground for resummation procedures that are invoked to remove divergences in the calculation of density-dependent transport coefficients.[9]

The diffusion process in this model is anomalous in the sense that the mean square displacement is not linear in time; hence, a fractal dimension of $d_f>2$ can arise. By means of computer simulations, we have studied this peculiar phenomenon in detail. In particular, we have investigated the dependence of d_f on the length of the trajectories and the correlation with the tree density. Most importantly, our calculations cover the whole range of abnormal diffusion; i.e., they extend from the so-called diffusion-localization transition to the regime that marks the crossover to normal diffusion ($d_f < 2$).

Model

In the wind-tree model by Ehrenfest and Ehrenfest,[10] noninteracting point particles (the wind particles) move between square-shaped scatterers of side length σ (the trees) that are randomly distributed over a two-dimensional plane. The wind particles can adopt no more than four different velocities, via specular reflection at the edges of the nonpenetrable trees.

If $Q^N = \{Q_1, \ldots, Q_N\}$ denotes the positions of the N scatterers and $\underline{x} = (\underline{r}, \underline{v})$ is the phase space of the wind particles, the probability distribution for a given wind particle and N scatterers is given by

$$\varrho(\underline{r}, Q^N) = Z_N^{-1} \exp[-H(\underline{r}, Q^N)] \qquad (3)$$

with the partition function

$$Z_N = \int \ldots \int d\underline{r}\, dQ^N \exp[-H(\underline{r}, Q^N)] \quad . \qquad (4)$$

The Hamiltonian can be written

$$H(\underline{r}, Q^N) = \sum_k W(\underline{r} - Q_k) + \sum_{k<l} W^*(Q_k - Q_l). \qquad (5)$$

In particular, we have

$$W(\underline{r} - Q_k) = \begin{cases} \infty, & \text{if } \underline{r} \text{ is inside of scatterer k} \\ 0, & \text{otherwise} \end{cases} \qquad (6)$$

for the particle-tree interaction and

$$W^*(Q_k - Q_l) = \begin{cases} \infty, & \text{if scatterer k and l overlap} \\ 0, & \text{otherwise} \end{cases} \qquad (7)$$

for the tree-tree interaction.

FRACTAL DIMENSION OF PARTICLE TRAJECTORIES

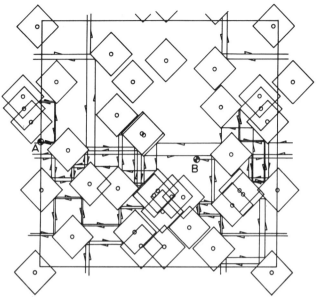

Fig.1 Typical particle trajectory for a wind-tree model with periodic boundary conditions. The system contains 32 trees, and A and B are the initial and final particle positions, respectively, for a trajectory involving 92 collisions.

Here we are mainly interested in models where the wind-trees are allowed to overlap. A typical particle trajectory for this case is shown in Fig. 1. Some time ago it was shown by Hauge and Cohen[9] that the diffusion coefficient for such models with overlapping trees vanishes because of the slow, less than linear growth of the mean square displacement with time. This unusual feature of the wind-tree model (previously confirmed by the computer simulations of Wood and Lado[11]) can be related to the anomalous behavior of the d_f as discussed below.

Trajectory Analysis

Consider some spatial curve $\underline{r}(t)$ that represents the microscopic path of a particle in a fluid. It is reasonable to determine the length of this spatial curve by means of a sequence of dividers of constant length ϵ that map out the actual trajectory of the particle. The total number of segments is $n_P(\epsilon)$; hence, the trajectory length is $L^P = n_P(\epsilon)\epsilon$. Averaging over an ensemble of trajectories, we obtain

$$L(\epsilon) = \langle L^P(\epsilon) \rangle = \langle n_p(\epsilon) \rangle \epsilon. \tag{8}$$

The particle gets displaced over an area of ϵ^2 during a time interval of $\tau_P(\epsilon)$. The mean square displacement is then given by

$$\langle |\underline{r}(\tau(\epsilon)) - \underline{r}(0)|^2 \rangle = \epsilon^2 \qquad (9)$$

Because of $\tau(\epsilon) = \langle \tau_P(\epsilon) \rangle$, we have $\langle n_P(\epsilon) \rangle = T/\tau(\epsilon)$, where T denotes observation time. By insertion of Eq.(8), we subsequently have

$$L(\epsilon) = \frac{T}{\tau(\epsilon)} \epsilon . \qquad (10)$$

It is illuminating to consider two limiting cases of this scenario. First, for dividers of infinitesimal length $d\epsilon$, the mean contour length is

$$L_c = \lim_{\epsilon \to d\epsilon} L(\epsilon) = \langle L^P(d\epsilon) \rangle.$$

For the mean square displacement we have in this case $(\tau(\epsilon) \to 0)$

$$\epsilon^2 \sim \tau^2(\epsilon).$$

Under substitution of Eqs. (1) and (10), we find that the fractal dimension of the particle trajectory is $d_f = 1$ ("free-flight limit").

In the other limiting case, the steps are large $(\epsilon \gg 0)$, and Einstein's diffusion equation is fulfilled:

$$\epsilon^2 = 2dD\tau(\epsilon),$$

where D is the usual diffusion constant and d denotes the euclidean dimension of the system of interest. Together with Eqs. (10) and (1), this yields $d_f = 2$, which was invoked by Mandelbrot for description of ideal Brownian motion ("diffusion limit").

Thus, by virtue of the varying step size ϵ, measurements of trajectory lengths establish a transition from a dynamic regime ($d_f = 1$) to a stochastic regime ($d_f = 2$). In the case of the wind-tree model, we now have

$$\epsilon^2 \sim \tau^\gamma(\epsilon)$$

with $\gamma < 1$ for long times, and it is immediately clear that $d_f > 2$.

In general, the d_f can be defined as a function of step size via

$$1 - d_f(\varepsilon) = \frac{d \log[L(\varepsilon)]}{d \log \varepsilon} \quad . \tag{11}$$

This "finite fractal analysis" allows to establish a connection with the fractal dimension that is directly obtainable from a computer experiment. Using the identity

$$\langle |\underline{r}(t) - \underline{r}(0)|^2 \rangle = \int_0^t dt' \int_{-t'}^{t-t'} ds \, \langle \underline{v}(s) \cdot \underline{v}(0) \rangle \tag{12}$$

with $\langle \underline{v}(s)\underline{v}(0) \rangle$ being the velocity autocorrelation function (vaf), it is easy to show that

$$\varepsilon^2 = 2\tau \int_0^\tau ds \, (1-s/\tau) \, \langle \underline{v}(s) \cdot \underline{v}(0) \rangle \quad . \tag{13}$$

From Eqs. (10), (11), and (13), we than have for finite time spans τ the exact scaling relation

$$d_f = \frac{d \log \tau}{d \log \varepsilon} = $$
$$= 2\left[1 - \int_0^\tau ds \, s \, \langle \underline{v}(s) \cdot \underline{v}(0) \rangle / \tau \int_0^\tau ds \, \langle \underline{v}(s) \cdot \underline{v}(0) \rangle \right] \quad . \tag{14}$$

If the vaf is known, it is hence possible to determine the d_f for any intermediate time scale. Moreover, Eq. (14) provides a measure for the approach of the diffusion limit because of the finite temporal range of the vaf. For dilute gases in the Boltzmann approximation, the vaf is given by $\exp(-\text{constant} \cdot t)$, and the fractal dimension becomes

$$d_f = 2(t^* + \exp(-t^*) - 1)/[t^*(1 - \exp(-t^*))] \tag{15}$$

with t^* denoting dimensionless time. This is a monotonous function between 1 and 2 indeed. Anomalous behavior can arise not only with respect to the transition from free-flight to the diffusion regime but also with respect to the asymptotic value in the diffusion limit. This is shown in the next section.

Results

To evaluate and test the trajectory analysis procedure, we first performed a standard molecular dynamics simulation of a three-dimensional Lennard-Jones system of 216 particles at a reduced density 0.83 and a temperature of 108°K. The trajectory length is computed by stepping along the particle path from its start point to that particular point on the path that is separated from the true endpoint by a distance of less than ϵ. The sum of all steps plus the fraction of ϵ at the end of the trajectory yields the total length $L^p(\epsilon)$. The function $L(\epsilon)$ is shown in Fig. 2 in bilogarithmic representation (lower curve in figure). The indicated slopes of the curve correspond to the α for the two limiting cases discussed above: $\alpha=0$ (or $d_f=1$) and $\alpha=0.84$ (or $d_f=1.84$) for small and large ϵ, respectively.

The latter d_f is in good accord with the result of Powles and Quirke,[1] who obtained a value of $d_f=1.65$ for a Lennard-Jones system that was less dense. As mentioned, the deviation from $d_f=2$ has to be attributed to long-time memory effects in either case. The crossover of α for intermediate ϵ is not pronounced and makes it difficult to assess the precise transition point between the dynamic and stochastic regimes.

We now turn to simulations of the wind-tree model. We introduce a Boltzmann mean free path of $l=1/(\sqrt{2}n\sigma)$, where σ is the side length of the wind-trees and $n=N/V$ is the number density within an area V. We further define a dimensionless area variable

$$\eta = V/(N\sigma^2) . \qquad (16)$$

For a sequence of C_k scattering configurations $Q^N{}_\mu=\{Q_{\mu 1},\ldots,Q_{\mu N}\}$, ($\mu=1,\ldots,C_k$), the tree positions are sampled randomly, uniformly, and independently and after the square areas V are transformed into unit squares by means of $q_i=Q_i/\sqrt{V}$. After assigning some random initial position $r_{\mu\delta}$ and initial velocity $v_{\mu\delta}$ to the scattering particle δ ($\delta=1,\ldots,C_l$) in a given configuration μ, the trajectory is propagated from collision to collision according to the specific Hamiltonian and under proper account of periodic boundary conditions.

We first investigate the influence of the trajectory length by performing two simulation series at an inverse density of $\eta=2$ and for $N=512$ scatterers and for 20 trajectories each. The number of collisions is 5000 in the one case and, actually, by extending the same simulations, 20000 in the other. The results of the fractal analysis,

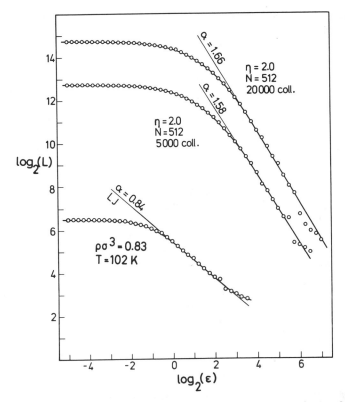

Fig.2 Log-log-plot of trajectory length L vs. step size ϵ for a 3-D Lennard-Jones system with $\rho\sigma^3=0.83$ and $T=102°K$ (lower curve), and for two wind-tree systems with 512 trees at $\eta=2.0$ and simulation lengths of 5000 and 20000 collisions (upper curves).

done along the lines described for the Lennard-Jones system, is shown in Fig. 2.

The obtained fractal dimensions are $d_f=2.58$ ($\alpha=1.58$) and $d_f=2.66$ ($\alpha=1.66$) for 5000 and 20000 collisions, respectively. The slight increase reflects the slow approach of the diffusion limit.[3] The result is a long way from the Mandelbrot value of $d_f=2$ and the results of some other recent computer simulations that caused considerable stir.[3-7]

The vaf is the key to a proper understanding of this phenomenon and will be discussed on the basis of the exact scaling relation Eq. (14). The trajectories that were examined in Refs. 3-7 all belong to systems whose fractal dimension grows continuously between $d_f=1$ and $d_f=2$. If the vaf is positive for all possible times, exactly this behavior is expected from Eq. (14). This is the case, e.g., for the ideal gas. In dense fluids, on the other hand, the main decay of the vaf is superimposed by rapid oscillations, partially into the negative, that are due to backscattering of reference particle in the confining cage of its neighbors. The balance between positive and negative contributions to the integrals in Eq. (14), connected with the specific shape of the vaf, thus determines the behavior of the fractal dimension that is encountered here. Under most conditions, however, the positive contributions are expected to dominate, last but not least because the vaf is positive in the regime of the well-known long-time tails with its extremely slow decay functions of $t^{-d/2}$ (where d is the dimensionality of the system). Nevertheless, for intermediate times it may still be possible to observe a fractal dimension of $d_f>2$. We note that the transition between $d_f=1$ and $d_f=2$ is in this case not monotonous. This is also discussed by Brooks III for systems of simple dense liquids.[7]

A somewhat different situation is posed by molecular trajectories in liquid water. Because of the peculiar nature of the hydrogen bond network, these trajectories are especially space filling, and one finds a fractal dimension of $d_f>2$.[12] In the diffusion limit, however, even this kind of system approaches the Mandelbrot value of $d_f=2$. As outlined, this is not the case for the wind-tree model where we have $d_f>2$ even in the diffusion limit. This is only possible if in Eq. (14)

$$\lim_{\tau \to \infty} \int_0^\tau ds \, \langle \underline{v}(s) \cdot \underline{v}(0) \rangle$$

overcompensates the contribution of $\lim_{\tau \to \infty} \tau^{-1}=0$. Since we can here invoke the definition

$$D = \lim_{\tau \to \infty} \frac{1}{2} \int_0^\tau ds \, \langle \underline{v}(s) \cdot \underline{v}(0) \rangle \qquad (17)$$

and since the diffusion coefficient vanishes for the wind-tree model, this is indeed the case. However, because of lack of knowledge concerning the functional dependence of $D(\tau)$, this limiting value cannot really be calculated.

We performed additional simulations to study the exact limits of these abnormalities in the wind-tree model. These simulations, again for 512 scatterers and 5000 collisions, were initiated in the vicinity of the percolation transi-

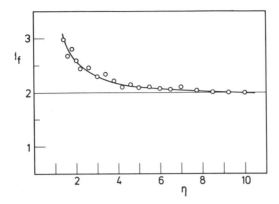

Fig.3 Fractal dimension d_f of trajectories in the wind-tree model from the diffusion-localization transition (inverse density $\eta \approx 1.3$) to the regime of normal diffusion ($\eta > 9$).

tion of the tree network that defines the end of the diffusive regime and that results in particle trapping. For a hard disk system such a transition is expected at an inverse density of about $\eta = 1.3$.[11] A precise calculation of this transition point for the wind-tree model is currently underway on the basis of critical percolation exponents.[13] For the purposes of the present study, we did the first simulations at $\eta = 1.4$ and then gradually increased η up to a value of $\eta = 10$. For wind-tree systems at low density (large η), there is a finite chance to encounter pathological (collisionless) trajectories where a channel of minimum width $\sqrt{2}\sigma$ spans the area V in x- or y-direction without hitting any trees. Wood and Lado[11] calculated this probability to be $\exp(-\sqrt{2N/\eta})$; i.e., for sufficiently large particle number N these collisionless trajectories can practically be neglected.

The fractal dimensions found for this series of simulations are depicted in Fig. 3. Close to the percolation transition the fractal dimension is almost $d_f = 3$. Deeper penetration into percolation region is naturally difficult because the wind particles get easily trapped in cages of trees. For increasing η the fractal dimension decreases approximately exponentially. Although we find a fractal dimension of slightly more than $d_f = 2$ even at $\eta = 9$, it can be estimated from the figure that the transition from the regime of abnormal diffusion ($d_f > 2$) to the regime of normal diffusion ($d_f < 2$) occurs in fact at about this density. This finding is in excellent agreement with the previous rough estimate of Wood and Lado of $\eta = 10$.[11]

Conclusions

The present study underlines the great value of the concept of fractal geometry for the understanding of the dynamics of many particle systems. It is evident that there is no universal scaling exponent (Richardson coefficient) as originally conjectured by Powles and Quirke.[1] In general, the relevant scaling factor is also not an integer. Rather, this quantity depends strongly on the characteristic microdynamics of the system under study. In fact, the fractal dimensionality of particle trajectories exhibits an astounding diversity: In addition to systems with $d_f < 2$[1,3-6,8] and systems that approach the limiting value of $d_f = 2$ from above,[7,12] there are also trajectories that display $d_f > 2$ even in the diffusion limit. This is a clear-cut manifestation of long-time memory effects. Interrelations with hydrodynamic effects are certainly an issue to be pursued.

The fractal behavior on intermediate time scales can be studied on the basis of vafs as given by Eq. (14). In this context, Toxvaerd recently derived from Eq. (14) another important relation [Eq. (8) in Ref. 14] that was subsequently criticized by Powles and Fowler.[15] It was claimed that Toxvaerd's relation violates the required behavior in the free-flight limit.[15] Apparently, this reflects nothing but a misunderstanding. In the denominator of Eq. (14), Toxvaerd substitutes the integral by the definition for the diffusion coefficient from Eq. (17). Not surprisingly, the obtained relation can be successfully applied only in the diffusion limit. The criticism of Powles and Fowler, however, does not affect at all the basic validity of Eq. (14) also in the free-flight limit.

Finally, our analysis illustrates that the transition from abnormal to normal diffusion occurs at an inverse density of about $\eta = 9$.

Study of the scaling behavior of particle trajectories in fluids thus contributes significantly to a better comprehension of molecular dynamics phenomena. The approach can be profitable even in the life sciences as is illustrated by a recent investigation of the Brownian motion of bacteriophage T4.[16]

References

[1] Powles, J. G. and Quirke, N., "Fractal Geometry and Brownian Motion," <u>Physics Review Letters</u>, Vol. 52, 1984, pp. 1571-1574.

[2] Mandelbrot, B. B., <u>The Fractal Geometry of Nature</u>, Freeman, San Francisco, CA, 1982.

[3]Powles, J. G., "Fractal Analysis of Molecular Motion in Liquids," Physics Letters, Vol. 107A, 1985, pp. 403-405.

[4]Rapaport, D. C., "Fractal Dimensionality of Brownian Motion," Physics Review Letters, Vol. 53, 1984, p. 1965.

[5]Rapaport, D. C., "The Fractal Nature of Molecular Trajectories in Fluids," Journal of Statistical Physics, Vol. 40, 1985, pp. 751-758.

[6]Erpenbeck, D. C. and Cohen, E. G. D., "Note on the Fractal Dimension of Hard Sphere Trajectories," Journal of Statistical Physics, Vol. 43, 1986, pp. 343-347.

[7]Brooks III, C. L., private communication, 1987.

[8]Toxvaerd, S., "Solution of the Generalized Langevin Equation," Journal of Chemical Physics, Vol. 82, 1985, pp. 5658-5661.

[9]Hauge, E. H. and Cohen, E. G. D., "Normal and Abnormal Diffusion in Ehrenfest's Wind-Tree Model," Journal of Mathematical Physics (New York), Vol. 10, 1969, pp. 397-414.

[10]Ehrenfest, P., Collected Scientific Papers, North-Holland, Amsterdam, 1959, p. 229.

[11]Wood, W. W. and Lado, F., "Monte Carlo Calculation of Normal and Abnormal Diffusion in Ehrenfest's Wind-Tree Model," Journal of Computational Physics, Vol. 7, 1971, pp. 528-546.

[12]Mausbach, P., "Teilchentrajektorien mit fraktalen Dimensionen $d_f > 2$", Zeitschrift für Angewandte Mathematik und Mechanik, Vol. 68, 1988, pp. T465-T467.

[13]Mausbach, P., private communication, 1988.

[14]Toxvaerd, S., "Fractal Dimension of Classical Mechanical Trajectories," Physics Letters, Vol. 114A, 1986, pp. 159-160.

[15]Powles, J. G. and Fowler, R. F., "Fractal Properties of Model Trajectories with Exponential Velocity Autocorrelation Functions," Physica, Vol. 141A, 1987, pp. 318-334.

[16]Matsuura, S., Tsurumi, S., and Nobuhisa, I., "Cross-over Behavior for Brownian Motion," Journal of Chemical Physics, Vol. 84, 1985, pp. 539-540.

Scaling Rules and Time Averaging in Molecular Dynamics Computations of Transport Properties

I. Greber*
Case Western Reserve University, Cleveland, Ohio
and
H. Wachman†
Massachusetts Institute of Technology, Cambridge, Massachusetts

ABSTRACT

A scheme of dimensional analysis, coupled with time averaging, has been applied to molecular dynamics calculations, for the purpose of exploring the lower limits of particle number and collision number required to model the behavior of a gas. Although geometric similarity cannot be preserved, the Knudsen number satisfies essential scaling criteria, and spacing ratios (e.g., particle volume to container volume) ensure that the model emulates a dispersed medium. The scheme is tested on models of heat conduction, Couette flow, and diffusion. Numerical results obtained compare satisfactorily with kinetic theory.

INTRODUCTION

The major emphasis of this paper is to determine how application of dimensional analysis and time averaging in molecular dynamics calculations can be used to render results obtained with small numbers of particles, applicable to the behavior of bodies of gases. The calculations examine extreme cases of very small numbers of closely spaced particles, and are intended to establish useful limits which can allow practical gas values to be obtained. We have used transport phenomena as examples of the types of problems to which the approach taken here should be appli-

---

Copyright © 1989 by the American Institute of Aeronautics and Astronautics, Inc. All rights reserved.
*Professor, Department of Mechanical and Aerospace Engineering.
†Professor, Department of Aeronautics and Astronautics.

cable, and have tested the validity of the approach in a determination of property profiles and transport coefficients in one-dimensional heat transfer, and two-dimensional Couette flow, and in a few preliminary calculations in one-dimensional diffusion.

BASIC MODEL

The basic approach in molecular dynamics calculations is that of kinetic theory, in which collisions between gas atoms determine the average behavior of a gas. A general review of work on molecular dynamics calculations has been given by Evans and Hoover[1].

In the present computations the gas is modelled as a set of interacting hard elastic spheres of specified masses and diameters. The initial locations and velocities of the spheres are given. The rules for their mutual interactions are those of classical mechanics with only binary collisions being considered. Interactions with surfaces are specified in ways which are consistent with the phenomenon under consideration. In cases where the interactions with surfaces are also elastic the system is entirely deterministic.

We have treated only cases with a hard sphere potential, hence all interactions are instantaneous with no interaction between collisions. Once initial positions and velocities are chosen, one can determine when the next collision will occur and identify which partners (which can be particle-particle or particle-wall) will participate in it, by computing the times of binary collision and particle-wall collision for all particles in the "gas" and selecting the collision that occurs soonest. The outcome of the next collision is determined, and the velocities of the colliding partners are revised to establish a new set of velocities of the system of particles. The new position of each particle in the system is then computed from its velocity and the time that it travels at that velocity. The process is repeated for as many collisions as desired. The calculations are shortened considerably because new calculations are needed only for collisions involving the just-colliding pair, and much of the time there is no need to record intermediate locations and unchanged velocities.

In all of the calculations presented here the computational domain is a cube. In terms of the calculations the differences in the transport phenomena that are treated are differences in boundary conditions; hence, the calculations are essentially identical, except for the information which is extracted from the results. Fig. 1 is a sketch depicting the computational domain and the boundary conditions for the several problems treated here.

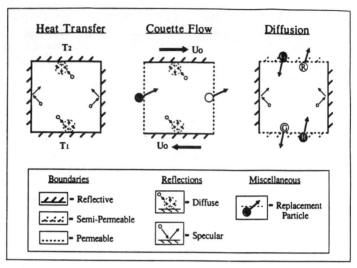

Fig. 1 Boundary condition for transport process computations.

In the heat transfer and Couette flow calculations two parallel walls ("upper" and "lower") represent the surfaces that participate in the transport processes. They may be fixed or moving surfaces at equal or different temperatures. Particles that collide with the upper or lower wall are re-emitted diffusely and are fully accommodated to the properties of the wall.

In the mass diffusion calculations the upper and lower walls are treated as sources of two different kinds of distinguishable particles, red and green. The lower wall reflects red particles specularly, but is porous to green particles; when the lower wall passes a green particle out of the computational domain, a green particle with identical dynamic properties is admitted into the computational domain through the upper wall. The action of these two walls towards green particles is the reverse of that described for red particles.

In the Couette flow calculations the upper and lower walls move in opposite directions at the same speeds. A symmetry condition is imposed at the upstream and downstream computational boundaries so that a particle leaving at either boundary is replaced by a particle of identical properties at the opposite boundary, hence the gas is essentially infinite in the directions of motion. Collisions at the other two walls of the cube are specular to model an infinite lateral region in these directions also.

TIME AVERAGING

Time averaging is a process by which the properties of the gas are determined from the fraction of the total time of an "experiment" during which specified volume elements in coordinates and velocity space are occupied by particles. Time averaging extends the ergodic idea that in equilibrium the instantaneous velocity distribution of all particles and the temporal velocity distribution of a single particle approach each other as observation time tends to infinity. Accordingly, if we let t_i be the time that the ith particle spends in a volume of phase space (coordinate space, velocity space, or a combination), q_i be the value of the property carried by the particle, and τ the total time of observation, then we define an extensive time average, Q, as

$$Q = (1/\tau) \sum (t_i q_i) \qquad (1)$$

The corresponding intensive average is obtained on dividing Q by the total mass, volume, or number of particles as needed.

For a gas in equilibrium the distribution function as usually defined indeed represents the time-average probability of the occupation of volume elements in velocity and coordinate space. For a steady-state process, time averaging is applied locally in the recognition that the asymptotic behavior is time-independent on the microscopic scale. In these cases time averaging is straightforward, and results in a computationally shorter effective approach to the steady state. If one is interested in the time evolution of a process, or if the process is inherently unsteady, then time averaging requires more care. The averages should be carried out over time intervals that are long enough to give useful average information and short enough so that the global properties do not change much over the time averaging interval.

SCALING

Dimensional analysis provides a useful basis for scaling, even though it is not rigorously applicable because one cannot maintain geometric similarity between the computational model and the gas. Exact geometric similarity would require that the number of particles within the computational domain be the same in the model as in the gas. With geometric similarity any gas property Q can be expressed functionally as

$$Q = f(n,d,L,m,u,etc.) \quad \text{or} \quad Q = f(s,d,L,m,u,etc.) \qquad (2)$$

where n is the number of particles per unit volume, d a characteristic particle diameter, L a characteristic length (for example, vehicle size or enclosure size), m the particle mass, u a characteristic velocity, s the spacing between centers of particles, and etc. refers to other variables such as additional masses for a mixture. Non-dimensionally, the functional relationship becomes

$$Q^* = f(nd^3, d/L, etc.) \quad \text{or} \quad Q^* = f(s/d, s/L, etc.) \qquad (3)$$

where Q^* is the non-dimensional Q. Dynamically similar behavior occurs if the non-dimensional variables are the same in the model as they are in the gas.

If one uses a small number density to model a large one, then the ratios s/d and s/L cannot both be kept the same in the model and in the gas; that is, geometric similarity with respect to particle diameter or spacing cannot both be maintained. For some properties this modeling deficiency is not important; for others, the results must be interpreted in the light of this deficiency.

For many problems the crucial non-dimensional parameter is the Knudsen number, Kn. As usually defined, the Knudsen number is the ratio of the molecular mean free path (dimensionally $1/nd^2$) to some relevant characteristic length, L. For scaling purposes the Knudsen number may be considered in terms of other properties, e.g., a ratio of the frequency of gas-surface to gas-gas collisions, a ratio of the total surface area of the gas particles to that of the surfaces that surround the gas and with which the molecules may collide, or a measure of the probability that a particle that is emitted from a body moving through a gas will return to that body after a single gas-gas collision.

For problems in which Kn is the major parameter, equation (3) expressed as a function of Knudsen number, simplifies to

$$Q^* = f(Lnd^2) \quad \text{or} \quad Q^* = f(L^2/Nd^2) \qquad (4a)$$

where $n = N/L^3$, and

$$Q^* = f(Kn) \qquad (4b)$$

If one chooses a value for the number of particles in the computational domain, then maintaining the Knudsen number, as defined in eqn. (4), the same in the model as in the gas, prescribes how one should decrease the body-diameter ratio, L/d, to compensate for the smaller number of particles in the model than in the equivalent region of the gas. In this process the particle linear spacing ratio, s/d (s is center-to-center distance), also decreases. When the Knudsen number is the same for the model and the gas, then these length ratios are related to the relative particle numbers by

$$(s/d)/(s/d)_0 = (N/N_0)^{1/6} \qquad (5)$$

$$(L/d)/(L/d)_0 = (N/N_0)^{1/2} \qquad (6)$$

where the subscript 0 denotes the gas value.

Although the spacing ratio, s/d, is thus determined by the Knudsen number scaling criterion, it is important to recognize that an arbitrary value of spacing ratio is not acceptable. The spacing between molecules in the model must be sufficient to allow molecular motion within the computational boundary to occur in a way that is consistent with the general model of a gas. The requirement is that the unoccupied volume in the model be sufficiently greater than the volume occupied by the particles themselves to maintain mobility associated with a gas (rather than with a liquid). A useful way to characterize the free volume within the computational boundaries is as a ratio, VR, between the total volume L^3 and the occupied volume $(N\pi d^3)/6$; the occupied volume represents the co-volume, b, in the van der Waals equation. From the definition of Kn, and the mean free path we find for VR,

$$VR = (6/\pi)(L^3/Nd^3) = 17.9(NKn^3)^{0.5} \qquad (7)$$

Maintaining NKn^3 sufficiently larger than unity preserves the gaseous character of the model and provides a useful way of ascribing gas properties to a computational model in which geometric similarity has not been preserved. Fortunately, numerical experiments show that this is not as restrictive a requirement as one might conjecture intuitively.

The adequacy of Knudsen number as the dominant scaling parameter can be partially tested by numerical experiments. One measure is the preservation of the particle-particle to particle-wall collision frequency ratio, notwithstanding the fact that s/d is unscaled. Perfect scaling implies that

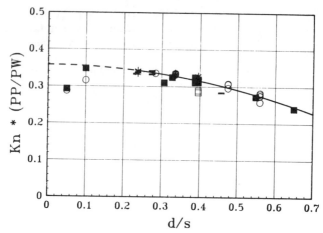

Fig. 2 Effect of spacing ratio on scaled collision frequency ratio. The ordinate is the product of the Knudsen number based on the Maxwell mean free path, and computed values of the ratio of particle-particle/particle-wall collision frequency. Values shown are from all three transport processes. The several symbols are for different processes and different numbers of particles.

he collision frequency ratio should vary inversely as the Knudsen number, independently of the spacing ratio. Numerical experiments were made, purposely carrying the scaling criteria to extremes. As few as 8 particles were used, with a smallest spacing ratio of 1.5! This value of spacing ratio should be considered in the context of the motion of a particle through the space formed by four particles in a square array, in which case the minimum spacing ratio must $s/d = \sqrt{2}$. Results of these numerical experiments are shown in Fig. 2, as a plot of the collision frequency ratio multiplied by the Knudsen number vs. d/s, so that extrapolation to d/s = 0 gives the limiting value for infinite spacing ratio (point particles). The points plotted in Fig. 2 are from computations of all three transport processes. They indicate that the extrapolated limit is approached quite closely even for modest spacing ratios. For example, s/d = 2.5 gives a value within 10% of the limit, as long as the number of particles is not excessively small, that is, as long as the Knudsen number is not too large. A few data points in the figure, at small d/s lie below the indicated curve. These were for small numbers of particles and small d/s, with corresponding nominal Knudsen numbers larger than ten. The scaling fails for large Knudsen numbers because for an enclosed set of particles, wall collisions then dominate and the absolute rate of particle-particle collisions becomes vanishingly small (consider the limit of one particle in a box) and inconsistent with the specified value of the Knudsen number. In this limit the product of Knudsen number and collision ratio tends to zero rather than to the value obtained under conditions where the collision frequency ratio is intact.

These numerical experiments suggest that models, costructed of very small numbers of rather closely packed particles and scaled according to Knudsen number, behave in a manner which is consistent with the behavior of a gas. This will also be seen in the transport phenomena results.

HEAT CONDUCTION

The model is of heat conduction between two parallel plates at temperature T_1 and T_2 (Fig. 1). Molecules rebound from the plates diffusely, with complete accommodation. This is pictured as though the outer surface of the plate were exposed to an infinite gas in equilibrium at the plate temperature, therefore rebounding molecules are assigned velocity components parallel to the surface randomly within a Maxwellian distribution, and normal components randomly within a Maxwellian distribution weighted by the normal component of velocity. Reflections from the computational boundaries of the model are specular.

Information collected in these computations includes the spatial distribution of properties (temperature and density) and the heat transfer rate at the upper and lower plates. The heat transfer rate is computed from the time averaged energy exchange at the wall as follows:

$$q = (1/At) \sum (E_r - E_i); \quad E = (m/2)V^2 \qquad (8)$$

when subscripts r and i denote re-emitted and incident, respectively. The conductivity coefficient can be deduced by taking the ratio of the heat transfer to the overall temperature gradient (the temperature difference divided by the distance between the walls with no correciton for temperature jump) at different Knudsen numbers, and finding the limit as the Knudsen number tends to zero.

The computations were all performed for a non-dimensional model, using as the length scale the enclosure side, and as the velocity scale the most probable speed of the Maxwellian distribution having the same total energy as the gas. The initial conditions were uniform speed and random directions. The conductivity coefficient of a gas can be found from:

$$\bar{K} = K d^2 (m/Tk^3)^{0.5} = 1.8 K^* / NKn \qquad (9)$$

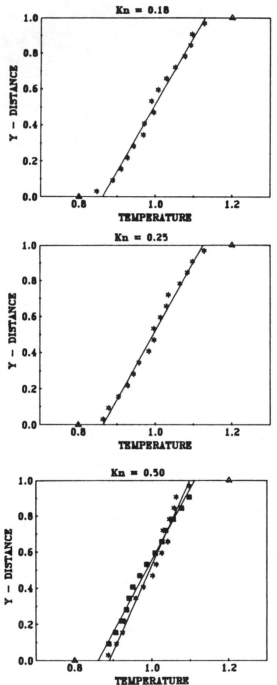

Fig. 3 Representative non-dimensional temperature profiles for several Knudsen numbers ■ 27 particles   * 512 particles

where K^* is the conductivity coefficient as calculated in the non-dimensional computations, d is the molecular diameter, m is the molecular mass and k is Boltzmann's constant. Fig. 3 shows temperature profiles at different Knudsen numbers. Each point shown represents the local temperature computed from the average velocity within a slab parallel to the plates at the location of that point. The figures also show least squares straight line fits to the numerical results. For Kn = 0.5 profiles are shown computed with 27 and with 512 particles. The close agreement is an indicator of the adequacy of the scaling criteria.

Fig. 4 contains a plot of the apparent conductivity parameter, $\bar{K}$, obtained using the overall temperature gradient, as a function of Knudsen number. Its limiting value as Knudsen number goes to zero represents the non-dimensional conductivity coefficient of the model gas. The solid curve describing the numerical results is a least squares fit using the simple kinetic theory form:

$$\bar{K} = \bar{K}_o/(1 + BKn) \qquad (10)$$

The current results are $\bar{K}_o$ = .766 and B = 3.84. Simple kinetic theory[2] yields $\bar{K}_o$ = .674. Also shown on Fig. 4 are recent kinetic theory results of Sone, Ohwada and Aoki[3], based on numerical solution of the linearized Boltzmann

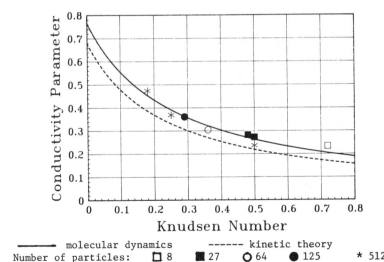

Fig. 4 Heat conductivity parameter, $Kd^2(m/Tk^3)^{0.5}$, vs. Knudsen number.

equation. The molecular dynamics results are approximately 12% higher. This level of agreement implies that the molecular dynamics and kinetic theory results are equally reasonable.

A more definitive statement about the comparative results is not warranted at this time because the molecular dynamics computations have not been subjected to a number of tests that might influence the numerical results. These include the effects of changes in the aspect ratio of the computational region, time averaging over a period of steady state behavior rather than from the beginning of the process, more critical examination of the achievement of steady state, and the influence of computations at lower Knudsen numbers on the least squares curve fit to the results.

VISCOSITY

The model consists of a gas between two parallel plates moving at the same speeds in opposite directions (Fig. 1). Molecules rebound from the plates diffusely, with complete accommodation. The velocity of a rebounding particle is the plate velocity plus a thermal velocity assigned in the same manner as in the heat conduction computation. A particle that leaves the computational region through one of the two computational boundaries lying perpendicularly to the flow direction is brought into the computational region through the opposite boundary. At the other two boundaries of the computational region particles are reflected specularly. The information collected is the spatial distribution of velocity component parallel to the moving plates at the shear stress at the plates. The shear stress, τ, is computed from the time average tangential impulse to the wall, as

$$\tau = (1/At) \sum (m(u_r - u_i)) \qquad (11)$$

where u is the velocity component in the direction of motion in the upper wall. As in the case of the conductivity coefficient, the viscosity coefficient can be deduced from the limiting value of the ratio of the shear stress to the overall velocity gradient as Knudsen number goes to zero. The viscosity coefficient of a gas can be deduced from:

$$\bar{\eta} = \eta d^2/(mkT)^{0.5} = (1/\pi)\eta^*/N*Kn \qquad (12)$$

where η^* is the viscosity coefficient as calculated in the non-dimensional computations.

Fig. 5 shows velocity profiles at different Knudsen numbers, where each point corresponds to the average velo-

SCALING RULES IN MOLECULAR DYNAMICS COMPUTATIONS 205

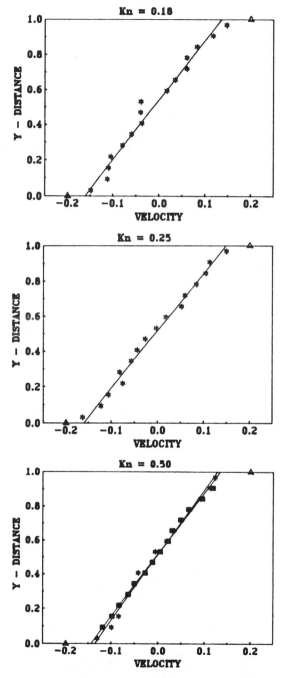

Fig. 5 Representative non-dimensional velocity profiles for several Knudsen numbers.   ■ 27 particles   * 512 particles

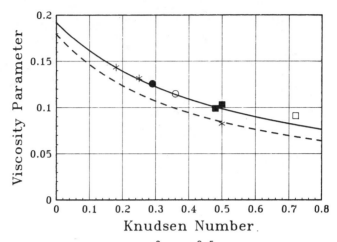

Fig. 6 Viscosity parameter, $\eta d^2/(mkT)^{0.5}$ vs. Knudsen number.
———— molecular dynamics   ------ kinetic theory
Number of particles: □ 8   ■ 27   ○ 64   ● 125   * 512

city in a corresponding slab, as described in connection with thermal conductivity. The profiles are essentially linear in the gas and exhibit marked velocity slip at the plates. The figures also display least squares straight line fits to the numerical results.

Fig. 6 is a plot of the apparent viscosity parameter, $\bar{\eta}$, obtained using the overall velocity gradient, as a function of Knudsen number. Its limiting value as Knudsen number goes to zero represents the non-dimensional viscosity coefficient of the gas. As with the thermal conductivity results, the solid curve describing the viscosity results is a least squares fit of the form:

$$\bar{\eta} = \bar{\eta}_0/(1 + CKn) \qquad (13)$$

The current results are $\bar{\eta}_0 = 0.192$ and $C = 1.90$. Kinetic theories[4] yield $\eta_0 = .180$ and a range of C values, with $C = 2.26$ being a good approximation. A curve using these values is shown on the figure for comparison. The current results are about 6% higher than the kinetic theory result. Again, this level of agreement implies that the molecular dynamics and kinetic theory results are equally reasonable.

CONCLUSIONS

The results of these numerical experiments suggest that when spacing ratios are selected so as to ensure that gaseous properties are preserved in a model of a gas, then a scaling criterion that maintains the Knudsen number in the model at the value prevailing in the gas appears to be a useful way of deducing gas behavior from molecular dynamics computations. Scaling coupled with time averaging significantly reduces the number of computations required and the number of particles needed per computation.

ACKNOWLEDGMENTS

Much of the computer programming was done by Thomas DiLiddo and Myeung Jouh Woo at C.W.R.U. and Greg Kass and Timothy Stone at M.I.T. The work was supported by a grant from NASA Lewis Research Center and by Project Athena at M.I.T., and a grant of computer time at the Pittsburgh Supercomputing Center.

REFERENCES

[1] Evans, D. J., and Hoover, W. G., "Flows Far from Equilibrium via Molecular Dynamics", Annual Review of Fluid Mechanics, Vol. 15, 1986, pp. 243-264.

[2] Chapman, S., and Cowling, T. G., The Mathematical Theory of Non-Uniform Gases, Cambridge University Press, 1958, pp. 100-106.

[3] Sone, Y., Ohwada, T., and Aoki, K., "Heat Transfer and Temperature Distribution in a Rarefied Gas between Two Parallel Plates - Numerical Analysis of the Boltzmann Equation for Hard Sphere Molecule", Abstracts of Sixteenth International Symposium on Rarefied Gas Dynamics, Pasadena, July 1988, pp. 48-50.

[4] Willis, R. D., "Comparison of Kinetic Theory Analyses of Linearized Couette Flow", Physics of Fluids, Vol. 5, 1962, pp. 127-135.

Chapter 3. Direct Simulations

Perception of Numercial Methods in Rarefied Gasdynamics

G. A. Bird*
University of Sydney, Sydney, New South Wales, Australia

Abstract

The relationships between the various numerical methods that have been applied to problems in rarefied gasdynamics are discussed. Particular emphasis is given to the sometimes conflicting viewpoints and computational requirements that are associated with "physical simulation" v "the numerical solution of the Boltzmann equation." The structure of a normal shock wave is used to illustrate these points, both historically and through new calculations for both very weak and very strong waves. The basic differences between the molecular dynamics and direct simulation methods are shown to affect their applicability to dense and rarefied flows. Finally, the methods for the probabilistic selection of representative collisions in the direct simulation Monte Carlo method are reviewed. A new method that combines the most desirable features of the earlier methods is presented.

Introduction

For the first two thirds of this century, the principal advances in all the flow regimes of fluid mechanics were based on parallel experimental and analytical studies. The classical kinetic theory for the transport coefficients[1] was one of the most impressive achievements of that era. This theory enabled the original empirical basis of the Navier-Stokes equations to be replaced by a solid analytical base. It also established the essential link between kinetic

---

Presented as an Invited Paper
Copyright 1989 by the American Institute of Aeronautics and Astronautics, Inc. All rights reserved.
*Professor of Aeronautical Engineering, Department of Aeronautical Engineering

theory and continuum fluid mechanics, but did not provide solutions for problems with a high degree of translational nonequilibrium. This mainstream effort was not, therefore, of direct help in the 1950's when engineering problems that involved rarefied gas flows attained some prominence. Also, many of these engineering problems arose in areas where laboratory experiments were either impossible or prohibitively expensive. The analytical methods for rarefied gasdynamic flows were more difficult to apply than those for continuum gasdynamics, so it was not long before the analytical efforts passed the point of diminishing returns. These difficulties have been largely bypassed by numerical methods that depend on modern digital computers.

Computational fluid mechanics for continuum gas flows is based almost exclusively on numerical methods for the solution of the mathematical equations that model the flows. For rarefied flows, the particulate nature of the gas must be recognized and one consequence is that the mathematical model, expressed through the Boltzmann equation, is far less amenable than the Navier-Stokes equations to either analytical or numerical approaches. The difficulties with the mathematical approach are exemplified by RGD being one of the few subject areas in which it has been respectable to replace the real equation by "model" equations that are mathematically more tractable. While these give the correct solutions in the continuum and free-molecule limits, the errors associated with solutions in the transition regime have always been difficult to assess. One problem has been that the model equations are at least as difficult as the Navier-Stokes equations so that numerical methods are required for complex flows. When a solution is clearly incorrect, it may be either the numerical method or the model itself that is at fault.

Numerical solutions of the "full" Boltzmann equation have been largely confined to one-dimensional steady flow studies using the Hicks-Yen-Nordsieck (HYN) method.[2] The restriction to relatively simple flows is primarily due to the computational requirements of any numerical method that has to work in phase space. However, even if unlimited computer resources were available, this approach would encounter the problem that the Boltzmann equation has not yet been formulated for chemically reacting and thermally radiating flows that are now the subject of engineering studies.

While the particulate nature of the gas has been responsible for the mathematical difficulties, it also permits the circumvention of these difficulties through the development of physically based numerical methods. The digital computer is essential for the physical approach and the first attempts at direct simulation date from the earliest days of the computer. These were "test-particle" methods,[3] which were quickly superseded by the molecular dynamics (MD) method[4] and the direct simulation Monte Carlo (DSMC)[5] method in which a large number of simulated molecules move simultaneously. These methods have become well established, but the strength of the mathematical tradition is such that a numerical solution of the approximate mathematical model would still be more generally acceptable than a physically based method that is free of these approximations.

A general problem with numerical methods is that there are many variations of a given method, either with time when the application is by a single individual or between implementations by separate individuals or groups. The physically based methods pose particular problems in that, while the overall procedures are generally based on mathematical expressions, some details such as default actions are defined at the level of the computer code. It is an extremely difficult task to properly include these in an analytical representation of the method and it becomes a quite impossible task when the code is not available. Because it is not practical to fully define a method each time it is applied, criticisms and "improvements" can be based on misunderstandings about the details of the method that was actually used. Also, there are computational variables such as grid size and time step associated with numerical methods so that, unless the requirements associated with these are met, poor results can be obtained and incorrect conclusions can be drawn.

Given the above difficulties, it is not surprising that there is confusion about the relative capabilities and merits of the various approaches. The following sections deal with topics that bring up, and hopefully throw some new light on, some of the more common areas of uncertainty.

The Shock Wave Revisited

The structure of the normal shock wave has served as a major test case for almost all of the numerical approaches. Most of the direct simulation calculations[6-9] have been for strong shock waves for which the continuum Navier-Stokes equations are invalid, but validation has been provided by comparisons with experiment.[10-12] It has been claimedy[13] that "the Monte Carlo techniques are not suited to very low Mach numbers because the solution is near equilibrium, and small deviations from equilibrium would be lost in the statistical fluctuations." This is incorrect; the structure of weak shock waves emerges from the scatter as the sample size increases, but weak shocks require much greater care with the boundary conditions and the details of the method.

Boundary Conditions

A normal shock wave simulation involves a one-dimensional flow with a supersonic upstream boundary and a subsonic downstream boundary. The boundary conditions may be:

1) The generation of the entering molecules from the equilibrium distributions corresponding to the Rankine-Hugoniot theory. Exiting molecules are removed from the flow. This "flux" condition has been used in the more recent studies.[9,14]

2) Specularly reflecting plane pistons with velocities determined from the Rankine-Hugoniot theory. If the boundaries are to retain a fixed distance from the wave, the piston must "jump" back to its original location at the end of the time step with additional equilibrium molecules being inserted at the upstream end and molecules outside the boundary being removed at the downstream end. Note that the piston boundary requires the specification of the downstream velocity only, while the flux boundary requires the additional specification of the density and temperature.

3) The "stagnation streamline" condition[15] with a specularly reflecting wall at the downstream end and a type 1 boundary at the upstream end. In the final steady state, the total number of molecules is held steady by appropriate molecule removal near the downstream end. This scheme is useful for reacting gases when a Rankine-Hugoniot solution is not available.

For type 1 boundaries, the initial conditions generally represent the shock as a discontinuity and the profile evolves with time. The piston boundary condition allows the shock to be formed in a physically realistic manner.[7]

Poisson Distribution or Random Walk?

When the boundary conditions are such that the flow geometry at large times is fixed, the fluctuation in the total number of simulated molecules follows the Poisson distribution and the standard deviation is approximately equal to the reciprocal of the square root of the average number. This is the case in most applications of the DSMC method, but does not apply to shock calculations in which the shock is free to move. The total number of molecules and the shock position then execute a random walk that can smear the time-averaged profile. Methods which have been used to overcome this problem have included a small normal velocity downstream of the desired shock position to act as a "shock holder,"[9] the movement of the sampling frame of reference to coincide with the shock center,[14] and the adjustment of the number of molecules entering the downstream boundary to keep the number of molecules constant.

The scatter may also show some characteristics of a random walk when an exactly conserved quantity is subdivided. For example, separate temperatures based on translational and internal modes may show departures from the mean that are greater than would be expected for a Poisson distribution based on the sample size. The same applies to the velocities and temperatures of the separate species in a gas mixture.

Weak Shock Calculations

The first attempts to calculate a Mach 1.1 wave were made with type 1 boundaries both upstream and downstream and with the wave stabilized by keeping the total number of molecules constant. The scatter in the results was far larger than expected and the wave was affected by the location of the boundaries. The removal of the shock stabilization improved the behavior of the program, but good results were not obtained until the downstream boundary was replaced by a type 2 "jumping" piston. It appears that, while the type 1 boundary is correct on the average, weak disturbances are produced by the mismatch between the exiting flux (which is subject to scatter) and the constant entering flux. These disturbances are unimportant for strong shocks, but affect the profile of weak shocks. The piston boundary is phsically more realistic and reacts instantaneously to the scatter so that no disturbances are generated. The profile was stabilised by limits on the fluctuation in the total number of molecules. Whenever this number moved outside an imposed limit, all molecules were moved a short distance in one direction and were discarded if they moved outside the boundary. The "gap" at the opposite end was filled with molecules generated from the appropriate equilibrium distribution.

The calculated density and temperature profiles for a Mach 1.2 shock wave in a hard sphere gas are shown in Fig. 1. The nondimensional results are the ratio of the difference between the local value and the upstream value to the difference between the downstream and upstream values.

These quantities were time averaged over a very long time in order to build up a sample of the order of 10^6 at each plotted point. The statistical scatter is consistent with the sample size and there is no sign of an "accumulation of error" that Yen[16] has suggested would occur with any DSMC method that does not follow the "Boltzmann-like" procedures set out by Nanbu.[17]

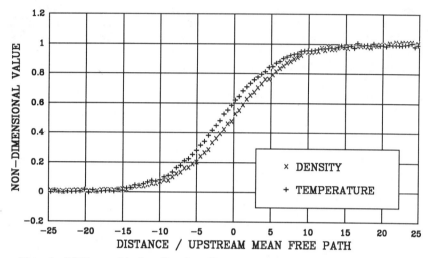

Fig. 1 DSMC result for the density and temperature profiles across a Mach 1.2 shock in a hard-sphere gas.

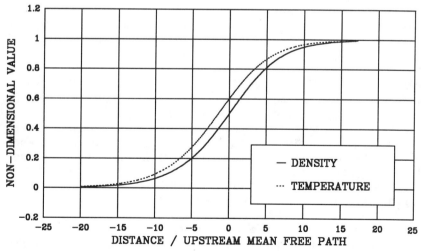

Fig. 2 Navier-Stokes result for the density and temperature profiles across a Mach 1.2 shock in a hard-sphere gas.

The theory of Gilbarg and Paolucci[18] has been used to obtain the Navier-Stokes profiles for the Mach 1.2 shock in a hard-sphere gas. These are shown in Fig. 2 and are in very good agreement with the DSMC results in Fig. 1. This result is at variance with the HYN results of Hicks et al[19] who, for a shock Mach number of 1.2, obtained a profile that is about 30% wider than the Navier-Stokes profile. The Navier-Stokes equations are valid[20] as long as the Chapman-Enskog theory for the transport properties is valid. This requires that the local Knudsen numbers based on the scale lengths of the macroscopic gradients should be small compared with unity. The breakdown in the Navier-Stokes equations is expected to commence when these local Knudsen numbers reach about 0.1. For the Mach 1.2 shock wave, the local Knudsen numbers are less than 0.02 and the Navier-Stokes solution should be valid.

An experimental study[21] of weak shock also led to shock wave thicknesses greater than the Navier-Stokes values. The thickness of the Mach 1.2 wave was in agreement with the hard sphere HYN result, while the variation of the thickness with Mach number was consistent with a Maxwell gas. This inconsistency is compounded by the fact that the HYN calculation[22] for this strength shock in a Maxwell gas was in agreement with the Navier-Stokes result. The extrapolation of the experimental curve back to a shock Mach number of 1.01 indicates an error in the Chapman-Enskog result of 10% at a local Knudsen number of 0.00005. Such an error should have been apparent in many other studies such as boundary layer measurements. The most likely explanation is that the shock was not fully formed in the shock tube experiments and the early HYN calculation for the hard sphere case was in error.

"Nanbu Procedures"

Nanbu[23] has adopted the point of view that the direct simulation procedures should be derived from the Boltzmann equation. Since the dependent variable in the Boltzmann equation is the velocity distribution function for a single molecule, the Nanbu procedures alter the velocity for only one of the molecules involved in a collision. This has the obvious disadvantage that twice as many collisions must be calculated. More seriously, momentum and energy are no longer exactly conserved and, while they are conserved on the average, there will be a random walk in these quantities at each collision. Random walks have the undesirable characteristic that the mean deviation and the mean time between zero crossings (of zero deviation) increase with the square root of the number of steps.

The program that led to Fig. 1 was modified so that the number of collisions was doubled and the velocity of only one molecule in the collision was altered. The results from a similar calculation to that which led to Fig. 1 are shown in Fig. 3. The modifications led to a serious degradation of the solution, the nature of which was consistent with the characteristics of a random walk. It is not rational to unnecessarily introduce undesirable effects into the simulation of a physical process merely to make the procedures more consistent with a particular mathematical model of that process.

Strong Shock Waves

The DSMC calculation of a strong shock wave with a sample size equal to that for the above weak shock calculations leads to profiles with a standard deviation of the statistical scatter as low as 0.1%. Figure 4 shows the result for a Mach 8 shock wave in a monatomic gas with viscosity proportional to the 0.68 power of temperature. This shows a small overshoot of less than 1% in the temperature that had not been apparent in the less accurate earlier results. Such an overshoot had been predicted earlier.[16]

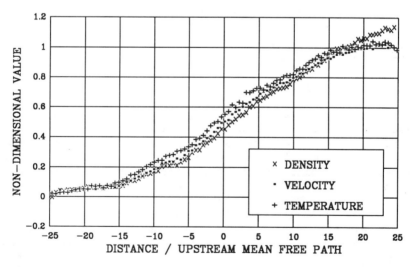

Fig. 3 Density and temperature profiles across a Mach 1.2 shock in a hard-sphere gas given by the DSMC method when the Nanbu procedures are incorporated.

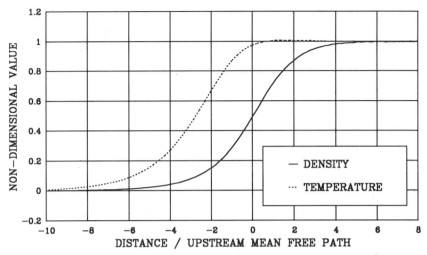

Fig. 4 DSMC result for the density and temperature profiles across a Mach 8 shock in a gas with viscosity coefficient proportional to temperature to the power 0.68.

Distinction between MD and DSMC Methods

Molecular Size in MD Calculations

The initial configuration of the molecules in the MD method is set probabilistically, but the subsequent calculation is deterministic. A collision occurs whenever the spacing between a pair of molecules is such that their force fields overlap. Apart from the approximate molecular model and the related difficulty in setting a limiting range for the field, this is realistic. There are practical limits on the number of simulated molecules set by the speed and storage capacity of computers. The speed limitation enters because the computing time in a straightforward application of the method is proportional to the square of the number of molecules. The flow to be modeled occupies a volume in physical space that may be specified in terms of cubic mean free paths.

Figure 5 illustrates the constraints that are imposed by the physics of real gases when one is faced with a limitation on the total number of molecules. A typical problem would involve 10,000 to 100,000 model molecules in a volume of 1000 to 10,000 cubic mean free paths. It can be seen that these conditions occur in real gases at densities well above standard density and most MD calculations have been for dense gases.[24,25] However, the method has recently been used by Meiburg[26] in a rarefied gas context and this can be

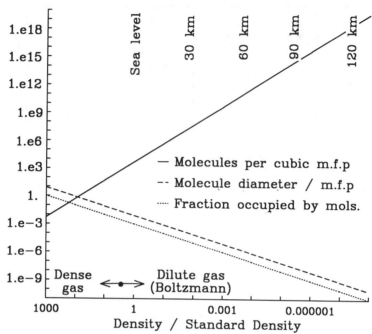

Fig. 5 Consequences of molecular size in a real gas.

done only if the molecules are regarded as having diameters that are many orders of magnitude larger than real molecules. Meiburg's calculation was for the flow past an inclined flat plate with a chord length of 75 mean free paths. The molecular diameter was 0.75 of a mean free path or 1% of the chord of the plate. These conditions would apply at a density of 125 times the atmospheric density. The Boltzmann equation would not apply under these conditions, the equation of state would be different from that in a dilute gas, and the miss distance impact parameter for collisions in the critical region near the ends of the plate would be one-sided in the direction to promote the formation of vortices.[27] These factors were evidently not appreciated either in the original report[28] of Meiburg's work or in the review by Evans and Hoover.[24]

Cell Size Effects in the DSMC Method

In contemporary applications[20] of the DSMC method, each simulated molecule is regarded as representing a set number of real molecules and the physically correct molecular size is used. Each of these molecules is subjected to the correct collision rate and the reference in the MD literature by Rapaport[25] that the DSMC method samples "only some of the collisions the molecules experience" is incorrect. However, while the division of the flow into cells in physical space is necessary in the MD method only for the output of results, it is used in the DSMC method for the probabilistic selection of collision pairs.

The DSMC method regards the molecules in a cell as representative of those at the location of the cell and their relative locations in the cell are disregarded when collision pairs are chosen. The motion of the molecules acts to create gradients in the cells and collisions between molecules on opposite sides of a cell tend to destroy them. The effect is most pronounced if the impact parameters are such that the velocities are interchanged in collision. Meiburg[26,28] drew attention to the possible effect on angular velocity and carried out some DSMC calculations for comparison with the MD results discussed above. Unfortunately, he used a cell size with a linear dimension of three mean free paths which is at least an order of magnitude greater than is needed for such a flow. This, coupled with the lack of recognition of molecule size effects in the MD calculations, meant that the comparison between the methods was not meaningful; it is unfortunate that the original conclusions have been uncritically repeated in the MD literature.[24,25]

Despite these shortcomings, the warning with regard to the cell size was valid and timely. It is essential that the DSMC cell dimensions be very small in directions where there are large gradients in the macroscopic flow properties. This condition is very easily met in one-dimensional flows and the shock wave calculations in the preceding section show that there is agreement with exact theory for sufficiently small cell sizes. It might be noted that, if one wished to dramatize the deleterious effect of excessive cell size in the shock wave context and again chose the worst possible coincidence of impact parameters, it could be said that "collisions act to reverse the flow gradients in the shock wave." Calculations that involve viscous boundary layers have been shown[29] to require the cell height adjacent to the surface to be small compared with the local mean free path. This is more readily achieved if sub cells[27] are used for the sampling of collision pairs and these can be sufficiently small for all collisions to be between near neighbors.

Recent calculations[27,30] have shown that the DSMC can be applied successfully to vortical flows. Those for the Oseen vortex decay problem[30] are particularly useful because of the availability of an exact solution for comparison. However, these are cases in which the physical location and extent of the vortex is set by the boundary conditions. The requirements for the successful simulation of the formation of an unconstrained vortex remain uncertain and this is an area where the DSMC method might have problems.

Collision Sampling Techniques in DSMC Method

The procedures for the probabilistic selection of a representative set of collisions at each discrete time step have been responsible for much of the comment on the DSMC. It is still useful to follow Derzko[31] and define three distinct methods to serve as the basis of the discussion, which are:

1) The "collision rate" method in which the non-equilibrium collision rate is calculated from the standard kinetic theory result that involves the average of all possible relative velocities. The required number of collisions is then given by the product of this rate and the time step δt. The collision pairs must be chosen with probability proportional to the product of the relative velocity v_r and collision cross section σ. The disadvantage is that the computation time is proportional to the square of the number of molecules.

2) The "time-counter" method in which the collision pairs are accepted with the same probability as in method 1 and, at each collision, a "cell time" is advanced by

$$2 / (N \langle n \rangle \sigma v_r) \qquad (1)$$

where N is the number of molecules in the cell and $\langle n \rangle$ is a time or ensemble averaged number density. Sufficient collisions are calculated to keep the cell time concurrent with the flow time.

The advantage of this method is that the computation time can be linearly proportional to the number of molecules.

3) The "direct" or Kac method in which all possible pairs in each cell are considered and the probability of collision within the time step is equal to the ratio of the volume swept out by the cross section (moving with the relative velocity) to the cell volume V_c. The disadvantage is again that the computation time is very nearly proportional to the square of the number of molecules.

Because of its efficiency, the time counter method has been the most used, but there are problems in highly nonequilibrium regions such as the front of a very strong shock. The acceptance of an unlikely collision can advance the cell time by an interval that is much larger than the time step δt and the overall collision rate can be distorted. The "direct" method can be modified by reducing the number of sampled pairs by some factor and increasing the collision probabilities by the same factor. If the factor is such that the maximum collision probability of any pair is unity, the number of pairs to be sampled is

$$0.5 \ N \ \langle N \rangle \ F_N \ (\sigma \ v_r)_{max} \ \delta t \ / \ V_c \qquad (2)$$

and the collision probability for each selection is

$$(\sigma \ v_r) \ / \ (\sigma \ v_r)_{max} \qquad (3)$$

These new procedures are, in fact, similar to those in the time counter method, except that the time counters are effectively replaced by the explicit relation for the number of molecule pair choices. The computing time remains directly proportional to the number of simulated molecules and the method is called the "no time counter" or NTC method. The maximum value of the product of the collision cross-section and the relative velocity may be set interactively during the running of the program. It should be noted that this quantity appears in the numerator of Eq. (2) and in the denominator of Eq. (3). This means that the collision rate is not affected by the fact that the interactively set value generally increases during a run. The number of choices at each time step must be an integer value and the remainder is, of course, carried forward to the next step and added to the value given by Eq. (2).

The operation of the collision procedures depends on the details of the coding and mathematical representations are generally incapable of providing an adequate model of these. Also, criticism may be based on imagined rather than the actual procedures. For example, the "null-collision technique" of Koura[32] addresses itself to a hypothetical problem associated with chemically reacting flows that had not, in fact, been present in the previous applications of the DSMC method to these flows.[15,20,29] Similarly, the critical study of the now obsolete "time counter" method by Belotserkovskii et al.[33] is based on an oversimplified mathematical model of its operation. The cell time does not have to be synchronised with the flow time at the end of each time step and an averaged number density can be used in Eq. (1) for both unsteady and steady flows. The claimed "accumulation of error" can be (and has been) readily avoided by sensible coding.

For gas mixtures, the details of the coding are particularly important in cases where the samples are such that

Table 1 The ratio of the simulated to the theoretical collision rates in a homogeneous equilibrium gas mixture.

Gas species	1	2	3	4	5
1	0.9982	1.0010	0.9998	0.9986	0.9980
2	1.0010	0.9933	1.0000	1.0050	1.0090
3	0.9998	1.0000	0.9875	1.0020	1.0060
4	0.9986	1.0050	1.0020	0.9961	1.0080
5	0.9980	1.0090	1.0060	1.0080	0.9838

two molecules of a trace species are present in a particular cell only part of the time. Table 1 presents the results from one of the test cases for the NTC method that dealt with a homogeneous gas mixture in equilibrium. There are five species and the fifth is only 2% of the total. There were 1000 molecules in the test volume, which was divided into 50 equal cells. This means that two type 5 molecules were present in any particular cell only 16% of the time. Despite the poor sample, Table 1 shows that the procedures led to the correct number of type 5-5 collisions between the molecules of the trace species. The sample size on which the ratio is based ranged from 5,000,000 for the 1-1 collisions down to 5,000 for the 5-5 collisions.

Acknowledgment

This work was supported by the NASA Langley Research Center under Grant NAGW-728 to the University of Sydney.

References

[1]Chapman, S. and Cowling T. G., *The Mathematical Theory of Non-Uniform Gases*, Cambridge University Press, London, 1952.

[2]Nordsieck A. and Hicks, B. L., "Monte Carlo Evaluation of the Boltzmann Collision Integral," *Rarefied Gas Dynamics*, edited by C. L. Brundin, Academic Press, New York, 1967, pp. 695-710.

[3]Haviland, J. K. and Lavin, M. L., "Application of the Monte Carlo Method to Heat Transfer in a Rarefied Gas," *The Physics of Fluids*, Vol. 5, 1962, pp. 1399-1405.

[4]Alder, B. J. and Wainwright, T. E., "Molecular Dynamics by Electronic Computers," *Transport Processes in Statistical Mechanics*, edited by I. Prigogine, Interscience, New York, 1958, pp 97-131.

[5]Bird, G. A., *Molecular Gas Dynamics*, Oxford University Press, London, 1976.

[6]Bird, G. A., "Shock Wave Structure in a Rigid Sphere Gas," in *Rarefied Gas Dynamics*, Vol. 1, edited by J. H. de Leeuw, Academic, New York, 1965, pp. 216-222.

[7]Bird, G. A., "The Formation and Reflection of Shock Waves," *Rarefied Gas Dynamics*, Vol. 1, edited by L. Trilling and H. Y. Wachman, Academic Press, New York, 1969, p. 85.

[8]Bird, G. A., "Aspects of the Structure of Strong Shock Waves," *The Physics of Fluids*, Vol. 13, May 1970, pp. 1172-1177.

[9]Bird, G. A., "The Structure of Shock Waves in Gas Mixtures,", *Rarefied Gas Dynamics*, Vol. 1, edited by H. Oguchi, University of Tokyo Press, Tokyo, 1984, pp. 175-182.

[10]Schmidt, B. "Electron Beam Density Measurements in Shock Waves in Argon," *Journal of Fluid Mechanics*, Vol. 39, 1969, pp. 361-371.

[11]Sturtevant, B. and Steinhilper, E. A., "Intermolecular Potentials from Shock Structure Experiments," *Rarefied Gas Dynamics*, edited by K. Karamcheti, Academic, New York, 1974, pp. 159-166.

[12]Alsmeyer, H., "Density Profiles in Argon and Nitrogen Shock Waves Measured by the Absorption of an Electron Beam," *Journal of Fluid Mechanics*, Vol. 74, 1976, pp.497-513.

[13]Sod, G. A., "A Numerical Solution of Boltzmann's Equation," *Communications on Pure and Applied Mathematics*, Vol.30, 1977, pp. 391-419.

[14]Nanbu, K. and Watanabe, Y., "Analysis of the Internal Structure of Shock Waves by Means of the Exact Direct-Simulation Method," *Rarefied Gas Dynamics*, Vol. 1, edited by H. Oguchi, University of Tokyo Press, Tokyo, 1984, pp. 183-190.

[15]Bird, G. A., "Direct Simulation of Typical AOTV Entry Flows," AIAA Paper 86-1310, June 1986.

[16]Yen, S. M., "Numerical Solution of the Nonlinear Boltzmann Equation for Nonequilibrium Gas Flow Problems," *Annual Reviews of Fluid Mechanics*, Vol. 16, 1984, pp. 67-97.

[17]Nanbu, K., "Direct Simulation Scheme Derived From the Boltzmann Equation. I. Monocomponent Gases," *Journal of the Physical Society of Japan*, Vol. 45, 1980, pp. 2042-2049.

[18]Gilbarg, D. and Paolucci, D., "The Structure of Shock Waves in the Continuum Theory of Fluids," *Journal of Rational Mechanics and Analysis*, Vol. 2, 1935, pp. 617-642.

[19]Hicks, B. L., Yen, S. M., and Reilly, B. L., "The Internal Structure of Shock Waves," *Journal of Fluid Mechanics*, Vol. 53, 1972. pp. 85-111.

[20]Bird, G. A., "Direct Simulation of Gas Flows at the Molecular Level," *Communications in Applied Numerical Methods*, Vol. 4, 1988, pp. 165-172.

[21] Garen, W., Synofzik, R., and Wortberg, G., "Experimental Investigations of the Structure of Weak Shock Waves in Noble Gases," Progress in Astronautics and Aeronautics: Rarefied Gas Dynamics, Edited J. L. Potter, Vol. 51, 1977, pp. 519–527.

[22] Yen, S. M. and Ng, W., "Shock Wave Structure and Intermolecular Collision Laws," Journal of Fluid Mechanics, Vol. 65, 1974, pp. 127–144.

[23] Nanbu, K., "Theoretical Basis of the Direct Simulation Monte Carlo Method," Rarefied Gas Dynamics, Vol. 1, Edited by V. Boffi and C. Cercignani, Vol.1, B.G. Tuebner, Stuttgart, Federal Republic of Germany, 1986, pp.369–383.

[24] Evans, D. J. and Hoover, W. G., "Flows Far from Equilibrium via Molecular Dynamics," Annual Reviews of Fluid Mechanics, Vol. 18, 1986, pp. 243–264.

[25] Rapaport, D. C., "Microscale Hydrodynamics: Discrete-particle Simulation of Evolving Flow Patterns,", Physical Review A, Vol. 36, 1987, pp.3288–3299.

[26] Meiburg, E., "Comparison of the Molecular Dynamics Method and the Direct Simulation Technique for Flows Around Simple Geometries," The Physics of Fluids, Vol. 29, 1986, pp. 3107–3113.

[27] Bird, G. A., "Direct Simulation of High-Vorticity Gas Flows,", The Physics of Fluids, Vol. 30, 1987, pp. 364–366.

[28] Meiburg, E., "Direct Simulation Technique for the Boltzmann Equation," Report DFVLR-FB 85-13, Göttingen, FRG, 1985.

[29] Moss, J. N. and Bird, G. A., "Direct Simulation of Transitional Flow for Hypersonic Re-entry Conditions," Progress in Astronautics and Aeronautics: Thermal Design of Aeroassisted Orbital Transfer Vehicles, Vol. 96, 1985, pp. 113–139.

[30] Wetzel, W. and Oertel, H. jr, "Gas-Kinetical Simulation of Vortical Flows," Acta Mechanica, to be published.

[31] Derzko, N. A., "Review of Monte Carlo Methods in Kinetic Theory," UTIAS Review 35, University of Toronto, 1972, 78 pp.

[32] Koura, K., "Null-collision Technique in the Direct Simulation Monte Carlo Technique," The Physics of Fluids, Vol. 29, 1986, pp. 3509–3511.

[33] Belotserkovskii, O. M., Erofeev, A. I., and Yanitskii, V. E., "A Non-stationary Method of Direct Statistical Modeling of Rarefied Gas Flows," USSR Computational Mathematics and Mathematical Physics, Vol. 20, 1980, pp. 82–112.

Comparison of Parallel Algorithms for the Direct Simulation Monte Carlo Method: Application to Exhaust Plume Flowfields

Thomas R. Furlani* and John A. Lord†

Calspan Advanced Technology Center, Buffalo, New York

Abstract

In this paper, results are presented to demonstrate that the execution time for the direct simulation Monte Carlo (DSMC) method can be improved significantly by the use of parallel processing. Specifically, we compare two distinct parallel algorithms and show that they both generate significant reductions in execution time. The first algorithm is one in which the collision processes within a cell and the molecular motions are done in parallel. Each of these processes is done for the molecules within a number of cells simultaneously, with each processor being assigned a new cell after completing the required computations from the previously assigned one. A ninefold reduction in execution time was obtained on 32 nodes using this algorithm. In the second algorithm, the computational domain is partitioned among the processors so that each processor carries out a distinct but coupled DSMC method calculation on a unique subspace. The calculations are not completely independent because the molecules may traverse subspace boundaries. A fivefold decrease in execution time was obtained on a 32 node system using this algorithm.

I. Introduction

A number of significant problems in fluid mechanics involve transitional flows, i.e., flows for which the the mean free path is of the same order of magnitude as a characteristic dimension. The most successful numerical technique for modeling complex transitional flows has been the direct simu-

---

Copyright © 1989 by the American Institute of Aeronautics and Astronautics, Inc. All rights reserved.
*Senior Chemist, Physical Sciences Department.
†Head, Physical Gasdynamics Group, Physical Sciences Department.

lation Monte Carlo (DSMC) method. This method, first proposed by Bird in the 1960's,[1] has evolved over the last 20 years from a technique originally applied only to artificial or idealized flows[2] to one applied to complex engineering situations.[3-8] Although this method has met with considerable success, execution times can become prohibitively long, especially for three-dimensional and low altitude (dense) flows. This paper is concerned with the reduction of execution times through the use of multiple (concurrent) processors.

The great advantage in concurrent computing lies in its ability, given the appropriate algorithm, to achieve super-o computer-like performance at a relatively low cost. Although the design of computer algorithms for a parallel environment has been an active area of research among computer scientists for a number of years, it is only since the introduction of commercially available general purpose concurrent computers that scientists in other disciplines have become actively involved. The fact that there are almost a dozen manufacturers in this area appears to signal the long heralded revolution in parallel computing. Indeed, all concurrent calculations reported here were carried out on the commercially available Intel Personal Super Computer - System 2 (IPSC/2).

In a previous paper,[9] we demonstrated that the execution time for the DSMC method can be improved significantly by the use parallel processing. In this paper, the application of parallel processing to reduce the execution time of the DSMC method using two distinct parallel algorithms is demonstrated for the transitional flowfield produced by the interaction of a high-altitude exhaust plume with the ambient stream. The specific flowfield computation performed is for the so-called enhancement region that occurs at altitudes typically above 100 km. At these altitudes, the mixing region of the ambient flow about a rocket and its exhaust plume extends to distances in which the characteristics scale of the interaction is comparable to the mean free path of the undisturbed stream, which in turn is much larger than the scale of the rocket. The resulting transitional flowfield produces a strong optical signature by which the rocket can be detected and tracked and has therefore been the subject of much current interest.[3,4]

A number of DSMC method solutions for such flows have been carried out previously, notably by Elgin[10] and by Hermina.[7] In the study reported here, a computer code developed at Calspan, which uses the DSMC method to model the axisymmetric interaction of the ambient stream with a single component exhaust plume, has been rewritten to operate on an Intel IPSC/2.

The DSMC method treats transitional flows on a molecular basis, statistically representing the molecules of the fluid interaction by a relatively small number of representative (computational) molecules. Proper number densities are achieved by assigning a molecular weight factor to each computational molecule. The flowfield is divided into cells that are smaller than the mean free path and the flow properties are determined by sampling the appropriate molecular properties within each cell. The motion of the molecules through the flowfield is computed separately from the evaluation of the effects of the collision processes on the velocity distribution function within each cell. The distinguishing feature of the DSMC method is that collision partners are selected at random from within a cell; the positions of the molecules are ignored when evaluating the collisions within a cell. This feature has made the DSMC method more efficient than other Monte Carlo methods because it avoids evaluations that are proportional to the square of the number of molecules.

There are a number of algorithms that may be employed to implement the DSMC method in a parallel architecture. As mentioned previously, in this paper, we discuss two such methods. In the first method, the collisions and molecular motions (as well as some indexing and sampling) are carried out in parallel while the remainder of the calculation is carried out sequentially on the host. Each processor computes the collisions for a given cell. As soon as one of the processors has finished its computations, it is assigned a new cell to work on. The collision process is therefore being distributed over all available processors.

In the second algorithm, we partition the flowfield into a series of subspaces and then distribute these subspaces over the available processors so that there is a one-to-one correspondence between processors and subspaces. Each processor then carries out a distinct, but coupled, DSMC method computation on its subspace. The calculations are coupled because molecules may leave the space spanned by one processor and enter the space spanned by another. Calculations have been carried out to determine the dependence of execution time on the number of nodes or processors using both algorithms. These results are presented along with discussions of the performance limitations for both algorithms.

The remainder of this paper is divided into five sections. In the next section, we describe the DSMC method in general and its application to our problem, namely the modeling of exhaust plume flowfields. A brief description of the algorithm for the sequential version of the code is also provided in this section. The following section provides a description of the Intel IPSC/2 concurrent computer. In the fourth section, we describe both parallel algorithms in detail. The results of our study are presented in the fifth section. Finally, we summarize our findings and describe future directions for our research.

II. Application of DSMC Method to Exhaust Plume Flowfields

We begin this section by way of a brief review of the direct simulation Monte Carlo method. In the DSMC method, the flowfield is partitioned into regions or cells that are small in size when compared to the distance over which significant changes in the flow properties occur (on the order of the mean free path). Molecules are moved, without changing their velocity, over a time interval dt that is small compared to the mean collision time. The distance and direction a given molecule travels is determined entirely by its initial velocity and the time interval. The molecular velocity, during this phase of the computation, changes only when a molecule strikes a boundary. After all the molecules have moved, a representative set of collisions (appropriate to time dt) is computed for each cell in the flowfield. Molecular collisions alter the velocities of the colliding species, they do not, however, change their positions. The collision process is probabilistic rather than deterministic in nature as collision partners are selected at random from within a cell. The above sequence of molecular motion followed by collision is repeated until a steady state in flow properties is obtained, that is, until flow properties such as number density remain constant (within statistical fluctuations) for several cycles. Unsteady flows can be computed by taking an ensemble average of a series of trials.

Since this method uses a relatively small number of simulated molecules to describe an intractably large number of real molecules, it can be subject to large statistical fluctuations in its macroscopic predictions. These variations can be reduced either by increasing the execution time of a single run or by taking an ensemble average of a series of runs. In either event the magnitude of the statistical uncertainty in the sampled properties decreases only as the inverse square root of the sample size (execution time).[2,10] A tenfold increase in execution time therefore results in only a threefold decrease in uncertainty.

As mentioned previously, we applied the DSMC method to the problem of modeling the enhancement region of high-altitude rocket plumes. Since the transitional flow regime occurs at altitudes where the enhancement region is several orders of magnitude larger than the nozzle exit diameter, it is possible to represent the nozzle exit plane by a point source within the computational grid. Introduction of a point source, however, requires that the grid be carefully selected so as to insure accurate results. The radial-like grid structure (see

Fig. 1), first proposed by Hermina,[7] was employed in our model. This grid is preferable to a rectangular mesh because the cell boundaries more closely follow along the streamlines of the undisturbed exhaust flow. Although the grid structure described above is desirable from a computational point of view, its complexity makes interpretation of results difficult. The computed results are therefore mapped onto a rectangular mesh.

From Fig. 1, we see that the computational domain is divided into four pie-shaped regions. The radial distance and the angular spacing between radial lines can be adjusted separately for each region. For all calculations presented here, the angular spacing between radial lines in Fig. 1 is 2 deg. Radial lengths are adjusted so as to concentrate cells near the source. There are a total of 450 cells in the flowfield. Exhaust molecules enter the domain at the source, while ambient molecules are introduced at the leftmost boundary of Fig. 1. The program allows for distinct plume and stream species. In all cases considered in this paper, the exhaust species was H_2O and the ambient (stream) species N_2. We made this choice because the physical properties of H_2O are representative of those of the exhaust species and N_2 is the major constituent of the atmosphere at these altitudes (157 km).

A flow chart for the sequential version of the code is given in Fig. 2. The program begins by initializing variables, calculating cell boundaries, and reading input data such as nozzle exit area and exhaust number density. Exhaust molecules, which move radially outward into the domain, are added to the initially empty domain at the origin (point source). Ambient molecules are introduced into the flow from the leftmost boundary of Fig. 1 only after the fluctuation in the number of exhaust molecules is zero, i.e., when as many exhaust molecules are leaving the domain as are entering it. Once a steady state in exhaust and ambient molecules occurs, a restart file that contains molecular position and velocity

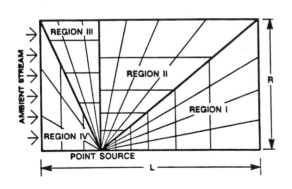

Fig. 1 Computational domain.

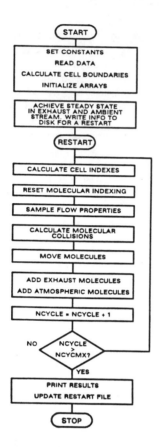

Fig. 2 Flow chart for sequential version of DSMC method code.

information is written. The calculation is continued from the restart file by specifying a value for NCYCMX (the maximum number of iterations). Flow properties, such as number density, are sampled during each iteration in this phase of the calculation. The program terminates by writing a new restart file after the appropriate number of iterations have taken place.

In the next section, we discuss the Intel IPSC/2 concurrent computer. Much of the material presented in this section is taken from Ref. 9 and is included here for the reader's convenience.

III. Intel IPSC/2

The Intel IPSC/2 consists of a multiuser host computer (resource manager), which serves as the front end to an array of processors (nodes). The host itself is composed of an Intel 80386 central processing unit (CPU), a 80387 math coprocessor, 8 Mbytes of random access memory (RAM), and a 140 Mbyte hard disk. The operating system is AT&T's Unix based System V. Each node is contained on a single board. Upon this board resides an Intel 80386 CPU, a 80387 math coprocessor, a communication chip, and a variable amount of RAM. The computer is a distributed memory system as opposed to a shared memory system. This means that each node can access only the memory contained on its own board; in our case, this was 4 Mbytes. One feature of a distributed memory system is that any data a given node operates on must have been generated by the node or received, in the form of a message, from either a fellow node or the host. A node, at present, cannot read data from a common (to all nodes) memory location or a disk.

Since the host processor is connected to the array of processors through a single node and since the nodes can communicate with each other only via message passing, the way in which the nodes are physically connected together is an important issue. There are many different ways of connecting the nodes together, they can be strung together to form loops, meshes, trees, tori, etc. In the Intel IPSC/2, the nodes are connected together in what is known as a hypercube architecture. A hypercube is an n-dimensional cube having 2^n vertices. In a computer using this construct, each processor can be thought of as occupying one of the vertices of the hypercube with the edges of the cube serving as the communication links. In this architecture, the total number of processors is given by the formula, $N = 2^n$, where n is the number of processors each node is directly connected to. For example, Fig. 3 presents node connectivity schemes for values of n equal to 1, 2, and 3 (2, 4, and 8 node systems).

One of the great advantages of the hypercube architecture is that through software one is able "mimic" other topologies such as meshes or loops. For example, an eight node hypercube can be thought of as an eight node loop by assuming that the processors are connected solely in the manner shown in Fig. 4. The programmer simply ignores each node's other connection when writing software.

One of the drawbacks to the hypercube architecture, in our opinion, is the fact that one can increase the size of the

Fig. 3 Processor connectivity for two-, four-, and eight-node systems (n refers to the number of other processors a given processor is connected to).

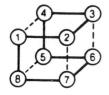

Fig. 4 Eight-node loop topology "superimposed" upon an eight-node hypercube topology.

cube only by doubling the number of processors and thereby doubling the price. Conversely, whenever a node board becomes inoperative, for whatever reason, the size of the machine is decreased by a factor of two.

In the next section, we describe the two algorithms we have elected to implement on the IPSC/2. We find it convenient to divide this section into two subsections. In the first, we describe the parallel algorithm in which the collisions, molecular motions, sampling, and some indexing are done concurrently. We will refer to this algorithm from herein out as method 1. The subsequent subsection describes what we refer to as the "completely" parallel algorithm in which each processor carries out a "distinct" but coupled DSMC method computation on a unique subspace of the computational domain.

IV. Parallel DSMC Method Algorithms

Method 1

In the first algorithm, only the molecular motions, collisions, and some indexing are carried out in parallel, the rest of the computation proceeds in the usual sequential manner. The motivation for this algorithm comes from the fact that the process of calculating a representative set of collisions for each cell in the flowfield is by far the most time-consuming step, accounting for some 80-90% of the execution time. Any initial efforts aimed at reducing execution time should be directed there. In addition, as each cell in the flowfield is treated independently of all others during the collision and molecular motion processes in the DSMC method, these processes are well suited to a parallel architecture.

Fig. 5 presents the flow chart for method 1. Two distinct sections have been made concurrent. First of all, the task of determining the cell index for each molecule in the flow is now being carried out by all (N) processors. Specifically, each node receives, from the host, molecular position information (two coordinates for each molecule) for its share of the simulated molecules in the flowfield (NMOL/N). Here, NMOL represents the total number of simulated molecules in the flow. Upon receiving the data, each node performs the appropriate calculations and then returns the cell indexes for its share of the simulated molecules back to the host.

APPLICATION TO EXHAUST PLUME FLOWFIELDS

Excluding the overhead involved in sending information to and from the nodes, one would expect the cell index array to be formed in 1/Nth the time.

The second area targeted for parallelization in this method was the process of calculating a representative set of collisions for each cell in the flowfield. In order for a node processor to compute collisions for a cell, it must have access to information such as molecular velocities (three components) and weight factors for all molecules within the cell. Since, as mentioned previously, the hypercube is a distributed memory machine, the node processors cannot access this information directly but rather must receive it in the form of a message from the host. By including molecular position information (two coordinates) in the message, one is also able to compute cell number densities and molecular motions in parallel. In the sequential version of the code (see Fig. 2), collisions are calculated one cell at a time for each cell in the flowfield. In the parallel version, we still cycle over

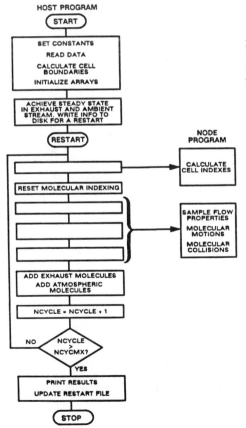

Fig. 5 Flow chart for parallel version (Method 1) of DSMC method code.

cells. This time, however, the appropriate molecular properties for a given cell are written to a buffer and then sent to an available (idle) node for processing. When a node has finished calculating the collisions, densities, and molecular motions, it writes the updated velocity, position, and density data to a buffer. These data are then sent to the host, which, upon receiving the information, places the data into its proper location in the appropriate arrays. The host then sends information on a new cell to the now idle processor, thereby assuring that the work is being distributed over all N processors (load balancing). This process continues until all cells in the flowfield have been cycled through.

There are two distinct advantages to this algorithm. First of all, memory requirements for a given processor are relatively small as the collision and molecular motion processes require information only for those "molecules" within the cell. This number is typically less than 30 and usually gets no larger than 100. In addition, the method usually achieves load balancing, that is, it keeps all the processors busy while calculating collisions and motions. As soon as a processor has finished with a given cell, it is sent information for a new cell, in this way, the collision and molecular motion processes are truly being distributed over N processors.

The disadvantage to this algorithm is that it very quickly reaches a point where an increase in the number of processors results in no significant decrease in execution time. Since the collision process accounts for 80-90% of the overall execution time, the best possible reduction in run time one could hope for (if only the collision process was made concurrent) would be a factor of five to ten; the host processor would still be required to carry out the remaining 10-20%. Execution times for calculations done using this algorithm are therefore limited by the rate at which the host processor carries out its computations.

Method 2

In the second, completely parallel algorithm, each processor carries out a "distinct" but coupled DSMC method computation on a unique subspace of the computational domain. The calculations are coupled because molecules, during the molecular motion process, may leave the space spanned by one processor and either enter the space spanned by another processor or leave the domain entirely. The processors must therefore exchange information after the molecular motions occur.

The boundaries for each subspace are required lie along the radial divisions of the computational domain (Fig. 1).

APPLICATION TO EXHAUST PLUME FLOWFIELDS

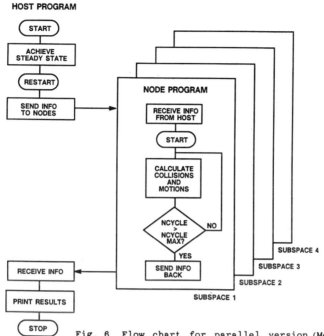

Fig. 6 Flow chart for parallel version (Method 2) of DSMC method code.

Many cirteria, such as number of molecules or cells, could be employed to determine the subspace boundaries. Our boundaries were selected so that each subspace contained, on average, the same number of cells.

After a steady state in exhaust and stream molecules is achieved in the usual sequential manner (see Fig. 6) the host initiates the concurrent computation by distributing the flowfield molecules among the processors according to cell location. Each processor, upon receiving its share of molecules, begins its computation by adding source and stream molecules to the flow. Upon completion of this task, a set of collisions appropriate to the time step dt is computed for each cell in the subspace. Once the appropriate set of collisions has been computed for a given subspace, the molecular motions occur. Until this point, each processor's DSMC method calculation has been uncoupled from the others. Now, however, each processor must wait until all other processors have reached the same point in the calculation in order to receive, in the form of messages from the other processors, the molecules that have entered its subspace. Once all the messages have been received, each processor adds the new molecules to its flowfield and then repeats the above processes until the appropriate number of iterations have taken place.

We make no assumptions concerning the movement of molecules between subspaces. By this we mean that the calculation is completely coupled in that a molecule, based upon its velocity, may travel to any subspace from a given subspace. A nearest neighbor approximation, in which only communication and hence movement between nearest neighbors in the flowfield is allowed, does not, in this case, conserve molecules. In addition, the nearest neighbor approximation becomes worse as one increases the number of processors because more molecules traverse multiple (non-nearest neighbor) subspace boundaries as the size of each subspace becomes smaller.

The advantages of this algorithm are twofold: 1) it allows one to contemplate solving problems that are currently too large to solve on conventional computers since the number of cells and hence the number of "molecules" are being distributed over N processor; and 2) the possibility of large decreases in execution time exists as one may reduce the amount of work any one processor has to do by decreasing the number of cells in its space, i.e., by increasing the number of processors. The major disadvantage of the second algorithm is that there is no guarantee the method will achieve load balancing as each processor must wait for the other processors to finish before it can continue.

In the next section, we present specific results, in terms of execution time, for the parallel and sequential algorithms described in the previous sections.

V. Results and Discussion

Before presenting specific results, we find it convenient to establish a context for the discussions that follow. In general, the maximum possible decrease in execution time a process can achieve, when run concurrently on N processors, is

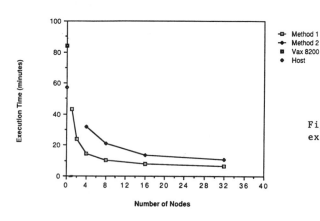

Fig. 7 Comparison of excecution times.

given by the following relationship:

$$t_p = t_s / N$$

where t_s is the execution time of the sequential version on one processor and t_p the theoretical execution time of the parallel version on N processors. The actual (observed) execution time t_{pa} is usually greater then t_p for the following reasons: 1) the communication overhead may be high; 2) the particular algorithm may not achieve load balancing, that is, not keep all processors busy; and 3) the optimal parallel algorithm itself may be less efficient than the optimal sequential one. Multiprocessor efficiency is then defined as the ratio of t_p to t_{pa}. Another measure of effectiveness often encountered when discussing parallel algorithms is the speed-up of a given algorithm. It is defined here as the ratio of the execution time of an algorithm on a single node to that on multiple nodes. Speed-up is, in essence, a measure of how ideally suited a given algorithm is to a parallel architecture. For example, in an ideal algorithm, one should decrease the execution time and hence increase the speed-up by a factor of two whenever one doubles the number of processors. We find it convenient to use both of the terms discussed above when describing the results of our study.

The main results of the present study are based upon ten iterations of both the sequential and parallel DSMC method codes described in Secs. II and IV. The iterations were run from a point in the solution where a steady state had been attained in both the exhaust and ambient stream. During these runs there were, on average, 38 molecules per cell, corresponding to a total of about 17,000 molecules. Although the resource manager (host) is a multiuser computer, we were the sole users of the system for all of the cases reported in this paper.

The variation in execution time with the number of nodes for both parallel algorithms is plotted in Fig. 7. The value of execution time given for zero nodes in Fig. 7 refers to the execution time for the sequential version of the program when run on the host (57.5 min). Similarly, any values reported for one node refer to the execution time for the parallel version (method 1) when run on a single node. For the sake of comparison, we included the execution time (84 min) for an equivalent run of the sequential program on a Vax 8200. It appears that the 80386-based host is some 1.5 times faster than the 8200. At present, we are unable to run the parallel code that employs method 2 on anything smaller than four nodes because of certain assumptions that were made when developing the code. While we intend to remove this constraint in future versions of the code, its presence in the current version makes calculating speed-up for the second algorithm difficult (remember that speed-up is defined as the ratio of execution time on a single node to that on multiple nodes). Rather than base the speed-up curve for the second algorithm on the execution time for a four-node system, we choose to use the execution time for method 1 (when run on a single node) as the basis of the speed-up curve. We believe this to be a fair approximation because, excluding some communication overhead, the execution time of method 1 on a single node is essentially that of the sequential version when run on a single node, which is exactly what method 2 would be when run on a single node.

At first, because of communication overhead in the parallel version, we were quite surprised to find that the execution time of the sequential version when run on the host was greater than that for method 1 when run on a single node. After some thought, however, we were able to reconcile the longer execution time on the host with the fact the host's processor is not a dedicated processor as are the nodes' processors, which means that the host must attend to system tasks in addition to running a program. The nodes' performance is further enhanced by the fact that each node also possesses a high speed instruction cache that the host does not.

We turn our attention now to the subject of this paper, namely, the comparison in terms of performance of the two parallel algorithms described in Sec. IV. From Fig. 7, we see that for the first parallel algorithm the execution time decreases by a factor of 9 in going from a sequential run on the host to a parallel run on a 32 node cube with the majority of the decrease (85%) coming in the first 4 nodes. The second parallel algorithm achieves only a fivefold reduction in execution time on a 32 node cube with 55% of the decrease coming in the first 4 nodes.

When comparing two algorithms, it is perhaps more informative to look at the speed-up that occurs in going from one node to multiple nodes. From the data contained in Fig. 7 we calculated speed-up as a function of number of nodes by dividing the execution times for runs on 2, 4, 8, and 16 node configurations into the execution time for the corresponding run on a single node; the results are given in Fig. 8. Ideally, one would like to see a plot that is linear over a large number of nodes, that is, the time required to run the program on 16 nodes should be one-half of that for 8 nodes. Unfortunately, the curve for the first method does not display this behavior, but instead levels off with increasing number of nodes. The rapid decrease in multiprocessor efficiency with the increasing number of nodes for the first algorithm indicates that this algorithm is not ideally suited to a parallel architecture. This is not to say that the algorithm we have selected is a poor one. It is quite possible that, when compared to other algorithms for the DSMC method, it will achieve the largest reductions in execution time.

Even though the first algorithm achieves load balancing, it suffers because the host is still required to perform some of the computation. In the sequential calculation that formed the basis for Figs. 7 and 8. the collision process accounted for about 90% of the overall execution time (51.75 out of 57 min). Even if our parallel algorithm were to reduce this fraction to 0%, the host would still be required to carry out the remaining 10% worth of computations. Our algorithm is thereby bounded by the rate at which the host can carry out these computations.

One advantage of the first algorithm is the ability to obtain significant reductions in execution time with a small number of nodes and hence a relatively small investment. A decrease in execution time on the order of a factor of two or three often translates into a difference of a day in turnaround time. Considering that, as mentioned previously, the statistical uncertainty in the output of DSMC method calculations fall off only as the inverse square root of sample size,[10] the reduction in execution time achieved on a two- or four node system is noteworthy. It is also quite impressive to note that the time spent calculating collisions has been reduced from 51.75 min (90%) for the sequential version of the code to about 4 min (63%) for the parallel version (method 1) of the code when run on 32 nodes. Included in this 4 min is, in addition to the collision time, the time required to move the molecules and sample flow properties.

Predictions concerning the second algorithm are a bit more difficult to make because it is no longer obvious how the

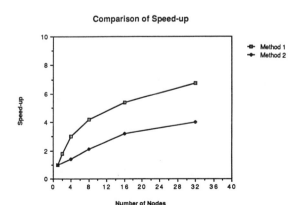

Fig. 8 Comparison of speed up.

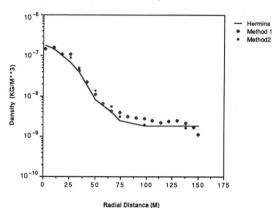

Fig. 9 Comparison of density profiles.

addition of more processors will effect load balancing. For example, the execution time decreased by a factor of only 1.3 in going from 4 to 8 nodes, while increasing the number of nodes from 8 to 16 resulted in a decrease much closer to the theoretical limit of a factor of 2 (1.7). We believe that the variations in multiprocessor efficiency evident in the second algorithm are indicative of the different levels of load balancing being achieved.

We should also point out that in the second algorithm we made no attempt to optimize the load balance other than to partition the cells equally among the processors. One could certainly obtain significant improvements by repartitioning the cells according to the work load.

Figure 9 presents density profiles at an altitude of 157 km and an axial position 50 m downstream from the source for 100 iterations of the parallel versions of the code. The density profile from a similar calculation carried out by Hermina[7] is also included. We include this plot for two reasons: 1) to illustrate the equivalence, in terms of results, between both parallel algorithms described within this paper and 2) to show that the DSMC method code used in this study compares well with codes developed elsewhere. Similar agreement between Hermina's code and our own was obtained at other altitudes.

VI. Summary and Recommendations for Future Research

In this paper, results have been presented which demonstrate that the execution time for the DSMC method can be improved significantly by the use of parallel processing. Specifically, we compared two distinct parallel algorithms and showed that they both generate significant reductions in execution time.

The first algorithm chosen for study was one in which the collision processes within a cell and the molecular motions are done in parallel. Each of these processes is done for the molecules within a number of cells simultaneously, with each processor being assigned a new cell after completing the required computations from the previously assigned one. Even though the first algorithm was limited in that speed-up curves obtained using it leveled off rather quickly, it still resulted in a ninefold decrease in execution time for 32 nodes. We feel quite certain that an optimized version of this algorithm could easily approach a tenfold decrease. One simple modification that would enhance the current algorithm would be to keep track of the execution time required to calculate collisions in each cell and to then distribute the cells to the processors on the next iteration such that the cells requiring the most time are sent first. This would avoid situations that occur at the end of the collision process in which you have only a few of the processors calculating collisions because they were sent "difficult" cells late in the collision process.

In the second, fully parallel algorithm, the computational domain is partitioned among the processors so that each processor carries out a distinct but coupled DSMC method calculation within its subspace. The calculations are not completely independent because molecules may traverse subspace boundaries. Although the algorithm achieves a fivefold decrease in execution time when run on 32 nodes, it, in general, does not achieve the performance increases the previous algorithm because it fails to achieve load balancing—that is, it fails to distribute the work load evenly among the processors. We feel, however, that significantly greater reductions in execution time could be achieved by "fine tuning" the current algorithm so that a more equitable distribution of the work load is obtained.

The second algorithm, however, does have the advantage that it allows one to contemplate solving DSMC method calculations that are currently too large to solve on conventional computers since the number of cells and hence the number of simulated molecules are being distriubted over N processors.

One word of caution is order here. As noted, we studied the effect of parallel processing on execution times only after the onset of a steady state. At present, the benefit concurrent computing would present before the onset of a steady state has not been evaluated, but the amount of time spent achieving a steady state is not significant when compared to the overall execution time for the example treated here.

In conclusion then, for calculations on a relatively small number of processors, it appears that greater decreases in execution time can be achieved by the first algorithm. If, however, one has access to a machine with a large number of processors (16 or more) one should consider the feasibility of employing the second algorithm with special emphasis given to the problem of achieving a load balance.

One other possibility, which we have not discussed previously, would be to perform a completely independent calculation for the entire domain in each processor and then compute the flowfield properties by taking an ensemble average of the results from each processor. Since the calculations are completely independent, this algorithm would have the advantage of showing almost perfect speed-up. The only disadvantage at present would be that the size of the calculation would limited by the amount at random access memory available on the node.

References

[1] Bird, G.A., "Article Title," *Physics of Fluids*, Vol. 6, 1963, p. 1518.

[2] Bird, G.A., *Molecular Gas Dynamics*, Clardendon Press, Oxford, England, U.K., 1976.

[3] Sheffield, J.S., "Monte Carlo Direct Simulation of High Altitude Plume Irradiance," Paper presented at 13th JANNAF Plume Technology Meeting, Houston, April 1982.

[4] Lee, R.H.C. and Nelson, D.A., "Enhancement Radiation from Rocket Plumes in the Transitional Flow Regime," Air Force Rocket Propulsion Laboratories, AFRPL TR-83-076, 1984.

[5] Moss, J.N. and Bird, G.A., "Direct Simulation of Transitional Flow for Hypersonic Re-entry Conditions," *Progress in Astronautics and Aeronautics: Thermal Design of Aeroassisted Orbital Transfer Vehicles*, Vol. 96, edited by H.F. Nelson, AIAA, New York, 1985.

[6] Cuda, V. and Moss, J.N., "Direct Simulation of Hypersonic Flows Over Blunt Slender Bodies," AIAA Paper 86-1348, June 1986.

[7] Hermina, W.L., "Monte Carlo Simulation of High Altitude Rocket Plumes with Nonequilibrium Molecular Energy Exchange," Paper presented at 15th JANNAF Plume Technology Meeting, San Antonio, 1985.

[8] Bird, G.A., "Monte-Carlo Simulation in an Engineering Context," *Progress in Astronautics and Aeronautics: Rarefied Gas Dynamics*, Vol. 74, edited by S.S. Fisher, AIAA, New York, 1988, Pt. 1, pp. 239-255.

[9] Furlani, T.R. and Lordi, J.A., "Implementation of Direct Simulation Monte Carlo Method for an Exhaust Plume Flowfield in a Parallel Computing Environment," *AIAA Thermophysics Plasmadynamics and Lasers Conference*, AIAA Paper 88-2736, June 27-29, 1988, San Antonio, TX.

[10] Elgin, J.B., "Getting the Good Bounce: Techniques for Efficient Monte Carlo Analysis of Complex Reacting Flows," Spectral Sciences Inc., Rept. SSI-TR-28.

Statistical Fluctuations in Monte Carlo Calculations

I. D. Boyd* and J. P. W. Stark†
University of Southampton, Southampton, England, United Kingdom

Abstract

The Direct Simulation Monte Carlo method has grown in status in recent years and is now a popular tool for the solution of rarefied flow problems. Several different simulation schemes now exist; so, in an effort to establish which of these is to be preferred, a series of investigations into the implementation and performance of the various methods has been undertaken. It is the intention of the present work to report on the results of a quantitative analysis of the Time Counter and Modified Nanbu simulation schemes. Specifically, the convergence of the calculations to a steady macroscopic state is examined. The effect of time step employed and the number of simulated molecules per cell is discussed with reference both to macroscopic and microscopic behavior.

Introduction

The Direct Simulation Monte Carlo method (DSMC) has been developed by Bird[1,2] over a number of years. The principal aim of this numerical technique is to allow acceptable simulation of rarefied flow phenomena while keeping computational expense to a minimum. In this capacity, Bird's Time Counter method has been quite successful and finds application in increasingly complex flow problems. For example, the results of a three-dimensional simulation of a chemically reacting gas mixture with inclusion of rotational and vibrational energy modes, flowing over diffusely reflecting, intersecting wedges, has recently been reported in Ref. 3.

Copyright © 1989 by the American Institute of Aeronautics and Astronautics, Inc. All rights reserved.
*Graduate Student, Department of Aeronautics and Astronautics; currently at NASA Ames Research Center, Moffett Field, California.
†Senior Lecturer, Department of Aeronautics and Astronautics.

This application highlights the variety of important physical phenomena that may be investigated with DSMC calculations.

Despite the fact that the Time Counter technique has been applied to a variety of flow problems by a number of workers, doubt still remains as to the validity of the method. In the solution of rarefied gas flow problems, the Boltzmann equation is generally accepted as the most appropriate mathematical model. While results obtained with Bird's method are found to be consistent with Boltzmann equation solutions, it is not possible for this method to be obtained directly from this model. Indeed, Nanbu[4] has shown that the Time Counter method is derived from the Kac master equation.

An alternative simulation scheme derived by Nanbu[5] directly from the Boltzmann equation would therefore seem to be preferred to that of Bird, at least from a theoretical standpoint. However, a previous investigation by the authors[6] concluded that for engineering solutions, the TC method is preferred both in terms of flexibility and of computational efficiency. In particular, Nanbu's scheme was found to be much more sensitive to the choice of the time step Δt over which molecular motion and intermolecular collisions are decoupled. In addition, no significant differences in the solutions provided by the two methods were discernible.

However, the debate over the DSMC techniques took a new turn with the advent of the Modified Nanbu technique, which has been developed at the University of Kaiserslautern, FRG. These improvements described by Ploss[7] immediately overcame the objections to the original Nanbu scheme that were reported in Ref. 6. Under these modifications, a major section of the collision routine may be vectorized, thus leading to significant improvements in computational performance. In addition, the introduction of the idea of executing the collision algorithm several times over each time increment overcomes the problem relating to the initial choice of Δt. The Modified Nanbu method is found to be just 20% more expensive than that of Bird when implemented on a computer with special vector facilities (see Ref. 7).

The new method was compared to the TC scheme and good qualitative agreement found for both unsteady one-dimensional and steady two-dimensional flows[8]. The number of times which the collision algorithm is called over each time step Δt

is denoted by L and must be determined prior to simulation. Included in Ref. 8 is a method for estimating this parameter from the initial flow conditions. The major conclusion derived from this work study is that the Modified Nanbu scheme should be seriously considered as an alternative solution technique to Bird's Time Counter method.

Having established the fact that the Modified Nanbu and TC schemes produce consistent macroscopic flow solutions, an investigation into the process by which these averaged solutions are seen to converge has been undertaken. In the previous comparisons, the sample size from which the final results are obtained is assumed to be the same for each technique. This assumption is unqualified. As the Modified Nanbu method calculates the collision probability for each and every simulated molecule in the flowfield, the scheme may be regarded as more deterministic in nature than the TC scheme. This attribute of the scheme may then be expected to produce results exhibiting less statistical scatter. If this is observed to be true, then use of this scheme would be preferred as acceptable error margins would be attained over a reduced statistical sample size. However, additional fluctuations may be associated with the Modified Nanbu method since energy and momentum are not conserved at the calculation of each individual collision.

The investigation reported here was made with reference to such variables as translational and rotational temperature, and flow velocity, which are sampled at several points in the flowfield. Both macroscopic averages and molecular distribution functions were analyzed. The calculation of inelastic collisions, in which transfer of energy between translational and internal energy modes is performed, was achieved through the use of the Larsen-Borgnakke phenomenological model[9]. In the case of the Modified Nanbu simulation scheme, this model was implemented in a manner that was in keeping with the fundamental philosophy of the Nanbu method. Thus, only one molecule undergoes change due to each collision.

Results and Discussion

The flowfield modelled is the sonic expansion of hot nitrogen into a vacuum and is similar to that reported in Refs. 6 and 8. The flow is two-dimensional and the large changes in flow density associated with the expansion are dealt with through variation of cell volume. In the following, all distributions and macroscopic values

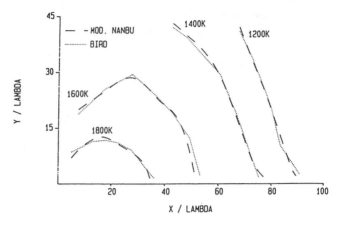

Fig. 1 Rotational temperature contours for sonic expansion of nitrogen.

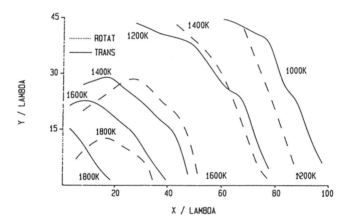

Fig. 2 Comparison of translational and rotational temperature contours.

are obtained by sampling the flow quantities over a suitable sample size once the flowfield has reached a steady state.

Good agreement between the methods for such macroscopic properties as temperature, density, and velocity vectors has been found previously. The inclusion of inelastic collisions has not been formally incorporated before into the Nanbu simulation methods. Therefore, in Fig. 1, rotational temperature contours for the two methods are presented. Good agreement is observed for these calculations. Rotational and translational temperature contours for the flowfield are compared in Fig.

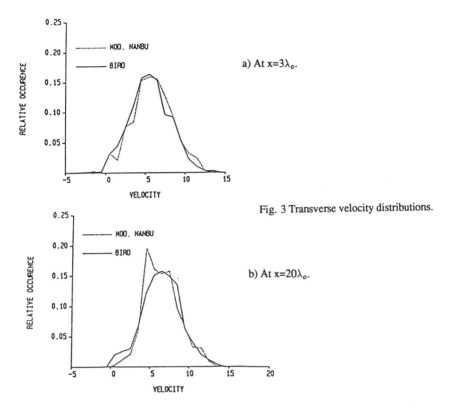

Fig. 3 Transverse velocity distributions.

2. The departure of these contours clearly highlights the regions of intense thermal nonequilibrium as we proceed away from the nozzle exit plane.

In addition to predicting comparable macroscopic properties, it is essential that detailed agreement exists for the molecular quantities simulated by the two schemes. Therefore, in the present study, distribution functions for velocity and rotational energy have been calculated at various points in the flowfield. The manner in which these distributions are seen to converge to a steady state is also investigated.

In Figs. 3a and 3b, transverse velocity distributions are shown at two separate points in the flowfield for the two simulation schemes. The first is at a point near to the nozzle exit that is characterized by high density and thermal equilibrium. The distribution in Fig. 3b was derived at the point near the axis at a distance of 20 exit mean free paths ($20\lambda_o$) downstream of the nozzle exit plane. Significant rarefaction effects are present in this region of the flowfield. The macroscopic velocities are found to be in good agreement for the two methods at both of these flowfield sampling locations. However, it is clear that, in the case of Fig. 3b, the distribution provided by Bird's method is smoother and more closely resembles those shown in Fig. 3a. The profile derived from the Modified Nanbu scheme indicates that significant statistical fluctuations exist. It is to be noted that all distributions are calculated from identical sample sizes.

The manner in which the Modified Nanbu method converges to a final distribution at $20\lambda_o$ is shown in Fig. 4 for rotational energy. The solid line represents the final distribution obtained after completion of the averaging process, while the other data were derived at the halfway point of this process. Similar plots for Bird's method are shown in Fig. 5 and indicate that a smaller degree of scatter and greater agreement exists for these two cases.

In Fig. 6, the convergence to steady state of translational temperature is plotted for the two methods. Once again it is clear that the calculations made with the Time Counter method are more satisfactory in that the values calculated throughout the averaging process are clustered around the final value.

Figures 3-6 indicate that the Modified Nanbu scheme exhibits an increased amount of statistical scatter when compared with calculations made with the TC method. In order to quantify such observations, a series of simulations has been undertaken in which the decoupling time step Δt and the average number N of simulated molecules per cell was varied. In the case of the Modified Nanbu calculations, the number of times that the collision algorithm is called over each time increment, (i.e., L) is also varied. Table 1 contains details of the 11 sets of conditions analyzed.

In Fig. 7 are shown values of translational temperature derived at the same point in the flowfield as used in Fig. 3b for each of the runs listed in Table 1. The results obtained from run 2 are assumed to represent the best DSMC predictions. This assumption is based on a number of previous investigations, all of which were found to be consistent with the calculations resulting from the use of these particular parameters. The translational temperature provided by the results of run 2 are bounded above and below by the generally accepted margin of 1σ error for DSMC

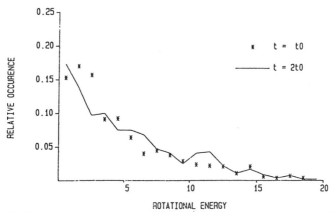

Fig. 4 Rotational energy distributions for the Modified Nanbu method, $x=20\lambda_o$.

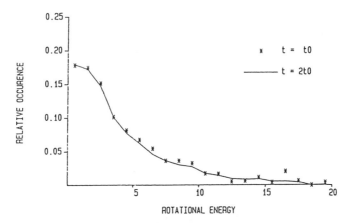

Fig. 5 Rotational energy distributions for the Time Counter method, $x=20\lambda_o$.

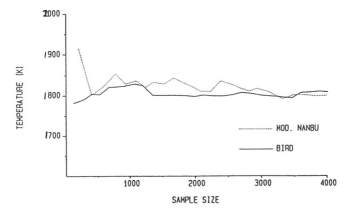

Fig. 6 Convergence of translational temperature to a steady state, $x=20\lambda_o$.

Table 1 Conditions of Simulations

Run	Method	N	Δt	Symbol
1	Time Counter	10	0.5	*(1)
2	Time Counter	20	0.5	*(2)
3	Time Counter	20	0.6	*(3)
4	Time Counter	20	0.7	*(4)
5	Modified Nanbu	10	0.5(2)	+(1)
6	Modified Nanbu	20	0.5(2)	+(2)
7	Modified Nanbu	20	0.6(2)	+(3)
8	Modified Nanbu	20	0.7(2)	+(4)
9	Modified Nanbu	20	0.7(4)	O
10	Modified Nanbu	20	0.7(8)	X
11	Modified Nanbu	20	0.7(16)	X

calculations, i.e., the percentage error is given by the inverse of the square root of the sample size. These errors are represented by the two horizontal broken lines. Each point in Fig. 8 is generated by storing the value of translational temepratures obtained in the cell after each time increment throughout the averaging procedure. Then, using the final value obtained at the end of the simulation, the deviation of these interim temperatures is calculated. The horizontal line represents the theoretical value.

Several important features are immediately discernible from Figs. 7 and 8. First, as expected, more accurate results are obtained by increasing the average number of simulated molecules per cell for a given time step. It should be remembered that all results reported are extracted from the same sample size, so that by doubling the number of molecules in each cell, the total number of time increments is thereby halved. This result is enforced in Fig. 8 where the observed statistical scatter about the mean value is also reduced as N is increased. These trends are observed for both simulation schemes. It is also apparent that, under the same conditions, the Modified Nanbu scheme shows a larger standard deviation about the mean. In addition, it may be seen that by assuming that run 2 provides the true solution also provides us with the value showing the least statistical scatter.

Examination of Fig. 7 for runs 2-4 reveals that for the TC method, as Δt is increased, the results obtained begin to depart from the true value. The amount of

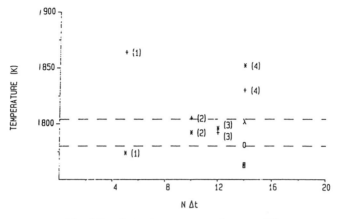

Fig. 7 Translational temperature values at $x=20\lambda_o$.

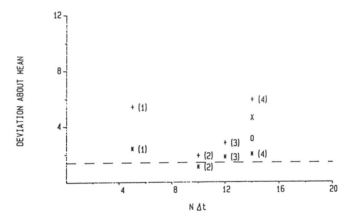

Fig. 8 Percentage deviation about the mean translational temperature, $x=20\lambda_o$.

statistical instability in these solutions is confirmed by the increases observed in the standard deviation, shown in Fig. 8.

Let us now consider the results obtained with the Modified Nanbu technique. The study of runs 6-8 allows analysis of the effect of the choice of Δt for a fixed value of L. It is clear from the figures that the conditions employed in run 7 offer the best results both in terms of consistency with the TC solution and with respect to the statistical scatter observed. This result is rather surprising when it is noted that the collision probability in the Modified Nanbu scheme is given by

$$P = \frac{\Delta t}{L} \frac{n}{N} \sigma_T c_r. \tag{1}$$

i.e., directly proportional to Δt.

From the previous investigations of the Nanbu schemes, it would be expected that smaller collision probabilities would lead to better calculation of collisional processes. However, similar behavior is observed in the results of runs 8-11. In these investigations, Δt is fixed while L is increased. Again, from Eq.(1), it is expected that the collision probability is best dealt with through a large value of L.

In run 8, a high deviation about the mean is associated with a largely inaccurate value for the translational temperature. The margin of these errors indicate that it is necessary to employ a larger value of L in order to improve the Modified Nanbu calculations, i.e., the collision probability has been allowed to exceed unity. Hence, as L is increased, the accuracy and scatter of the calculated results would be expected to improve. While this is immediately observed in the cases where L is increased to values of 4 and 8, a further rise to 16 leads to a higher degree of scatter and a less accurate result. It is concluded that increased statistical errors are associated with the Modified Nanbu method for both very high and very low collision probabilities.

Finally, it should be noted that the standard deviations about the mean of the values obtained with the TC method are certainly lower than those for the Modified Nanbu scheme. The sample size employed in each of the investigations 1-11 was 5000 giving a statistical fluctuation level of about 1.4%. This value is very close to that exhibited by the calculations performed in run 2. The smallest fluctuation level observed with the Modified Nanbu scheme is 1.9% for run 6. If it is assumed that the deviation about the mean may be expressed as a single power of the sample size N, then it may be deduced that

$$\sigma = N^{-0.465} \tag{2}$$

With this expression, the sample size required to give a deviation of 1.4% is found to be about 9700, i.e., almost double that required for the Bird method. In a further investigation, it was found that for a sample size of 10,000, the deviation for the Modified Nanbu scheme had been reduced to only 1.7%. It is therefore concluded that an increased amount of statistical noise is always present in the Modified Nanbu calculations. These additional fluctuations must result from differences between the two simulation schemes. They are therefore attributable either to the manner in which collision partners are selected or to the way in which post collision values are assigned. It is unlikely that the selection procedure for colliding molecules will have a large effect. In general, it is the pairs of molecules having the largest collision probabilities which will be accepted for collision.

The fact that energy and momentum are only conserved in the Nanbu schemes for the single particle system will be unimportant when the number of simulated molecules is large. In such cases, a large number of collisions will be calculated and the fluctuations in energy and momentum will be damped out. However, for a smaller number of simulated molecules, giving rise to only one or two collisions over each time increment, failure to conserve these quantities will become significant. There appear to be two limitations placed on the practical use of the Modified Nanbu scheme. If the flow parameters are such that the collision probability of a significant number of molecules exceeds unity, then the type of poor result observed in Ref. 6 may occur. Alternatively, if the collision probability is too small (small Δt or large L) then increased statistical fluctuations occur due to the lack of conservation of momentum and energy.

Other flow properties examined in this study include rotational temperature and flow velocity. In the case of the internal mode temperature, the trends observed were consistent with those reported above for the translational modes. When the velocity distributions such as that shown in Fig. 3b were analyzed in detail, it was found that despite differences in higher moments, excellent agreement existed for the mean value and for the deviation about the mean. Under the worst conditions, the mean values were found to differ by no more than 4%.

While the above discussion refers to flow properties at one particular point, similar trends were observed throughout the flowfield and have been interpretted through application of computer graphics output in Ref. 10. In regions near to the nozzle exit, the solutions provided by the various runs showed better agreement and less scatter. In regions where the local collision rate was low, the statistical scatter for both schemes was generally larger. However, it was found that the calculations made with the TC method were in each case observed to provide solutions with less statistical scatter.

Conclusions

It has been observed with reference to translational temperature that the Time Counter method of Bird shows less statistical scatter than that found with the Modified Nanbu simulation scheme. Nanbu[11] has previously shown that in his original method the statistical fluctuations are indeed inversely proportional to the square root of the sample size. However, these investigations were carried out in a model cell environment and not in the more useful context of an engineering application. It is important to note that, in Bird's scheme, the fluctuation level is maintained within the theoretical limit.

It has also been noted that large statistical fluctuations are associated with the Modified Nanbu scheme when the number of subdivisions of the decoupling time step is taken to be too large. This is a previously unexpected result. It is concluded that these fluctuations are due to the failure of the method to conserve energy and momentum in regions of the flowfield characterized by a low collision rate. This aspect of the DSMC technique has previously been noted by Hermina[12] with reference to the Time Counter method. It appears that the problem is more acute in the case of the Modified Nanbu scheme.

It is concluded that, while the Modified Nanbu scheme may have a better theoretical basis, it cannot compete with Bird's Time Counter method in terms of computational efficiency. While it has just a 20% computational overhead when comparison is made on the basis of the collision algorithms alone, the Modified Nanbu method requires a larger sample size to reach a particular fluctuation level. Thus when the two schemes are compared in the context of obtaining particular results, the Modified Nanbu scheme may require as much as 100% more computational effort than the Time Counter method.

References

[1] Bird, G.A., Molecular Gas Dynamics, Clarendon Press, Oxford, England, 1976.

[2] Bird, G.A., "Monte-Carlo Simulation in an Engineering Context", Progress in Astronautics and Aeronautics: Rarefied Gas Dynamics, AIAA, New York, 1980, pp. 239-255.

[3] Celenligil, M.C., Bird, G.A., and Moss, J.N., "Direct Simulation of Three Dimensional Flow about Intersecting Blunt Wedges", AIAA Paper-88-0463, AIAA 26th Aerospace Sciences Meeting, Reno, NV, 1988.

[4] Nanbu, K., "Interrelations Between Various Direct Simulation Methods for Solving the Boltzmann Equation", Journal of the Physical Society of Japan, Vol. 52, Oct. 1983, pp. 3382-3388.

[5] Nanbu, K., "Direct Simulation Scheme Derived from the Boltzmann Equation: I. Monocomponent Gases", Journal of the Physical Society of Japan, Vol. 49, May 1980, pp. 2042-2049.

[6] Boyd, I.D. and Stark, J.P.W., "A Comparison of the Implementation and Performance of the Nanbu and Bird Direct Simulation Monte Carlo Methods", Physics of Fluids, Vol. 30, Dec. 1987, pp. 3661-3668.

[7] Ploss, H., "On Simulation Methods for Solving the Boltzmann Equation", Computing, Vol. 38, 1987, pp. 101-115.

[8] Boyd, I.D. and Stark, J.P.W., "On the Use of the Modified Nanbu Direct Simulation Monte Carlo Method", Journal of Computational Physics, to be published.

[9] Borgnakke, C. and Larsen, P.S., "Statistical Collision Model for Monte Carlo Simulation of Polyatomic Gas Mixtures", Journal of Computational Physics, Vol. 18, 1975, pp. 405-420.

[10] Boyd, I.D., "Monte Carlo Simulation of an Expanding Gas", Computers in Physics, to be published.

[11] Nanbu, K., "Direct Simulation Scheme Derived from the Boltzmann Equation: V. Effect of Sample Size, Number of Molecules, Step Size, and Cutoff Angle upon Simulation Data", Institute for High Speed Mechanics, Tohoku University, Tohoku, Japan, Rept. 348, 1981.

[12] Hermina, W.L., "Monte Carlo Simulation of Transitional Flow Around Simple Shaped Bodies", Proceedings of 15th International Symposium on Rarefied Gas Dynamics, Teubner, Stuttgart, FRG, 1987, pp. 452-460.

Applicability of the Direct Simulation Monte Carlo Method in a Body-fitted Coordinate System

T. Shimada*
Nissan Motor Company, Ltd., Tokyo, Japan
and
T. Abe†
Institute of Space and Astronautical Science, Kanagawa, Japan

Abstract

A formulation of the direct simulation Monte Carlo method in a body-fitted coordinate system of multidimensional arbitrary configuration is presented and is also extended to unsteady moving boundary applications. A few sample calculations of two- and three-dimensional external flow problems of geometries, including singular points, multiply-connecting regions, and actual aerobrake configuration are conducted. From the computed results, the versatility of the present method and its applicability to the multidimensional problem of arbitrary bodies are confirmed.

Introduction

Recently, aerothermodynamic problems of hypersonic rarefied gas flows past bodies have drawn much attention in relation to newly planned re-entry vehicles such as space planes and aeroassisted orbital transfer vehicles (AOTV). In such a flow regime the molecular gasdynamics approach is required to investigate the aerothermodynamics of problems. So far, there are not many methods of solution other than the Monte Carlo methods, and especially, the direct simulation Monte Carlo (DSMC) method developed by Bird[1] has been most widely used.

As those application needs become larger, chances of the DSMC method being used to solve flows around a multidi-

---
Copyright © American Institute of Aeronautics and Astronautics, 1988.
*Aerospace Technologist, Aerospace Division.
†Associate Professor of Gasdynamics.

mensional complex boundary geometry, such as vehicle shape, will increase. A typical DSMC calculation usually consists of procedures for flow initialization, molecular transfer, molecular indexing, molecular collision, and flowfield sampling. Among these procedures, molecular transfer and molecular indexing strongly relate to boundary geometries of specific problems. Since the Cartesian coordinates are usually used, procedures become complicated for identifying the cell where a particle belongs as well as for detecting the particle-surface interactions as the geometry becomes complex. Some users might be discouraged by this complexity in programming DSMC codes for three-dimensional flow calculation, or by the program production cost, since the program must be tailored to each different vehicle shape.

One solution to this inconvenience was given by Bird[1] introducing the point-network approach, which enables one to construct an algorithm for handling arbitrary geometry of multidimensions. In that approach each particle is considered as in a cell when it is closer to the cell centroid than to any other. A particle-surface interaction is detected when its nearest cell centroid is an imaginary point lying inside the wall. The extension to three-dimensions is not simply straightforward because additional concepts of subdivided tetrahedral elements must be incorporated. They are still complicated for programmers to handle.

A new approach for freeing this difficulty is proposed by one of the authors.[2] The scheme is unique in the sense that it utilizes the body-fitted coordinate system so that an arbitrary geometry can be treated. In the field of computational continuum fluid dynamics, the body-fitted coordinate system has been used for years and has proved to be a powerful technique for multidimensional problems. It was developed by Thompson et al.,[3] followed by various applications and extensions by many researchers since the late 1970's.[4] Its fundamental concept is that all of the calculations are conducted in a rectangular plane or solid region which is transformed to a complex-shaped physical space through a general transformation. The transformation is discretely defined through metrics calculated at the grid points where their (x,y,z) coordinates are given. These grid points are made to correspond to points uniformly distributed in the computational space. This makes application of a finite-difference scheme to a complex geometry very simple and allows arbitrary shapes to be handled by a single program.

In this paper we first describe the original formulation of the DSMC method in a body-fitted coordinate system[2] in order to review its merits and demerits. As will be mentioned, since in the original formulation derivatives called metrics are used to define the contravariant velocity, or particle velocity in the computational space, it cannot directly apply to the grid geometry including singular points or points where metrics cannot be defined, e.g., at a sharp leading edge of a plate and along the centerline. We propose, then, a modified formulation that utilizes cell boundary surface area vectors to describe the transformation and show a few applications results for geometries including singular points. Next, the modified formulation is extended to three dimensions and is applied to a three-dimensional flow around a lifting-brake configuration. Finally, the applicability of this method to unsteady calculations of moving or deforming boundaries is discussed.

Basic Formulation

The basic concept and formulation of the DSMC method in a body-fitted coordinate system are given in Ref. 2 and are reviewed briefly here.

The body-fitted coordinate system (ξ, η) is defined by the following general transformation:

$$x = x(\xi, \eta) \tag{1a}$$

$$y = y(\xi, \eta) \tag{1b}$$

which transforms a rectangular region in the (ξ, η) space into a region around bodies in the physical space (x,y). A schematic explanation of the transformation is given in

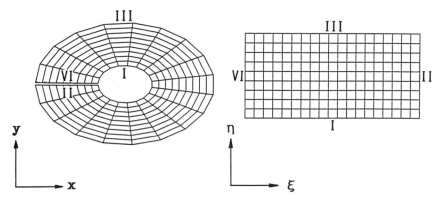

a) Body-fitted coordinate system. b) Rectangular coordinate system.

Fig. 1 Schematic illustration of the general transformation between the body-fitted coordinate system and the rectangular coordinate system.

Fig. 1. Note that this is only an example of the transformation, in this case for an O grid, and any other grid topology that can be expressed by Eqs. (1) can replace it. To discretize the physical space into finite-volume cells, a body-fitted grid system is used that is automatically generated, e.g., by solving elliptic partial differential equations with appropriate boundary conditions. Such a grid consists of two groups of lines: lines along the body contour and lines emitted from the body surface. In the (ξ,η) space, these lines are expressed as η=constant and ξ=constant of arbitrary values, which usually are integers so that $\Delta\xi$ and $\Delta\eta$ are unities. Note that the relations between the (x,y) space and the (ξ,η) space are only known at the grid points where both coordinate values are given.

Each molecule is assigned the physical velocity components (u,v,w) and the computational space coordinates (ξ,η). The major difference between this method and the standard DSMC method is that the molecular transfer and molecular indexing are conducted in the (ξ,η) space. The merits of the method are that 1) the molecular-surface interaction is easily detected since the surface is identical to an η=constant line; 2) the molecular indexing is conducted quickly since each cell in the (ξ,η) space is a unit square; and 3) a program for arbitrary geometry is constructed in a simple, systematic manner. Molecular transfer calculation is performed by integrating the equations of motion of a particle under no external forces expressed in the (ξ,η) space:

$$d\xi/dt = (\partial y/\partial \eta \; u - \partial x/\partial \eta \; v \;)/J \qquad (2a)$$

$$d\eta/dt = - (\partial y/\partial \xi \; u - \partial x/\partial \xi \; v \;)/J \qquad (2b)$$

$$J = \partial x/\partial \xi \; \partial y/\partial \eta - \partial x/\partial \eta \; \partial y/\partial \xi \qquad (2c)$$

The molecular collisions are counted in a cell over a time interval Δt_c the same as that in the standard method. Then, Eq. (2) must be integrated from $n\Delta t_c$ to $(n+1)\Delta t_c$, which is conducted numerically as follows:

$$(\xi^{k+1}-\xi^{k})/\Delta t = [(\partial y/\partial \eta u - \partial x/\partial \eta v)/J]^{k} \qquad (3a)$$

$$(\eta^{k+1}-\eta^{k})/\Delta t = [-(\partial y/\partial \xi u - \partial x/\partial \xi v)/J]^{k} \qquad (3b)$$

where Δt is the time step and ξ^{k} represents $\xi(n\Delta t_c+k\Delta t)$, etc. The metrics $\partial x/\partial \xi$, $\partial x/\partial \eta$, $\partial y/\partial \xi$, and $\partial y/\partial \eta$ are evaluated at each grid point by the central finite-difference scheme. The superscript k of the right-hand side of Eqs. (3)

indicates that this term is evaluated at $\xi=\xi^k$ and $\eta=\eta^k$. The metrics necessary for the evaluation can be obtained from the interpolation of those values known at the neighboring grid points.

Although the preceding basic formulation can be applied to various body configurations, it cannot handle the grid including singular points. Another point to be considered is how to choose the time step for the molecular transfer calculation. In the basic formulation it is chosen as a small constant independent of ξ and η. This works well as long as Δt is small enough to resolve the particle trajectory even at the severest space distortion due to the transformation. It is therefore more reasonable to define Δt corresponding to the local space distortedness.

Present Method

Molecular Transfer

An alternative and improved approach is to express the contravariant velocity in terms of cell boundary surface area vectors. The method can be recognized as an extension of the basic method where the appropriate evaluation is made of the one-sided differential coefficients at the singular point. These cell boundary surface area vectors are defined in Fig. 2 and can be evaluated from the node point locations as

$$\vec{S}_{Iij} = (y_{i,j+1} - y_{i,j})\vec{e}_x - (x_{i,j+1} - x_{i,j})\vec{e}_y \qquad (4a)$$

$$\vec{S}_{Jij} = -(y_{i+1,j} - y_{i,j})\vec{e}_x + (x_{i+1,j} - x_{i,j})\vec{e}_y \qquad (4b)$$

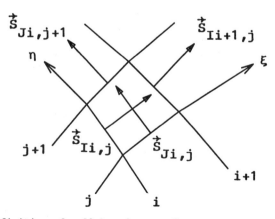

Fig. 2 Definition of cell boundary surface area vectors and nodal point indices of the two-dimensional cell (i,j).

where $\vec{e}_x$ and $\vec{e}_y$ are x- and y-directional unit vectors, respectively. For an infinitesimally small cell, Eqs. (4) can be written as follows:

$$\vec{S}_I = \partial y/\partial \eta \, \vec{e}_x - \partial x/\partial \eta \, \vec{e}_y \tag{5a}$$

$$\vec{S}_J = -\partial y/\partial \xi \, \vec{e}_x + \partial x/\partial \xi \, \vec{e}_y \tag{5b}$$

Using Eqs. (5), the contravariant velocity components Eqs. (2) are expressed with $\vec{S}_I$ and $\vec{S}_J$ as

$$U = d\xi/dt = \vec{S}_I \cdot \vec{q} \, / |\vec{S}_I \times \vec{S}_J| \tag{6a}$$

$$V = d\eta/dt = \vec{S}_J \cdot \vec{q} \, / |\vec{S}_I \times \vec{S}_J| \tag{6b}$$

where $\vec{q}$ is the physical velocity vector of a particle and the operators • and x represent taking a scalar product and a vector product of vectors, respectively.

To obtain a particle path in the computational plane, Eqs. (6) are integrated numerically employing, e.g., a simple forward difference scheme as in Eqs. (7):

$$(\xi^{n+1} - \xi^n) / \Delta t^n = \vec{S}_I^n \cdot \vec{q} \, / |\vec{S}_I^n \times \vec{S}_J^n| \tag{7a}$$

$$(\eta^{n+1} - \eta^n) / \Delta t^n = \vec{S}_J^n \cdot \vec{q} \, / |\vec{S}_I^n \times \vec{S}_J^n| \tag{7b}$$

In Eqs. (7), $\vec{S}_I^n$ and $\vec{S}_J^n$ represent those vectors at $(\xi,\eta) = (\xi^n,\eta^n)$ and are generally unknown. A simple, effective interpolation scheme for the evaluation of these vectors can be written as follows, assuming $\Delta\xi = \Delta\eta = 1$:

$$\vec{S}_I^n = (i+1-\xi^n) \, \vec{S}_{Ii,j} + (\xi^n - i) \, \vec{S}_{Ii+1,j} \tag{8a}$$

$$\vec{S}_J^n = (j+1-\eta^n) \, \vec{S}_{Ji,j} + (\eta^n - j) \, \vec{S}_{Ji,j+1} \tag{8b}$$

where i and j are the largest integers but are not greater than ξ^n and η^n, respectively. Note that variables i and j are the surface indices, and at the same time, they represent the coordinate values, in other words, a location in the (ξ,η) space.

During the course of this study the time step for the particle trace integration is found to be critical for the accuracy. In this formulation the time step is estimated locally at the particle location at each step from the contravariant velocity of the particle (U,V):

$$\Delta t^n = \text{MIN}(\Delta t_c, \Delta t_\xi, \Delta t_\eta) \qquad (9a)$$

$$\Delta t_\xi = \delta \Delta \xi / |U| \qquad (9b)$$

$$\Delta t_\eta = \delta \Delta \eta / |V| \qquad (9c)$$

where δ is a small constant and $\delta < 0.05$ is found to give good results. Since the time step varies from particle to particle, we use a time-keeping array that contains the time of all particles.

Boundary Procedures

The solid boundary contour is chosen as the line of $\eta=1$. Therefore, a particle impingement is detected when $\eta<1$ is encountered.

Before conducting the full DSMC simulation we checked the particle path traceability of the method, especially in the vicinity of a singular point. From this test it was found that, as long as the time step estimated by Eqs. (9) is used, this method can trace a particle accurately enough for the Monte Carlo simulation. The only case when a special treatment becomes necessary is when a particle hits the singular point. The mathematical probability of a molecule to hit a point is zero, but in the simulation, due to the finite time step, such impingement occurs, though it is very rare. The frequency of such impingement becomes lower as the time step becomes smaller. We recommend the constant coefficient δ in Eqs. (9) to be 0.01, in which case the experienced rate of the impingement is practically zero (less than $1/10^7$). Since such impingement seldom occurs, we simply eliminate the particle when it should hit a singular point.

Molecular Indexing

The indexing scheme we use is the standard Bird's algorithm. Then identification of the cell where each particle belongs is required. This is done simply as follows. When a particle is at (ξ^*, η^*), the cell indices (i,j) containing this particle are $i=\text{INT}(\xi^*)$ and $j=\text{INT}(\eta^*)$, where the function INT(X) represents the largest integer but is not greater than X.

Extension to Three-Dimensions

The extension of the present method to three-dimensions is quite straightforward. The particle contravariant veloci-

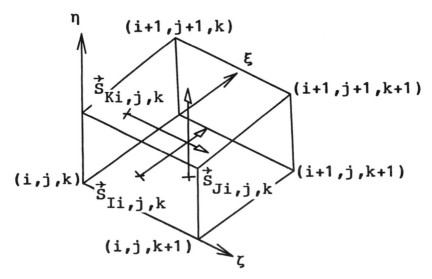

Fig. 3 Definition of cell boundary surface area vectors and nodal point indices of the three-dimensional cell (i,j,k).

ty components are expressed as follows, employing cell boundary surface area vectors $\vec{S}_{Ii,j,k}$, $\vec{S}_{Ji,j,k}$, $\vec{S}_{Ki,j,k}$ defined as depicted in Fig. 3:

$$U = d\xi/dt = (\vec{S}_I \cdot \vec{q})/J \qquad (10a)$$

$$V = d\eta/dt = (\vec{S}_J \cdot \vec{q})/J \qquad (10b)$$

$$W = d\zeta/dt = (\vec{S}_K \cdot \vec{q})/J \qquad (10c)$$

$$J = [\,(\vec{S}_J \times \vec{S}_K) \cdot \vec{S}_I\,]^{1/2} \qquad (10d)$$

and $\vec{q} = (u,v,w)$. In Eqs. (10) vectors $\vec{S}_I$, $\vec{S}_J$, $\vec{S}_K$ are evaluated at particle locations from the interpolation similar to Eqs. (8).

Results and Discussions

Two-Dimensional Flows

Three two-dimensional rarefied flows are simulated with a single program of the DSMC simulation coded along the present method. Simulated two-dimensional geometries are a vertical flat plate, a lifting-brake AOTV, and a pair of cylinders. The cell network used for these simulations are shown in Fig. 4.

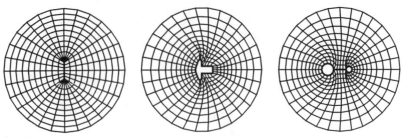

a) Vertical flat plate.   b) Lifting-brake AOTV.   c) Pair of cylinders.

Fig. 4 Cell networks for two-dimensional bodies.

The current program can handle the gas mixture and internal degree of freedom, as well as an arbitrary geometry. For the modeling of the polyatomic flows, the Larsen-Borgnakke model is employed.[5] The frequency of molecular collisions are controlled employing Bird's time counter. The hard sphere model is used as the molecular model.

The surface is assumed to be specularly reflective, and the freestream conditions used are the following: molecular species dealt include nitrogen, atomic oxygen, and molecular oxygen; their number densities ($1/m^3$) are 3.8×10^{17}, 1.4×10^{17}, and 5.4×10^{16}, respectively; the kinetic temperature is 335 K; and the stream velocity is 8000 m/s. These correspond to flight conditions in the Earth's atmosphere at 120 km altitude. Estimated mean free path is about 3 m. For two-dimensional simulations, the number of real molecules f_n represented by one simulation molecule is set to 5×10^{15}.

To check the ability of the present method of handling the geometry including singular points, a problem of a flow past a vertical flat plate shown in Fig. 4a is chosen. The length of the plate is 2.6 m; then the overall Knudsen number is 1.2. Computed molecule number density and the velocity field are shown in Fig. 5. Impingement of a molecule on one singular point during Δt_c was observed once out of more than 10^7 particle's movements on the average. We simply eliminate these molecules, but no bad influence has been found.

The same program is used to simulate a flow past a two-dimensional lifting-brake AOTV. No change in the program is necessary, and only grid coordinates specified as input data are changed. The diameter of the brake is 4.3 m, and then the corresponding Knudsen number is 0.7. The flow attack angle α is 20 deg. The computed number density and the velocity are shown in Fig. 6. The peak number density is found in the stagnation region and is about 3.3 times higher than the freestream value. About 4×10^6 simulation molecules

DSMC METHOD IN A BODY-FITTED COORDINATE SYSTEM 267

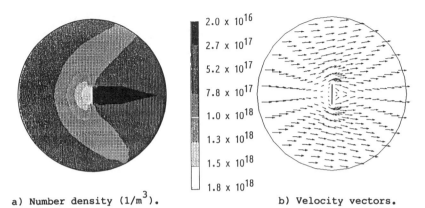

a) Number density (1/m^3). b) Velocity vectors.

Fig. 5 Computed results for a flow past a vertical flat plate (freestream speed = 8 km/s, 120 km altitude Earth's atmosphere, Kn = 1.2).

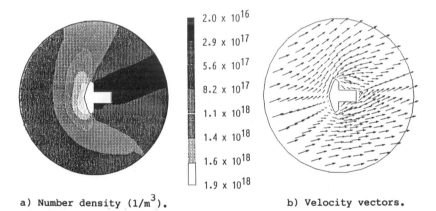

a) Number density (1/m^3). b) Velocity vectors.

Fig. 6 Computed results for a flow past a lifting-brake AOTV (freestream speed = 8 km/s, 120 km altitude Earth's atmosphere, Kn = 0.7, angle of attack = 20 deg).

are sampled in total where the time average is used after the flowfield becomes steady. The total CPU time is about 5000 s in CRAY X-MP12 supercomputer, and the core storage used is about 600K.

To demonstrate the applicability of the method to the problem of the multiply-connected geometry, a flow around a cylinder pair is chosen as an example. The grid is given in Fig. 4c. The diameters of the big and the small cylinders are 2.0 m and 1.0 m, respectively, and the distance between their centers are 3.5 m. The freestream enters from the southeast (α=135 deg) direction. For this application the

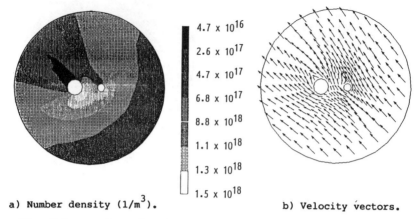

a) Number density (1/m^3). b) Velocity vectors.

Fig. 7 Computed results for a flow past a pair of cylinders (freestream speed = 8 km/s, 120 km altitude Earth's atmosphere, Kn = 0.7, flow enters from down, right).

program is modified in order to incorporate the boundary procedures at the common cut (a portion of the line of $\eta=1$ that connects two circles). This is simply done by replacing the ξ coordinate by $(\xi_{MAX}+1-\xi)$ when a molecule crosses the common cut. The computed density contours and the velocity vectors shown in Fig. 7 clearly give the nature of the flow interactions.

Three-Dimensional Flow

A three-dimensional rarefied flow is simulated in order to test the feasibility of the present method in three dimensions. As a testing geometry, a three-dimensional lifting-brake AOTV configuration is chosen, though only the half-space is considered for saving the computer storage. The cell system used is generated by revolving the two-dimensional grid system (upper half) of Fig. 4b around its centerline. In the program the grid point coordinates are specified as input data and the surface area vectors and the cell volumes are calculated from these grid point location data. The attack angle is set to 20 deg and diffuse model of surface reflection is assumed. The freestream conditions used are the following: number densities (1/m^3) of N_2, O, and O_2 are 8.7×10^{18}, 4.1×10^{17}, and 2.0×10^{18}, respectively; the temperature is 199 K; and the stream velocity is 8000 m/s. These correspond to flight conditions in the Earth's atmosphere at 100 km altitude. The parameter f_n is changed to 3.9×10^{16} due to the storage limitation of the computer system. The computed number density and the veloci-

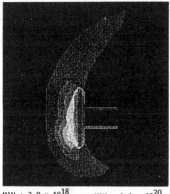
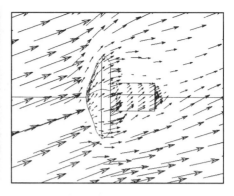

MIN : 3.0 x 10^18   MAX : 1.1 x 10^20

a) Number density (1/m³).   b) Velocity vectors.

Fig. 8 Computed results for a flow past a three-dimensional lifting-brake AOTV (freestream speed = 8 km/s, 100 km altitude Earth's atmosphere, angle of attack = 20 deg).

ty vectors on the body surface and the symmetrical plane are shown in Fig. 8.

Extension to Unsteady Problems

As is known, through the use of the body-fitted coordinate system, the deformation and acceleration of the geometry can be treated easily.³ Let us discuss such application with the present DSMC method. To simulate flows past deforming or accelerating bodies, we actually deform and move the physical body-fitted coordinate system in an inertial frame of reference. Both the direction and the magnitude of the surface area vectors change due to such time-dependent deformation of the grid, whereas, in the case in which the grid system accelerates or moves without deformation, they remain constant.

The general transformation is defined so that at any moment it transforms the moving cell system in the physical space to a fixed rectangular region in the computational space. Using such transformation, the particle velocity in the computational space can be written as follows:

$$U = d\xi / dt = \vec{S}_I \cdot \vec{q}* / J \qquad (11a)$$

$$V = d\eta / dt = \vec{S}_J \cdot \vec{q}* / J \qquad (11b)$$

$$W = d\zeta / dt = \vec{S}_K \cdot \vec{q}* / J \qquad (11c)$$

$$J = [\ (\vec{S}_J \times \vec{S}_K) \cdot \vec{S}_I\ ]^{1/2} \qquad (11d)$$

$$\vec{q}^* = \vec{q} - \vec{q}_c \qquad (11e)$$

where $\vec{q}_c$ is the velocity of the moving body-fitted coordinate system and then q^* is the relative velocity of the particle to the moving coordinates.

In the numerical integration of Eqs. (11), $\vec{S}_I$, $\vec{S}_J$, and $\vec{S}_K$ in the right-hand side are evaluated at the particle location at the current time. The velocity of the body-fitted coordinate system $\vec{q}_c$ is also evaluated at the particle location at the same time level. This can be done by interpolating current values of $\vec{q}_c$'s known at the nodal points of the cell.

Conclusions

The present method is highly versatile in applications of the DSMC method to multidimensional flows of arbitrary geometry, including singular points and multiply-connected regions, and furthermore, possibly to problems of unsteadily moving boundaries of any shape. By the use of this method a DSMC program obtains strong universality, and its production and the running costs will be reduced.

References

[1] Bird, G. A., *Molecular Gas Dynamics*, Clarendon Press, Oxford, 1976, pp. 118-200.

[2] Abe, T., "Rarefied Gas Flow Analysis by Monte Carlo Direct Simulation Method in Body-Fitted Coordinate System," *Journal of Computational Physics*, 1989 (in press).

[3] Thompson, J. F., Thames, F. C., and Mastin, C. M., "Automatic Numerical Generation of Body-Fitted Curvilinear Coordinate System for Field Containing Any Number of Arbitrary Two-Dimensional Bodies," *Journal of Computational Physics*, Vol. 15, July 1974, pp. 299-319.

[4] Steger, J. L., "Implicit Finite Difference Simulation of Flow about Arbitrary Two-Dimensional Geometries," *AIAA Journal*, Vol. 16, July 1978, pp. 679-686.

[5] Larsen, P. S. and Borgnakke, C., "Statistical Collision Model for Simulating Polyatomic Gas with Restricted Energy Exchange," *Rarefied Gas Dynamics*, Vol.1, edited by M. Becker and M. Fiebig, Deutsche Forschungs und Versuchsanstalt für Luft und Raumfahrt (DFVLR) Press, Porz-Wahn, FRG, 1974; see also Borgnakke, C. and Larsen, P. S., "Statistical Collision Model for Monte Carlo Simulation of Polyatomic Gas Mixture," *Journal of Computational Physics*, Vol. 18, Aug. 1975, pp. 405-420.

Validation of MCDS by Comparison of Predicted with Experimental Velocity Distribution Functions in Rarefied Normal Shocks

Gerald C. Pham-Van-Diep* and Daniel A. Erwin†
University of Southern California, Los Angeles, California

Abstract

Velocity distribution functions in normal shock waves in argon and helium are calculated using Monte Carlo direct simulation. These are compared with experimental results for argon at $M = 7.18$ and for helium at $M = 1.59$ and 20. For both argon and helium, the variable-hard-sphere (VHS) model is used for the elastic scattering cross section, with the velocity dependence derived from a viscosity-temperature power-law relationship in the way normally used by Bird. Argon results are presented for several values of the viscosity-temperature exponent S; it is found that, under the present conditions, where the freestream is quite cold (16 K), a value of 1 is appropriate for S; i.e., argon closely resembles a Maxwell gas here. For helium, the value $S = 0.647$ is used. In both cases, agreement between computed and measured distribution functions is excellent.

Introduction

It has been accepted for some time now that Monte Carlo direct simulation[1] (MCDS) is reliable for prediction of macroscopic properties of rarefied flowfields. However, little attention has been paid to validation of detailed MCDS predictions such as the velocity distribution functions. This is probably due to the difficulty in performing experimental measurements of such quantities and to the amount of

---

Copyright ©1989 by the American Institute of Aeronautics and Astronautics, Inc. All rights reserved.
* Research Assistant, Department of Aerospace Engineering.
† Assistant Professor, Department of Aerospace Engineering.

computation required to build up sufficient statistics in phase space.

This work presents calculations of velocity distribution functions in argon and helium normal shock waves, with conditions chosen to match existing experimental data:[2-4] argon at $M = 7.18$ and helium at $M = 1.59$ and 20. The flowfield parameters are summarized in Tables 1-3. The experimental works cited give detailed distribution functions, in both the streamwise (parallel) and transverse (normal) directions, at several positions in the argon and $M = 20$ helium shocks; for the $M = 1.59$ helium shock, the half-width of the parallel distribution is given. In all cases, the data were taken at several stations within the shocks. The present results were

Table 1 Conditions of simulated shock wave (argon, $M = 7$) chosen to match the experimental conditions of Holtz and Muntz[2]

	Freestream	Downstream
Mach number	7.183	0.461
Density, cm^{-3}	1.144×10^{15}	4.324×10^{15}
Temperature, °K	16	276
Velocity, m/s	539.58	142.74

Table 2 Conditions of simulated shock wave (helium, $M = 1.59$) chosen to match the experimental conditions of Muntz and Harnett[3]

	Freestream	Downstream
Mach number	1.59	0.7034
Density, cm^{-3}	2.88982×10^{15}	5.2808×10^{15}
Temperature, °K	160	244.92
Velocity, m/s	1184.87	648.09

Table 3 Conditions of simulated shock wave (helium, $M = 20$) chosen to match the experimental conditions of Muntz[4]

	Freestream	Downstream
Mach number	20	0.44961
Density, cm^{-3}	2.88982×10^{15}	5.2808×10^{15}
Temperature, °K	2.2	285.1452
Velocity, m/s	1770	447.3901

obtained using a one-dimensional MCDS code written by Graeme Bird, with modification (described later) to permit recording of the distributions. As originally written, the variable-hard-sphere (VHS) collision model is used, with the diameter of the colliding particles derived from the viscosity curve; the viscosity μ is assumed to be a power function of temperature:

$$\mu(T) \propto \left(\frac{T}{T_{\text{ref}}}\right)^S \qquad (1)$$

where T_{ref} is a reference temperature at which the viscosity is known. ($S = 0.5$ describes hard spheres, whereas $S = 1.0$ describes Maxwell molecules.) The actual temperature-viscosity curves for argon and helium[5] are shown in Fig. 1. For argon at $T \lesssim 273°$K a value of $S = 0.81$ is appropriate; for much higher T, $S \cong 0.68$, and at low T, $S \cong 0.98$.

When the viscosity is assumed to vary as T^S, the elastic scattering cross section varies roughly as $(1/V_R^2)^{S-1/2}$, where V_R is the relative velocity of the colliding particles.

The change in slope of Fig. 1 occurs in the neighborhood of the temperature corresponding to the depth of the interatomic potential,

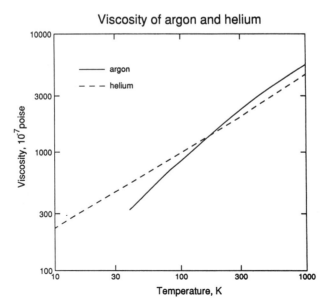

Fig. 1 Viscosity of argon and helium as functions of temperature, after Ref.5 .

which for argon is about 120 K. By contrast, that for helium is about 10 K; thus, the entire log-log viscosity curve for helium is straight with a slope $S \cong 0.65$.

Method of Computation

As mentioned, a one-dimensional MCDS code written originally by Bird was used. In this code, the shock develops normal to the x axis, which is divided into several hundred cells corresponding to an interval of a few centimeters containing the shock. To record velocity distribution functions during the simulation, two integer arrays, each containing 200 values ("bins"), are set up for each cell in physical space. One array is for the parallel (x) distribution, and the other for the perpendicular distribution. Each bin represents a region of velocity space; the arrays cover the interval between -2.4 and 3.2 times the freestream velocity. During each sampling of the flowfield properties, the x and y components of the velocity of each atom are used to increment the appropriate bins of the parallel and perpendicular arrays, respectively, in the cell occupied by the atom.

At the end of a run, the arrays hold discretized, non-normalized parallel and perpendicular velocity distributions. It should be mentioned that the parallel velocity is sampled without regard to the perpendicular, and vice versa; thus, the measured parallel distribution is actually $f_x(v_x) = \int \int f(v_x, v_y, v_z) \, dv_y \, dv_z$. This is appropriate for comparison with experiment, since there the distributions are obtained from Doppler-broadened spectral lines of electron beam stimulated atoms; the procedure measures velocity shifts in the direction of observation without regard to motion in the other directions.

The distribution functions vary with position within the shock. The simulation gives results for each physical cell; the location of each cell may be related to the normalized number density,

$$\hat{n}(x) \equiv \frac{n(x) - n_{\text{FS}}}{n_{\text{DS}} - n_{\text{FS}}} \qquad (2)$$

where the subscripts FS and DS denote freestream and downstream respectively.

The distribution functions were recorded for each cell in physical space. To relate these to the values of the normalized number density $\hat{n}$ at which the experimental distributions were taken, the distributions at the discrete cells were interpolated using a five-point cubic fit to the desired normalized number densities, the interpolation based on the values of $\hat{n}$ in the discrete cells.

Results and Discussion

Argon at $M = 7$

The shock density profiles for argon at $M = 7$ are shown in Fig. 2. Profiles are given for several values of S; a viscosity of 2.11×10^{-5} poise at 273 K was used in all calculations. The profile is only slightly sensitive to S, and all the curves fall close to the experimental data.

The computed distribution functions for Maxwell molecules are shown in Fig. 3 and 4 for the parallel and perpendicular components of velocity, respectively; the distributions are given at several points

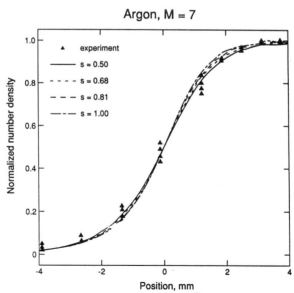

Fig. 2 Shock density profile for argon at $M = 7$: Computed results, for several values of S, and measurements. The shock thickness is seen to be roughly 3 mm.

within the shock. The perpendicular distributions are centered at zero velocity; their spreading as the shock is traversed represents the temperature increase. The parallel distributions become bimodal in the interior of the shock, blending the freestream and downstream distributions.

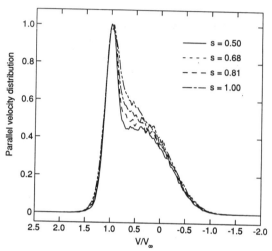

Fig. 3 Argon, $M=7$: Computed parallel velocity distributions for Maxwell molecules at several positions within the shock.

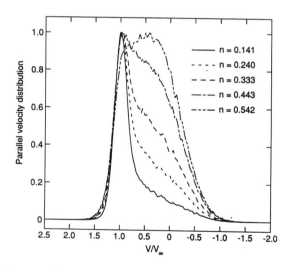

Fig. 4 Argon, $M=7$: Computed perpendicular velocity distributions for Maxwell molecules at several positions within the shock.

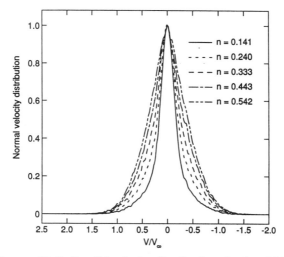

Fig. 5 Argon, $M=7$: Parallel velocity distributions for $\hat{n} = 0.333$ for several values of S.

Figure 5 gives the parallel distributions at $\hat{n} = 0.333$ computed for several values of S. An increase in S causes an increase in the collision cross section at low relative velocities, so that the distribution "fills in" between the peaks. By contrast, the perpendicular distributions do not change appreciably with S.

To properly compare the computations with the experimental results of Holtz and Muntz, the computed distributions were convolved with the measured instrument profiles, which are given in the dissertation of Holtz.[6] These profiles are due to the finite transmission linewidth of the Fabry-Perot etalon used in the measurements. The parallel velocity distributions, both computed and measured, are given in Fig. 6 for several values of $\hat{n}$ corresponding to those at which the measurements were performed.

The convolution has a smoothing effect, which unfortunately removes some of the detail, making comparison with experiment less meaningful. Nevertheless, it is clear that, although agreement is fair in all cases, the Maxwell molecules provide the best fit to experiment.

The perpendicular distributions, both the computed ones, convolved with the appropriate instrument responses, and the experimental measurements, are shown in Fig. 7. Agreement is excellent, except in the case of $\hat{n} = 0.333$; as suggested by Muntz, this measurement may actually have been made somewhat downstream of $\hat{n} = 0.333$.

Helium at $M = 1.59$

The MCDS for a normal shock in helium at a Mach number of 1.59 required modification of the code previously used for the argon shock simulations discussed earlier. Since for $M = 1.59$ the shock in helium is thicker than in argon at $M = 7.18$, it was necessary to increase the size of the computation window from 2 cm to 4 cm. Note

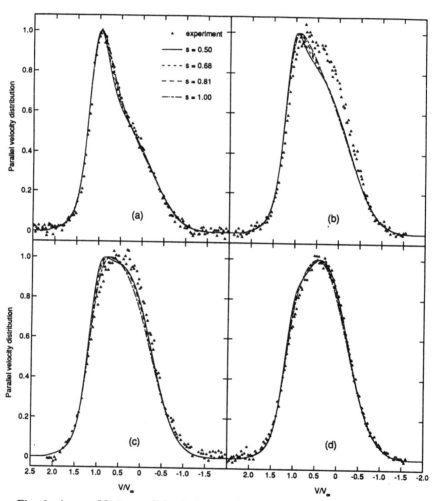

Fig. 6 Argon, $M=7$: parallel velocity distributions, computed and measured, for several values of S. The computed curves have been convolved with measured instrument functions for comparison with experiment: a) $\hat{n}=0.240$; b) $\hat{n}=0.333$; c) $\hat{n}=0.443$; d) $\hat{n}=0.542$.

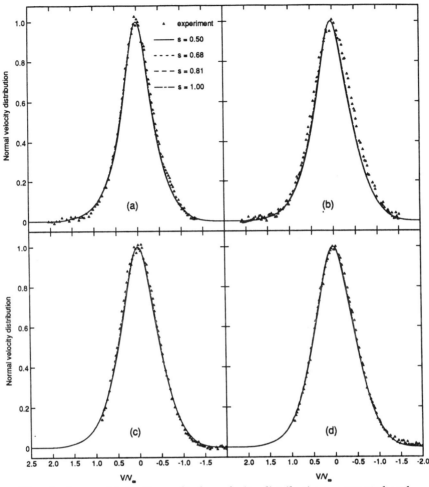

Fig. 7 Argon, $M=7$: Perpendicular velocity distributions, computed and measured, for $S=1.0$ (Maxwell molecules). Details as in Fig. 6.

that, in this case, the upstream mean free path is approximately 1.33 mm, which is about five times the cell size. A value of $S = 0.647$ (see Fig. 1) was chosen for the simulation. To compare the computations with the results of Muntz et al.[3] the non dimensional half-widths of the velocity distributions vs non dimensional number density were calculated. The non dimensional half-width is defined as

$$\hat{W}(x) \equiv \frac{W(x) - W_{\text{FS}}}{W_{\text{DS}} - W_{\text{FS}}} \qquad (3)$$

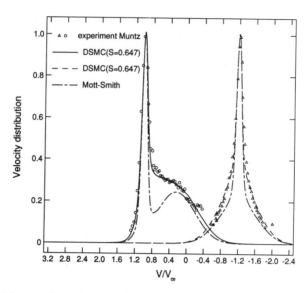

Fig. 8 Helium, $M=1.59$: Non dimensional half width vs non dimensional density.

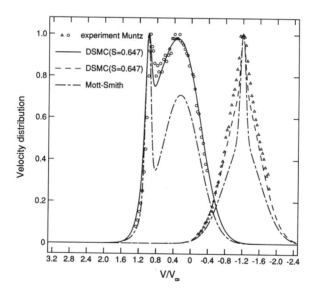

Fig. 9 Helium, $M=20$: Parallel and perpendicular velocity distributions for $\hat{n} = 0.585$, computed and experimental. The perpendicular distributions are shown shifted to the right for clarity; they are actually centered at 0.0.

where the subscripts FS and DS, as defined earlier, denote freestream and downstream. Note that $\hat{W}(x)$ was computed with the velocity distributions convolved with the appropriate instrument profile.

Figure 8 shows the computed and measured non dimensional half-widths, along with those given by the Mott-Smith and Chapman-Enskog distributions.[1] It can be seen that the MCDS as well as the Mott-Smith predictions agree fairly well with the experiment. This is particularly true for the perpendicular velocity distributions. The agreement between the parallel velocity distribution half-widths and the data is not as good, however, it can be seen that the discrepancies are not very significant and could be explained by experimental errors as suggested by Muntz.[4] The statistical characteristic of the MCDS computation is illustrated by the presence of fluctuations in the half-width curves. These tend to disappear as the simulation sampling time increases. The Chapman-Enskog model is less accurate than either MCDS or Mott-Smith.

Helium at $M = 20$

For the simulation of a normal shock in helium at $M = 20$, the window chosen for computation spanned from $x = -4$ cm to $x = +2$ cm, with the shock roughly centered at $x = 0$; the temperature influence propagates quite far upstream for strong shocks, necessitating a wide simulation window.

To match the experimental conditions, in which the shock was formed through a free expansion, a Mach number of 20 corresponding to an upstream temperature of 2.2 K was used only for the parallel velocity distribution prediction; a simulation of a $M = 28$ shock for a freestream temperature of 1.1 K was computed to predict the perpendicular velocity distributions. A temperature exponent of $S = 0.647$ was used for these simulations. The parallel and perpendicular velocity distributions, both computed and measured, are shown in Figs. 9 and 10 for several values of $\hat{n}$. Again, the computed distribution functions were convolved with the appropriate instrument functions. It can be seen that the agreement between experiment and prediction is quite good for both parallel and perpendicular distributions. One must also note that the blending of the freestream and downstream distributions is simulated with the MCDS considerably better than with the Mott-Smith prediction.

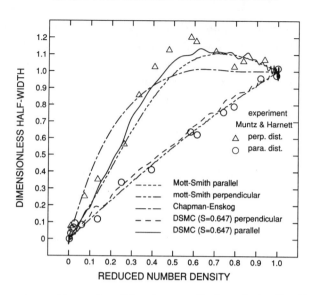

Fig. 10 Helium, $M=20$: Parallel and perpendicular velocity distributions for $\hat{n} = 0.30$, computed and experimental (perpendicular distribution shifted for clarity).

Conclusions

We may now state definitively that Monte Carlo direct simulation is completely adequate in its treatment of elastic interatomic collisions, to the best of our ability to measure. In particular, the variable-hard sphere collision model appears to work well, quantitatively reproducing measured distribution functions in strong shock waves. This detailed agreement is not obtained with either the Mott-Smith or Chapman-Enskog solutions to the Boltzmann equation.

Acknowledgments

We are grateful to Professor Graeme Bird for providing his MCDS source code to us and to E. P. Muntz for valuable comments and insights. This work was supported by NASA/DOD Hypersonics Research and Training Grant NAGW-1061.

References

[1] Bird, G. A., Molecular Gas Dynamics (Clarendon Press, Oxford, 1976).

[2]Holtz, T. and Muntz, E. P., "Molecular velocity distribution functions in an argon normal shock wave at Mach number 7," Physics of Fluids, Vol. 26, 1984, pp. 2425-2436.

[3]Muntz, E. P. and Harnett, L. N., "Molecular velocity distribution function measurements in a normal shock wave," Physics of Fluids, Vol. 12, 1969, pp. 2027-2035.

[4]Muntz, E. P., private communication, 1988.

[5]Hirschfelder, J. O., Curtiss, C. F., and Bird, R. B., Molecular Theory of Gases and Liquids, Wiley, New York, 1964.

[6]Holtz, T., "Measurements of Molecular Velocity Distribution Functions in an Argon Normal Shock Wave at Mach Number 7," Ph.D. Thesis, Univ. of Southern California, Los Angeles, CA, 1974.

Direct Monte Carlo Calculations on Expansion Wave Structure Near a Wall

F. Seiler*
Deutsch-Französisches Forschungsinstitut Saint-Louis (ISL),
Saint-Louis, France
and
B. Schmidt†
University of Karlsruhe, Karlsruhe, Federal Republic of Germany

Abstract

The influence of the boundary layer on the structure of a centered rarefaction wave moving along a wall was numerically investigated by using the gaskinetic direct simulation Monte Carlo method in comparison with a one-dimensional isentropic continuum solution given by the method of characteristics. In the simulation study the real gas flow is modeled statistically at atomic level for a monatomic, perfect gas with three degrees of freedom. The centered two-dimensional rarefaction wave is generated by a piston, which is suddenly set in motion and then moves with constant velocity. Calculations have been done for piston Mach numbers $M_p = u_p/a_0$ 1) with gas particles following the piston speed ($M_p < f$) and 2) at which the gas flow is detached from the piston ($M_p > f$), where f designates the degrees of freedom. If the gas temperature is initially set equal to the wall temperature, only small density and temperature gradients appear toward the wall. In the case of a cooler wall, the weak expansion wave structure far away from the wall changes next to the wall into a very steep wave behavior. An interesting phenomenon develops when the piston speed exceeds the maximum possible flow velocity. Toward the wall no density gradients are observed, whereas for the temperature and velocity distribution the boundary-layer effect is clearly visible.

Copyright © 1989 by ISL. Published by the American Institute of Aeronautics and Astronautics, Inc. with permission.

* Scientist, Department for hypersonic gasdynamics
† Professor, fluid mechanics, Institut für Strömungslehre und Strömungsmaschinen

Introduction

A piston, which is suddenly set in motion with a constant velocity inside a channel, being filled with gas, induces two different gasdynamic phenomena (Fig. 1). In front of the piston the gas is compressed by a "quasistationary" shock-wave. Contrary to the shock-wave compression in this case the gas behind the piston is expanded, forming a "time-dependent" centered rarefaction wave. One-dimensional gaskinetic calculations of the shock- wave[1] and the expansion wave[2] structure have been done in the past in order to show the internal structures of both wave phenomena on a molecular level.

In case the flowfield is bounded by a solid wall, the interaction of the gas particles with the wall becomes very important. At least, the flow becomes two-dimensional with large gradients of the state variables perpendicular to the wall. Results of gaskinetic modeling of the shock structure close to a wall were presented by Seiler and Schmidt[3] several years ago. For these calculations the direct simulation Monte Carlo (DSMC) technique, as developed by Bird,[4] was used. Since the DSMC method was found to be a successful gaskinetic procedure for calculating rarefied gas flows, this method was also applied to the two-dimensional expansion-wave flow pattern in the region near a wall. The suspicion that the conservation of angular momentum is not satisfied[5] is not true in general. It depends on the cell size in the mean free paths, the number of gas particles per cell, and the amount of vorticity that has to be modeled by the DSMC method. If cell sizes and the number of particles per cell are optimized the DSMC method should give an exact picture of the flow under consideration.

Gaskinetic Flow Formation

To apply the simulation technique, a numerical model was developed as shown in principle in Fig. 2, which can be used for both the shock-structure evaluation and the expansion-wave generation. The simulated flow region in this numerical computation is subdivided into a large number of cells of variable cell width. The flow behind the pis-

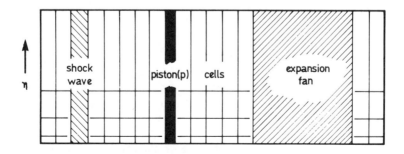

Fig. 1 Shock-wave and rarefaction-wave formation.

ton, respectively, in its front, is modeled in terms of a relatively large number of model particles. At the solid wall diffuse reflexion is used, according to a wall accomodation coefficient α, for simulating the boundary conditions at the wall. The rarefaction wave formation is generated by a piston, which is suddenly set in motion, thus expanding the flow at constant velocity.

The simulation procedure starts with particles initially uniformly distributed in space, which is given in two dimensions with the normalized x- and y-coordinate system of Fig. 2. For normalization the mean free path λ_0 of the particles in the undisturbed gas in front of the expansion wave is used ($\xi = x/\lambda_0, \eta = y/\lambda_0$). The initial velocity distribution is a Maxwellian one corresponding to thermodynamic equilibrium in the quiescent gas at the temperature T_0. The time $\tau = t \cdot c_{mp_0}/\lambda_0$, with c_{mp_0} being the most probable thermal speed of the particles at initial conditions (0), proceeds in discrete time steps. During the time intervals the particles move in phase space according to their individual velocities and are reflected at the boundaries. At each time step representative collision pairs are selected randomly in each cell according to their relative speed. The collision process has been calculated classically with the hard sphere model.

The value of the macroscopic quantities, i.e., density, temperature, and flow velocity, can be extracted in each cell at given time points τ by averaging over appropriate molecular quantities in each cell. The fluctuations of the macroscopic quantities in the flowfield are reduced by sampling and averaging the computational results of about 100 independent calculation runs.

The initial number of model particles within a cell was 10 for all runs. The initial cell size in the ξ direction was 0.5 λ_0. In the η direction the cell width next to the wall was set to be 1 λ_0 and increases with growing distance from the wall. Calculations have been carried out for piston Mach numbers $M_p = u_p/a_0 < f$ and for the case $M_p > f$, respectively, providing a survey of the two-dimensional expansion flow pattern of both flow domains: 1) the gas flow can follow the piston speed, and 2) the piston base is detached from the gas behind it. The

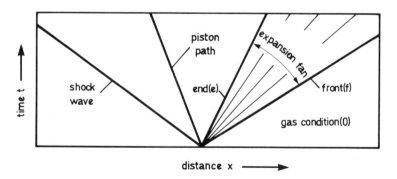

Fig. 2 Expansion and shock-wave generation model.

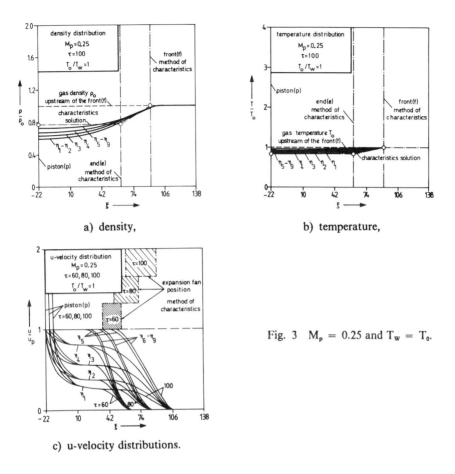

a) density,

b) temperature,

c) u-velocity distributions.

Fig. 3 $M_p = 0.25$ and $T_w = T_0$.

transition happens for $M_p = 3$, for which the velocity of the end (e) of the expansion wave equals exactly the piston speed.[2] All calculations have been done for a perfect monatomic gas, i.e., a gas with three thermodynamic degrees of freedom f. The interaction of the gas particles with the wall surface is simulated with the full accommodation at the wall, $\alpha = 1$.

Calculation Results

Attached Expansion Flow

The numerically obtained density, temperature, and u-velocity distributions for $M_p < f$ along the wall coordinate ξ at increasing distances from the surface $\eta_1 = 0.5$, $\eta_2 = 1.5$, $\eta_3 = 3$, $\eta_4 = 6$, $\eta_5 = 15$, $\eta_6 = 35$, $\eta_7 = 75$, $\eta_8 = 300$, and $\eta_9 = 1250$ are presented in Figs. 3-6. Two distinct boundary conditions with regard to the wall temperature are considered: initially 1) the gas temperature T_0 is equal to the wall

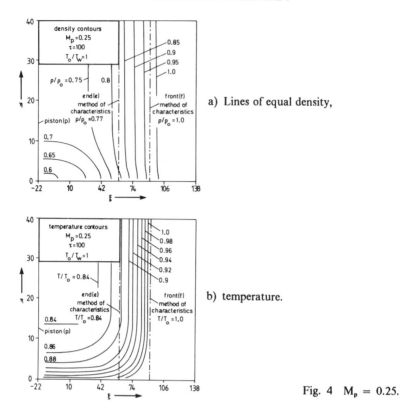

a) Lines of equal density,

b) temperature.

Fig. 4 $M_p = 0.25$.

temperature T_W (Fig. 3), and 2) the wall is cooler than the gas particles (Fig. 5). In case 1), small deviations from the one-dimensional rarefaction wave structure (approximately η_9) can be found for the density and temperature profiles at distinct distances from the wall. Regarding the u-velocity distribution of Fig. 3, the expansion fan widening in time τ is obvious. Because of the boundary condition at the wall (u = 0), a substantial velocity decrease from the outer flow region (u = u_p) toward the wall is present.

From the results of the DSMC calculations, as shown in Fig. 3, contours of equal density and temperature have been extracted and are plotted in Fig. 4. The boundary-layer influence is clearly evident with the deviations of the density and temperature contours from the one-dimensional flow pattern with vertically arranged lines of constant density and temperature, respectively. The small gradients in the flow quantities inside the boundary layer perpendicular to the wall are caused by taking the wall temperature T_W equal to the gas temperature T_0 in front of the expansion fan.

For the "hot" expansion flow development gathered from the DSMC calculations with a gas temperature $T_0 = 3\,T_W$, strong changes in the flow quantities toward the wall have been extracted from the si-

DIRECT MONTE CARLO CALCULATIONS

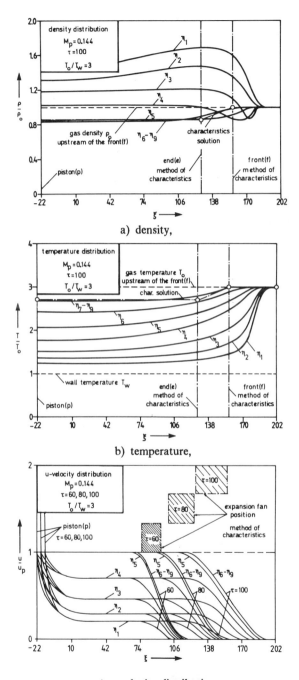

Fig. 5 $M_p = 0.144$ and $T_0 = 3 T_w$.

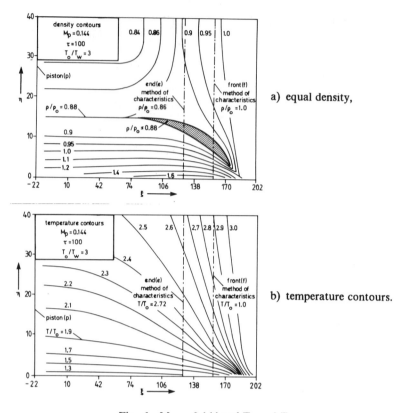

Fig. 6 $M_p = 0.144$ and $T_0 = 3\,T_w$.

mulation computations (Fig. 5). For the density and temperature profiles, the weak expansion wave structure far away from the wall (η_6 to η_9) changes into a very strong and steep wave pattern closer to the wall surface (η_1). The gradient is comparable with the flow variations inside a pressure wave. The reason for this expansion flow pattern next to the wall is that the heat-conduction effect in this simulation calculations is more important than for the results given in Figs. 3 and 4 with $T_w = T_0$. The u-velocity variations of Fig. 5 are similar to those shown in Fig. 3. Responsible for this u-velocity similarity is that the flow boundary conditions are equal for both cases: $u = 0$ for $\eta = 0$. Furthermore, the time dependence of the expansion wave development will be clearly shown with the results of Fig. 5c by means of the expansion wave widening at successive time points $\tau = 60$, 80, and 100.

An interesting pattern for lines of constant density and temperature develops for the wall affected expansion wave structure shown in Fig. 6. These lines are extracted from the density and temperature distribution profiles along the wall at several distances η of Figs. 5a and 5b. At the front of the expansion wave for both, i.e., for the density and the temperature flow pattern, a very surprising forward-

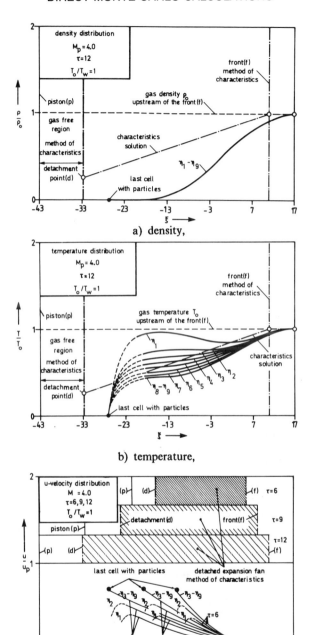

Fig. 7 $M_p = 4$ and $T_0 = T_w$.

facing turn appears close to the wall as already seen from DSMC calculations in the density field of the shock structure near a wall.[6] But, in contrast to the shock curvature with a backward-bent shock in the outer part of the flow region, the expansion wavefront of Fig. 6a always seems to be turned forward close to the wall. A similar picture develops for the density contours in Fig. 4a ($T_w = T_0$). They also show a similar but smaller inclined forward-facing front. Concerning the lines of equal temperature the different boundary conditions are obvious (see Fig. 4b: $T_0 = T_w$ and Fig. 6b: $T_0 = 3 T_w$). In the case $T_0 = T_w$ the gas temperature upstream of the end (e) of the expansion fan in the outer flow region is lower than the wall temperature. For $T_0 = 3 T_w$ the gas always gets "hotter" than the boundaries, forming the differing temperature pictures of Fig. 4b, respectively, Fig. 6b.

Detached expansion fan

If the piston speed exceeds $u_p = f a_0$, the gas will be detached from the piston base. A particle-free region develops closely behind the piston, as shown in Fig. 7. At the edge of this gas-free region the pressure is extremely low and especially here only the gaskinetic solution should be valid. In agreement with the prediction of the continuum theory, a well-developed gas-free region is also given by the Monte Carlo results for $M_p = 4$. The simulation produces a larger particle-free space at the back side of the piston, as found already for the one-dimensional DSMC calculations.[2] The reason for the particles being detached too far may be founded in the use of the hard sphere model and in the application of possibly too large cell sizes, as discussed by Seiler.[2]

An unforeseen picture forms for the density distribution as seen in Fig. 7a. Toward the wall no density gradients can be recognized, probably caused by the boundary conditions in front of the expansion wave ($\rho = \rho_0$) and at the detachment point ($\rho = 0$), valid at the wall as well as far away.

In recognizing the temperature variation of Fig. 7b, a sudden temperature fall to zero is present. Because of the strong particle decrease in the cells near the gas front, the statistical scatter there increases and the smoothing error gets larger in the dashed line region of the temperature distribution. The influence of the wall boundary condition ($T_w = T_0$) produces a temperature increase toward the wall, as can be clearly seen by the lines of equal temperature as shown in Fig. 8.

The beginning of the temperature boundary layer at the front part of the expansion wave flow pattern just as it is growing downstream can be illustrated with the curved contour lines of Fig. 8. Of particular interest is the fact that at the onset of the particle-free area the front contour is formed straight and arranged perpendicular to the wall; i.e., there the wall effect vanishes to zero.

Especially for the piston Mach number $M_p = 4$ with a detached gas flow, the continuum solution with regard to the DSMC results for the profiles far away from the wall should fail due to the strong dilute-

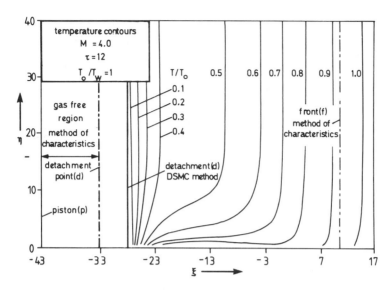

Fig. 8 Equal temperature contours for $M_p = 4$ and $T_w = T_0$.

gas phenomena present at the detachment (d) to vacuum. Contrary to the results presented in Fig. 3 ($M_p = 0.25$) and Fig. 5 ($M_p = 0.144$), strong deviations from the characteristics solution are shown in Fig. 7 for the density and temperature distributions at far wall distances, e.g., η_9. A remarkable density shift considering the characteristics calculations exists (Fig. 7a). Also, the temperature DSMC solution (Fig. 7b) shows strong differences concerning the continuum considerations. Surprisingly, a good agreement in the velocity structure has been found for the distances η_3 to η_9 from the wall, as already found by the one-dimensional calculation of Seiler.[2]

Concerning the front (f) of the expansion wave the agreement of both, the DSMC technique and the method of characteristics is fairly good. As expected, the simulation Monte Carlo method gives a continuous wave profile in which the discontinuity at the front (f) of the characteristics solution vanishes.

Conclusion

The results obtained from the DSMC computations give a survey of the two-dimensional expansion wave formation done for piston Mach numbers $M_p = 0.144$, 0.25, and 4. The wall temperature T_w has been set to $T_w < T_0$ and $T_w = T_0$, respectively.

In the cases of the lower piston Mach numbers with $M_p < 1$ and for $T_w = T_0$, the density and temperature gradients inside the boundary layer perpendicular to the wall are very small due to the low temperature change from the outer flow part up to the wall. If the gas is hotter than the wall ($T_0 = 3\, T_w$), the weak one-dimensional expansion fan density and temperature profiles being present far away from the wall develop into a steep wave behavior next to the wall. Concerning the u-velocity profiles at several distances: η_1, η_2, ... η_9, from the wall, the velocity decrease toward the wall (u = 0) is visible. The characteristics solution can describe fairly good the position of the front (f) and the end (e), especially for the DSMC calculations with the boundary condition $T_w = T_0$. For $T_w < T_0$, nearby the wall, the DSMC results show a forward bent of the equal density and temperature contours, whereas in the wall region large deviations from the continuum solution obtained with the method of characteristics are present.

If the piston Mach number exceeds $M_p = f$, a well-developed gas-free region, predicted by the continuum theory, is given by the DSMC calculations. The simulation produces a larger particle-free space at the back of the piston. The reason for this discrepancy is not yet evident. It may be found in 1) the intermolecular potential used and/or 2) the cell sizes being not determined optimally. Furthermore, it might be possible that due to the dilute gas effects at the border to vacuum the continuum theory does not correctly describe the particle behavior. With regard to the front (f), the DSMC technique and the method of characteristics are in good agreement, comparable with the results obtained for $M_p < f$ and $T_w = T_0$. An interesting result was found for the density flow pattern. No boundary-layer influence appears, probably due to the boundary conditions upstream ($\rho = \rho_0$) and downstream ($\rho = 0$) of the expansion wave. Concerning the temperature and velocity distribution in the flowfield for $M_p = 4$, the effect of the boundary condition at the wall can be seen.

References

[1] Bird, G. A., "The Velocity Distribution Function Within a Shock Wave," Journal of Fluid Mechanics, Vol. 30, No. 3, 1976, p. 479.

[2] Seiler, F., "Direct Monte Carlo Calculations on Expansion Wave Structure," Proceedings of the 15th International Symposium on Rarefied Gas Dynamics, Edited by V. Boffi and C. Cercignani, B. G. Teubner Verlag, Stuttgart, 1986.

[3] Seiler, F. and Schmidt, B., "Shock Structure Near a Wall," Proceedings of the 12th International Symposium on Rarefied Gas Dynamics, Edited by S. S. Fisher, Published by AIAA, New York, 1980.

[4] Bird, G. A., Molecular Gas Dynamics, Clarendon, Oxford, UK, 1976.

[5] Wetzel, W. and Oertel, H., "Gas-kinetical Simulation of Vortical Flows," Acta Mechanica, Vol. 70, 1987, pp. 127-143.

[6] Seiler, F., "Boundary Layer Influenced Shock Structure," Proceedings of the 13th International Symposium on Shock Tubes and Waves, Edited by C. E. Treanor and J. G. Hall, State University of New York Press, Albany, New York, 1981.

Chapter 4. Numerical Techniques

Numerical Analysis of Rarefied Gas Flows by Finite-Difference Method

Kazuo Aoki*
Kyoto University, Kyoto, Japan

Abstract

Recent results of accurate numerical analysis of rarefied gas flows on the basis of finite-difference methods are reported. Three different approaches: 1) the integral equation method for the linearized BGKW model, 2) the finite-difference method for the nonlinear BGKW model, and 3) the finite-difference method for the linearized Boltzmann equation are explained, with emphasis on their applications.

I. Introduction

In connection with the modern development of aerospace engineering and vacuum technology, clear understanding of the properties of rarefied gas flows for all degrees of gas rarefaction (i.e., all values of the Knudsen number) has become increasingly important. In the two extreme cases of slightly and highly rarefied gas, analytical approaches are effective, and many successful results have been obtained. On the other hand, for the moderately rarefied gas, imprecise methods, such as the moment method, have often been employed without reasonable discussions about their accuracy because of the difficulty in solving the Boltzmann equation analytically. In this case, however, direct numerical approaches seem to be advantageous for the purpose of accurate analysis. In fact, concerning the model Boltzmann equations such as the Bhatnagar-Gross-Krook-Welander (BGKW) model,[1,2] reliable results were obtained in the early stages for some geometrically simple problems. At the present time, the recent progress of high-speed

---
Presented as an Invited Paper.
Copyright © 1989 by Kazuo Aoki. Published by the American Institute of Aeronautics and Astronautics, Inc. with permission.
*Associate Professor, Department of Aeronautical Engineering.

computers is making a remarkable contribution to the development of a new stage in direct numerical analysis.

The aim of the present paper is to introduce accurate numerical analysis of some basic problems in rarefied gas dynamics, based on finite-difference approximation of the Boltzmann equation and its model equation (BGKW model), carried out recently in the research group that the author belongs to.

II. Integral Equation Method for Linearized BGKW Equation

Let us consider linearized steady boundary-value problems in which the deviation from an equilibrium state at rest is small. Under the diffuse reflection boundary condition, the BGKW model of the Boltzmann equation, being integrated along its characteristics, gives the expression of the velocity distribution function in terms of the macroscopic variables, such as the density, velocity, and temperature of the gas. By taking the appropriate moments of the result, we obtain a system of linear integral equations for the macroscopic quantities (e.g., Refs. 2-8). This system, the independent variables of which are the space coordinates only, is suitable for precise analysis. In fact, it has been widely used for both analytical and numerical studies of rarefied gas flows since the 1960's. However, except for very few examples, the existing works are restricted to one-dimensional problems (Knudsen layers,[2,6-14] plane and cylindrical Couette flows,[4,15] heat transfer between parallel plates,[16] etc.) and the problems reducible to one-dimensional problems by appropriate similarity solutions[17] (plane and cylindrical Poiseuille flows,[5,18] thermal creep flow,[7,18-20] flow past a sphere,[21,22] thermophoresis,[17] etc.). Nowadays, we can readily carry out numerical analysis of these problems as accurately as we wish for any Knudsen number. Furthermore, the precise numerical analysis of the above-mentioned system has been extended to some essentially two-dimensional problems (with two space coordinates as independent variables). We will demonstrate two such examples here.

A. Poiseuille and Thermal Transpiration Flows for Pipes with Various Cross Sections[23,24]

Let us consider a rarefied gas in an infinitely long straight pipe with a uniform cross section (parallel to the X_3 axis, where X_i is the space rectangular coordinate system) and investigate Poiseuille flow (flow due to a uniform pressure gradient along the axis of the pipe) and thermal transpiration (flow due to a uniform temperature gradient of the temperature of the pipe wall along the axis) for various cross sections.

In this problem, the solution can be obtained in the form in which the velocity has only the X_3 component v_3 and the pressure and temperature are uniform in the cross section perpendicular to the X_3 axis. Thus, the integral equations are greatly simplified. That is, if we put

$$v_3 = (\frac{RT_w}{2})^{1/2}(- \frac{L}{p} \frac{dp}{dX_3} u_P + \frac{L}{T_w} \frac{dT_w}{dX_3} u_T) \tag{1}$$

where p is the pressure, T_w the temperature of the pipe wall, L the reference length, and R the specific gas constant, the nondimensional velocities u_P and u_T, corresponding to Poiseuille and thermal transpiration flows, respectively, are the solutions of the following integral equations:

$$u_P(\mathbf{r}) = \frac{1}{\pi k} \int_{(\bar{S})} u_P(\hat{\mathbf{r}}) \frac{J_0(|\hat{\mathbf{r}}-\mathbf{r}|/k)}{|\hat{\mathbf{r}}-\mathbf{r}|} d\hat{\mathbf{r}} + \frac{1}{\pi} \int_{(\bar{S})} \frac{J_0(|\hat{\mathbf{r}}-\mathbf{r}|/k)}{|\hat{\mathbf{r}}-\mathbf{r}|} d\hat{\mathbf{r}} \tag{2}$$

$$u_T(\mathbf{r}) = \frac{1}{\pi k} \int_{(\bar{S})} u_T(\hat{\mathbf{r}}) \frac{J_0(|\hat{\mathbf{r}}-\mathbf{r}|/k)}{|\hat{\mathbf{r}}-\mathbf{r}|} d\hat{\mathbf{r}}$$

$$+ \frac{1}{\pi} \int_{(\bar{S})} \frac{J_0(|\hat{\mathbf{r}}-\mathbf{r}|/k) - J_2(|\hat{\mathbf{r}}-\mathbf{r}|/k)}{|\hat{\mathbf{r}}-\mathbf{r}|} d\hat{\mathbf{r}} \tag{3}$$

$$k = \frac{\sqrt{\pi}}{2} \frac{\ell}{L} = \frac{\sqrt{\pi}}{2} Kn, \qquad J_n(\eta) = \int_0^\infty t^n \exp(-t^2 - \frac{\eta}{t}) dt$$

where $\mathbf{r} = (X_1/L, X_2/L)$, ℓ is the mean free path of the gas molecules, Kn is the Knudsen number, and the domain of integration $(\bar{S})$ is the part of the cross section visible from $\mathbf{r}$ (thus, it is the whole cross section when its boundary is concave).

The mass flow rate M, which is obtained by integrating Eq. (1) over the entire cross section (S), is written as

$$M = p(2RT_w)^{-1/2} S(- \frac{L}{p} \frac{dp}{dX_3} Q_P + \frac{L}{T_w} \frac{dT_w}{dX_3} Q_T) \tag{4}$$

$$Q_{P,T} = \frac{L^2}{S} \int_{(S)} u_{P,T}(\hat{\mathbf{r}}) d\hat{\mathbf{r}}$$

where S is the cross-sectional area, and Q_P, Q_T are the dimensionless flow rates corresponding to u_P, u_T, respectively.

In Refs. 23 and 24, Eqs. (2) and (3) have been solved for various cross sections (rectangle, regular hexagon, ellipse, semicircle) for the entire range of the Knudsen number by constructing the Neumann series numerically. That is, if Eqs. (2) and (3) are written symbolically as

$$f = \bar{A}[f] + b \tag{5}$$

where f stands for u_P or u_T, $\bar{A}[\cdots]$ the integral operator, and b the inhomogeneous term, a sequence $\{f^{(n)}\}$ is constructed with the recursion formula

$$f^{(n+1)} = \bar{A}[f^{(n)}] + b \tag{6}$$

with an initial function $f^{(0)}$. In the scheme for an arbitrary cross section in Ref. 24, the domain of integration is divided into triangular elements, and $f^{(n)}$ is expanded in terms of a system of basis functions,

$$f^{(n)} = \sum_m f_m^{(n)} \Psi_m \tag{7}$$

where $f_m^{(n)}$ is the value of $f^{(n)}$ at the mth nodal point, and Ψ_m is the associated basis function. Ψ_m is so chosen that Eq. (7) takes the exact value at each nodal point and is linear in both X_1/L and X_2/L in each triangular element. Thus, provided that the integral $\bar{A}[\Psi_m]$ is computed for all the nodal points beforehand, the computation of $\bar{A}[f^{(n)}]$ in Eq. (6) is reduced to a sum of simple products of known quantities. In this way, very efficient computation is achieved.

In this paper, we give the results only for the regular hexagon and ellipse shown in Fig. 1 (the reference length L is taken as in the figure). The velocity distributions u_P, u_T for typical Kn are shown in Fig. 2. The corresponding Q_P, Q_T are shown in Fig. 3 as functions of the Knudsen number.

The Q_P and Q_T for rectangular cross sections are also obtained in Ref. 25, where u_P and u_T are, however, not given.

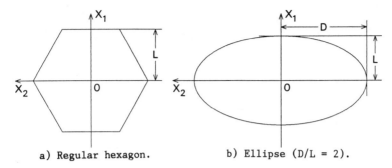

a) Regular hexagon. b) Ellipse (D/L = 2).

Fig. 1 Cross section of the pipe (Ref. 24).

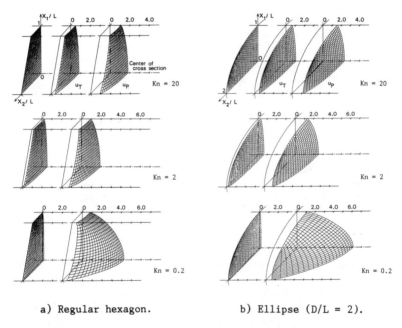

a) Regular hexagon. b) Ellipse (D/L = 2).

Fig. 2 Nondimensional flow velocities u_P, u_T (Ref. 24).

B. Flow Induced in a Gas between Noncoaxial Circular Cylinders[26]

Next we consider a rarefied gas contained between two noncoaxial circular cylinders at rest with different uniform temperatures. We investigate the flow induced in the gas, the force acting on the cylinders, and the heat transmitted to them for the entire range of the Knudsen number.

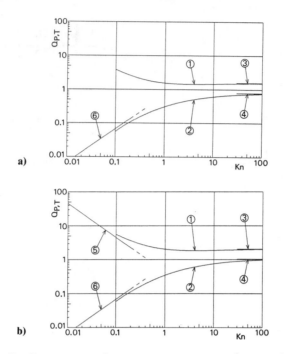

Fig. 3 Nondimensional mass flow rates Q_P, Q_T (Ref. 24).
a) Regular hexagon. b) Ellipse ($D/L = 2$).
①, numerical result (Q_P)[24]; ②, numerical result (Q_T)[24];
③, free molecular flow (Q_P); ④, free molecular flow (Q_T);
⑤, Navier-Stokes flow without slip (Q_P); ⑥, analytical result for small Kn based on the asymptotic theory[9] (Q_T).

We use the following notations: L is the radius of the inner cylinder, Lr that of the outer cylinder, Ld the distance between the axes of the cylinders, T_0 the temperature of the inner cylinder, $T_0(1 + \Delta\tau)$ that of the outer cylinder, p_0 the reference pressure, ℓ_0 the mean free path at the reference equilibrium state at rest with temperature T_0 and pressure p_0; the X_3 axis is taken parallel to the axes of the cylinders, and the positive X_2 axis toward the center of the outer cylinder from that of the inner (Fig. 4).

In this problem, we have to deal with the complicated system of integral equations for the density, velocity, and temperature of the gas, and the function defined on the boundary, corresponding to the density of the reflected molecules. In principle, the numerical construction of the Neumann series also applies to this system. We encounter, however, several difficulties in the actual computation,

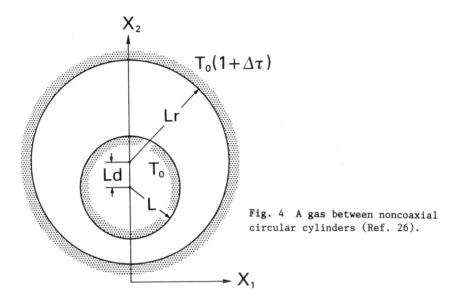

Fig. 4 A gas between noncoaxial circular cylinders (Ref. 26).

for example:

1) Since the induced velocity field is small compared with the variation of the density and temperature, very accurate computation is required to obtain the reliable velocity field.

2) As the velocity and temperature of the Neumann series converge, its density and boundary function, mentioned earlier, exhibit a uniform shift by a small constant at each iterative step. Thus, the Neumann series does not converge.

The unfavorable behavior in item 2 is attributed to the fact that the homogeneous equations of the integral equations possess a nontrivial solution (a stationary equilibrium solution with an arbitrary density) and to the error introduced by numerical approximation (discretization). In fact, the shift is of the order of the conceivable error of discretization. This can be resolved by imposing another condition associated with the mass of gas between the cylinders at each step of iteration.

The typical flow patterns in the X_1X_2 plane for $r = 2$, $d = 0.5$ are shown in Fig. 5, where $k = (\sqrt{\pi}/2)Kn = (\sqrt{\pi}/2)(\ell_0/L)$ and the arrows stand for the nondimensional velocity vector $(2RT_0)^{-1/2}v_i/\Delta\tau$ [v_i: the velocity of the gas ($v_3 = 0$)] at their starting points. This flow, as well as the thermal transpiration flow discussed in Subsec. A, is induced by the temperature field. This type of flow is peculiar to rarefied gas and vanishes when k = 0 (continuum flow). But differently from the thermal transpiration, the present flow vanishes also in the limit k → ∞ (free molecular flow).[27-29] For small k, the flow is attributed to thermal stress slip flow[10,30,31] caused by the velocity slip on the boundary, proportional to the tangential thermal stress there.[32] For moderate and large k, the flow is induced by a different mechanism, which is explained in Ref. 26.

The force F_i = (0, F_2, 0) acting on the inner cylinder per unit length and the energy H transmitted to it per unit length per unit time are shown as functions of k in Figs. 6 and 7, respectively, for the same r and d. When the outer cylinder is heated ($\Delta\tau$ > 0), the force is toward the widest gap for large and moderate values of k, whereas it is in the opposite direction for small k. The reversal occurs at k ∼ 0.15. When the inner cylinder is heated ($\Delta\tau$ < 0), the direction of the force is reversed. The magnitude of $F_2/(p_0 L\Delta\tau)$ attains to the maximum at k ∼ 1.5.

Kogan et al.[33] pointed out that, for small Knudsen number and nonsmall temperature difference of the cylinders, another type of flow is induced by the effect of thermal stress in the gas.

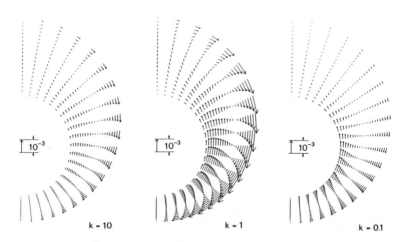

Fig. 5 Flow induced in the gas (r = 2, d = 0.5) (Ref. 26). The arrows stand for the nondimensional velocity vector $(2RT_0)^{-1/2} v_i/\Delta\tau$ (v_3 = 0) at their starting points. The magnitude of 10^{-3} is shown in the figures. The inner and outer cylinders are located at the innermost and outermost arrows.

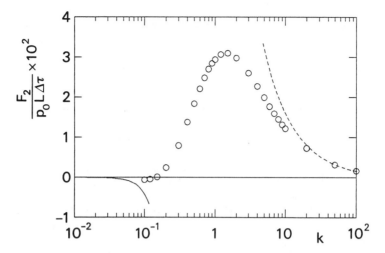

Fig. 6 Force acting on the inner cylinder F_i ($r = 2$, $d = 0.5$) (Ref. 26). O, numerical result[26]; ———, analytical result[32] based on the asymptotic theory[9,10]; ----, 1/k-order correction to the free molecular flow solution (S. Tanaka, unpublished paper).

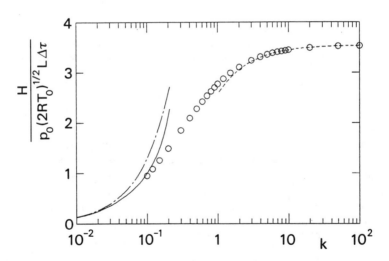

Fig. 7 Energy transmitted to the inner cylinder H ($r = 2$, $d = 0.5$) (Ref. 26). O, numerical result[26]; ———, analytical result based on the asymptotic theory[9,10] (S. Tanaka, unpublished paper); —·—, Navier-Stokes result without temperature jump; ----, 1/k-order correction to the free molecular flow solution (S. Tanaka, unpublished paper).

III. Finite-Difference Method for Nonlinear BGKW Equation

If we restrict ourselves to the spatially one-dimensional problems where the independent variables are the time t, one of the space rectangular coordinates (say, X_1), and the molecular velocity ξ_i, the nonlinear BGKW equation is reduced to a system of integro-differential equations, with t, X_1, ξ_1 as the independent variables, by taking the appropriate moments with respect to ξ_2 and ξ_3.[34] This system is useful for analyzing one-dimensional time-dependent problems of basic importance (e.g., shock wave formation,[35] piston problem,[36] Rayleigh problem[37]). In fact, if the ξ_1 variable is discretized (discrete ordinate), the usual finite-difference method in the $X_1 t$ space conveniently applies to the system, and time-dependent boundary-value problems are easily analyzed. In this section, we will report the recent results on a strong evaporation-condensation problem based on this approach.[38-41]

Let us consider a semi-infinite expanse of a gas ($X_1 > 0$) bounded by its plane condensed phase ($X_1 = 0$) at temperature T_w. Suppose that a uniform flow [velocity (u_∞, 0, 0), pressure p_∞, temperature T_∞] is realized in the half-space ($X_1 > 0$) at time $t = 0$ and is always imposed at infinity for $t > 0$. We investigate the time development of the disturbance produced by the interaction of the flow and the condensed phase, with special emphasis on clarifying the relation among the parameters that provide a steady solution and on obtaining the accurate profiles of the steady solutions (Knudsen layer).

We analyze the initial-boundary-value problem defined in the preceding paragraph under the assumption that the gas molecules leaving the surface of the condensed phase have the stationary Maxwellian distribution corresponding to the saturated gas at the surface temperature. Since the problem is characterized by the three dimensionless parameters, $M_\infty = |u_\infty|(\frac{5}{3} RT_\infty)^{-1/2}$ (Mach number at infinity), p_∞/p_w, and T_∞/T_w, where p_w is the saturation gas pressure at temperature T_w, it is convenient to consider the three-dimensional space of these parameters (M_∞, p_∞/p_w, T_∞/T_w) for summarizing the main results.

A. Strong Condensation on a Plane Condensed Phase[38,39,41]

We first consider the case $u_\infty < 0$. The time development of the solution is classified into the following four types:

(I) The gas is compressed on the condensed phase, and

the compression region propagates or diffuses upstream. The speed of propagation, however, slows down and finally vanishes. A steady state with the prescribed condition at infinity (M_∞, p_∞/p_w, T_∞/T_w) is established.

(II) The gas is compressed on the condensed phase, and a compression wave (shock wave) propagates up to upstream infinity. The region behind the wave approaches a steady state with a new subsonic state at infinity. (For sufficiently small p_∞/p_w, steady evaporation takes place instead of condensation.)

(III) A rarefaction region develops on the condensed phase and diffuses as time goes on. Finally a steady state with the prescribed condition at infinity is established.

(IV) A rarefaction region develops on the condensed phase, and an expansion wave propagates up to upstream infinity. The region behind the wave approaches a steady state with a new subsonic or sonic state at infinity.

The three-dimensional space of the prescribed data (M_∞, p_∞/p_w, T_∞/T_w) is divided into four regions according to the types of solution the data give. The schematic view of its cross section at T_∞/T_w = const is shown in Fig. 8. The steady solution with the prescribed data at infinity exists in regions I and III and on the boundary between regions II and IV. In other words, for given T_w, the subsonic steady solution is determined by two parameters at infinity (e.g., p_∞, T_∞), whereas the supersonic steady solution is determined by the three parameters there (u_∞, p_∞, T_∞).

The actual map of the cross section at $T_\infty/T_w = 1$ based on the numerical data is shown in Figs. 1 and 2 of Ref. 38. However, the position of the III-type solution in Fig. 2 of Ref. 38 is not accurate enough because of the numerical error (see Postscript of Ref. 38). Here we give

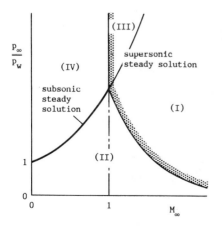

Fig. 8 Schematic view of the cross section of the (M_∞, p_∞/p_w, T_∞/T_w) space at T_∞/T_w = const (Ref. 38).

the more accurate map[39] around the intersection of the boundaries in Fig. 9, where the points giving the solution of type I, II, III, and IV are marked by ○, ●, △, and ▲, respectively.

Let the boundary surface between regions II and IV be $M_\infty = f_s(p_\infty/p_w, T_\infty/T_w)$ and that between regions I and II $M_\infty = f_b(p_\infty/p_w, T_\infty/T_w)$. We now try to construct these surfaces numerically.

As mentioned earlier, the state behind the shock (expansion) wave in the type II (IV) solution approaches a steady state with a new subsonic (subsonic or sonic) state at infinity. Therefore, if we read the new u_∞, p_∞, T_∞ (i.e., the values at the uniform portion behind the shock or expansion wave) for the case $u_\infty < 0$, $M_\infty < 1$ and plot the corresponding point in the (M_∞, p_∞/p_w, T_∞/T_w) space, it lies on the surface $M_\infty = f_s(p_\infty/p_w, T_\infty/T_w)$. Thus, taking this procedure for many solutions of type II and IV, we can construct the surface $M_\infty = f_s(p_\infty/p_w, T_\infty/T_w)$.

In Ref. 38, the asymptotic behavior of the I- and II-type solutions for large time is explained physically with the aid of the combination of a subsonic steady solution

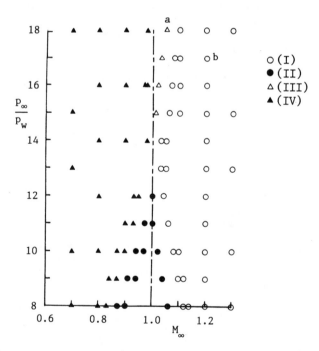

Fig. 9 Cross section of the (M_∞, p_∞/p_w, T_∞/T_w) space at $T_\infty/T_w = 1$ (Ref. 39). Numerical result around the intersection of the boundaries.

and a shock wave. This discussion and the detailed
numerical computation show that a point on the surface
$M_\infty = f_b(p_\infty/p_w, T_\infty/T_w)$ is given by the upstream condition of
a standing shock wave whose downstream condition is the
state at infinity of a subsonic steady solution. Therefore,
applying the jump condition of a standing shock wave to
each of the data on the subsonic steady solution surface
$M_\infty = f_s(p_\infty/p_w, T_\infty/T_w)$, obtaining the upstream state u_∞', p_∞',
T_∞', and plotting the corresponding point in the $(M_\infty,$
$p_\infty/p_w, T_\infty/T_w)$ space, we can construct the boundary surface
$M_\infty = f_b(p_\infty/p_w, T_\infty/T_w)$.

The projections of the surfaces $M_\infty = f_s(p_\infty/p_w, T_\infty/T_w)$
and $M_\infty = f_b(p_\infty/p_w, T_\infty/T_w)$ thus constructed on the plane
T_∞/T_w = const are shown in Figs. 10 and 11, respectively.
In Fig. 10, the analytical result based on the weakly
nonlinear theory[42,43,14] is also shown for $T_\infty/T_w = 1$.
These surfaces are almost perpendicular to the plane T_∞/T_w
= const, except in the region where M_∞ is close to unity.

The profiles of the supersonic steady solutions
corresponding to a, b in Fig. 9 and those of the subsonic
steady solutions corresponding to c, d in Fig. 10 are shown
in Figs. 12a, 12b, 12c, and 12d, respectively, where p is
the gas pressure, $(u_1, 0, 0)$ the gas velocity, T the gas
temperature, ℓ_w the mean free path of the saturated gas in
equilibrium at rest at temperature T_w, and $a_\infty = [(5/3)RT_\infty]^{1/2}$ the sound speed at the upstream infinity.

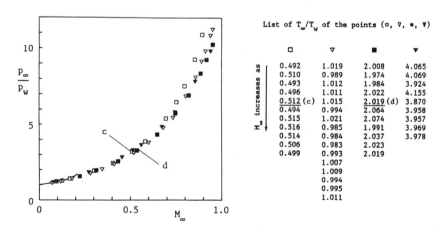

Fig. 10 The projection of the subsonic steady solution surface
$M_\infty = f_s(p_\infty/p_w, T_\infty/T_w)$ on the plane T_∞/T_w = const (Refs. 39, 41).
The values of T_∞/T_w are classified into four groups (□, ∇, ■, ▼)
and are listed in order of M_∞ in each group. ———, analytical
result ($T_\infty/T_w = 1$) by the weakly nonlinear theory.[42,43,14]

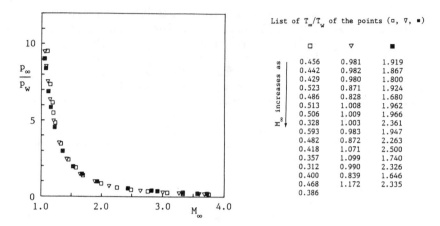

Fig. 11 The projection of the boundary surface $M_\infty = f_b(p_\infty/p_w, T_\infty/T_w)$ on the plane T_∞/T_w = const (Refs. 39, 41). The values of T_∞/T_w are classified into three groups (□, ▽, ■) and are listed in order of M_∞ in each group.

The half-space steady strong condensation has been studied by many authors (e.g., Refs. 44-49). The range of existence of the supersonic steady solution just described is completely different from that predicted by the moment method in Ref. 45, where it is concluded that the supersonic steady solution exists on a surface in the (M_∞, p_∞/p_w, T_∞/T_w) space. Supersonic steady solutions that are not on this surface have also been reported in Refs. 48 and 49.

B. Strong Evaporation on a Plane Condensed Phase[40,41]

As in the case of condensation, we can investigate the property of the steady strong evaporation, which has interested many authors (e.g., Refs. 44, 50, 51), by pursuing the long-time behavior of the initial-boundary-value problem for $u_\infty > 0$. The numerical result shows that, unlike condensation, the steady solution exists only on a curve, say, $p_\infty/p_w = h_1(M_\infty)$, $T_\infty/T_w = h_2(M_\infty)$, in the ($M_\infty$, p_∞/p_w, T_∞/T_w) space and, moreover, does not exist for $M_\infty > 1$. The projections of the curve, $p_\infty/p_w = h_1(M_\infty)$ and $T_\infty/T_w = h_2(M_\infty)$, are shown in Figs. 13a and 13b, respectively, where the analytical result based on the weakly nonlinear theory[42,43,14] and the result by the moment method[50] are also shown. In Fig. 13a, the result by the moment method does not deviate from the numerical result for the whole range of M_∞ while, in Fig. 13b, the analytical result

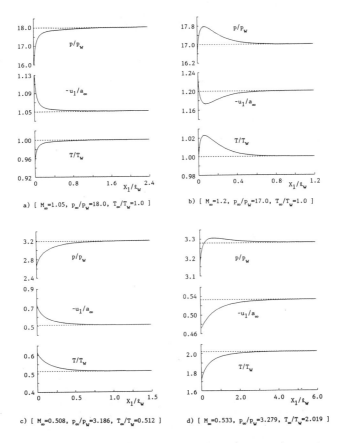

Fig. 12 Profiles for steady condensation (Ref. 39).
a) $M_\infty = 1.05$, $p_\infty/p_W = 18.0$, $T_\infty/T_W = 1.0$ (cf. Fig. 9).
b) $M_\infty = 1.2$, $p_\infty/p_W = 17.0$, $T_\infty/T_W = 1.0$ (cf. Fig. 9).
c) $M_\infty = 0.508$, $p_\infty/p_W = 3.186$, $T_\infty/T_W = 0.512$ (cf. Fig. 10).
d) $M_\infty = 0.533$, $p_\infty/p_W = 3.279$, $T_\infty/T_W = 2.019$ (cf. Fig. 10).

agrees completely with the latter for $M_\infty < 0.25$. The profiles of the steady solution for $M_\infty = 0.4500$ and 0.9892 are shown in Fig. 14.

The nonexistence of a steady solution for $M_\infty > 1$ has been proved on the basis of the BGKW equation linearized around the Maxwellian distribution at the downstream infinity.[52-54] This linearization, however, is not valid near the condensed phase, where the state of gas significantly deviates from the equilibrium state at the downstream infinity.

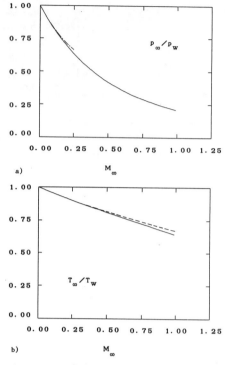

Fig. 13 The projections of the steady solution curve (Refs. 40, 41). a) $p_\infty/p_w = h_1(M_\infty)$. b) $T_\infty/T_w = h_2(M_\infty)$. ———, numerical result[40,41]; —·—, analytical result based on the weakly nonlinear theory[42,43,14]; ----, moment method.[50]

IV. Finite-Difference Method for Linearized Boltzmann Equation

Finally, we consider one-dimensional boundary-value problems of the linearized Boltzmann equation for hard-sphere molecules. A finite-difference analysis similar to that employed in the previous section can also be applied. In this case, however, the two components of the molecular velocity (normal and parallel components to the boundary) remain as the independent variables, though the reduction of the number of the independent variables with the aid of similarity solutions is available, and thus we have to devise the computation of the complicated collision integral in the space of the two components of the molecular velocity. Recently we constructed a finite-difference scheme with a very efficient computation of the collision integral and succeeded in the precise analysis of some basic boundary-value problems in rarefied gas

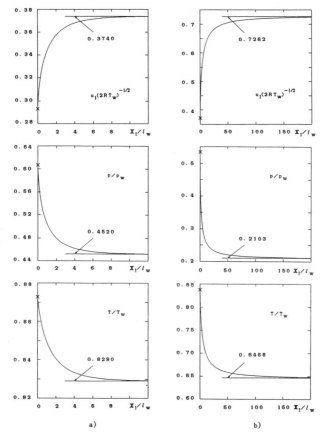

Fig. 14 Profiles for steady evaporation (×: the value at $X_1 = 0$) (Ref. 40). a) $M_\infty = 0.4500$. b) $M_\infty = 0.9892$.

dynamics.[55,56] In what follows, we describe the analysis of the temperature jump problem,[55] which is one of the most fundamental problems and has been investigated by various authors (e.g., Refs. 2, 12, 57-60).

Consider a semi-infinite expanse ($X_1 > 0$) of a rarefied gas at rest bounded by a plane wall ($X_1 = 0$, temperature T_0). Suppose that there is a constant heat flow normal to the wall from infinity. Investigate the steady behavior of the gas (velocity distribution function, temperature and density distributions, temperature jump at the wall, etc.).

We analyze the problem on the basis of the linearized Boltzmann equation for elastic hard-sphere molecules with the diffuse reflection boundary condition.

Let us denote the velocity distribution function by $\rho_0(2RT_0)^{-3/2}E(1+\phi)$, where $E = \pi^{-3/2}\exp(-\zeta_i^2)$, $(2RT_0)^{1/2}\zeta_i$ is the molecular velocity, $\rho_0 = p_0(RT_0)^{-1}$, and p_0 is the pressure at infinity. We seek the solution of the present half-space boundary-value problem in the form

$$\phi = \gamma(\Phi_H + \Phi_K), \quad \Phi_H = (\zeta_i^2 - \frac{5}{2})(x_1 + \beta) - \zeta_1 A(|\zeta_i|) \quad (8)$$

where γ and β are arbitrary constants, $x_1 = (2/\sqrt{\pi})(X_1/\ell_0)$; $\gamma\Phi_H$ is a solution of the linearized Boltzmann equation expressing a stationary state with a constant heat flow, and $\gamma\Phi_K$ is the correction term, called Knudsen layer, appreciable in the region (with thickness of the order of the mean free path) adjacent to the wall; $A(|\zeta_i|)$ is the solution of the following integral equation and subsidiary condition:

$$L[\zeta_1 A(|\zeta_i|)] = -\zeta_1(\zeta_i^2 - \frac{5}{2}), \quad \int_0^\infty \zeta^4 A(\zeta) e^{-\zeta^2} d\zeta = 0 \quad (9)$$

where $L[\cdots]$ is the linearized collision operator, the kernel of which is expressed in a concise form[61,3] (see Eq. (2) in Ref. 55). Φ_K is then governed by the following equation (linearized Boltzmann equation) and boundary condition:

$$\zeta_1 \frac{\partial \Phi_K}{\partial x_1} = L[\Phi_K] \quad (10)$$

$$\Phi_K = -\beta(\zeta_i^2 - 2) + \zeta_1 A(|\zeta_i|) - 2\sqrt{\pi}\int_{\xi_1 < 0} \xi_1 \Phi_K E d\xi, \quad (\zeta_1 > 0, \ x_1 = 0) \quad (11)$$

$$\Phi_K \to 0, \quad (x_1 \to \infty) \quad (12)$$

The boundary-value problem, Eqs. (10-12), has a unique solution if and only if β takes a special value. This theorem, which was conjectured by Grad,[62] was proved recently.[63,64] Corresponding to Eq. (8), the temperature T and the density ρ are expressed as

$$(T-T_0)/T_0 = \gamma[x_1 + \beta + \Theta(x_1)], \quad (\rho-\rho_0)/\rho_0 = \gamma[-x_1 - \beta + \Omega(x_1)] \quad (13)$$

$$\Theta(x_1) = \frac{2}{3}\int (\zeta_i^2 - \frac{3}{2})\Phi_K E d\zeta, \quad \Omega(x_1) = \int \Phi_K E d\zeta \qquad (14)$$

If we neglect the Knudsen-layer correction Φ_K, the difference between the temperature of the gas at the wall and that of the wall itself is given as

$$T_{X_1=0} - T_0 = \frac{\sqrt{\pi}}{2}\beta\ell_0 (\frac{dT}{dX_1})_{X_1=0} \qquad (15)$$

This relation and the constant β [or $(\sqrt{\pi}/2)\beta$] are called the temperature jump condition and the temperature jump coefficient, respectively.

In slightly rarefied gas flows around solid bodies, the overall behavior of the gas is described by a system of fluid dynamic type of equations with slip or jump boundary conditions.[9,10,62,65] The correction, called the Knudsen-layer correction, to the fluid dynamic type of solution is required in a thin layer adjacent to the boundary. As far as the first-order correction to the classical fluid dynamic solution is concerned, the temperature jump condition and the Knudsen-layer correction for the temperature and density distributions are given, respectively, by Eq. (15) and
$(\sqrt{\pi}/2)\ell_0(dT/dX_1)_{X_1=0}[\Theta(x_1), \Omega(x_1)]$ if the x_1 axis is considered as the local coordinate axis, normal to the boundary and pointed to the gas, with the origin at the boundary.

In Ref. 55, the boundary-value problem, Eqs. (10-12), is solved in the following way. Taking a large constant d for which Φ_K at $x_1 = d$ is negligibly small, we consider the function ϕ_ε:

$$\phi_\varepsilon = \varepsilon[\Phi_H + \Phi_K - (\zeta_i^2 - \frac{5}{2})(d + \beta)] + \Phi_K \qquad (16)$$

where ε is a constant. Then ϕ_ε satisfies the following equation and boundary condition:

$$\zeta_1 \frac{\partial \phi_\varepsilon}{\partial x_1} = L[\phi_\varepsilon] \qquad (17)$$

$$\left.\begin{array}{l}
\text{[Eq. (11) with } \Phi_K \text{ and } \beta \text{ replaced by } \phi_\varepsilon \text{ and} \\
\beta^* = (1 + \varepsilon)\beta + \varepsilon d, \text{ respectively]}, \quad (\zeta_1 > 0, \; x_1 = 0) \\
\phi_\varepsilon(d, \zeta_1, \zeta_2, \zeta_3) = -\phi_\varepsilon(d, -\zeta_1, \zeta_2, \zeta_3), \\
\qquad\qquad\qquad\qquad\qquad\qquad (\zeta_1 < 0, \; x_1 = d)
\end{array}\right\} \qquad (18)$$

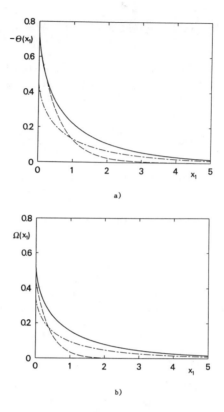

Fig. 15 Knudsen-layer functions Θ and Ω (Ref. 55). a) $\Theta(x_1)$. b) $\Omega(x_1)$. ———, numerical result[55]; – – –, the result for hard-sphere molecules by moment method[57]; —·—, the result for BGKW equation.[12]

For a properly chosen β^*, we solve Eq. (17) under the boundary condition (18). Since $(T - T_0)/T_0$ corresponding to $\phi = \phi_\varepsilon$ is given as $\varepsilon(x_1 - d)$ around $x_1 = d$, we can determine ε by reading the temperature gradient around $x_1 = d$ of the temperature distribution computed from ϕ_ε. Then β and ϕ_K are obtained from $\beta = (\beta^* - \varepsilon d)/(1 + \varepsilon)$ and Eq. (16), respectively.

It is shown that the solution of the form $\phi_\varepsilon(x_1, \zeta_1, \zeta_r)$ $[\zeta_r = (\zeta_2^2 + \zeta_3^2)^{1/2}]$ is compatible with the boundary-value problem, Eqs. (17) and (18). The solution ϕ_ε is obtained by pursuing the long-time behavior of the solution of the initial-boundary-value problem composed of the time-dependent Boltzmann equation [Eq. (17) with $\partial \phi_\varepsilon / \partial \bar{t}$ term added to the left-hand side, where $(\sqrt{\pi}/2)\ell_0(2RT_0)^{-1/2}\bar{t}$ is the time], the boundary condition (18), and an initial condition (e.g., $\phi_\varepsilon = 0$). The time-dependent problem is solved by a finite-difference method. In the computation of the collision integral, we expand $E\phi_\varepsilon$ in terms of a system of basis functions associated with the lattice points in the $\zeta_1\zeta_r$ space. The collision integral is then reduced to a product of a square matrix of the values of the universal collision integrals of the basis functions and a column vector of the values of $E\phi_\varepsilon$ at the lattice points. This form enables very efficient computation of the collision integral by the use of a vector computer.

The numerical value of the temperature jump coefficient β thus determined in Ref. 55 is $\beta = 2.3994$. The values of β obtained by other methods are: $\beta = 2.382$ [moment method (hard sphere)[57]], 2.371 [variational method (hard sphere)[59]], 1.30272 (BGKW[12]). The Knudsen-layer functions Θ and Ω, defined by Eq. (14), are shown in Fig. 15, together with the results based on the moment method[57] and the BGKW equation.[12] The moment method, which gives a fairly good result for the temperature jump coefficient, fails to describe the Knudsen-layer structure.

By using the relation between the thermal conductivity k_g and the mean free path ℓ_0, $\beta\ell_0$ in Eq. (15) can be written as $(\sqrt{\pi}/2)\beta\ell_0 = \tilde{\beta}T_0 p_0^{-1}(2RT_0)^{-1/2}k_g$, where $\tilde{\beta} = 4\beta/(5\kappa)$ and $\kappa = 1.922284$ (hard sphere) or $\kappa = 1$ (BGKW).[65] This $\tilde{\beta}$ is sometimes called the temperature jump coefficient. The value of $\tilde{\beta}$ for the BGKW equation ($\tilde{\beta} = 1.04217$) is fairly close to that for the hard-sphere molecules ($\tilde{\beta} = 0.9986$).[55]

References

[1] Bhatnagar, P. L., Gross, E. P., and Krook, M., "A Model for Collision Processes in Gases, I," *Physical Review A*, Vol. 94, 1954, pp. 511-525.

[2] Welander, P., "On the Temperature Jump in a Rarefied Gas," *Arkiv för Fysik*, Vol. 7, 1954, pp. 507-553.

[3] Cercignani, C., *Theory and Application of the Boltzmann Equation*, Scottish Academic, Edinburgh, U.K., 1975, Chap. 4, Secs. 12 and 5.

[4] Willis, D. R., "Comparison of Kinetic Theory Analysis of Linearized Couette Flow," *The Physics of Fluids*, Vol. 5, 1962, pp. 127-135.

[5] Cercignani, C. and Daneri, A., "Flow of Rarefied Gas between Two Parallel Plates," *Journal of Applied Physics*, Vol. 34, 1963, pp. 3509-3513.

[6] Sone, Y., "Kinetic Theory Analysis of Linearized Rayleigh Problem, Journal of the Physical Society of Japan, Vol. 19, 1964, pp. 1463-1473.

[7] Sone, Y., "Thermal Creep in Rarefied Gas," Journal of the Physical Society of Japan, Vol. 21, 1966, pp. 1836-1837.

[8] Sone, Y., "Some Remarks on Knudsen Layer," Journal of the Physical Society of Japan, Vol. 21, 1966, pp. 1620-1621.

[9] Sone, Y., "Asymptotic Theory of Flow of Rarefied Gas over a Smooth Boundary I," Rarefied Gas Dynamics, edited by L. Trilling and H. Y. Wachman, Academic Press, New York, 1969, pp. 243-253.

[10] Sone, Y., "Asymptotic Theory of Flow of Rarefied Gas over a Smooth Boundary II," Rarefied Gas Dynamics, edited by D. Dini, Editrice Tecnico Scientifica, Pisa, 1971, pp. 737-749.

[11] Pao, Y. P., "Some Boundary Value Problems in the Kinetic Theory of Gases," The Physics of Fluids, Vol. 14, 1971, pp. 2285-2290.

[12] Sone, Y. and Onishi, Y., "Kinetic Theory of Evaporation and Condensation," Journal of the Physical Society of Japan, Vol. 35, 1973, pp. 1773-1776.

[13] Sone, Y. and Onishi, Y., "Kinetic Theory of Evaporation and Condensation — Hydrodynamic Equation and Slip Boundary Condition," Journal of the Physical Society of Japan, Vol. 44, 1978, pp. 1981-1994.

[14] Onishi, Y. and Sone, Y., "Kinetic Theory of Slightly Strong Evaporation and Condensation — Hydrodynamic Equation and Slip Boundary Condition for Finite Reynolds Number," Journal of the Physical Society of Japan, Vol. 47, 1979, pp. 1676-1685.

[15] Cercignani, C. and Sernagiotto, F., "Cylindrical Couette Flow of a Rarefied Gas," The Physics of Fluids, Vol. 10, 1967, pp. 1200-1204.

[16] Bassanini, P., Cercignani, C., and Pagani, C. D., "Comparison of Kinetic Theory Analysis of Linearized Heat Transfer between Parallel Plates," International Journal of Heat and Mass Transfer, Vol. 10, 1967, pp. 447-460.

[17] Sone, Y. and Aoki, K., "A Similarity Solution of the Linearized Boltzmann Equation with Application to Thermophoresis of a Spherical Particle," Journal de Mécanique Théorique et Appliquée, Vol. 2, 1983, pp. 3-12.

[18] Sone, Y. and Yamamoto, K., "Flow of Rarefied Gas through a Circular Pipe," The Physics of Fluids, Vol. 11, 1968, pp. 1672-1678; Erratum: The Physics of Fluids, Vol. 13, 1970, p. 1651.

[19] Niimi, H., "Thermal Creep Flow of Rarefied Gas between Two Parallel Plates," Journal of the Physical Society of Japan, Vol. 30, 1971, pp. 572-574.

[20] Loyalka, S. K., "Thermal Transpiration in a Cylindrical Tube," The Physics of Fluids, Vol. 12, 1969, pp. 2301-2305.

[21] Cercignani, C., Pagani, C. D., and Bassanini, P., "Flow of Rarefied Gas past an Axisymmetric Body II. Case of a Sphere," The Physics of Fluids, Vol. 11, 1968, pp. 1399-1403.

[22] Lea, K. C. and Loyalka, S. K., "Motion of a Sphere in a Rarefied Gas," The Physics of Fluids, Vol. 25, 1982, pp. 1550-1557.

[23] Sone, Y. and Hasegawa, M., "Poiseuille and Thermal Transpiration Flows of a Rarefied Gas through a Rectangular Pipe," Journal of the Vacuum Society of Japan, Vol. 30, 1987, pp. 425-428 (in Japanese).

[24] Hasegawa, M. and Sone, Y., "Poiseuille and Thermal Transpiration Flows of a Rarefied Gas for Various Pipes," Journal of the Vacuum Society of Japan, Vol. 31, 1988, pp. 416-419 (in Japanese).

[25] Loyalka, S. K., Storvick, T. S., and Park, H. S., "Poiseuille Flow and Thermal Creep Flow in Long, Rectangular Channels in the Molecular and Transition Flow Regimes," Journal of Vacuum Science and Technology, Vol. 13, 1976, pp. 1188-1192.

[26] Aoki, K., Sone, Y., and Yano, T., "Numerical Analysis of a Flow Induced in a Rarefied Gas between Noncoaxial Circular Cylinders with Different Temperatures for the Entire Range of the Knudsen Number," The Physics of Fluids A, Vol. 1, 1989, pp. 409-419.

[27] Aoki, K. and Ishizuka, H., "A Force on a Cylinder in a Highly Rarefied Gas Bounded by a Plane Wall," Rarefied Gas Dynamics, edited by O. M. Belotserkovskii et al., Plenum, New York, 1985, pp. 413-420.

[28] Sone, Y., "Highly Rarefied Gas around a Group of Bodies with Various Temperature Distributions I — Small Temperature Variation," Journal de Mécanique Théorique et Appliquée, Vol. 3, 1984, pp. 315-328.

[29] Sone, Y., "Highly Rarefied Gas around a Group of Bodies with Various Temperature Distributions II — Arbitrary Temperature Variation," Journal de Mécanique Théorique et Appliquée, Vol. 4, 1985, pp. 1-14.

[30] Sone, Y., "Flow Induced by Thermal Stress in Rarefied Gas," The Physics of Fluids, Vol. 15, 1972, pp. 1418-1423.

[31] Sone, Y., "Rarefied Gas Flow Induced between Non-Parallel Plane Walls with Different Temperatures," Rarefied Gas Dynamics, edited by M. Becker and M. Fiebig, DFVLR, Porz-Wahn, FRG, 1974, D. 23.

[32] Sone, Y. and Tanaka, S., "Thermal Stress Slip Flow Induced in Rarefied Gas between Noncoaxial Circular Cylinders," Theoretical and Applied Mechanics, edited by F. P. J. Rimrott and B. Tabarrok, North-Holland, Amsterdam, 1980, pp. 405-416.

[33] Kogan, M. N., Galkin, V. S., and Fridlender, O. G., "Stresses Produced in Gases by Temperature and Concentration Inhomogeneities," Soviet Physics Uspekhi, Vol. 19, 1976, pp. 420-430.

[34] Chu, C. K., "Kinetic-Theoretic Description of the Formation of a Shock Wave," The Physics of Fluids, Vol. 8, 1965, pp. 12-22.

[35] Chu, C. K., "Kinetic-Theoretic Description of Shock Wave Formation. II," The Physics of Fluids, Vol. 8, 1965, pp. 1450-1455.

[36] Wu, Y. and Lee, C. H., "Kinetic Theory of the Piston Problem by the Bhatnagar-Gross-Krook Equation," The Physics of Fluids, Vol. 13, 1970, pp. 2222-2229.

[37] Chu, C. K., "The High Mach Number Rayleigh Problem According to the Krook Model," Rarefied Gas Dynamics, edited by C. L. Brundin, Academic, New York, 1967, pp. 589-605.

[38] Sone, Y., Aoki, K., and Yamashita, I., "A Study of Unsteady Strong Condensation on a Plane Condensed Phase with Special Interest in Formation of Steady Profile," Rarefied Gas Dynamics, edited by V. Boffi and C. Cercignani, Teubner, Stuttgart, FRG, 1986, Vol. 2, pp. 369-383.

[39] Aoki, K., Sone, Y., and Yamada, T., Proceedings of the 19th Fluid Dynamics Symposium (Sendai, Japan, 1987), Japan Society for Aeronautical and Space Sciences, Tokyo, Japan, 1987, pp. 98-101 (in Japanese).

[40] Sone, Y. and Sugimoto, H., "Strong Evaporation from a Plane Condensed Phase," Journal of the Vacuum Society of Japan, Vol. 31, 1988, pp. 420-423 (in Japanese).

[41] Sone, Y., Aoki, K., Sugimoto, H., and Yamada, T., "Steady Evaporation and Condensation on a Plane Condensed Phase," Theoretical and Applied Mechanics, Bulgarian Academy of Sciences, Sofia, Bulgaria, Year XIX, No. 3, 1988, pp. 89-93.

[42] Sone, Y., "Kinetic Theory of Evaporation and Condensation — Linear and Nonlinear Problems," Journal of the Physical Society of Japan, Vol. 45, 1978, pp. 315-320.

[43] Onishi, Y., "Kinetic Theory Treatment of Nonlinear Half-Space Problem of Evaporation and Condensation," Journal of the Physical Society of Japan, Vol. 46, 1979, pp. 303-309.

[44] Kogan, M. N. and Makashev, N. K., "Role of the Knudsen Layer in the Theory of Heterogeneous Reactions and in Flows with Surface Reactions," Fluid Dynamics, Vol. 6, 1971, pp. 913-920.

[45] Oguchi, H. and Hatakeyama, M., "One-Dimensional, Steady Supersonic Condensation," Rarefied Gas Dynamics, Progress in Astronautics and Aeronautics, Vol. 74, edited by S. S. Fisher, AIAA, New York, 1981, pp. 321-329.

[46] Ytrehus, T. and Alvestad, J., "A Mott-Smith Solution for Nonlinear Condensation," Rarefied Gas Dynamics, Progress in Astronautics and Aeronautics, Vol. 74, edited by S. S. Fisher, AIAA, New York, 1981, pp. 330-345.

[47] Yen, S. M., "Numerical Solutions of the Boltzmann and Krook Equation for a Condensation Problem," Rarefied Gas Dynamics, Progress in Astronautics and Aeronautics, Vol. 74, edited by S. S. Fisher, AIAA, New York, 1981, pp. 356-362.

[48] Abramov, A. A. and Kogan, M. N., "Conditions for Supersonic Condensation of Gas," Soviet Physics Doklady, Vol. 29, 1984, pp. 763-765.

[49] Kryukov, A. P., "One-Dimensional Steady Condensation of Vapor Velocities Comparable to the Velocity of Sound," Fluid Dynamics, Vol. 20, 1985, pp. 487-491.

[50] Ytrehus, T., "Theory and Experiments on Gas Kinetics in Evaporation," Rarefied Gas Dynamics, Progress in Astronautics and Aeronautics, Vol. 51, edited by J. L. Potter, AIAA, New York, 1977, pp. 1197-1212.

[51] Soga, T., "On Arbitrary Strong One-Dimensional Evaporation Problem," Transactions of the Japan Society for Aeronautical and Space Sciences, Vol. 21, 1978, pp. 87-97.

[52] Arthur, M. D. and Cercignani, C., "Non-existence of a Steady Rarefied Supersonic Flow in a Half-Space," Zeitschrift für angewandte Mathematik und Physik, Vol. 31, 1980, pp. 634-645.

[53] Siewert, C. E. and Thomas, J. R. Jr., "Strong Evaporation into a Half Space. II. The Three-dimensional BGK Model," Zeitschrift für angewandte Mathematik und Physik, Vol. 33, 1982, pp. 202-218.

[54] Greenberg, W. and van der Mee, C. V. M., "An Abstract Approach to Evaporation Models in Rarefied Gas Dynamics," Zeitschrift für angewandte Mathematik und Physik, Vol. 35, 1984, pp. 156-165.

[55] Sone, Y., Ohwada, T., and Aoki, K., "Temperature Jump and Knudsen Layer in a Rarefied Gas over a Plane Wall: Numerical Analysis of the Linearized Boltzmann Equation for Hard-Sphere Molecules," The Physics of Fluids A, Vol. 1, No. 3, 1989.

[56] Ohwada, T., Aoki, K., and Sone, Y., "Heat Transfer and Temperature Distribution in a Rarefied Gas between Two Parallel Plates: Numerical Analysis of the Boltzmann Equation for a Hard Sphere Molecule," This volume.

[57] Gross, E. P. and Ziering, S., "Heat Flow between Parallel Plates," The Physics of Fluids, Vol. 2, 1959, pp. 701-712.

[58] Sone, Y., "Effect of Sudden Change of Wall Temperature in Rarefied Gas," Journal of the Physical Society of Japan, Vol. 20, 1965, pp. 222-229.

[59] Loyalka, S. K. and Ferziger, J. H., "Model Dependence of the Temperature Slip Coefficient," The Physics of Fluids, Vol. 11, 1968, pp. 1668-1671.

[60] Siewert, C. E. and Thomas, J. R. Jr., "Half-Space Problems in the Kinetic Theory of Gases," The Physics of Fluids, Vol. 16, 1973, pp. 1557-1559.

[61] Grad, H., "Asymptotic Theory of the Boltzmann Equation II," Rarefied Gas Dynamics, edited by J. A. Laurmann, Academic, New York, 1963, pp. 26-59.

[62] Grad, H., "Singular and Nonuniform Limits of Solutions of the Boltzmann Equation," Transport Theory, edited by R. Bellman, et al., American Mathematical Society, Providence, RI, 1969, pp. 269-308.

[63] Bardos, C., Caflisch, R. E., and Nicolaenco, B., "The Milne and Kramers Problems for the Boltzmann Equation of a Hard Sphere Gas," Communications on Pure and Applied Mathematics, Vol. 39, 1986, pp. 323-352.

[64] Cercignani, C., "Half-Space Problems in the Kinetic Theory of Gases," Trends in Applications of Pure Mathematics to Mechanics, edited by E. Kröner and K. Kirchgässner, Springer-Verlag, Berlin, FRG, 1986, pp. 35-50.

[65] Sone, Y. and Aoki, K., "Steady Gas Flows past Bodies at Small Knudsen Numbers—Boltzmann and Hydrodynamic Systems," Transport Theory and Statistical Physics, Vol. 16, 1987, pp. 189-199; "Asymptotic Theory of Slightly Rarefied Gas Flow and Force on a Closed Body," Memoirs of the Faculty of Engineering, Kyoto University, Vol. 49, 1987, pp. 237-248.

Application of Monte Carlo Methods to Near-Equilibrium Problems

S. M. Yen*
University of Illinois at Urbana-Champaign, Urbana, Illinois

Abstract

Both random and systematic errors have to be minimized before any Monte Carlo method can be applied to solve the flow problem for a near-equilibrium gas. The concept of transport coefficients is no longer valid in a near-equilibrium gas. The collision integral is basic not only to the solution of the Boltzmann equation but also to the calculation of transport properties. The accurate evaluation of the collision integral is essential to the solution of the Boltzmann equation for a near-equilibrium gas. A correction technique to increase the fidelity of the Monte Carlo solution and a sampling technique to control the random error of the solution are outlined in this paper. The distinct near-equilibrium behavior in a weak shock wave obtained from the Boltzmann solution and its disagreement with the prediction by the linear kinetic theory are given. Numerical experiments have been performed to study the application of Bird's direct simulation Monte Carlo method to the Rayleigh problem under the near-equilibrium condition. It was found that the random error can be controlled by using large collision samples and that the solution at different levels of nonequilibrium conditions may be obtained. The study of the application of Bird's method to a flow problem over a sphere indicates that the assessment of random error distribution for multidimensional problems and the development of ways to control it require further experiments because they depends on more factors and

---

Copyright © 1989 by the American Institute of Aeronautics and Astronautics, Inc. All rights reserved.
* Professor, Aeronautical and Astronautical Engineering.

parameters. For example, the statistical scatter depends strongly on the location in the flowfield, particularly in the wake in which fewer molecules are scattered.

Introduction

It is necessary to apply kinetic theory approach to problems under those near-equilibrium conditions in which the concept of transport coefficients on the basis of linear kinetic theory is no longer valid. Since the flowfield of modern gasdynamics problems may contain local near-equilibrium regions, it would be of interest to treat the entire flow using the kinetic theory approach in order to avoid the computational difficulties of using a hybrid method.[1]

In the kinetic theory approach, a Monte Carlo method is used to simulate the intermolecular collision phenomenon and the gas surface interactions, and it is necessary to use a large collision sample to evaluate accurately the collisional effect on the flow and to minimize both the statistical scatter and the systematic error in order to detect the small but distinct effect on the flow, especially in a near-equilibrium gas. We shall review in this paper the techniques used in Monte Carlo methods to increase the fidelity of calculations under near-equilibrium conditions and the numerical experiments that may be performed to study the influence of computational parameters on the solution.

There are two distinct Monte Carlo methods: one is to solve directly the nonlinear Boltzmann equation; the other, the direct simulation. Both methods are considered in this paper.

Monte Carlo Method to Solve the Boltzmann Equation

Nordsieck's method[2] for solving the Boltzmann equation directly involves the calculation of the velocity distribution function, as well as the collision integral at each velocity cell in a quantized velocity space. This must be done for each of the selected positions in physical space using an intermolecular collision model and a gas surface interaction model appropriate for a given problem. From the moments of the distribution function and the collision integral, we may calculate the macroscopic properties that exhibit the distinct nonequilibrium behavior.

Extensive studies of the Monte Carlo, as well as the systematic errors of the calculations of the distribution function, the collision integral, and their moments, have been made for two problems and under a wide range of nonequilibrium conditions.[3] These error studies have led to the development of techniques to increase the fidelity of calculations under a wide range of nonequilibrium conditions. We shall describe some of the results of error studies relevant to the study of near-equilibrium flows and the techniques to increase the fidelity of calculations under the near-equilibrium condition.

The distinctive feature of numerical methods for solving the Boltzmann equation is to embed in a method of integration a Monte Carlo method for evaluating the collision integral. The numerical method consists of two explicit stages: the evaluation of the collision integral and the integration of the differential equation. The Boltzmann solution includes the distribution function, the collision integral, and their moments. All these functions are related and have practical significance. For example, the collision integral is the rate of change of distribution function, and the moments of the collision integral are the transport properties. It is thus important to assess the errors of all these functions and find ways to minimize them. The change of computational parameters influences many errors that should be controlled to increase the fidelity of calculations. Several techniques have been used to minimize the errors. We shall describe briefly only those that are directly relevant to the calculations in near-equilibrium conditions.

Studies have been made of the two kinds of errors in the Monte Carlo evaluation of the Boltzmann collision integral.[4] Systematic errors are caused by the use of a finite number of cells in the phase space. Random errors are associated with the finite size of each collision sample used. Both kinds of errors can be reduced, for a gas near equilibrium, by the MB (Maxwell-Boltzmann) technique.[4] We customarily use the MB corrections for all calculations, that is, for gases both far from and near equilibrium for the sake of consistency and uniformity of treatment. Near equilibrium, the corrections produce a significant increase of accuracy in the results; far from equilibrium, the corrections are small and do no harm. With the MB corrections, it has been possible to make calculations for gases much closer to equilibrium, whether the gas is in a very weak shock or is near the boundary of a strong shock. We conclude that this technique increases the fidelity of calculation of the flow in a Knudsen layer of a near-equilibrium gas.

The MB technique amounts to correcting the calculated values of the gain and loss terms of the collision integral in such a way that their differences (gain minus loss) equal zero for a gas in equilibrium that has the same values of density, temperature, and gas velocity as the nonequilibrium gas. The MB correction adds about 15% to the computing time. The shock wave calculations using the MB correction were compared with the theory of Baganoff and Nathenson[5] and the experiment of Holtz et al.[6]

Another technique that is directly relevant to the Boltzmann calculations of the distribution function and the two parts of the collision integral for a near-equilibrium gas is the use of one sample (1S).[4] In this technique, the _same_ collision sample was used for each computation node in the physical space and iteration so that the _same gas_ was therefore situated at each node. The Boltzmann solution is obtained from the average of several solutions of different collision samples. Before the 1S algorithm was introduced, we used the FS algorithm in which a fixed but different set of independent collision samples was used at each node (this set was repeated for each iteration of a given run). Figure 1 shows the implementation of FS and 1S sampling techniques. The FS algorithm, in effect, describes the

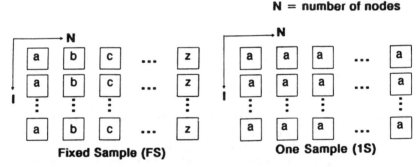

Fig. 1 Implementation of the FS (fixed sample) and 1S (one sample) sampling techniques to evaluate the Boltzmann collision integral in the velocity space at the N computation nodes in the physical space. In the FS technique, the collision samples at the nodes are different and statistically independent but are repeated at subsequent iterations. In the 1S technique, the same collision sample is used at all nodes and subsequent iterations. The Boltzmann solution is obtained from the average of several solutions of different collision samples.

interactions of slightly different nonequilibrium gases because the collision samples at various stations are statistically independent. We observed fluctuations in the Monte Carlo calculations of the collision integral for a given velocity cell using this technique. The 1S technique reduced these fluctuations in the physical space and separated even more cleanly than before the variations of both velocity distribution functions and collision integrals in the physical space from their variations in velocity space. In the case of the important function density gradient dn/dx in a shock wave, the dn/dx vs $\hat{n}$ curves are nested, as shown in Figure 2; that is, the curves for different 1S samples (i.e., different runs) are similar but displaced along the dn/dx axis as though a different x scale corresponds to each sample. The curves in Figure 2 show the variation of dn/dx in a shock wave of M = 1.2 obtained for five different collision samples. It should be noted that the density gradient dn/dx is calculated as a moment of the collision integral.

Another virtue of the 1S algorithm is that it permits an increase in computing time by a factor of as much as 3. This increase is achieved by calculating, for each randomly chosen collision, the contributions to the collision integral for each of the nodes in the physical space before going on to similar calculations for the next randomly chosen collision.

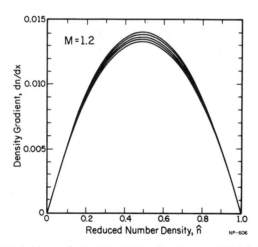

Fig. 2 Variation of the density gradient dn/dx (obtained from the collision integral) vs the reduced density $\hat{n}$ in a shock wave of M = 1.2. Results were obtained from five different collision samples. $\hat{n} = (n - n_1)/(n_2 - n_1)$.

Our calculations for the near-equilibrium gas using the two techniques described earlier exhibit some distinct behavior of interest. Figure 3 shows the departure of the Boltzmann calculation of the viscosity coefficient from the linear prediction in a shock wave of Mach number 1.2. The departure is expressed as the ratio of the viscosity coefficient μ to that of the linear prediction μ_ℓ. The density profile is included in the figure to show the relative nonequilibrium position at which the computed departure occurs. We observe that the linear kinetic theory underestimates appreciably the transport properties even near the cold and hot sides of the shock wave. This finding suggests that the gas departs from, and arrives at, an equilibrium condition abruptly and that neither the continuum approach nor methods based on linear kinetic theory can be used to solve problems of near-equilibrium gas.

The shock wave may be considered as a flow consisting of two back-to-back Knudsen layers. A Knudsen layer, in which the gas relaxes from a nonequilibrium condition to one of equilibrium, occurs in practically all rarefied gas flowfields. The study of shock waves is therefore basic to that of rarefied gasdynamics problems in general, and the observation made here is equally valid for other

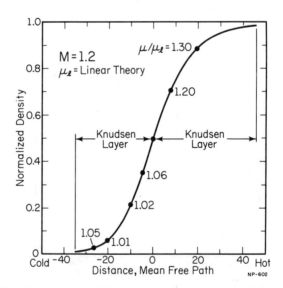

Fig. 3 Density profile in a shock wave of M = 1.2, showing the Boltzmann calculations of the viscosity coefficient ratio μ/μ_ℓ at several positions in the two Knudsen layers.

Knudsen layers that may occur in a flowfield of a near-equilibrium gas, e.g., near the stagnation point of a hypersonic flow at high altitudes.[7]

Direct Simulation Method

There are four different direct simulation methods.[8] We shall consider only the tracer method. This method, first developed by Bird[9] and referred to as the DSMC method, dynamically follows a subset of molecules, and interactions between the molecules occur only within the randomly selected subset. There are two explicit steps in the calculation: to evaluate the interactions among the molecules and to move them. Since only a small number of molecules and, thus, a small number of collision samples are used, we are interested in studying the random error in the Monte Carlo calculations and its ability to make calculations of a near-equilibrium gas, which requires large collision samples. We consider the application of the DSMC method to two problems, the one-dimensional Rayleigh problem and the problem around a sphere. We have performed several numerical experiments to study the influence of change of the computational parameters on the random error and the range of nonequilibrium conditions that the DSMC method can simulate.

Three sets of numerical experiments have been performed by changing individually the following computational parameters: NCYC, number of cycles (i.e., number of collision samples); MC, number of molecules per cell; and Kn, Knudsen number. The objectives of these experiments are to obtain an understanding of the random error of the Monte Carlo solution, the nonequilibrium condition it can simulate, and the methods to control the error and to increase the fidelity of calculations. All the experiments were performed for the Rayleigh problem with the boundary conditions that the dimensionless wall velocity = 2 and the dimensionless wall temperature = 1. The basic feature of this problem is the influence of the molecules reflected from the surface on the gas velocity above the plate; therefore, the principal result observed in our experiments is the velocity profile.

We are interested in knowing the method for selecting computational parameters in order to control the random error and to increase the fidelity and resolution of Monte Carlo calculation at all levels of departure of equilibrium condition. For each case, irrespective of Knudsen number, we also obtained the free molecular solution through bypassing the collision loop in the program. The comparison with the analytical free molecule solution gives a measure of the resolution that can be obtained from the number of molecules used in the simulation. The comparison of the velocity profile for each case of Knudsen number with the continuum analytical solution gives a measure of the departure of the equilibrium condition for this Knudsen number. Two parameters were used to measure, respectively, the smoothness and statistical scatter of the Monte Carlo calculation of the velocity profile. The definition and the method of calculation of these two parameters are given in Appendix A.

In the first set of experiments, the effect of changing the collision sample, NCYC, was studied. For these experiments, we set the number of molecules per cell MC = 20 and NCYC = 1, 10, 20, 50, 100, and 200. In the second set of experiments, we set NCYC = 20 and MC = 10, 20, 50, 100, and 200. In Figures 4a and 4b, we compare the velocity profiles for two cases: (a) MC = 200 and NCYC = 20 and (b) MC = 20 and NCYC = 200. We observe that the

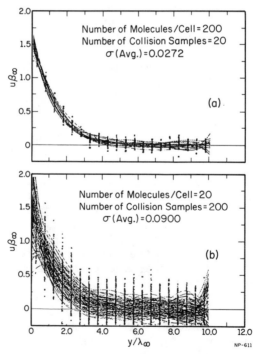

Fig. 4 Rayleigh problem velocity profiles obtained for each collision sample for two numerical experiments. Kn = 1; normalized wall velocity = 2; normalized time = 5.

velocity profiles for case (a) nest within a smaller band, as observed in our Boltzmann calculations for the density gradient dn/dx, shown in Figure 2. The scatter parameter (AVG) is reduced from 0.0815 to 0.0272, when MC increases from 20 to 200. Similar scatters were observed for the case without collisions (free molecular). The comparison of the average velocity profile for the two cases is shown in Figure 5. We observe that the results are in agreement since the product (MC)(NCYC) is the same for the two cases.

The objective of the third set of experiments is to study the effect of Knudsen number on the solution. For this study, we chose MC = 20 and NCYC = 20. The results for two lower Kn of 0.10 and 0.01 are shown in Figure 6. In each case, the velocity profile is compared with the

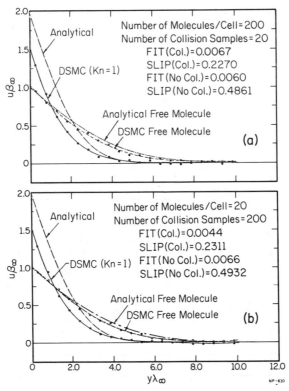

Fig. 5 Comparison of the average velocity profile obtained from the results of the two cases shown in Fig. 4. The continuum analytical solution, the DSMC free molecule solution, and the free molecule analytical solution are also shown.

continuum analytical solution in order to show the effect of the departure from equilibrium condition. The velocity slip at the surface, also given in each figure, is a measure of the departure from equilibrium. The Monte Carlo and analytical free molecules solutions are also compared. The value of the smoothness parameter FIT is also given in the figure for each Monte Carlo solution.

It may be concluded that the random error of the solution of the Rayleigh problem using the DSMC method may be assessed and controlled by using a relatively large collision sample and that accurate near-equilibrium, as well as free molecule, solutions can be obtained by choosing an appropriate set of computational parameters.

The second problem studied was that of flow over a sphere. Results were obtained for 10 cycles (collision samples) for freestream velocity = 775 and 7750 m/s.

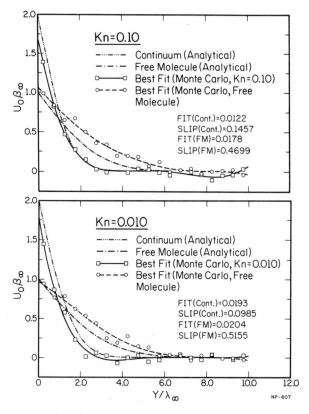

Fig. 6   Comparison of DSMC solutions of the Rayleigh problem for Kn = 0.10 and 0.010. Normalized time = 5; number of molecules per cell = 20; number of cycles = 20.

Figures 7 and 8 show the results on the velocity profile for the two freestream velocities in the radial direction for 10 collision samples at four different angular positions measured counterclockwise from the horizontal axis: 1) wake: 0-10 deg, 2) wake: 40-50 deg, 3) front: 130-140 deg, and 4) near stagnation point: 170-180 deg. We observe that the scatter is considerable in the wake region for both cases. The scatter for the case of high speed (7750 m/s) is much smaller in the front of the sphere. Further studies have to be made to assess the random error and to control the scatter in the wake.

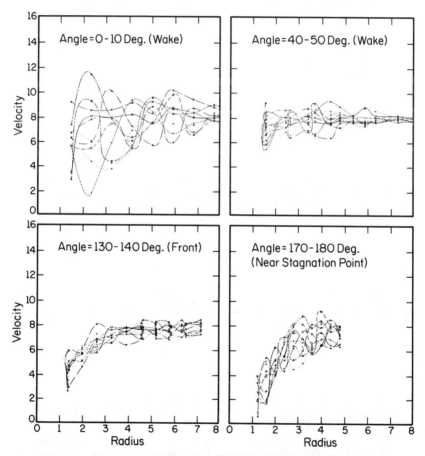

Fig. 7 Velocity profiles obtained from a DSMC solution in the radial direction over a sphere at four angular positions measured in the counterclockwise direction from the horizontal axis. Freestream velocity = 775 m/s.

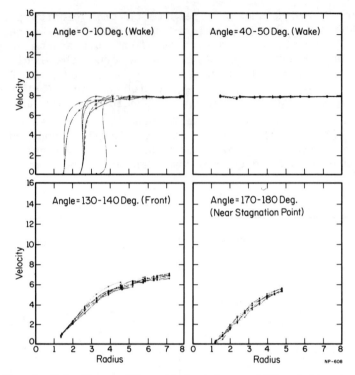

Fig. 8 Velocity profiles obtained from a DSMC solution in the radial direction over a sphere at four angular positions measured in the counterclockwise direction from the horizontal axis. Freestream velocity = 7750 m/s.

Acknowledgment

The numerical experiments on Bird's DSMC method for the Rayleigh problem and the problem over a sphere were made by E. Bermingham.[10]

Appendix A

The parameters used to evaluate the smoothness and statistical scatter of the results are a curve-fit parameter FIT and the average standard deviation $\sigma(AVG)$ in the Rayleigh problem and average standard deviation σ_{avg} in the sphere problem. The curve-fit parameter gives an indication of the "smoothness" of the curve, whereas the average standard deviation gives an indication of the statistical scatter in the results.

These parameters are calculated as follows:

$$\text{FIT} = \{[\sum_i (V_{BF} - V_{CELL})_i^2]^{1/2}/U_w\}/NC \qquad (A1)$$

where

V_{CELL} = velocity in cell i
V_{BF} = velocity indicated by the best-fit curve at the corresponding cell i
U_w = wall velocity
NC = number of cells

$$\sigma(\text{AVG}) = \sum_i \{[\sum_j (V_{AVG_i} - V_{i,j})^2/NCYC]^{1/2}/U_w\}/NC \qquad (A2)$$

where V_{AVG_i} = average velocity in cell i, and $V_{i,j}$ = velocity in cell i of cycle j.

$$\sigma_{avg} = \sum_i \{[\sum_j (V_{AVG_i} - V_{i,j})^2/NS]^{1/2}/FV\}NR \qquad (A3)$$

where FV = freestream velocity, NS = number of samples, and NR = number of radii in one sample.

References

[1] Aristov, V. V. and Tcheremissine, F. G., "The Kinetic Numerical Method for Rarefied and Continuum Gas Flows," Proceedings of the 13th International Symposium on Rarefied Gas Dynamics, Vol. 1, Plenum Press, New York, 1985, pp. 269-276.

[2] Nordsieck, A. and Hicks, B. L., "Monte Carlo Evaluation of the Boltzmann Collision Integral," Rarefied Gas Dynamics, edited by C. L. Brundin, Academic Press, New York and London, 1967, pp. 695-710.

[3] Hicks, B. L. and Smith, M. A., "On the Accuracy of Monte Carlo Solution of the Nonlinear Boltzmann Equation," Journal of Computational Physics, 1968, pp. 58-79.

[4] Hicks, B. L., Yen, S. M., and Reilly, B. J., "Numerical Studies of the Nonlinear Boltzmann Equation. Part II: Studies of New Techniques for Error Reduction," Coordinated Science Laboratory, Univ. of Illinois, Urbana, Rept. R-412, 1969.

[5] Baganoff, D. and Nathenson, M., "Constitutive Relations for Stress and Heat Flux in a Shock Wave," Physics of Fluids, 1970, pp. 596-607.

[6] Holtz, T., Muntz, E. P., and Yen, S. M., "Comparison of Measured and Predicted Velocity Distribution Functions in a Shock Wave," Physics of Fluids, 1971, pp. 545-548

[7] Yen, S. M., "Thermal Nonequilibrium Shock Layer Near the Stagnation Point," Physics of Fluids, 1986, pp. 9-10.

[8] Eaton, R. R., Fox, R. L., and Touryan, K. J., "Isotope Enrichment by Aerodynamic Means: A Review and Some Theoretical Considerations," Journal of Energy, 1977, pp. 229-236.

[9] Bird, G. A., Molecular Gas Dynamics, Oxford University Press, London, 1976, pp. 118-132 and 151-153.

[10] Bermingham, E., "Application of a Direct Simulation Monte Carlo Method to Rarefied Gas Flow Problems," M.S. Thesis, Aeronautical and Astronautical Engineering Department, Univ. of Illinois, Urbana, 1987.

Direct Numerical Solution of the Boltzmann Equation for Complex Gas Flow Problems

S. M. Yen* and K. D. Lee†
University of Illinois at Urbana-Champaign, Urbana, Illinois

Abstract

The parallel processing and multiprocessing capability of supercomputers ideally accommodates the computation strategy for solving the Boltzmann equation for rarefied gas flow problems. This strategy is to decouple the calculations in the velocity space from those in the physical space. The increase in computational efficiency resulting from the use of a supercomputer enhances the possibility of solving the Boltzmann equation for more complex problems. Our plans for revising our method to solve increasingly more complex problems are: 1) apply a method of integration to solve the Boltzmann equation for multidimensional problems, 2) implement our Monte Carlo method to evaluate the collision integral for both self-collisions and cross-collisions, and 3) incorporate inelastic collision into our Monte Carlo method. The proposed methods and their implementations are outlined in this paper.

Introduction

The advent of supercomputers has spurred our interest in applying the Monte Carlo method to solve directly the nonlinear Boltzmann equation for complex multidimensional problems. We shall review in this paper our plans to incorporate the complexities in our method and our strategies to implement on the supercomputer the revisions to increase systematically the degree of complexity in our method.

Copyright © 1989 by the American Institute of Aeronautics and Astronautics, Inc. All rights reserved.
* Professor, Aeronautical and Astronautical Engineering.
† Associate Professor, Aeronautical and Astronautical Engineering.

Our numerical method for solving the nonlinear Boltzmann equation embeds a Monte Carlo method to evaluate the collision integral in a method of integration. It consists of two explicit stages: 1) the evaluation of the collision integrals in the quantized and discretized velocity space at each computation node, using the velocity distribution function obtained from the previous iterative step, and 2) the update of the velocity distribution function at all the computation nodes by integrating the Boltzmann equation for each of the velocity cells in the velocity space.

Our plans for revising our method to solve the Boltzmann equation numerically for increasingly complex problems are: 1) apply a method of integration for multi-dimensional problems, 2) implement our method for gas mixtures, and 3) incorporate inelastic collisions in our method. We shall describe these revisions and their implementations on the computer.

Parallel Processing and Multiprocessing

The computation strategy is shown in Figure 1. The important aspect of the strategy is the decoupling of the calculations in the velocity space from those in the physical space. The parallel processing and multi-processing capability of a supercomputer is ideal for accommodating our strategy, increasing the overall computation efficiency, and allowing the implementation of more complexities.

The time-consuming phase of the numerical solution of the Boltzmann equation is in the Monte Carlo evaluation of the collision integral in the velocity space. Any increase in computation efficiency in this phase of computation greatly enhances the possibility of solving the nonlinear Boltzmann equation for complex problems. We use an iterative scheme to solve the Boltzmann equation numerically; therefore, a total of N x I x J independent collision samples have to be processed for each solution. (N = number of computational nodes in the physical space, I = number of iterations, and J = number of runs or sample sets.) If there are N number of processors in a computer, then each processor can accommodate each node, and the collision integral at all nodes can be evaluated simultaneously for each iteration. We have used two techniques to reduce the random error and increase the efficiency of the Monte Carlo calculation: the fixed sample set (FS) and the one sample set (1S). In the FS

DIRECT NUMERICAL SOLUTION OF THE BOLTZMANN EQUATION 339

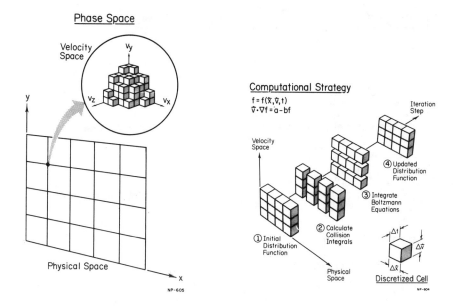

Fig. 1 Schematic presentation of the computation strategy to make Boltzmann calculations in the phase space. Evaluation of the collision integral in the velocity space for the velocity distribution function at the previous iterative step are performed concurrently at each of the computation nodes in the physical space. The next step is to integrate the Boltzmann equation in the physical space to update the velocity distribution function in the velocity space. This numerical integration is carried out concurrently for all the velocity cells in the velocity space.

technique, a fixed set of collision samples is used at each node, this set is repeated for each iteration; therefore, the number of collision samples to be processed = N x J. In the 1S technique, the same set of sample is used for all the nodes and iterations; therefore, the number of samples to be processed is reduced to J. In the implementation of this technique, the contributions of each of the randomly chosen collisions to the collision integral is made and may be paged to the processor of all the interior nodes before going on to the next randomly chosen collision. This technique permits a significant increase in computational efficiency. Figure 2 shows the implementation of three different sampling techniques and the relative computational efficiencies.

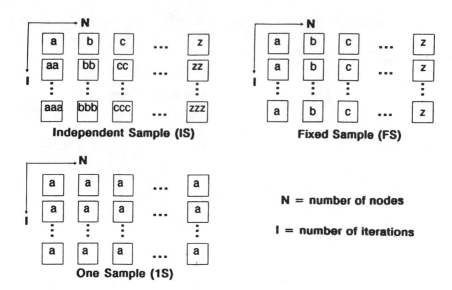

Fig. 2 Implementation of three Monte Carlo sampling techniques to evaluate the Boltzmann collision integral at the computation nodes in the physical space. In the IS (independent sample) technique, the collision samples at each node and iteration are different and statistically independent, and the number of samples to be processed = N x I for each run. In the FS (fixed sample) technique, the collision sample at each node is repeated at subsequent iterations, and the number of samples to be processed = N. In the 1S (one sample) technique, the same sample is used at each node and iteration, and the sample is changed for each run.

Multidimensional Problems

We shall describe briefly the numerical scheme to integrate the Boltzmann equation proposed by Tcheremissine.[1,2] For simplicity, we consider the Boltzmann equation for the steady two-dimensional flow:

$$v_x \frac{\partial f^I}{\partial x} + v_y \frac{\partial f^I}{\partial y} = a^{(I-1)} - b^{(I-1)} f^I \qquad (1)$$

where f is the distribution function, a-bf the collision integral, and I the iteration number.

In this scheme, the finite difference approximation in divergence form is used for the transport operator for each computation cell as follows:

$$\int (\bar{v} \cdot \bar{n}) f^I \, d\ell = \iint (a^{(I-1)} - b^{(I-1)} f^I) dA \qquad (2)$$

where ℓ is the boundary surface, $\bar{n}$ a vector normal to $d\ell$, and A the area of the cell. Equation (2), in its discretized form, implies a flux balance of the distribution function f with a source term due to molecular collision.

Gas Mixtures

We shall now describe briefly the implementation of our method to problems of binary gas mixtures.[3] There are four collision integrals to be evaluated independently. The complication that lies in the quantization and discretization of velocity spaces for two different gases may be resolved by choosing a common velocity space and proper units for the velocity and density. The formulation of the two sets of Boltzmann equation will reflect these revisions. There will be no computation difficulties other than increased computation time.

Inelastic Collisions

The implementation of inelastic collisions is being considered. It will increase the complexity of Monte Carlo computations; however, the basic strategy of decoupling the Monte Carlo calculations from the integration scheme and their implementation on the supercomputer will be the same. Many collision models have been proposed for gases with internal degree of freedom. The statistical model proposed by Holway[4] and the formulation are of interest. In his collision model, the collision term is broken into two parts: 1) the elastic term representing collisions in which no change occurs in the internal energy of the colliding molecules and 2) the inelastic term representing collisions in which at least one of the colliding molecules changes its internal energy state.

An advantage of breaking up the collision term is that elastic collisions normally occur much more frequently than inelastic collisions. For elastic collisions, the form of the collision term is unaffected by the existence of internal states, which only determine the labeling of particles. The application to a diatomic gas would be of interest. For the case in which the rotational energy can be treated classically, we introduce the notation $n_s(e, \bar{r}, t) f_s(\bar{v}, e, \bar{r}, t) de d\bar{v} d\bar{r}$, which represents the number of particles in vibrational state s with rotational energy between e and e + de in the volume element $d\bar{r}$ about $\bar{r}$ with velocities in the range $d\bar{v}$ about $\bar{v}$ at time t. The collision term consists of three parts as follows:

$$\left(\frac{\partial f_s(e,v)}{\partial t}\right)_{el} + \left(\frac{\partial f_s(e,v)}{\partial t}\right)_{in} + \left(\frac{\partial f_s}{\partial t}\right)_{rot} \qquad (3)$$

The final term in the preceding expression is added to take into account inelastic collisions in which there is an exchange between rotational and translational degrees of freedom while the vibrational state is unchanged. The statistical model, as well as other collision models is being considered.

References

[1] Tcheremissine, F. G., "Solution of the Boltzmann Equation for Plane Rarefied Gas Dynamics Problems," Proceedings of the 13th International Symposium on Rarefied Gas Dynamics, Vol. 1, Plenum Press, New York, 1985, pp. 303-310.

[2] Yen, S. M., "Numerical Solution of Nonlinear Boltzmann Equation for Nonequilibrium Gas Flow Problems," Annual Review of Fluid Mechanics, 1984, pp. 67-97.

[3] Yen, S. M., Hicks, B. L., and Osteen, R. M., "Further Developments of a Monte Carlo Method for Evaluation of Boltzmann Collision Integral," Proceedings of the 9th International Symposium on Rarefied Gas Dynamics, Vol. 1, DFLVR-Press, Porz-Wahn, Germany, 1974, pp. A.12-1-A.12-10.

[4] Holway, L. H., "New Statistical Models for Kinetic Theory: Methods of Construction," Physics of Fluids, Vol. 9, September 1966, pp. 1658-1673.

Advancement of the Method of Direct Numerical Solving of the Boltzmann Equation

F. G. Tcheremissine*
USSR Academy of Sciences, Moscow, USSR

Abstract

This paper presents progress in the development and application of the direct numerical method for solving the fundamental kinetic equation for gases during the period after the XIII International Rarefied Gas Dynamics Symposium. It describes some of the main results of investigations directed toward the improvement of the computational efficiency of the method especially for numerically difficult parameters of gas flows (small Knudsen numbers, considerable changes in density or concentration among components of gas mixtures, and high Mach numbers.)

The interpretation of numerical algorithms that are given show their relation with a direct simulation method on one hand and a discrete velocities approach on the other. Such an interpretation permits an experimental observation that the method gives reasonably good results even with crude parameters of discretization. Yet most solutions presented in the paper were obtained on rather fine meshes and with a detailed control of the convergence of numerical results by internal parameters of the algorithm.

Introduction

Initially, numerical solutions of the Boltzmann kinetic equation were based on a special elegant technique of statistical simulation of elastic molecular collisions and obtained at the beginning of the sixties at the University of Illinois in the U. S. by Nordsieck and

---

Copyright ©1989 by the American Institute of Aeronautics and Astronautics, Inc. All rights reserved.
*Computing Center.

Hicks.[1] The work was then continued by Yen and his collaborators and had a stimulating influence on the development of other approaches.[2] One such approach was developed in the USSR by this author and his colleagues and was named the "method of direct numerical solution of the Boltzmann equation."[3] In developing our method we imposed the strict requirement that it must not include any assumption that does not follow from the Boltzmann equation. On this basis we constructed numerical algorithms that at the limit of convergency strictly approximate the kinetic equation, with the accuracy of each particular solution being determined only by the parameters of discretization. But the Boltzmann equation is rather complex and difficult for numerical solution; thus our main efforts were directed toward the optimization of general numerical algorithms for given ranges of characteristic flow parameters. For this we used different standard means of computational mathematics, including the heuristic postulate of computational fluid mechanics which gives preference to conservative finite-difference schemes. Some special techniques were also developed such as control of the accuracy of computations according to a considered point in phase space or a number of succesive iterations.

As a result of these improvements the class of problems that could be solved in exact formulation and with rather high accuracy was broadened. Now with a computer of about 1 MFL productivity it is possible to investigate different plane flows and one-dimensional flows of gas mixtures, including those with high differences in concentration or with physicochemical reactions. Some one-dimensional problems like that of shock-wave structure were solved with guaranteed accuracy for different intermolecular laws.[4,5] For flows close to the continuum limit (small Knudsen numbers), a special method was developed as well as a kinetic approach to solving Euler and Navier-Stokes equations; some examples of matching those solutions with the solution of the Boltzmann equation were demonstrated.

The computation of hypersonic flows stimulated the construction of a new Monte Carlo technique for the evaluation of collision integrals.

Finally, it was shown on some examples and theoretically justified that numerical algorithms of the proposed approach could be considered as a realization of a discrete model of rarefied gas composed of "phase particles" of finite size and describes rather well its main properties regardless of the parameters of discretization. That consideration permits an organization of very fast computations with reasonable confidence in the results.

General Structure of the Method

Write the Boltzmann kinetic equation in the form

$$\mathcal{L}(f) = I(f, f)$$

where $\mathcal{L}(f)$ is a linear differential transport operator, and $I(f, f)$ is a nonlinear integral collision operator. It is often convenient to write the collision integral in the form

$$I(f, f) = -\nu(f)f + N(f, f) \qquad (1)$$

The main elements of the method are 1) the discrete presentation of unknown solutions as a matrix $f_{i\beta}^l$ with indices $i = (i_1, i_2, i_3)$ and $\beta = (\beta_1, \beta_2, \beta_3)$ marking a node in coordinate and velocity spaces and index l corresponding to a time level for an unsteady problem or a number of iterations for a steady one; 2) finite-difference approximation of the operator $\mathcal{L}(f)$ on a fixed coordinate grid; 3) method of evaluation of collision integrals $I_{i\beta}^l$ or $\nu_{i\beta}^l$ and $N_{i\beta}^l$ at any node by the Monte Carlo technique; 4) an iterative procedure for advancing the index l; and 5) evaluation of macroscopic variables of the flow of each node x_i as a corresponding sum by the index β, e.g., $n_i^l = \Sigma_\beta f_{i\beta}^l$.

Hence, in this method the deterministic finite-difference approximation of the left side of the kinetic equation is used together with the stochastic estimation of the collision integrals.

As a finite-difference approximation of the operator $\mathcal{L}(f)$, we mainly used divergent approximations of the first order, but it may occur that for complex geometry some schemes of the method of characteristics would be preferable. As an example we show here the scheme that was used in the method of steady iterations for the one-dimensional problem and for $\xi_\beta \geq 0$:

$$\xi_\beta \frac{f_{i\beta}^l - f_{i-1\beta}^l}{\Delta x_i} = -\nu_{i\beta}^{l-1} f_{i\beta}^l + N_{i\beta}^{l-1} \qquad (2)$$

It was demonstrated that the application of this formula for flows with small Knudsen numbers permits the choice of the coordinate mesh Δx_i being superior to the local mean free path. The only restriction for the mesh is that it should be smaller than some characteristic length of change in the solution.

In unsteady conservative splitting methods the solution of the Boltzmann equation over a small time interval is performed in two stages. In the first stage the collisionless kinetic equation is solved, and in the second the equation is solved that describes relaxation in each node (or in each cell).

$$\frac{\partial f^*}{\partial t} + \xi \frac{\partial f^*}{\partial x} = 0 \qquad (3)$$

$$\frac{\partial f}{\partial t} = -\nu(f^*)f + N(f^*, f^*) \qquad (4)$$

For both Eqs. (3) and (4) the first-order finite-difference schemes were used, and for Eq. (3) the explicit divergent scheme with the stability condition $\xi \Delta t / \Delta x \leq 1$ was used.

It was proved that the schemes (3) and (4) approximate the Boltzmann equation with first-order accuracy for Δt and Δx. For practical use, however, the stability condition imposes rather inconvenient restrictions on time step Δt. Two methods for softening this restriction were used. The first one consists in subsplitting of a reasonably chosen time interval Δt on m fractions and successively doing m solutions of Eq. (3) with the fractional step. After that the solution of Eq. (4) over a total time interval has to be performed. The second method that can be applied in conjunction with the first one consists of considering the time step as an iterative parameter and choosing a locally dependent parameter on a coordinate mesh Δx_i.

To construct a strictly conservative algorithm, a correction of the solution after stage (4) should be added

$$\tilde{f} = f(1 + \mathcal{P}(\xi))$$

The polynomial $\mathcal{P}(\xi)$ generally contains five free parameters to satisfy density, momentum and energy conservation laws for the corrected function $\tilde{f}$. It was proved long ago that the correction does not reduce the order of approximation of the Boltzmann equation by the splitting scheme. On the contrary, the practice showed that the correction not only improves the stability of computations but raises the final accuracy of the results. With the use of the conservative splitting scheme fast computations on very crude meshes and with low numbers of random tests also became possible.[5]

Figure 1 shows a solution of a one-dimensional piston problem for total numbers of nodes in velocity spaces $K_1 = 50$ and 15, with the number of random tests for both calculations $K_0 = 10$. The density graphics are presented for a speculary reflecting piston with velocity $v = 2(kT/m)^{1/2}$, where T is the temperature in the unperturbed gas.

Note that the solution for $K_1 = 50$ deviates from that obtained with $K_1 = 500$ and $K_0 = 100$ less than 1.5%, but the execution was about 50 times faster. The solution on a "supercoarse" mesh is also quite reasonable, although the velocity mesh here was about 1.5 times bigger than the mean terminal velocity.

It was demonstrated on different examples that extremely low numbers of random tests (sometimes $K_0 = 1$) could be used in conservative algorithms. In Fig. 2 such a demonstration is given for a problem of steady one-dimensional heat flux between two parallel plates with $T_2/T_1 = 4$ and $Kn = 1$. However, it is important to note that in other examples a small systematic deviation was observed, which is quite natural because of the nonlinearity of the kinetic equation.

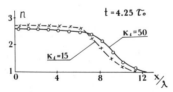

Fig. 1 Solution of a one-dimensional piston problem for two different K_1 values.

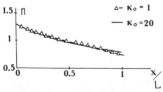

Fig. 2 Solution of the steady one-dimensional heat flux between two parallel plates.

The Principle of Hydrodynamic Consistence and The "Phase Particles" Model

Consider a finite-difference approximation of the Boltzmann equation on a grid t^j, x_i, ξ_β:

$$(f_{i\beta}^j)_{\bar{i}} + \xi_\beta A_k f_{i\beta}^j = -\nu_{i\beta}^{j-1} f_{i\beta}^j + N_{i\beta}^{j-1} \qquad (5)$$

Here the subscript $\bar{i}$ indicates a time derivative and A_k is any finite-difference operator on the x variable, $\nu_{i\beta}^{j-1}$ and $N_{i\beta}^{j-1}$ being collision integrals approximated as Monte Carlo sums with the distribution function at a previous iteration or at a previous time level.

It is obvious that hydrodynamic equations could be obtained by multiplication of the continuous Boltzmann equation by collision invariants $(1, \vec{\xi}, \vec{\xi}^2)$ and integrating over the total velocity subspace, and have a form

$$\frac{\partial R^{(\alpha)}}{\partial t} + \operatorname{div} \Phi^{(\alpha)} = 0 \quad \text{where} \quad R^{(\alpha)}(t,x) = \int_{-\infty}^{\infty} \vec{\xi}^\alpha f\alpha\vec{\xi},$$

$$\Phi^{(\alpha)} = \int_{-\infty}^{\infty} \vec{\xi}\vec{\xi}^\alpha f d\vec{\xi} \qquad (6)$$

After multiplying Eq. (5) by $(1, \vec{\xi}_\beta, \vec{\xi}_\beta^2)$ and summing by index β, we get

$$\hat{R}_{\bar{i}}^{(\alpha)} + \operatorname{div}_h \hat{\Phi}^{(\alpha)} + \delta_1 = \delta_2 \qquad (7)$$

Where $\hat{R}_{\bar{i}}^{(\alpha)}$ is a time derivative and div_h is the finite-difference operator of divergence. The variables $\hat{R}^{(\alpha)}$ and $\hat{\Phi}^{(\alpha)}$ are determined as sums by the index β analogous to Eq. (6). The disbalance δ_1 has its origin in the approximation on the left side of the equation, and δ_2 is an error of the right side due to the splitting of collision integrals, the numerical error in the Monte Carlo estimation, and the error of numerical integration by ξ_β.

We would say that the finite-difference scheme (5) together with procedures for the evaluation of integrals are conservative if $\delta_1 = \delta_2 = 0$

Indeed, it is important to formulate for a conservative scheme the appropriate boundary and initial conditions in a discrete form.

DIRECT NUMERICAL SOLVING OF THE BOLTZMANN EQUATION

The unperturbed flow on the external boundary is usually described by a Maxwellian function with n_∞, u_∞, and T_∞ as parameters. But integrating this function numerically we do not get exactly the hydrodynamic parameters of a free flow. There are two ways to obtain exact hydrodynamic parameters of a free flow in a discrete form: renormalization or a correction of the Maxwellian function analogous to that used in splitting algorithms.

Boundary conditions on a body surface contain some balance relations for mass, energy, and momentum fluxes. To satisfy those relations in a discrete form the correction could be used as well.

We would call the numerical algorithm hydrodynamically consistent if 1) the finite-difference scheme is conservative, and 2) boundary conditions in discrete form give exact hydrodynamic parameters on a free boundary and exact balance relations on a body surface.

It is evident that such an algorithm could be considered as a discrete model of rarefied gas. To describe this model in more detail let us introduce a grid ω_β in a limited velocity space Ω with the condition $U\omega_\beta = \Omega$ and introduce the basis of step function $\hat{\omega}_\beta$ defined as follows:

$$\hat{\omega}_\beta(\vec{\xi}) = \begin{cases} 1, \vec{\xi} \in \omega_\beta \\ 0, \vec{\xi} \in \omega_\beta \end{cases}, \quad \int_\Omega \hat{\omega}_{\beta_1}\hat{\omega}_{\beta_2}d\vec{\xi} = v_0\delta_{\beta_1\beta_2}$$

$$\int_\Omega \hat{\omega}_\beta d\vec{\xi} = v_0 \quad v_0^{-1}\int_\Omega \varphi(\vec{\xi})\hat{\omega}_\beta(\vec{\xi})d\vec{\xi} = \varphi(\vec{\xi}_\beta)$$

The last condition is equivalent to a rule of integration by a middle point $\vec{\xi}_\beta$ of a cell ω_β, the value v_0 being the volume of a cell. The solution could be presented as an expansion:

$$f(t, \vec{x}, \vec{\xi}) = \Sigma_\beta f(t, \vec{x}, \vec{\xi}_\beta)\hat{\omega}_\beta(\vec{\xi})$$

The hydrodynamic parameters could be expressed as sums containing densities of "phase particles" $n_\beta(t, \vec{x})$:

$$n_\beta = \int_\Omega f\hat{\omega}_\beta \, d\vec{\xi}, \quad n(t, \vec{x}) = \int_\Omega f \, d\vec{\xi} = \Sigma_\beta n_\beta(t, \vec{x})$$

$$n\vec{u} = \Sigma_\beta \vec{\xi}_\beta n_\beta, \quad nE = \Sigma_\beta(\vec{\xi}_\beta^2/2)n_\beta$$

By multiplication of the Boltzmann equation on $\hat{\omega}_\beta$ and integration over Ω, we will get

$$\frac{\partial n_\beta}{\partial t} + \operatorname{div}(\xi_\beta n_\beta) = -\nu_\beta n_\beta + N_\beta v_0 \qquad (8)$$

It is easy to demonstrate that integrals ν_β and $N_\beta v_0$ are expressed through $n_{\beta 1}, n'_\beta$ and $n'_{\beta 1}$ the same way as they were expressed through f_1, f' and f'_1 in the equation. It means that Eq. (8) may be considered a kinetic equation for "phase particles" n_β.

It was demonstrated by Aristov[6] that for hard-sphere molecules the integration in ν_β and $N_\beta v_0$ over angular impact parameters could be performed analytically, and collision integrals on the right side of Eq. (8) were expressed as finite sums by a three-dimensional index β_1.

Evaluation of Collision Integrals and the Statistical Error of the Method

For any central-symmetric intermolecular potential the collision integrals could be written in the form

$$\mathcal{V}(f) = \int_{-\infty}^{\infty} d\vec{\xi}_1 \int_0^{b_m} b\, db \int_0^{2\pi} f(\vec{\xi}_1)|\vec{\xi} - \vec{\xi}_1|\, d\epsilon$$

$$N(f,f) = \int_{-\infty}^{\infty} d\vec{\xi}_1 \int_0^{b_m} b\, db \int_0^{2\pi} f(\vec{\xi}')f(\vec{\xi}'_1)|\vec{\xi} - \vec{\xi}_1|\, d\epsilon$$

In the Monte Carlo method those integrals are expressed as sums:

$$\mathcal{V}_\beta^j = \sum_{\gamma=1}^{K_0} \frac{1}{P_\gamma}[f^j(\xi_{1\gamma})|\xi_\beta - \xi_{\beta\gamma}|]$$

$$N_\beta^j = \sum_{\gamma=1}^{K_0} \frac{1}{P_\gamma}[f^j(\xi'_\beta)f^j(\xi'_{\beta\gamma})|\xi_\beta - \xi_{\beta\gamma}|]$$

Here random five-dimensional vectors $(\xi_{\beta\gamma}, b_\gamma, \epsilon_\gamma)$ have density distribution $P_\gamma = K_0/V_0 \pi B_m^2$, where V_0 is the volume of a limited space. This formula was used at the initial stages of our investigation; however, later it was found that for small and moderate Mach numbers two other Monte Carlo techniques could reduce the statistical error. The first one consists of separation of the total volume $V_0 \pi b_m^2$ into K_0 equal parts and generation of one random vector in each subvolume. In the second method the volume of integration is divided into $K_0/2$ equal hypercubes, and inside each of these one antithetic pair of random vectors is generated. For high Mach numbers it is recommended to use the technique of essential choice and raise the density of random vectors in areas that correspond to most probable collisions. Such a technique was used for an example of hypersonic plane flow cited later in the paper.

The use of a fixed discrete grid in velocity space permits considerable reduction of the number of arithmetic operations if contributions in Monte Carlo sums are made simultaneously at a number of points x_i, ξ_β with the same random sample. It was found that, for one processor computer, this organization of calculations may reduce the total computational time per one collision (one contribution at the collision integrals) about $(10 - 50)$ times with the total cost per one collision about 5-10 arithmetic operations. We have previously mentioned the control of the accuracy of calculations that could be quite easily done by variation of the size of a random sample according to a phase point or a number of iterations. Note that for reducing statistical error in conservation laws (before correction) it is preferable to use different samples for different velocity nodes ξ_β. On the contrary, each pair of integrals $\nu_{i\beta}^j$ and $N_{i\beta}^j$ with identical indices should be evaluated on the same sample. Then for the Maxwellian distribution function f_M and for each random number such an estimation is valid:

$$f'_{M\beta} f'_{M\beta_1} - f_{M\beta} f_{M\beta_1} = O(\Delta \xi) \tag{9}$$

It is evident from Eq. (9) that near local thermodynamic equilibrium the accuracy of estimation of the total collision integral $I_{i\beta}^j$ is increased.

Under certain assumptions it is possible to estimate a statistical error in the solution of Eq. (5) with the use of K_0 independent random vectors at K_1 velocity nodes and with the iteration time step $\Delta t = \tau_0/j$, where τ_0 is a mean time of a molecular free path. Then a mean square error in hydrodynamic moments could be presented as follows:

$$G_M = G_0 (K_0 K_1 j)^{-1/2}$$

where G_0 is a constant.

This estimation explains the surprisingly accurate results in Fig. 2, because although $K_0 = 1$, $K_1 = 200$ and $j = 50$ so $G_M = G_0 \cdot 10^{-2}$.

Computations with Low Knudsen Numbers

It is well known that the main difficulty in solving the kinetic equation for low Knudsen numbers presents a small parameter at the left side of the equation. In dimensionless form we have

$$Kn\mathcal{L}(f) = I(f,f)$$

As a result, the number of iterations increases, but the accuracy may decrease and the convergent solution may not be found. The previously mentioned relation (9) is a key to solving the problem because it excludes the trivial part of the collision integral that is due to the Maxwellian function. Let us write the solution near equilibrium in the form

$$f = f_M + Kn\varepsilon + \cdots$$

Using Eq. (9) we get

$$I_{i\beta}^j(f,f) = I_{i\beta}^j(f_M, f_M) + 2Kn I_{i\beta}^j(f_M, \rho) + \cdots$$

$$= 2Kn I_{i\beta}^j(f_M, \rho) + O(\Delta\xi) + \cdots$$

It is evident that the small parameter Kn disappears from the kinetic equation. Note that the approximation error in Eq. (9) could be reduced near equilibrium to $O(Kn\Delta\xi)$ by the interpolation on the basis of local Maxwellian function near a node $\vec{\xi}_\gamma$ under the condition $|\vec{\xi} - \vec{\xi}_\gamma| < \Delta\xi$:

$$f(\vec{\xi}) = f(\vec{\xi}_\gamma) \exp\left(\frac{-(\vec{\xi} - \vec{u})^2 + (\vec{\xi}_\gamma - \vec{u})^2}{2T}\right)$$

It was noted that the splitting algorithm is convenient for matching the Boltzmann equation solution with the Euler and Navier-Stokes equations obtained by the kinetic approach.[7] Remember that to solve Euler and Navier-Stokes equations it is necessary to replace the stage of relaxation (4) with the construction of the Euler or Navier-Stokes distribution function.[7]

Extension on Gas Mixtures

The extension of our conservative splitting algorithm to gas mixtures was done by Raines.[8] Here the specific point is the construction of a correction procedure. The problem lies in the fact that an inert mixture with j components has $j+4$ conservation laws; thus, using polynomials in the previously mentioned form we will get only $j+4$ equations for $5j$ free coefficients. It was proposed before correction to separate the total numerical errors in energy and momentum between the components in proportion to their partial energies and momenta. This method was successfully applied to the investigation of the problems of uniform relaxation and shock-wave structure in a binary gas mixture. On the well-known problem of relaxation of a small eddy with the concentration ratio 10^{-3}, it was found that the method is quite effective and permits the computation of mixtures with very high differences in partial concentrations.[9] In this example the effectiveness of the considered method was found to be about 10 times higher than for the improved Monte Carlo simulation approach.[10] The calculation of the mixture with concentration ratio 10^{-4} did not show any difficulties as well.

An example of computation of the uniform relaxation in an active gas mixture was done by Aristov and the author.[11] The mixture of gases with essential physical parameters of CO_2 and N_2 with two vibrational levels was considered. The excitation process was described by the probability

$$\mathcal{P}_{AB}^{A'B'} = p_{AB}^{A'B'} X\left[\frac{\mu}{2}((\vec{\xi}_A - \vec{\xi}_B) \cdot \vec{K}) - \epsilon_{AB}^{A'B'}\right]$$

where $p_{AB}^{A'B'}$ is a steric factor, $X(z)$ the Heaviside function, $\epsilon_{AB}^{A'B'}$ the energy of excitation, and $\vec{K}$ the unity vector directed on a line of collision. It was supposed that velocities of colliding partners after an inelastic process are distributed uniformly in the center of mass system and a corresponding collision operator was presented as a seven-dimensional integral.[12] It was found that such detailed description and the computation of 10 different inelastic processes do not make the problem too difficult to solve.

Some Solutions of Plane Steady Problems

Presented in this section are some examples that illustrate the peculiarities of the method.

As an illustration of its sensitivity to the intermolecular potential, we show in Fig. 3 the distribution of a viscous stress along a plate of finite length for two-dimensional flow of a one-component monatomic gas with zero angle of incidence, $M = 4, Kn = 0.1$, temperature of the surface being equal to the temperature of unperturbed flow, with diffuse reflection and for two intermolecular potentials: hard-sphere molecules – continuous line, potential $\varphi = a\tau^{-12}$ – dashed line. It can be seen that the statistical error of calculations is rather small. The problem was solved by the method of steady iterations, and about 30 iterations were needed.

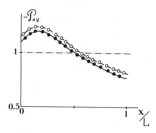

Fig. 3 Distribution of viscous stress along a finite length plate for hard-sphere molecules (continuous line) and for potential $\varphi = a\tau^{-12}$ (dashed line).

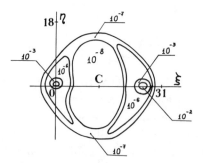

Fig. 4 Profile of the distribution function mear a flat plate leading edge.

Figure 4 shows results for a hypersonic flow over a plate with the same geometry as for the previous problem, same boundary condition on the plate, the potential $\varphi = a\tau^{-12}, M = 20, Kn = 2.5$, a profile of the distribution function $f(\xi, \tau, \zeta_0, x_0, y_0)$ near a leading edge with $\zeta_0 = 1.15 v_T, x_0 = -0.1\lambda, y_0 = 0.03\lambda$, where v_T is a thermal velocity and λ is the mean free path in unperturbed flow, is shown.

The hollow structure of the function is due to the domination of collisions between incident and reflected gas fluxes. To improve the Monte Carlo technique the essential choice of random vector was used in such a way that collisions corresponding to those fluxes were taken with higher probability. The collision integrals for velocity nodes inside a spherical layer of the radius $u_\infty/2$ were computed on a larger sample to raise the accuracy of computations. The problem was solved by the method of steady iterations, and six iterations were sufficient for the convergence.

Figures 5 and 6 present the structure of a plane supersonic jet with $M = 4$ through a split of $3\lambda_0$ size at $x = 0$. The molecular model of hard sphere gas was taken. Figure 5 shows isolines of density for a gas expansion into a vacuum. Low-density lines are well distinguished, and this is an important capability of the method. In fact, at $n \to 0$ the collision integral vanishes also; thus the statistical error of computations becomes negligible, the left side of the equation being approximated by a deterministic finite-difference scheme.

Figure 6 presents graphics of the flow velocity on a symmetry line $y = 0$ for three examples: flow into a vacuum, flow into a low density gas with $n_\infty/n_0 = 0.1$ and $T_\infty/T_0 = 1$, and a flow with $n_\infty/n_0 = 0.1$ and $T_\infty/T_0 = 2$. For the first graphic, a small rise of flow velocity is seen. For the two other examples the velocity diminishes faster for expansion into a heated gas. This problem was also solved by the method of steady iterations, and 10 iterations were

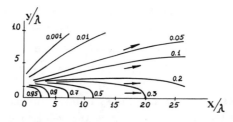

Fig. 5 Isolines of density for a gas expansion into vacuum.

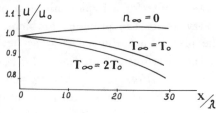

Fig. 6 Flow velocity on a symmetry line for flow into a vacuum and flow into a low density gas with two different temperature ratios.

needed for the convergence. Note that for $Kn < 0.1$ all problems were solved by the conservative splitting method.

Conclusion

The analysis and the practice of the method that has been discussed showed some advantages that we would like to summarize: 1) because the method is constructed only by the use of standard means of computational mathematics, its correlation to the continuous Boltzmann equation is quite evident; 2) the accuracy of the solution depends only on the parameters of the discretization and can be easily controlled; 3) the use of a fixed velocity grid considerably reduces computational expenses as simultaneous calculations of collision integrals become possible; 4) near the continuum flow limit, the main Maxwellian part of the distribution function can be eliminated from the collision integral considerably raising the total accuracy of the solution; 5) near the free molecular limit, the statistical error vanishes; and 6) the method is well adapted for flows with high variations of density or concentrations of components in a gas mixture. Finally, the numerical algorithms are quite simple and well adapted for parallel processing.

It was mentioned that the method could be used for rather accurate calculations and for those with crude parameters of discretization as well. Practice has shown that, for a one component gas and the accuracy in hydrodynamic parameters of 1.5-3%, it is necessary to have 20-40 nodes in velocity space for any isotropic problem, 200-500 nodes for a cylindrical symmetry in velocity space, and 1500-5000 nodes for a general case, the velocity mesh being taken constant. The average number of random vectors in Monte Carlo evaluation of collision integrals was about 100 per one node in phase space, the number of steady iterations varied from between 5 and 30, and the average number of time steps in the splitting method was on the order of 100. So, in this method the number of velocity nodes is usually higher than the number of cell particles in a direct simulation approach. On the other hand, the number of iterations is much less and the cost of one "collision" (one contribution in a collision integral) in arithmetic operations is considerably less as well. We believe that the last factor is the most important for the effectiveness of any method.

References

[1] Nordsieck, A. and Hicks, B. L., "Monte-Carlo Evaluation of the Boltzmann Collision Integral," *Proceedings of the V International Symposium on Rarefied Gas Dynamics*, Vol. 1, Plenum, New York, 1967.

[2] Hicks, B. L. and Yen, S. M., "Collision Integrals for Rarefied Gas Flow Problems," *Proceedings of the VII International Symposium on Rarefied Gas Dynamics*, Plenum, New York, 1971, pp. 845-854.

[3] Tcheremissine, F. G., "Numerical Solution of the Boltzmann Kinetic Equation for One-dimensional Steady Gas Flows," *J. of Comp. Math. and Math. Phys.*, Vol. 10, No. 3, 1970, pp. 645-655, (in Russian).

[4] Tcheremissine, F. G., "Numerical Methods of Direct Solution of the Boltzmann Kinetic Equation," *J. of Comp. Math. and Math. Phys.*, Vol. 25, No. 12, 1985, pp. 1840-1855, (in Russian).

[5] Aristov, V. V. and Tcheremissine, F. G., "Solutions of One and Two-dimensional Problems for the Boltzmann Equation," *Committee on Applied Mathematics – Reports of the Computer Center of the Academy of Science, USSR*, 1987 p. 47, (in Russian).

[6] Aristov, V. V., "On the Solution of the Boltzmann Equation by a Discrete Velocity Method," *Doklady Academii Nauk SSSR*, Vol. 283, No. 4, 1985, pp. 831-834, (in Russian).

[7] Aristov, V. V. and Tcheremissine, F. G., "The Kinetic Numerical Method for Rarefied and Continuum Gas Flows," *Proceedings of the XIII International Symposium on Rarefied Gas Dynamics*, Vol. 1, Plenum, New York, 1985, pp. 269-276.

[8] Raines, A. A., "Numerical Solution of the Boltzmann Equation for Binary Gas Mixtures," *Proceedings of the XIII International Symposium on Rarefied Gas Dynamics*, Vol. 2, Plenum, New York, 1985, pp. 1285-1293.

[9] Bird, G. A., *Molecular Gas Dynamics*, Clarendon, Oxford, UK, 1976.

[10] Korolev, A. E. and Yanitskii, V. E., "Direct Statistical Simulation of a Collision Relaxation in Gas Mixture with High Variation in Concentrations," *J. of Comp. Math. and Math. Phys.*, Vol. 23, No. 3, 1983, pp. 674-680, (in Russian).

[11] Aristov, V. V. and Tcheremissine, F. G., "Numerical Solution of the Boltzmann Equation for a Gas Mixture with Vibrational Degrees of Freedom," *Committee on Applied Mathematics – Reports of the Computing Center of the Academy of Science USSR*, 1985, p. 12 (in Russian).

[12] Rykov, V. A., "On Kinetic Equations of Chemically Reacting Gas Mixtures," *Izvestiya of the Academy of Sciences of USSR, Mechanics of Fluids and Gases*, No. 4, 1972 (in Russian).

New Numerical Strategy to Evaluate the Collision Integral of the Boltzmann Equation

Zhiqiang Tan,* Yih-Kanq Chen,† Philip L. Varghese,‡ and John R. Howell§

The University of Texas at Austin, Austin, Texas

Abstract

A new scheme for the calculation of the collision integral of the nonlinear Boltzmann kinetic equation is presented, based on a piecewise constant discrete approximation of the distribution function in phase space and corresponding precalculated collision rate coefficients. The computing effort required for evaluation of the fivefold collision integral is significantly reduced by the use of precalculated rate coefficients. The method was combined with the conservative splitting method to solve the Boltzmann equation for a number of test cases to examine the performance of this method and demonstrate its efficiency.

I. Introduction

The Boltzmann equation, the basic equation of kinetic theory, is required for detailed characterization of gas flows that are far from local thermal equilibrium. Direct numerical solution of the Boltzmann equation presents several computational difficulties. These difficulties arise from the nonlinearity and complexity of the collisional integral, together with the multidimensionality of the equation.

Groups led by Yen and Cheremisin have spearheaded the development of direct numerical solutions of the Boltzmann equation.[1-5] The basic approaches of the two groups are

---

Copyright © 1989 by the American Institute of Aeronautics and Astronautics, Inc. All rights reserved.

*Graduate student, Department of Mechanical Engineering.

†Postdoctoral Research Associate, Department of Mechanical Engineering.

‡Associate Professor, Department of Mechanical Engineering.

§Professor, Department of Mechanical Engineering.

similar and consist of two steps: 1) evaluation of the collision integral by conducting Monte Carlo simulations (random quadrature), and 2) approximation of the differential operators by the finite-difference method. For the second step, Aristov and Cheremisin's conservative splitting method[4] is the best currently available because it guarantees conservation of mass, momentum, and energy at every iteration. The Monte Carlo method for computing the collision integral developed by Nordsieck[6] has been systematically tested and refined for several years. The calculations are time-consuming because a large number of multifold integrations must be performed at every position, velocity, and time grid. The discrete velocity approach has also been studied,[7] but it does not provide quantitative predictions. Because the collision integral cannot be calculated efficiently, difficult problems involving high Mach numbers, small Knudsen numbers, large derivatives in the flow parameters, and multidimensional physical space cannot be solved within a reasonably short period of time, even on the fastest supercomputer.

In this paper, a new method for evaluating the nonlinear collision integral of the Boltzmann equation is presented. In Sec. 2, the formulation and corresponding numerical method for evaluating the integral are given. It is shown that some basic constants can be computed and stored in advance to reduce total computing time. The important properties of these constants are also discussed. In Sec. 3, the present method, combined with the conservative splitting method,[4] is applied to one- and two-dimensional cases to examine the method's performance.

II. Formulation

Consider the Boltzmann equation without external forces:

$$\frac{\partial f}{\partial t} + \xi \cdot \frac{\partial f}{\partial x} = \int_0^{4\pi} d\Omega \int_{-\infty}^{\infty} (f'f'_1 - ff_1) G(|\zeta-\xi|,\mathbf{n}) d\zeta \equiv I \qquad (1)$$

The position vector $\mathbf{x}$ has the Cartesian components (x,y,z), the velocity vector ξ (or ζ) has corresponding components (ξ_x, ξ_y, ξ_z), f is the velocity distribution function of the gas molecules normalized to the local gas density, Ω is the solid angle, and $\mathbf{n} = (n_x, n_y, n_z)$ is the unit direction vector along the apse line in the plane of the collision. The function $G(g,\mathbf{n}) = g\sigma(g,\mathbf{n})$, where $g = |\zeta-\xi|$ is the relative speed of the collision pair and σ the differential collision cross section, is determined by the form of the intermolecular potential. Subscript 1 indicates quantities for the collision partner, primed quantities indicate values after a collision, and the functional dependence of the distribution functions has been suppressed for compactness, i.e., $f = f(t,\mathbf{x},\xi)$, $f_1 = f(t,\mathbf{x},\zeta)$, etc. The velocities of the collision partners after a collision that results in a particular scattering can be determined from

$$\xi' \equiv \xi - \mathbf{nn}\cdot(\xi - \zeta) \qquad (2a)$$

$$\zeta' \equiv \zeta - \mathbf{nn}\cdot(\zeta - \xi) \qquad (2b)$$

The collision integral I in the Boltzmann equation can be separated into two parts:

$$I = A - Bf$$

where

$$A = \int_0^{4\pi} d\Omega \int_{-\infty}^{\infty} f'f'_1\, G(g,\mathbf{n})\, d\zeta$$

and

$$B = \int_0^{4\pi} d\Omega \int_{-\infty}^{\infty} f_1\, G(g,\mathbf{n})\, d\zeta$$

In order to solve the equation numerically, we may approximate the infinite velocity space by a finite cube $|\xi_i| < V$ (where V is a given constant, i = x, y, z) and then divide this cube into smaller cubic elements e_i (i = 1,..., N_e, where N_e is the total number of elements) of the same side length ΔV. Each velocity axis is divided into an odd number of elements so that the origin is at the center of one of the elements.

The velocity distribution function is discretized by setting

$$f(t,\mathbf{x},\xi) = \sum_{i=1}^{N_v} \phi_i(\xi) f_i \qquad (3)$$

where $f_i \equiv f(t,\mathbf{x},\xi_i)$, ξ_i are the discretized points in ξ space, N_v is the total number of ξ_i, and ϕ_i is an interpolating function. For simplicity, ϕ_i is chosen as a piecewise constant function in this paper, so that the number of elements N_e is equal to the number of discrete velocity points N_v:

$$N_e = N_v \equiv N$$

Substituting Eq. (3) into (1) and letting $\xi = \xi_k$ yields

$$\frac{\partial f_k}{\partial t} + \xi_k \cdot \frac{\partial f_k}{\partial \mathbf{x}} = A_k - B_k f_k \qquad (4)$$

where

$$A_k \equiv \sum_{i=1}^{N} \sum_{j=1}^{N} \alpha_{ij}^k f_i f_j, \qquad B_k \equiv \sum_{i=1}^{N} \beta_i^k f_i \qquad (5)$$

with *collisional rate coefficients* α_{ij}^k and β_i^k defined by

$$\alpha_{ij}^k = \int_0^{4\pi} d\Omega \int_{-\infty}^{\infty} \phi_i(\xi_k - \mathbf{n}\mathbf{n}\cdot(\xi_k-\zeta))\phi_j(\zeta - \mathbf{n}\mathbf{n}\cdot(\zeta-\xi_k))G(|\zeta-\xi_k|,\mathbf{n})d\zeta \quad (6a)$$

$$\beta_i^k = \int_0^{4\pi} d\Omega \int_{-\infty}^{\infty} \phi_i(\zeta) G(|\zeta-\xi_k|,\mathbf{n}) d\zeta \qquad (6b)$$

Because the collisional rate coefficients α_{ij}^k and β_i^k are independent of **x**, t, f, and initial and boundary conditions, they can be computed and stored in advance for later use on a variety of problems with different geometries or conditions. The collision rate coefficients depend on the collision cross section and hence vary for different gases. For hard-sphere or inverse-power gases, the coefficients are merely rescaled and need not be recalculated.

The limitation of the technique is the large number ($O(N^3)$) of α_{ij}^k and β_i^k to be evaluated and stored. Even when the α_{ij}^k are computed, the $N^2 + N$ multiplications to compute A_k and B_k are extremely time consuming. The key to practical implementation is to exploit some properties of α_{ij}^k that reduce the computing time required. The evaluation of β_i^k is much easier than α_{ij}^k; the properties of β_i^k are not discussed in detail but are mentioned where appropriate.

The first property of α_{ij}^k is that the coefficient remains unchanged when ξ_i, ξ_j, and ξ_k are shifted (but not rotated) to new positions (say, ξ_1, ξ_m, ξ_n) in velocity space, keeping the relative positions the same. The relation

$$\alpha_{ij}^k = \alpha_{mn}^l$$

follows directly from Eqs. (4) and (6a) because only the relative velocity affects the solution of the Boltzmann equation. Hence, one need only compute α_{ij}^k with $\xi_k=0$, and the collision rate coefficients for all other k are obtained by shifting. Thus, there are N^2 independent α_{ij}^k; hereafter, only the computation of $\alpha_{ij}^k|_{\xi_k=0} \equiv \alpha_{ij}$ will be considered. Similarly, $\beta_i^k|_{\xi_k=0} \equiv \beta_i$, and only N coefficients need be computed and stored.

Second, if the angular integral in 4π space is approximated by an M-point numerical integration formula

$$\int_0^{4\pi} F(\mathbf{n}) \, d\Omega = \sum_{t=1}^{M} w_t F(\mathbf{n}^t) \tag{7}$$

where w_t are the integration weights and $\mathbf{n}^t$ the integration points, then the number of nonzero α_{ij} is at most 9MN. The proof for this property can be outlined as follows. If $\alpha_{ij} \neq 0$ for some i and j, then, with $\xi_k = 0$, Eq. (6a) gives vector inequalities

$$|\xi_i - \mathbf{n}\mathbf{n}\cdot\zeta| < (\Delta V/2)\hat{\mathbf{1}}$$

$$|(\mathbf{I} - \mathbf{n}\mathbf{n})\cdot\zeta - \xi_j)| < (\Delta V/2)\hat{\mathbf{1}} \tag{8}$$

where $\mathbf{I}$ is the unit matrix, $\hat{\mathbf{1}}$ is the unit vector, and the vector magnitude of a vector $|\mathbf{v}| \equiv (|v_x|, |v_y|, |v_z|)$. To satisfy these two inequalities, the following must hold:

$$|(\xi_i \times \mathbf{n}^t)_r| < (S^t - |n_r^t|)\frac{\Delta V}{2},$$

$$i = 1,\ldots,N, \quad t = 1,\ldots,M, \quad r = x, y, z \tag{9a}$$

$$|\xi_j \cdot \mathbf{n}^t| < S^t \frac{\Delta V}{2}, \quad j = 1,\ldots,N, \quad t = 1,\ldots,M \tag{9b}$$

where $S^t = |n_1^t| + |n_2^t| + |n_3^t|$, and subscript r on the left side of Eq. (9a) denotes the rth component of a vector. Inequality (9a) represents a hexagon in velocity space. The maximum number of ξ_i in the hexagon is less than $3N^{1/3}$ and this occurs when $|n_x^t| = |n_y^t| = |n_z^t| = 1/\sqrt{3}$. Inequality (9b) represents a plate of thickness $S^t \Delta V$. The maximum number ξ_j that lie in this plate is less than $3N^{2/3}$, and this is obtained when $|n_x^t| = |n_y^t| = |n_z^t|/\sqrt{2} = 1/2$ (or any permutation of subscripts x, y, z). Therefore, the total number of pairs (ξ_i, ξ_j) $(i, j = 1,\ldots,N)$ that satisfy Eqs. (9a) and (9b) does not exceed 9N for any $\mathbf{n}^t$. Because there are M vectors $\mathbf{n}^t$, the maximum number of nonzero α_{ij} is 9MN. Numerical experiments were conducted to verify this for N ranging from 3^3 to 11^3, and the number of nonzero was found to be between 0.9MN and 2.1MN. The exact number depended on the discretization of velocity space and the choice of numerical scheme for angular integration. Because M << N typically, this property greatly reduces the number of nonzero α_{ij}.

Combining Eqs. (6a) and (7), the five-dimensional integrals α_{ij} may be written explicitly as

$$\alpha_{ij} = \sum_{t=1}^{M} w_t \int_{-\infty}^{\infty} \phi_i(\mathbf{n}^t \mathbf{n}^t \cdot \zeta) \, \phi_j(\zeta - \mathbf{n}^t \mathbf{n}^t \cdot \zeta) \, G(|\zeta|, \mathbf{n}^t) \, d\zeta \quad (10)$$

It is convenient to transform variables to permit evaluation of the integral by Gaussian quadrature. The following transformation was used:

$$\eta \equiv \begin{pmatrix} \eta_x \\ \eta_y \\ \eta_z \end{pmatrix} = \begin{pmatrix} n_x & n_y & n_z \\ -n_x n_y & 1-n_y^2 & -n_y n_z \\ -n_x n_z & -n_y n_z & 1-n_z^2 \end{pmatrix} \begin{pmatrix} \xi_x \\ \xi_y \\ \xi_z \end{pmatrix} \equiv R\xi$$

For convenience, it is assumed that the M points chosen for the angular integration are such that all n_r^t are nonzero in Eq. (7). (If any one of the components n_r^t is zero, the integration points can be rotated to new sets so that all n_r^t are nonzero. This does not change the order of accuracy of numerical integration because of the spherical symmetry of the domain of the integral.) It is easy to show that $|R| = n_x$ (| | denotes the determinant) and that the inverse of R exists. Substituting in Eq. (10)

$$\alpha_{ij} = \sum_{t=1}^{M} w_t \int_{a_3}^{b_3} d\eta_z \int_{a_2}^{b_2} d\eta_y \int_{a_1}^{b_1} G(|R^{-1}\eta|, \mathbf{n}^t) \frac{d\eta_x}{|n_x^t|} \quad (11)$$

where

$$a_1 = \max_s \left(\frac{(\xi_i)_s}{n_s^t} - \frac{\Delta V}{2|n_s^t|} \right), \quad b_1 = \min_s \left(\frac{(\xi_i)_s}{n_s^t} + \frac{\Delta V}{2|n_s^t|} \right),$$

$$s = x, y, z$$

$$a_2 = (\xi_j)_y - \frac{\Delta V}{2}, \quad b_2 = (\xi_j)_y + \frac{\Delta V}{2}$$

$$a_3 = \max \left((\xi_j)_z - \frac{\Delta V}{2}, \, -\frac{n_x^t(\xi_j)_x + n_y^t \eta_y}{n_z^t} - \frac{|n_x^t|\Delta V}{2|n_z^t|} \right)$$

$$b_3 = \min \left((\xi_j)_z + \frac{\Delta V}{2}, \, -\frac{n_x^t(\xi_j)_x + n_y^t \eta_y}{n_z^t} + \frac{|n_x^t|\Delta V}{2|n_z^t|} \right)$$

and the integral is set to zero if $b_r \le a_r$ for any r ($r = 1, 2,$ and 3). Here $(\xi_j)_s$ denotes the sth component of ξ_j.

Thus, the total number of independent nonzero collisional rate coefficients α_{ij}^k is of the order of MN instead of N^3. This reduction is very important because it determines the computation time and memory size required for data storage. The coefficients can be stored efficiently using one array to store the values of the nonzero elements and another to store the indices of these elements. When a flow is computed, the collision integral may be evaluated rapidly and efficiently using the stored collision rate coefficients α and β. In contrast, the Monte Carlo technique is relatively inefficient because it requires repeated random number generations, location of ζ' and ζ, and evaluation of G.

III. Numerical Examples

The technique was applied to solve the Boltzmann equation for three cases to demonstrate its capabilities. In these calculations, velocity space was approximated by a cube of side $4\sqrt{3}v_0$, where $v_0 = (kT_0/m)^{1/2}$ is a characteristic molecular velocity in the unperturbed gas. Each side of this domain was divided into 5, giving 125 small cubic elements in velocity space. The shifting property of the collision rate coefficient was utilized to consider collisions between molecules near the edges of the cube with partners outside this region, so that depleting collisions were computed on an enlarged domain ($9 \times 9 \times 9$). An M = 18 point formula for the angular integral was used in the numerical integrations. For simplicity, the gas molecules were modeled as hard spheres. The computed α and β were stored and used for all three problems. The discretized Boltzmann equation was solved using the conservative splitting method.[4] Steady-state problems were solved by an explicit transient approach.

Stationary Shock Wave

Figure 1 presents a comparison between the results obtained using the present method (solid line) and those of Aristov and Cheremisin[4] (symbols) for the structure of a stationary one-dimensional shock wave in a hard-sphere gas flow at Mach 2.5. The reduced density N, velocity U, and temperature T are plotted vs nondimensional distance

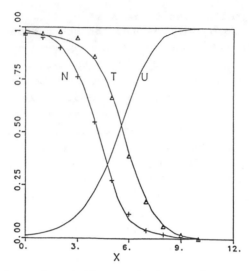

Fig. 1 Structure of a stationary shock wave in a hard-sphere gas at Mach 2.5. The solid lines are the results of this work; the symbols were taken from Aristov and Cheremisin.[4] N, U, and T denote reduced density, velocity, and temperature, respectively. X has been scaled by the mean free path in the unshocked gas.

X. Reduced quantities are defined by

$$N = \frac{N_x - N_0}{N_1 - N_0}, \quad U = \frac{U_x - U_0}{U_1 - U_0}, \quad T = \frac{T_x - T_0}{T_1 - T_0}$$

where subscripts 0 and 1 denote upstream and downstream values, respectively. Distances have been scaled by the mean free path $\lambda (= 1/\sqrt{2}\pi d^2 N_0)$ in the unshocked gas. The predictions of the two techniques are in very good agreement.

Piston Problem

The technique was then applied to a transient problem presented by Aristov and Cheremisin.[4] The half-space, $x > 0$, is filled with gas at uniform temperature and density T_0 and N_0, respectively, moving with uniform mean velocity $2v_0$ in the negative x direction. The molecular velocity distribution is Maxwellian when $t < 0$. At the instant $t = 0$, an infinite wall perpendicular to the x axis is placed at $x = 0$. The reflection of gas molecules on the wall is assumed to be specular.

The semi-infinite physical space was replaced by the finite segment [0, L]. The following characteristic parameters are used for nondimensionalization:

$$x^* \equiv \frac{L}{30}, \quad t^* \equiv \frac{L}{3\sqrt{2}v_0}, \quad f^* \equiv \frac{N_0}{(2\pi)^{3/2}v_0^3}, \quad \xi^* \equiv \sqrt{2}v_0, \quad Kn \equiv \frac{\lambda}{x^*}$$

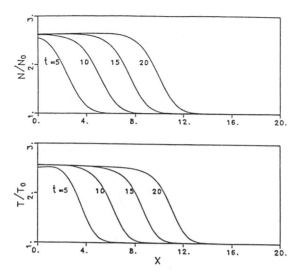

Fig. 2 Nondimensional density and temperature at various times for the reflecting piston problem when Kn = 0.02.

The only characteristic length in the problem is the shock-wave thickness, which is not known a priori, so that the choice of x^* and t^* is rather arbitrary. In Fig. 2, the dimensionless density and temperature profiles at various dimensionless times are presented for Kn = 0.02. Corresponding results for Kn = 0.2 are presented in Fig. 3. The results show that a sharp gradient propagates in the positive x direction corresponding to the formation of a shock because of collisions between the reflected and incident molecules. Comparing the results of the two cases, the shock thickness is seen to be inversely related to the Knuden number (as expected) and is approximately $4x^*$ at Kn = 0.02 and $12x^*$ at Kn = 0.2. At the lower Knudsen number, the shockwave profile rapidly approaches steady state and moves at constant speed thereafter. Steady-state propagation is delayed at the higher Knudsen number. Local departures from equilibrium are most clearly seen in the temperature profile because it is the highest moment of the distribution function. The maximum in the temperature profiles at early times, which is very pronounced at the higher Knudsen number, corresponds to the fact that reflected molecules traveling at $2v_0$ penetrate some distance into the incoming gas before being thermalized by collisions. Similar trends are seen in the results of Aristov and Cheremisin[4] obtained by Monte Carlo evaluation of the collision integral. It should also be noted that

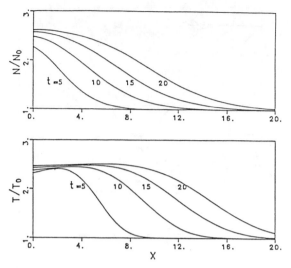

Fig. 3 Nondimensional density and temperature at various times for the reflecting piston problem when Kn = 0.2.

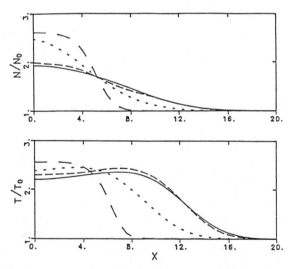

Fig. 4 Effect of Kn on the density and temperature profiles at nondimensional time t = 10 for the reflecting piston problem. The collisionless flow (Kn = ∞) was computed with a 13×13×13 grid in velocity space. All other cases were computed with a 5×5×5 grid. The distortion of the Kn = 2 case resulting from the coarse velocity grid may be noted. ——— Kn = ∞; ———— Kn = 2; ———— Kn = 0.2; —— —— —— Kn = 0.02.

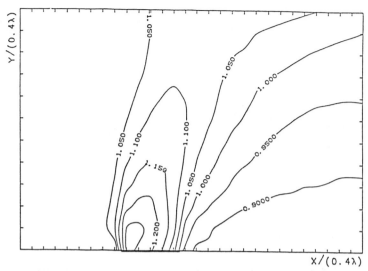

Fig. 5 Density contours calculated for supersonic flow of a monatomic gas over a flat plate. The incident Mach number is 2, and the plate length is 2λ, where λ is the mean free path in the freestream. Gas/surface interactions are modeled by a combination of specular reflection and diffusion, with a wall accommodation coefficient of 0.5; the wall temperature is specified to be the same as the freestream temperature. Distances have been scaled by 0.4λ so that the plate is 5 units long. Note that the y scale has been stretched for clarity. The density has been normalized by the freestream value.

these solutions were obtained in collision-dominated flows (low Knudsen numbers) where Boltzmann equation solutions have traditionally been difficult because of computational difficulty in evaluating the collision term to satisfactory precision.

Figure 4 shows the effect of Kn on the flowfield at nondimensional time t = 10. For negligible collision rates (Kn $\rightarrow \infty$), the reflected molecules propagate freely without interacting with incoming molecules to thermalize the kinetic energy. This leads to a lower temperature in the vicinity of the wall and a wide transition region between the zone of unperturbed gas and gas that has been brought to rest by the wall. Note that when collisions are relatively unimportant (Kn = 2), the profiles appear slightly distorted. The distortion arises because the division of velocity space into 125 elements is insufficient to describe the convection term in the Boltzmann equation. For collisionless flow (Kn $\rightarrow \infty$), the distortions were somewhat larger for the same discretization. These distorted results are not shown in Fig. 4, and the results

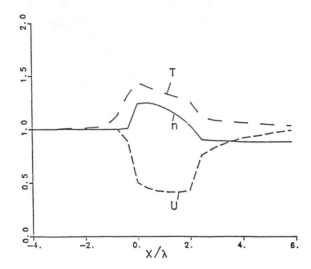

Fig. 6 Variation of density, mean velocity, and temperature along y = 0, for flow over a flat plate shown in Fig. 5. The plate extends from x = 0 to x = 2. All fluid properties have been normalized by the corresponding freestream values.

for collisionless flows were calculated using a denser grid of 13 ×13 ×13 for the convective and unsteady terms with the collision integral set to zero. As Kn decreases, the distortions disappear, indicating that the division into 125 elements is sufficient for accurate calculation of the collision integral.

A remedy for the distortion problem at large Kn may be to use a different number of velocity space divisions for the left- and right-hand sides of the Boltzmann equation. A small number of A_k (say, 125, as in this example) may be calculated and then interpolated to denser velocity nodes for the convective term of the Boltzmann equation.

Comparisons were also made between conservative and nonconservative solutions to this problem. The computations indicated that the present method required the conservative scheme for satisfactory solutions using the fairly coarse grid chosen in velocity space. The nonconservative results obtained were considerably better than those obtained by Monte Carlo evaluation of the collision integral.[4]

Two-Dimensional Flow

The steady flow of a rarefied gas over a finite flat plate was also solved using the new technique. Similar problems have been studied by Cheremisin using Monte Carlo evaluation of the collision integral.[8,9] In this test case, the wall accomodation coefficient was taken to be 0.5 and the wall temperature was set to the temperature of the unperturbed flow. The Mach number of the incident flow was set at 2, and the length of the plate was twice the mean free path in the unperturbed flow, giving a Knudsen number of 0.5. The physical space was divided into 260 unequal cells. Figure 5 presents the density contours in the flowfield. The locus of points of maximum density may be identified with an oblique shock, which appears to intersect the plate about 0.3λ behind the leading edge and extends at an angle of approximately 65 degree to the plate. (It should be noted that the scale in the figure has been stretched in the y direction for clarity.) These steady-state results are in qualitative agreement with Cheremisin's result[8] for a higher-speed flow (M = 4). The present calculations were run at conditions that matched his earlier paper,[9] but the results do not agree.

Figure 6 is a plot of density, mean x velocity, and temperature along the plane y = 0; the plate extends from 0 to 2. The density and temperature are increased just upstream (one or two mean free paths) of the plate and decline steadily over the length of the plate. The velocity drops sharply just upstream of the plate, declines slowly over the leading edge, and is almost constant over the trailing edge. In the wake, the velocity recovers rapidly to the freestream value, but there is a persistent density decrease and corresponding temperature increase relative to the freestream. The trends in density and temperature obtained in these results are similar to those obtained by Cheremisin for Mach 4 flow over heated and cooled plates.[9] The same trends are also obtained from direct simulation Monte Carlo calculations of higher-speed flows.[10]

The present solution technique could not be extended immediately to higher speed flows (M > 3) with good accuracy, because the present division of velocity space into five elements in each direction with piecewise constant interpolation cannot represent bimodal distribution functions satisfactorily. A higher-order interpolating function and a finer mesh for the left-hand side terms would produce more accurate results without increasing computing time significantly. Work on these areas is in progress.

IV. Conclusions

A new numerical method to evaluate the collision integral of the Boltzmann equation was presented. The technique was successfully applied to obtain solutions of the Boltzmann equation for some example problems. The follow-

ing conclusions may be drawn:
1) The present method can generate collision integrals efficiently since most of the work needed for numerical integration is done by precalculation. Because the precalculated collision rate coefficients are independent of physical space, the method is readily extended to multidimensional problems. The examples show its ability to deal with problems with strong collision effects. Good agreement was obtained with results obtained by the Monte Carlo method for the stationary shock-wave problem.
2) The efficiency and accuracy of the present technique may be improved by: using different velocity meshes for convection and collision terms; applying higher order interpolating functions; and using a better quadrature scheme for angular integrations.

Acknowledgment

This research was supported by the U.S. Department of Energy under grant DE-FG05-87ER13735.

References

[1] Yen, S. M., "Monte Carlo Solutions of Nonlinear Boltzmann Equation for Problems of Heat Transfer in Rarefied Gases," *International Journal of Heat and Mass Transfer,* Vol. 14, 1971, pp. 1865-1869.

[2] Yen, S. M., and Ng, W., "Shock Wave Structure and Inter-molecular Collision Laws," *Journal of Fluid Mechanics,* Vol. 65, 1974, pp. 127-144.

[3] Cheremisin, F. G., "Numerical Methods for the Direct Solution of the Kinetic Boltzmann Equation," *U. S. S. R. Computational Mathematics and Mathematical Physics,* Vol. 25, 1985, pp. 156-166.

[4] Aristov, V. V., and Cheremisin, F. G., "The Conservative Splitting Method for Solving Boltzmann's Equation," *U. S. S. R. Computational Mathematics and Mathematical Physics,* Vol. 20, 1980, pp. 208-225.

[5] Yen, S. M., and Tcheremissine, F. G., "Monte-Carlo Solution of the Nonlinear Boltzmann Equation," *Progress in Aeronautics and Astronautics: Rarefied Gas Dynamics,* Vol. 74, edited by S. S. Fisher, AIAA, New York, 1981, Pt. I, pp. 287-304.

[6] Nordsieck, A., and Hicks, B. L., "Monte Carlo Evaluation of the Boltzmann Collision Integral," *Rarefied Gas Dynamics,* edited by C. L. Brundin, Vol. 1, 1967, pp. 695-710.

[7] Cabannes, H., "Couette Flow for a Gas with a Discrete Velocity Distribution," *Journal of Fluid Mechanics,* Vol. 76, 1976, pp. 273-287.

[8]Tcheremissine, F. G., "Solution of the Boltzmann Equation for Rarefied Gas Dynamics," *Rarefied Gas Dynmics,* Vol. 1, edited by O. M. Belotserkovskii, M. N. Kogan, S. S. Kutateladze, and A. K. Rebrov, Plenum, New York, pp. 303-310.

[9]Cheremisin, F. G., "Solution of the Plane Problem of the Aerodynamics of a Rarefied Gas on the Basis of Boltzmann's Kinetic Equation", *Soviet Physics-Doklady,* Vol. 18, 1973, pp. 203-204.

[10]Hermina, W. L., "Monte Carlo Simulation of Rarefied Flow Along a Flat Plate," AIAA Paper 87-1547, 1987.

Comparison of Burnett, Super-Burnett and Monte Carlo Solutions for Hypersonic Shock Structure

Kurt A. Fiscko*
U.S. Army and Standford University
and
Dean R. Chapman†
Stanford University, Stanford, California

Abstract

The continuum Navier-Stokes, Burnett, and Super-Burnett equations are solved for one-dimensional shock structure in various monatomic gases. Solutions are obtained for a hard sphere gas, for argon, and for a Maxwellian gas from Mach 1.3 to Mach 50. A new numerical method is employed which utilizes the complete time-dependent continuum equations and obtains the steady-state shock structure by allowing the system to relax from arbitrary initial conditions. Included is a brief discussion of the numerical method used and difficulties encountered in solving the Burnett equations. Shock density, velocity, temperature, and entropy profiles are also obtained using the Direct Simulation Monte Carlo method, and these results are used as bases of comparison for continuum solution profiles. It is shown that the Burnett equations yield shock structure solutions in much closer agreement to both Monte Carlo and experimental results than do the Navier-Stokes equations. Solutions to the Super-Burnett equations with coefficients as presently derived, however, are inferior to those of the Burnett equations; but this may possibly be due to an algebraic error in the coefficient of one of the terms in these equations. Shock density thickness, density asymmetry, and density-temperature separation are all more accurately predicted by the Burnett equations than by the Navier-Stokes equations.

Nomenclature

C = constant of proportionality
c = peculiar (thermal) atom velocity
c_v = constant volume specific heat
e = total energy density
$\mathbf{e}$ = nondivergent velocity gradient tensor
e_i = internal energy density
E = total internal energy

Copyright ©1989 by the American Institute of Aeronautics and Astronautics, Inc. All rights reserved.

* Graduate Student, Department of Aeronautics and Astronautics; currently Captain, U.S. Army Space Program Office.

† Professor, Department of Aeronautics and Astronautics.

f = normalized velocity distribution function
$\mathbf{J}_s$ = nonequilibrium entropy flux
k = Boltzmann's constant
k = coefficient of thermal conductivity
n = atom number density
N = total number of atoms
m = mass of a single atom
p = pressure
$\mathbf{p}_1$ = Navier-Stokes contribution to viscous stress
$\mathbf{p}_2$ = Burnett contribution to viscous stress
q = total normal heat flux
q_1 = Navier-Stokes contribution to heat flux
q_2 = Burnett contribution to heat flux
Q = heat
Q_ρ = density asymmetry quotient
R = gas constant
s = entropy per unit mass
S = entropy
t = time
t_ρ = shock density thickness
T = temperature
T_n = normalized temperature
$\mathbf{u}$ = vector stream velocity
u = scalar stream speed
$\mathbf{v}$ = vector atom velocity
V = volume
V_c = velocity space
w = entropy production rate per unit volume
W = work
x = cartesian coordinate in physical space
α, β = constants defined in text
$\Delta_{\rho T}$ = half point rise separation distance
ϵ = Knudsen number
γ = ratio of specific heats
κ = intermolecular force constant of proportionality
λ = mean free path
μ = coefficient of viscosity
Ω = total number of microscopic states
ρ = mass density
ρ_n = normalized mass density
σ = total normal stress
ω, θ = quantities defined in text
∇ = divergence operator

Introduction

The study of shock wave structure in dilute gases has historically been the subject of much investigation.[1] There are mainly two reasons for this: 1) the shock wave represents a flow condition that is far from thermodynamic equilibrium, and 2) shock wave phenomena is unique in that it allows one to separate the continuum differential equations of fluid motion from the boundary conditions that must be stated to complete a well-posed problem. The boundary conditions for a shock wave are not in question, being determined by the Rankine-Hugoniot relations. Thus, in the study of shock structure one is able to isolate effects due to the differential equations themselves.

It is beyond the scope of this investigation to assess the limits of a continuum description of shock flow. However, the practical motivation for this investigation stems from the concept that an altitude regime exists in the atmosphere that is within the continuum domain, but wherein the conventional Navier-Stokes description ceases to be accurate. The altitude limits of this regime, referred to as the continuum transition regime, are not fixed, but depend on vehicle size and speed.[2,3] A measure of the lower altitude limit of this regime would be where the thickness of the bow shock wave is no longer negligible when compared to the shock stand-off distance from a hard body. Above this altitude, flow computations must be made *through* the shock wave structure. Various types of transatmospheric vehicles operate or will operate extensively in this continuum transition regime: All atmosphere entry and transatmospheric vehicles must pass through it. The Aero-assist Orbital Transfer Vehicle (AOTV) and manned Mars return vehicle will operate extensively within this regime; the inlet cowl lip on the National AeroSpace Plane (NASP) will be in the continuum transitional regime over a sizeable portion of its powered ascent trajectory; and the hard-body flowfield radiation from an ascending missile is dominantly in this regime.

When examining various mathematical models of shock structure, one needs to establish parameters that are indicative of the quality of the predicted shock structure. Because shock structure is primarily a function of Mach number, molecular mean free path, and intermolecular force model, it is instructive to examine not only the accuracy of a predicted flow variable profile through the shock wave, but also the accuracy of these parameters over a very wide Mach number range. Three parameters are frequently used in this investigation: the shock density thickness t_ρ, shock density asymmetry quotient Q_ρ, and shock temperature-density separation $\Delta_{\rho T}$. Two of these parameters are defined in Fig. 1 and the other is defined in Fig. 2. Thus, in addition to investigating the accuracy of flow variable spatial profiles at a given Mach number, these parameters will also be scrutinized.

One indication that the Navier-Stokes equations fail in strong shock waves is contained in Fig. 3. The density thickness parameter for argon appears to be inaccurate for all but very low Mach number shocks. Fig. 4, however, indicates that Navier-Stokes predictions of this parameter are also somewhat inaccurate even at low Mach numbers, as has been previously observed.[4] Furthermore, density asymmetry (Fig. 5) is in gross error

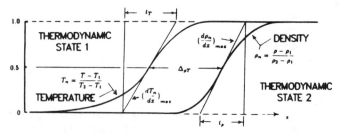

Fig. 1 Shock profile parameter definitions.

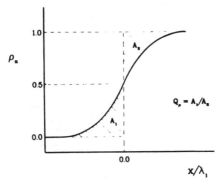

Fig. 2 Definition of shock asymmetry quotient.

at very low Mach numbers. Indeed, as noted in Ref. 5, the Navier-Stokes description of shock flow is valid only for monatomic gases at Mach numbers very near one. Reasons for the failure of the Navier-Stokes equations may be associated with any or all of the following assumptions embodied in these equations:

1) linear stress-strain tensor dependent only on velocity gradients
2) linear heat-flux vector dependent only on temperature gradients
3) zero bulk viscosity
4) small Knudsen number flow (continuum gas model)
5) no direct contribution of nonequilibrium internal molecular energy to viscous stress or heat flux.

This investigation focuses mainly on 1) and 2).

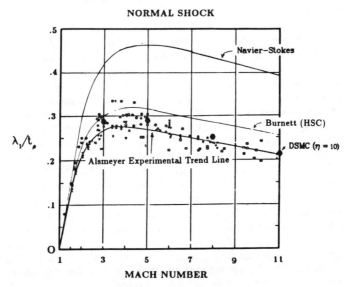

Fig. 3 Argon inverse density thickness.

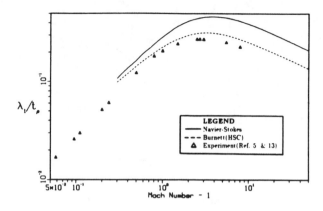

Fig. 4 Argon inverse density thickness.

Many attempts have been made to accurately model shock flow.[1] These range from pure particulate descriptions of the flow to complex continuum formulations. Naturally, the success and/or acceptance of any model usually depends on its ability to predict shock structure in agreement with experimentally measured shock flow, its basis in theory, and on its computational practicality. For all of these reasons the Direct Simulation Monte Carlo (DSMC) method of Bird has gained broad acceptance.[6] Yet, while this method is capable of good predictions of hypersonic flow in the continuum transition regime, it is quite expensive in terms of computer time. The work in this investigation has been conducted as part of efforts to expand the continuum description of the flow well into the continuum transition regime, with the hope that continuum flow solutions of acceptable accuracy in this regime can be obtained more economically than DSMC solutions.

If one adopts the idea that flow within shock waves falls within the range of applicability of the Boltzmann equation, then one may attempt solutions of that equation to model shock flow. One particular class of solutions to the Boltzmann equation is that obtained by the moment methods of Ref. 7 (e.g., Grad's 13-moment equations). A subclass of solutions is obtained by multiplying the local Maxwellian weighting function by a power series in the Knudsen number ϵ. As noted in Ref. 8, "since dissipation is the transfer of energy from macroscopic to microscopic length scales, ϵ is the natural expansion parameter for all those functions whose values are determined by dissipation." This subclass of solutions is more commonly referred to as those obtained by Chapman-Enskog expansion methods. They include, but are not limited to, the Navier-Stokes equations, Burnett equations, and Super-Burnett equations.

Prior to this investigation, no shock structure solutions of the Burnett equations could be obtained for hypersonic Mach numbers, and no solutions of the Super-Burnett equations could be obtained for shocks of any Mach number. Previous investigations have attempted to obtain steady-state solutions directly from a system of ordinary differential equations (ODE) involving strong singularities. The time-dependent solution method used herein avoids these singularities, and is apparently a more powerful mathematical method.

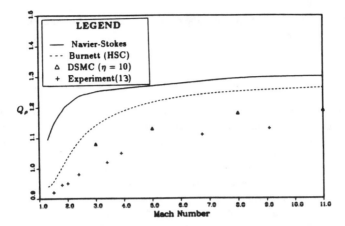

Fig. 5 Argon density asymmetry.

Direct Simulation Monte Carlo Solutions

Generally only shock density profiles have been determined experimentally. Exceptions are the low Mach number temperature profiles of Ref. 9 and the velocity distribution functions of Ref. 10. To obtain more information concerning the flow, the DSMC method[6] is used to compute particulate solutions. In addition to density profiles, these solutions yield velocity, temperature, and entropy profiles, as well as even higher order moments such as velocity distribution function skewness and kurtosis. Using this method one can also completely determine the velocity distribution function at all points in the shock structure.

A formulation based on the Boltzmann equation contains inherent assumptions concerning the flow,[11] including the dilute gas assumption of large spatial separation of molecules relative to molecular diameters and the hypothesis that only binary molecular collisions occur. Implicit in the first assumption is the fact that the potential energy of molecules is negligible compared to their kinetic energy, and that potential energy is only important during the brief encounters of molecular "collisions." While the DSMC method does not explicitly solve the Boltzmann equation, Bird has claimed that, with a sufficiently large number of particles, it is fully equivalent to solving this equation.[6] Furthermore, this method is consistent with the normal assumptions of a dilute gas. In this investigation molecules are represented as point centers of mass and point centers of repulsion, containing no internal structure. Studies of the effect of molecular model on shock structure[12] have confirmed the appropriateness of this type of molecular model for hypersonic shock flow. Naturally, to include real gas ionization effects, which are significant in monatomic shocks above Mach number 10, it would be necessary to enhance the molecular model.

An important conclusion of the Chapman-Enskog method of expansion–and fundamental link between continuum and particulate solutions–is that when the intermolecular repulsive force law is given by

$$F = \frac{\kappa}{r^\eta} \tag{1}$$

a direct relationship exists between the continuum coefficient of viscosity and the gas temperature given by

$$\mu = CT^{(\frac{1}{2} + \frac{2}{\eta - 1})} \tag{2}$$

Also, the coefficient of thermal conductivity is given by $k = c_p\mu/Pr$, with $Pr = 2/3$ for a monatomic gas.

The practical need to compute through shock wave structure is for hypersonic flows involving very high temperatures. In molecular collisions at such temperatures, the molecular attraction forces are negligible compared to repulsive forces. Hence, a repulsive force model can provide realistic simulations.

The molecular force model of Eq. (1) for $\eta = 5$ corresponds to a Maxwellian gas. The case $\eta = \infty$ represents a so-called hard sphere gas. By setting $\eta = 10$ for argon, good agreement is obtained with experimentally measured shock density thickness[13] and, as shown herein, also with density asymmetry. This corresponds to $\mu \approx T^{0.72}$, which is consistent with measurements of viscosity and conductivity for argon.

The following expression is used throughout as the reference length scale for shock structure:

$$\lambda_1 = \frac{16\mu_1}{5\rho_1(2\pi RT_1)^{0.5}} \tag{3}$$

This is the mean free path that would exist in the upstream unshocked region if the gas were composed of hard elastic spheres and had the same viscosity, density, and temperature of the gas being considered. The actual unshocked mean free path will be somewhat different for each gas considered other than hard spheres.

A typical particle simulation schematic is depicted in Fig. 6. Numbers used in this Fig. are only representative and change based on the particular case being examined. Steady-state shock waves are generated numerically by placing molecules (about 175,000) into the tube and driving a piston into the equilibrium gas. A shock is established that propagates down the tube at a velocity that is a function of the piston velocity. Molecules are moved in a deterministic manner, and molecular collisions are calculated on a probabilistic basis.[6] Macroscopic flow features, such as density and temperature, are determined as the large time scale averages of many isolated observations.

Continuum Formulation

For one-dimensional shock waves in a simple monatomic gas the fundamental continuum equations that describe the flow are as follows:

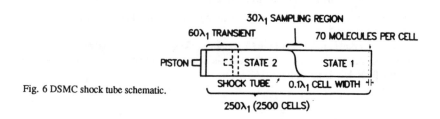

Fig. 6 DSMC shock tube schematic.

$$\frac{\partial \rho}{\partial t} + \frac{\partial (\rho u)}{\partial x} = 0 \quad \text{(mass)} \tag{4}$$

$$\frac{\partial (\rho u)}{\partial t} + \frac{\partial (\rho u^2 - \sigma)}{\partial x} = 0 \quad \text{(momentum)} \tag{5}$$

$$\frac{\partial e}{\partial t} + \frac{\partial [(e-\sigma)u + q]}{\partial x} = 0 \quad \text{(energy)} \tag{6}$$

$$p = (\gamma - 1)\rho e_i \quad \text{(state)} \tag{7}$$

$$e_i = c_v T \quad \text{(calorically perfect gas)} \tag{8}$$

$$e = \rho(e_i + \frac{u^2}{2}) \quad \text{(definition)} \tag{9}$$

Since these six equations involve eight unknowns, two more equations are needed to close the system. Conventional continuum theory closes this system by making assumptions that relate viscous stress to rate of strain and heat flux to temperature gradient. Truesdell[14] has speculated on possible forms for higher order non-linear continuum closure models. However, these forms contain large numbers of undetermined constants, and are of little practical use.

Starting with the known equilibrium velocity distribution function (Maxwellian), the Chapman-Enskog theory of rarefied gases closes the system of continuum equations by taking moments of the Boltzmann equation. Detailed procedures for doing this can be found in Ref. 6. When complete thermodynamic equilibrium exists, the following closure relations result:

$$\sigma = -p \tag{10}$$

$$q = 0 \tag{11}$$

The continuum equations with these closure relations are the Euler equations.

When first order departures from equilibrium occur, the closure relations for one-dimensional flow are those of Navier-Stokes:

$$\sigma = -p + \frac{4}{3}\mu \frac{\partial u}{\partial x} \tag{12}$$

$$q = -k \frac{\partial T}{\partial x} \tag{13}$$

The Chapman-Enskog theory not only closes the system of equations, but, for a given molecular force model, also specifies the functional relationship between μ, k, and T [see Eq. (2)].

When second order departures from equilibrium are considered in the Chapman-Enskog expansion procedure, the Burnett equations are obtained. Closure relations (as found in Ref. 15) in one-dimensional flow for this system are:

$$\sigma = -p + \frac{4}{3}\mu \frac{\partial u}{\partial x} - \frac{\mu^2}{p}\left[\left(\frac{2}{3}\omega_1 - \frac{14}{9}\omega_2 + \frac{8}{27}\omega_6\right)\left(\frac{\partial u}{\partial x}\right)^2 - \frac{2}{3}\omega_2 \frac{RT}{\rho}\frac{\partial^2 \rho}{\partial x^2} \right.$$
$$\left. + \frac{2}{3}\omega_2 \frac{RT}{\rho^2}\left(\frac{\partial \rho}{\partial x}\right)^2 - \frac{2}{3}(\omega_2 - \omega_4)\frac{R}{\rho}\frac{\partial \rho}{\partial x}\frac{\partial T}{\partial x}\right] \tag{14}$$

$$+ \frac{2}{3}(\omega_4 + \omega_5)\frac{R}{T}\left(\frac{\partial T}{\partial x}\right)^2 - \frac{2}{3}(\omega_2 - \omega_3)R\frac{\partial^2 T}{\partial x^2}\Bigg]$$

$$q = -k\frac{\partial T}{\partial x} + \frac{\mu^2}{\rho}\Bigg[\left(\theta_1 + \frac{8}{3}\theta_2 + \frac{2}{3}\theta_3 + \frac{2}{3}\theta_5\right)\frac{1}{T}\frac{\partial u}{\partial x}\frac{\partial T}{\partial x}$$
$$+ \frac{2}{3}(\theta_2 + \theta_4)\frac{\partial^2 u}{\partial x^2} + \frac{2}{3}\theta_3\frac{1}{\rho}\frac{\partial u}{\partial x}\frac{\partial \rho}{\partial x}\Bigg] \tag{15}$$

The preceeding expressions are valid for any dilute gas that is represented by the force model of Eq. (1). Values for the various ω's and θ's are pure numbers dependent on η in Eq. (1). Thus far, they have been derived only for $\eta = 5$ and $\eta = \infty$. The values in Table 1 are those determined from Ref. 15. Foch[16] has argued that some of these values for the Maxwellian Gas are slightly in error; however, this does not seem to impact significantly any of the results presented in this investigation. When results for the Burnett equations are shown in this investigation, indication of the set of coefficients used is made by specifying "Burnett (HSC)," implying "Burnett equations with Hard Sphere Coefficients," or "Burnett (MXC)," meaning "Burnett equations with MaXwellian Coefficients." From the above discussion it may appear as though the Burnett equations can only be derived based on Chapman-Enskog expansion concepts; however, an independent derivation of the Burnett equations based on mean free path considerations is found in Ref. 8. Woods[17] has also suggested a modification to the Burnett equations of Ref. 15 to ensure their frame independence, which has negligible impact on the results presented herein. A slight modification to the Burnett equations, termed the Burnett(-) equations is obtained when deleting the density second derivative term from the viscous stress. This form of the equations will be referred to subsequently.

It seems as though only one article[18] presents the complete Super-Burnett equations for one dimensional flow in a Maxwellian gas. For reference purposes, they are listed here:

Table 1: Burnett Coefficients for Two Theoretical Gases

	Hard Sphere	Maxwellian
η	∞	5
ω_1	4.056	3.333
ω_2	2.028	2.
ω_3	2.418	3.
ω_4	0.681	0.
ω_5	1.428	3.
ω_6	7.424	8.
θ_1	11.644	9.375
θ_2	−5.822	−5.625
θ_3	−3.090	−3.
θ_4	2.418	3.
θ_5	25.158	29.25

$$\sigma = -p + \frac{4}{3}\mu\frac{\partial u}{\partial x} - \frac{\mu^2}{p}\left[\frac{40}{27}\left(\frac{\partial u}{\partial x}\right)^2 - \frac{4}{3}\frac{RT}{\rho}\frac{\partial^2 \rho}{\partial x^2}\right.$$
$$+ \frac{4}{3}\frac{RT}{\rho^2}\left(\frac{\partial \rho}{\partial x}\right)^2 - \frac{4}{3}\frac{R}{\rho}\frac{\partial \rho}{\partial x}\frac{\partial T}{\partial x} + 2\frac{R}{T}\left(\frac{\partial T}{\partial x}\right)^2 + \frac{2}{3}R\frac{\partial^2 T}{\partial x^2}\right]$$
$$- \frac{\mu^3}{p^2}\left[\frac{47}{3}\frac{R}{\rho}\frac{\partial T}{\partial x}\frac{\partial \rho}{\partial x}\frac{\partial u}{\partial x} - \frac{64}{9}\frac{RT}{\rho^2}\left(\frac{\partial \rho}{\partial x}\right)^2\frac{\partial u}{\partial x} + \frac{40}{9}\frac{RT}{\rho}\frac{\partial^2 \rho}{\partial x^2}\frac{\partial u}{\partial x}\right. \quad (16)$$
$$- \frac{2}{3}\frac{RT}{\rho}\frac{\partial \rho}{\partial x}\frac{\partial^2 u}{\partial x^2} - \frac{21}{3}\frac{R}{T}\left(\frac{\partial T}{\partial x}\right)^2\frac{\partial u}{\partial x} - \frac{47}{9}R\frac{\partial^2 u}{\partial x^2}\frac{\partial T}{\partial x}$$
$$\left. - \frac{31}{9}R\frac{\partial u}{\partial x}\frac{\partial^2 T}{\partial x^2} + \frac{2}{9}RT\frac{\partial^3 u}{\partial x^3} + \frac{16}{27}\left(\frac{\partial u}{\partial x}\right)^3\right]$$

$$q = -k\frac{dT}{dx} + \frac{\mu^2}{\rho}\left[\frac{95}{8T}\frac{\partial u}{\partial x}\frac{\partial T}{\partial x} - \frac{7}{4}\frac{\partial^2 u}{\partial x^2} - \frac{2}{\rho}\frac{\partial u}{\partial x}\frac{\partial \rho}{\partial x}\right]$$
$$+ \frac{\mu^3}{p\rho}\left[-\frac{8035}{336}\frac{1}{T}\frac{\partial T}{\partial x}\left(\frac{\partial u}{\partial x}\right)^2 + \frac{166}{21}\frac{1}{\rho}\frac{\partial \rho}{\partial x}\left(\frac{\partial u}{\partial x}\right)^2 + \frac{949}{168}\frac{\partial u}{\partial x}\frac{\partial^2 u}{\partial x^2}\right.$$
$$+ \frac{917}{8}\frac{R}{\rho T}\frac{\partial \rho}{\partial x}\left(\frac{\partial T}{\partial x}\right)^2 - \frac{1137}{16}\frac{R}{\rho^2}\frac{\partial T}{\partial x}\left(\frac{\partial \rho}{\partial x}\right)^2 + \frac{397}{16}\frac{R}{\rho}\frac{\partial \rho}{\partial x}\frac{\partial^2 T}{\partial x^2} \quad (17)$$
$$+ \frac{701}{16}\frac{R}{\rho}\frac{\partial T}{\partial x}\frac{\partial^2 \rho}{\partial x^2} - \frac{813}{16}\frac{R}{T^2}\left(\frac{\partial T}{\partial x}\right)^3 - \frac{1451}{16}\frac{R}{T}\frac{\partial T}{\partial x}\frac{\partial^2 T}{\partial x^2}$$
$$\left. - \frac{157}{16}R\frac{\partial^3 T}{\partial x^3} - \frac{41}{8}\frac{RT}{\rho^2}\frac{\partial \rho}{\partial x}\frac{\partial^2 \rho}{\partial x^2} - \frac{5}{8}\frac{RT}{\rho}\frac{\partial^3 \rho}{\partial x^3} + \frac{23}{4}\frac{RT}{\rho^3}\left(\frac{\partial \rho}{\partial x}\right)^3\right]$$

The complete Super-Burnett equations for a Maxwellian gas with the viscous stress term

$$-\frac{\mu^3}{p^2}\frac{47}{3}\frac{R}{\rho}\frac{\partial T}{\partial x}\frac{\partial \rho}{\partial x}\frac{\partial u}{\partial x}$$

deleted, is referred to herein as the Super-Burnett(-) equations.

Continuum Numerical Method

When appropriate closure relations are included in the continuum equation set, the system may be solved by utilizing a numerical algorithm. The equation set is finite differenced[2] with all derivatives being treated implicitly, and the steady state solution is obtained by allowing the system to relax from arbitrary initial conditions. On the cold side of the shock, boundary conditions specified are temperature, density, and Mach number; on the hot side only the Rankine-Hugoniot pressure is specified. A smooth spatial distribution of flow variables is imposed on a mesh corresponding to the expected total shock thickness, and the system is discretely advanced in time until changes no longer take place. The solution is then checked to ensure that fluxes are conserved at all points within the shock.

All continuum solutions presented herein are for a spatial mesh consisting of 150 points. As a check on the algorithm, solutions have also been obtained for as many as 1000 spatial mesh points, yielding results that are nearly indistinguishable from the 150 point solutions. It was discovered that solutions of more than 150 points produce changes of less than 1% in the very sensitive density asymmetry quotient. As a further check on the numerical algorithm, the continuum equations have also been solved using a purely explicit method incorporating the MacCormack predictor-corrector scheme. These results are identical to those obtained by implicit methods, although much more computer time is used in obtaining the explicit solutions.

The algorithm presented in Ref. 2 utilized flux splitting procedures for the Euler terms. While this greatly accelerates convergence, the type of flux splitting used in that algorithm produces discontinuous first derivatives at the sonic point. This problem is amplified as one computes even higher order derivatives at the sonic point. Hence, for the results presented herein, no flux splitting is employed in order that the maximum solution accuracy might be obtained.

Convergence of the Chapman-Enskog Knudsen Number Series

Numerous authors have *speculated* quite boldly concerning the convergence characteristics of the velocity distribution function series expansion employed in the Chapman-Enskog method. Moreover, in a 1964 paper[19] Holway concludes that closure relations that are derived based on moments of the Boltzmann equation (the Burnett equations represent a sub-class of this class of solutions to the Boltzmann equation) will not yield solutions for hypersonic shock waves. However, in a 1967 paper[20] Butler and Anderson found that by changing the weighting function of the system of moment equations to something other than the local Maxwellian distribution, solutions for hypersonic shock structure could, in fact, be obtained. The Chapman-Enskog expansion method does change the weighting function of the system by multiplying the local Maxwellian distribution by the Knudsen number series.

In conjunction with this investigation, an attempt was made to solve Grad's 13-moment equations utilizing a similar time-dependent numerical algorithm. Results were obtained for no higher Mach number shocks than those obtained using conventional ODE methods–fully consistent with Holway's conclusions regarding Grad's equations. However, it now appears as though the Chapman-Enskog expansion procedure in itself permits solutions to exist to equations of at least third-order (Super-Burnett equations).

Furthermore, it appears that prior to this time, solutions of the Burnett equations for hypersonic shock structure could not be obtained numerically because of shortcomings in the numerical algorithm employed, not inherent failure of the equations themselves.

Entropy Considerations

That numerical solutions to the Burnett equations can be obtained for hypersonic shock structure creates very many interesting new possibilities regarding the flow of dilute gases. This investigation attempts to consider more than merely density, velocity, and temperature profiles through shock waves. The DSMC method enables complete determination of the velocity distribution function at any point within the shock structure. Thus, one can readily examine the features of any function of velocity and potentially draw conclusions regarding the flow of nonequilibrium gases.

For any property that depends only on molecular velocity in a simple monatomic gas $\phi = \phi(\mathbf{u})$, the general transport equation[21] describing the movement of this property through the fluid is:

$$\frac{\partial(n\overline{\phi})}{\partial t} + \nabla \cdot (n\overline{\mathbf{u}\phi}) = w \tag{18}$$

The term w is due to molecular collisions, and can be considered as the source strength, or rate of production of the property ϕ. For mass, momentum, and energy, w is identically 0 under the assumptions of classical mechanics. However, when considering the property entropy, the term, w, is not necessarily 0, unless we have specified an isentropic flow.

Boltzmann has stated that the total entropy S of a system consisting of Ω microstates is

$$S = k \ln \Omega \tag{19}$$

Invoking the Heisenberg uncertainty principle for a gas with Planck's constant represented by 'h', it can be shown[22] that this is equivalent to

$$S = kN - kV \int_{-\infty}^{+\infty} (nf) \ln\left(\frac{h^3}{m^3}nf\right) dV_c \tag{20}$$

If we are interested only in changes in S, then we can write the entropy per unit mass as

$$s = -\frac{k}{nm} \int_{-\infty}^{+\infty} (nf) \ln(nf) dV_c \tag{21}$$

So, for entropy transport, Eq. (18) can now be written as

$$\frac{\partial(\rho s)}{\partial t} + \nabla \cdot (\rho \mathbf{v} s + \mathbf{J}_s) = w \tag{22}$$

where

$$\mathbf{J}_s = -k \int_{-\infty}^{+\infty} \mathbf{c}(nf) \ln(nf) dV_c \tag{23}$$

Thus, from DSMC calculations, all quantities on the left hand side of Eq. (22) are known, enabling us to compute w, the rate of entropy production per unit volume, at all points within the shock.

Continuum concepts also enable the calculation of an entropy, entropy flux, and entropy production consistent with any given continuum approximation. For flow conditions not in thermodynamic equilibrium, these variables are expanded in a Knudsen number series, namely

$$s = s_0 + s_1 + s_2 + \cdots \tag{24}$$

$$\mathbf{J}_s = \mathbf{J}_{s_1} + \mathbf{J}_{s_2} + \cdots \tag{25}$$

$$w = w_1 + w_2 + \cdots \tag{26}$$

Note that w_0 and $\mathbf{J}_{s_0}$ are zero (thermodynamic equilibrium), whereas the corresponding entropy s_0 is

$$s_0 = \frac{3}{2} R \ln\left(\frac{RT}{\rho^{\frac{2}{3}}}\right) + \text{constant} \tag{27}$$

and satisfies the relation

$$T \, ds_0 = dQ = dE - \mathrm{d}W = dE + p \, d\left(\frac{1}{\rho}\right) \tag{28}$$

It is widely accepted[7] that $s_1 = 0$. This implies that the error term in Eq. (28) is $O(\epsilon^2)$ and not $O(\epsilon)$, and explains why the assumption of "local thermodynamic equilibrium" is usually a good hypothesis.

Woods[7] indicates that

$$\mathbf{J}_{s_1} = \frac{q_1}{T} \tag{29}$$

where $q_1 = -kT\mathbf{g}$, with $\mathbf{g} = \frac{\nabla T}{T}$, and that for a Maxwellian gas,

$$\mathbf{J}_{s_2} = \frac{q_2}{T} - \frac{3\mu^2 R}{pT} \nabla T \cdot \mathbf{e} \tag{30}$$

He also states that the following expression for s_2 is fully consistent with that of Ref. 23:

$$s_2 = \frac{1}{\rho T}\left(-\frac{\mu^2}{p}\right)\left[\alpha \mathbf{e} : \mathbf{e} + \beta \frac{R}{T} \nabla T \cdot \nabla T\right] \tag{31}$$

where for Maxwellian molecules, $\alpha = 1$ and $\beta = \frac{45}{16}$.

The entropy production rate per unit volume w is generally viewed to be due to irreversible effects. However, only w_1 can be proven to be greater than zero (normally referred to as dissipation). Woods has reasoned that the additional terms that the Bur-

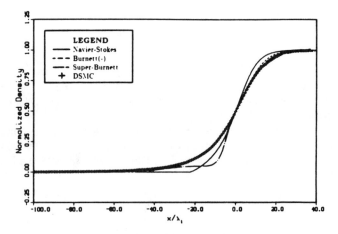

Fig. 7 Mach 35 density profiles for a Maxwellian gas.

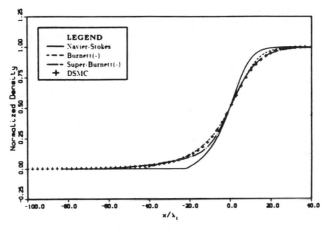

Fig. 8 Mach 35 density profiles for a Maxwellian gas.

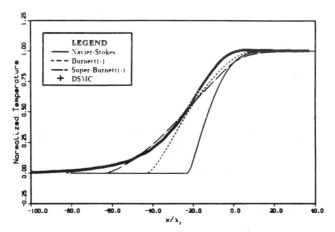

Fig. 9 Mach 35 temperature profiles for a Maxwellian gas.

nett equations add to the viscous stress and heat flux are due to reversible effects, so that, although they can change the entropy of a fluid element, they cannot increase the dissipation in the fluid element. He presents

$$w_1 = \frac{1}{T}\left[-\mathbf{p}_1 : \mathbf{e} - \mathbf{q}_1 \cdot \mathbf{g}\right] \tag{32}$$

$$w_1 + w_2 = \frac{1}{T}\left[-(\mathbf{p}_1 + \mathbf{p}_2) : \mathbf{e} - (\mathbf{q}_1 + \mathbf{q}_2) \cdot \mathbf{g}\right] \\ + \frac{4\mu^2}{p}\left[\mathbf{e} : \mathbf{e} \cdot \mathbf{e} + a\frac{R}{T}\nabla T \cdot \mathbf{e} \cdot \nabla T\right] \tag{33}$$

where $a = \frac{189}{32}$ for a Maxwellian gas.

Results

Solutions for hypersonic shock structure of the Burnett and Super-Burnett equations have been obtained for all gases and Mach numbers examined, with one exception. Only the Burnett(-) equations could be solved for hypersonic shock waves in a Maxwellian gas. Presently, it is unclear if this is due to deficiencies in the numerical algorithm or problems contained in the Burnett equations themselves. For a Maxwellian gas, the complete Burnett equations could be solved only for shocks up to Mach 3.8.

As depicted in Fig. 7 through 19, the Burnett equations consistently predict shock structure for all Mach numbers and shock parameters investigated in closer agreement with Monte Carlo solutions than do the Navier-Stokes equations. However, as can be seen in Fig. 7, the complete Super-Burnett solution does not appear to be better than the Burnett(-) solution, or even physical for that matter. This problem with the Super-Burnett equations (as given by Foch,) however, is due to just one term in the stress,

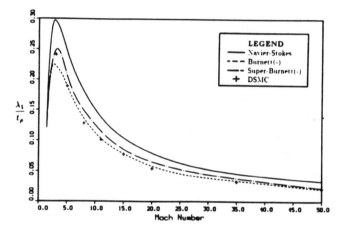

Fig. 10 Maxwellian gas inverse density thickness.

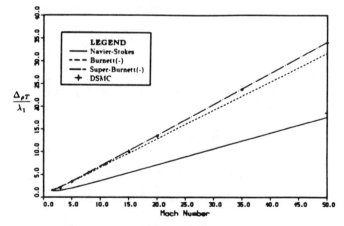

Fig. 11 Maxwellian gas temperature-density separation.

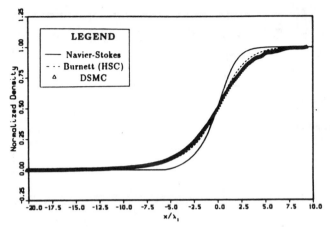

Fig. 12 Mach 35 density profiles for argon.

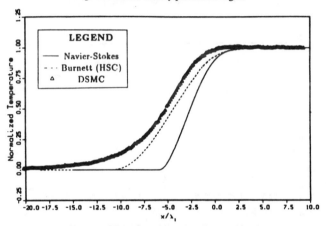

Fig. 13 Mach 35 temperature profiles for argon.

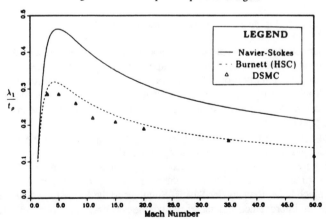

Fig. 14 Argon inverse density thickness.

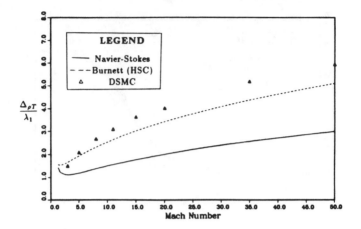

Fig. 15 Argon temperature-density separation.

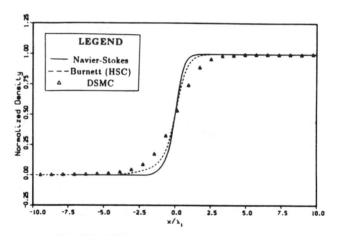

Fig. 16 Mach 35 density profiles for a hard sphere gas.

namely that involving the triple product of temperature, density, and velocity derivatives. With this one term omitted, solution of the Super-Burnett(-) equations yields results that are somewhat better over portions of the shock structure than results from the Burnett(-) equations. Thus, it is suspected that the term that has been deleted from the Super-Burnett equations to form the Super-Burnett(-) equations, may possibly have been derived in error. This possibility should be explored.

Density profiles of the Burnett, Burnett(-), and Super-Burnett(-) equations for soft-potential gases are in strong agreement with DSMC results at all Mach numbers, and temperature-density separation is consistently better than that predicted by Navier-Stokes. For harder potentials, like that of the hard sphere gas, agreement is not as good, but is still much better than Navier-Stokes. Solutions to the Burnett(-) equations represent substantial improvements in accuracy over the Navier-Stokes solutions. However, solutions to the Super-Burnett(-) equations are in some respects superior and in others inferior to solutions of the Burnett(-) equations.

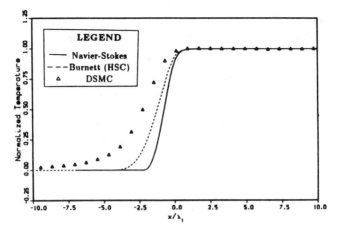

Fig. 17 Mach 35 temperature profiles for a hard sphere gas.

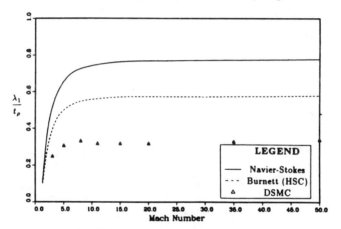

Fig. 18 Hard sphere gas inverse density thickness.

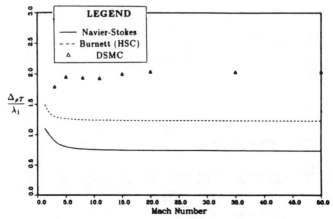

Fig. 19 Hard sphere gas temperature-density separation.

The ability to obtain numerical solutions to the Burnett equations seems to be very dependent on the values chosen for ω's and θ's. Obtaining solutions to the Burnett(MXC) equations for argon appears much more difficult (not possible with the numerical algorithm presently used) than obtaining solutions to the Burnett(HSC) equations for argon. *Each* term in the Burnett expressions for viscous stress and heat flux appears to represent a physically significant feature of the flow. For example, the first term in the Burnett heat flux is dominantly responsible for increasing the separation between the temperature and density profiles. Also, the density second derivative term in the viscous stress is dominantly responsible for making predicted density asymmetry agree with experimental density asymmetry in argon; yet, it has negligible effect on the density thickness.

Based on a singular point analysis, Ref. 4 predicted a temperature overshoot within the shock structure. In this investigation, a very small temperature overshoot (less than 1%) has been observed in DSMC, Burnett, and Super-Burnett solutions of shock structure, but not in Navier-Stokes solutions.

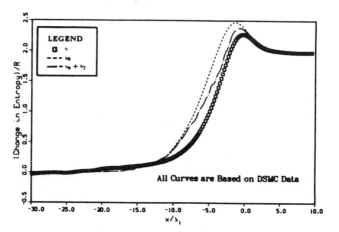

Fig. 20 Maxwellian gas Mach 5 entropy.

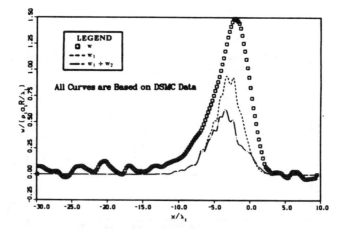

Fig. 21 Maxwellian gas Mach 5 entropy production rate.

Because DSMC data permits complete determination of the velocity distribution function, the Boltzmann entropy [Eq. (21)] can be determined at all points in the shock. Fig. 20 shows that the equilibrium entropy s_0 is considerably different from s at most locations in the shock structure of a Maxwellian gas. Using Eq. (22) the entropy production rate per unit volume has also been determined within the shock structure. Note that this quantity is nonnegative at all points in the shock (neglecting statistical error in the extreme upstream and downstream parts of the shock). If J_s were 0 in the shock (thermodynamic equilibrium), Eq. (22) and Fig. 20 indicate that negative entropy production would be predicted on the hot side of the shock. Thus, it is the nonequilibrium entropy flux J_s that keeps the entropy production rate positive in this region. The second order entropy production rate $w_1 + w_2$ has been determined from DSMC data and is positive at

all points in the shock structure (Fig. 21). Furthermore, calculation of the second order entropy production rate using data from continuum solutions is also positive at all points in the shock.

Concluding Remarks

Whereas previous investigations of hypersonic shock wave structure using the Burnett equations have been unable to obtain steady-state solutions directly from a system of ordinary differential equations containing strong singularities, the present investigation has been successful in obtaining solutions from a time-dependent system of partial differential equations. Some previous views about inadequacies of the Burnett equations are now believed due to inadequacies in the mathematical methods previously used.

With the present time-dependent method, solutions to hypersonic shock structure have also been obtained using the Super-Burnett equations. These correspond to third-order departures from equilibrium, whereas the Burnett equations correspond to second order, and the Navier-Stokes to first order.

Shock structure of three different gas models– hard sphere, argon, and Maxwellian– has been investigated from both continuum and particulate perspectives. Continuum solutions have been compared with solutions obtained using the DSMC particulate method, and with experimental data.

Solutions to the Burnett equations have been obtained for argon and hard-sphere gases at all Mach numbers investigated (1.3 through 50). Solutions for a Maxwellian gas, however, could be obtained only up to Mach 3.8. By eliminating one Burnett term, namely, the second derivative of density term in the viscous stress, solutions to the resulting Burnett(-) equations have been obtained up to Mach 50 for a Maxwellian gas as well as for argon and hard-sphere gases. The Burnett equations, and the Burnett(-) equations yield shock structure in much closer agreement with DSMC solutions than do the Navier-Stokes equations. All parameters examined–including density thickness, density asymmetry, and temperature-density separation– are more accurately predicted at all Mach numbers by the Burnett equations than by the Navier-Stokes equations. Results for "soft" potential models, which result in thicker and lower-Knudsen-number shocks, are in much better agreement with DSMC data than results for "hard" potential models. For molecules like argon, and presumably also nitrogen, the Burnett equations appear to be of acceptable accuracy for engineering flow-field computations in the continuum transitional regime at the higher altitudes at which Navier-Stokes computations are relatively inaccurate.

Based on the results obtained herein, a need exists to derive the precise Burnett coefficients (ω's and θ's) for potentials which represent real gases, such as argon and nitrogen. Present results for argon using Burnett(-) coefficients derived for hard sphere and for Maxwellian gases differ by only a small amount.

A curious result is that, although solutions for a Maxwellian gas could not be obtained with the full Burnett equations at Mach numbers greater than 3.8, they could readily be obtained with the higher-order Super-Burnett equations at all Mach numbers up to 50. Super-Burnett solutions, however, are much poorer than Burnett solutions due to one particular third-order term. With this term omitted, solutions to the Super-Burnett(-) equations for a Maxwellian gas are in some, but not all, respects slightly better than those of the Burnett(-) equations. It is emphasized that the term deleted from the Burnett equations to form the Burnett(-) equations is believed to be physically correct, but creates numerical difficulties. Conversely, the term deleted from the Super-Burnett equations to form the Super-Burnett(-) equations is believed possibly to have been derived with an incorrect pure number coefficient, but inclusion of this term does not cause numerical difficulties.

DSMC entropy calculations based on the Boltzmann definition are consistent with expectations concerning the second law of thermodynamics. Furthermore, in the case examined, the Burnett equations yield positive entropy production rates at all points within the shock structure.

Appendix: Note Added in Proof

Very recently some comparisons have been made of shock structures computed from the present DSMC code with those computed from a code provided by Graeme Bird. For argon the two DSMC codes agree closely for all Mach numbers, while for a Maxwellian gas small differences are observed. For a hard sphere gas the two DSMC codes differ significantly so that Burnett equation results are closer to Bird's DSMC results than to present DSMC calculations. Although the precise reasons for the differences are not yet understood, the present conclusions about the improved realism of Burnett relative to Navier-Stokes equations for all three gases would not be affected by which set of DSMC calculations is used for comparison.

Acknowledgments

This research is supported by SDIO/IST managed by the Army Research Office under Contract DAALO3-86K-0139, and by ONR/AFOSR/NASA Hypersonic Training and Research Grant NAGW-965. We would also like to acknowledge the Department of the Army for the graduate fellowship of the principal author, and the NASA Ames Research Center for providing computer time on the Cray XMP and Cray 2.

References

[1] Fiszdon, W., Herczynski, R., and Walenta, A., "The Structure of a Plane Shock Wave of a Monatomic Gas: Theory and Experiment," *Rarefied Gas Dynamics: Proceedings of the Ninth International Symposium*, Vol. 2, 1974, pp. Ax.-B.23-1 to Ax.-B.23-57.

[2] Fiscko, K. A. and Chapman, D. R., "Hypersonic Shock Structure with Burnett Terms in the Viscous Stress and Heat Flux," AIAA Paper 88-2733, 1988.

[3] Chapman, D. R., Fiscko, K. A., and Lumpkin, F. E., "A Fundamental Problem in Computing Radiating Flow Fields with Thick Shock Waves," *SPIE Proceedings on Sensing, Discrimination, and Signal Processing, and, Superconducting Materials and Instrumentation*, Vol. 879, 1988, pp. 106-112.

[4] Garen, W., Synofzik, R., and Wortberg, G., "Experimental Investigation of the Structure of Weak Shock Waves in Noble Gases," *Progress in Aeronautics and Astronautics: Rarefied Gas Dynamics*, edited by J. L. Potter Vol. 51, Pt. 1, AIAA, New York, 1977, pp. 519-528.

[5] Elliott, J. P. and Baganoff, D., "Solution of the Boltzmann Equation at the Upstream and Downstream Singular Points in a Shock Wave," *Journal of Fluid Mechanics*, Vol. 65, Pt. 3, 1974, pp. 603-624.

[6] Bird, G. A., *Molecular Gas Dynamics*, Clarendon, Oxford, 1976.

[7] Kogan, M. N., *Rarefied Gas Dynamics*, Plenum, New York, 1969.

[8] Woods, L. C., "On the Thermodynamics of Nonlinear Constitutive Relations in Gasdynamics," *Journal of Fluid Mechanics*, Vol. 101, Pt. 2, 1980, pp. 225-242.

[9] Sherman, F. S., "A Low-Density Wind-Tunnel Study of Shock-Wave Structure and Relaxation Phenomena in Gases," *NACA TN 3298*, 1955.

[10] Holtz, T. and Muntz, E. P., "Molecular Velocity Distribution Functions in an Argon Normal Shock Wave at Mach Number 7," *The Physics of Fluids*, Vol. 26, No. 9, 1983, pp. 2425-2436.

[11] Grad, H., "Theory of Rarefied Gases," *Rarefied Gas Dynamics: Proceedings of the First International Symposium held at Nice*, Pergamon, New York, 1960, pp. 100-138.

[12] Steinhilper, E. A., "Electron Beam Measurements of the Shock Wave Structure," Ph.D. Thesis, California Inst. of Tech., Pasadena, CA, 1971.

[13] Alsmeyer, H., "Density Profiles in Argon and Nitrogen Shock Waves Measured by the Absorption of an Electron Beam," *Journal of Fluid Mechanics*, Vol. 74, April 1976, pp. 497-513.

[14] Truesdell, C., "A new definition of a fluid. II. The Maxwellian Fluid;" *Journal de Mathematiques*, Vol. 30, 1951, pp. 111-151.

[15] Chapman, S. and Cowling, T. G., *The Mathematical Theory of Non-Uniform Gases*, Cambridge Univ. Press, London, 1970.

[16] Foch, J. D., Jr., "On Higher Order Hydrodynamic Theories of Shock Structure," *Acta Physica Austriaca*, Suppl. X., 1973, pp. 123-140.

[17] Woods, L. C., "Frame-Indifferent Kinetic Theory," *Journal of Fluid Mechanics*, Vol. 136, 1983, pp. 423-433.

[18] Simon, C. E., "Theory of Shock Structure in a Maxwell Gas based on the Chapman-Enskog Development Through Super-Super-Burnett Order," Ph.D. Thesis, Univ. of Colorado, CO, 1976.

[19] Holway, L. H., Jr., "Existence of Kinetic Theory Solutions to the Shock Structure Problem," *The Physics of Fluids*, Vol. 7, 1964, pp. 911-913.

[20] Butler, D. S. and Anderson, W. M., "Shock Structure Calculations by an Orthogonal Expansion Method," *Proceedings of the Rarefied Gas Dynamics Symposium*, Vol. 1, 1967, pp. 731-744.

[21] Woods, L. C., *The Thermodynamics of Fluid Systems*, Clarendon, Oxford, 1975.

[22] Vincenti, W. G. and C. H. Kruger, Jr., *Introduction to Physical Gas Dynamics*, Krieger Publishing, Malabar, FL, 1982.

[23] Shavit, A. and Zvirin, Y., *Technion Report TME-110*, Haifa, Israel, 1970.

Density Profiles and Entropy Production in Cylindrical Couette Flow: Comparison of Generalized Hydrodynamics and Monte Carlo Results

R. E. Khayat* and B. C. Eu†

McGill University, Montreal, Quebec, Canada

Abstract

In this paper we report on calculations of density profiles and some related flow properties for cylindrical Couette flow of a Lennard–Jones fluid. The flow is subject to a temperature gradient and thermoviscous effects are taken into consideration. We apply the generalized fluid dynamic equations provided by the modified moment method for the Boltzmann equation. The results are in good agreement with those obtained by the Monte Carlo direct simulation method for all Knudsen numbers for which the simulation data are available. The shear stress, heat flux, effective shear viscosity and heat conductivity, and entropy production are shown to decay with the increasing Knudsen number(Kn). Their Kn dependence indicates that the generalized hydrodynamic equations asymptotically approach the ideal fluid dynamic equations as Kn increases.

I. Introduction

The Boltzmann equation and a generalized Boltzmann equation have been studied in refs. 1–8 by one of the present authors in an effort to put theory of nonlinear irreversible processes in fluids on the foundations of kinetic theory. The study has yielded a theory[9] of macroscopic nonlinear irreversible processes in fluids, which is consistent with the thermodynamic laws. The theory turns out to be a generalization of the Navier–Stokes and Fourier (NSF) equations, which hold near equilibrium. We call this theory generalized hydrodynamics, because it reduces to classical hydrodynamics as the

---

Copyright © 1989 by the American Institute of Aeronautics and Astronautics, Inc. All rights reserved.

* Graduate Student, Department of Chemistry.

† Professor, Department of Chemistry and Associate Member, Department of Physics.

processes involved become steady in time and linear with respect to thermodynamic forces driving them. Based on the generalized hydrodynamic equations(GHE), we have carried out a series of studies in which non–Newtonian fluids and non–Fourier heat conductivity have been investigated successfully. In this paper we apply them to study density profiles and some other aspects of fluid properties that have not been reported in a previous paper(ref. 10) on flow properties of a Lennard–Jones fluid in cylindrical Couette flow.

We consider a Lennard–Jones gas contained between two concentric cylinders of radius R_i and R_o, respectively. The inner cylinder rotates at an angular velocity Ω while the outer cylinder is at rest. The steady state temperatures of the two cylinders are different by 8 deg. The temperature of the inner and outer cylinder will be denoted by T_i and T_o respectively. This is the experimental arrangement used by Alofs and Springer[11] who thereby obtained density profiles at various gas densities of argon. Nanbu[12] has calculated the density profiles by means of the Monte Carlo direct simulation(MCDS) method of Bird[13]. In this paper we shall make comparison of the density profiles obtained by experiment, the MCDS method, the NSF theory with slip boundary conditions, and the present theory of generalized hydrodynamics in order to judge the utility of the latter. This comparison also allows assessment of the MCDS method. We will also report on the Kn dependence of the shear stress, heat flux, effective shear viscosity and effective thermal conductivity at the inner cylinder wall. Their Kn dependence will enable us to examine the approach of the generalized hydrodynamic equations to the ideal hydrodynamic equations in the limit of Kn → ∞. It will be shown that the shear stress and heat flux decrease as Kn increases in contrast to the NSF theory prediction that they should remain constant at a given temperature. This behavior assures the asymptotic approach to the ideal fluid behavior by the generalized hydrodynamic equations. The entropy production is also calculated as a function of Kn. The result shows that it also decreases with increasing Kn. This behavior is different from the behavior predicted by the NSF theory, since the latter predicts a constant entropy production for all values of Kn. The decreasing entropy production is brought about in the present theory as a result of the nonlinear transport processes that give rise to the shear stress and heat flux decreasing with increasing Kn.

II. Reduced Steady Generalized Hydrodynamic Equations

We will present only the steady generalized hydrodynamic equations in reduced form which are appropriate for the geometry of flow in question. More details can be found in the previous papers (refs.1–10 and 14).

We first define the following reduced variables scaled by a suitably chosen set of reference variables. With the definitions $\Delta = T_o$

$-T_i$, $D = R_o - R_i$ and denoting the reference set of variables by T_r, p_r, ρ_r, U_r, η_r, and λ_r for temperature T, pressure p, mass density ρ, velocity u, viscosity η_0, and heat conductivity λ_0, respectively, we define the reduced variables

$$T^* = T/T_r, \qquad p^* = p/p_r, \qquad u^* = u_\theta/U_r,$$

$$\xi = r/D, \qquad h^* = T\hat{C}_p/T_r\hat{C}_p(T_r), \qquad \alpha^* = pT\hat{C}_p/p_rT_r\hat{C}_p(T_r),$$

$$\eta^* = \eta_0/\eta_r, \qquad \lambda^* = \lambda_0/\lambda_r, \qquad \gamma^* = \gamma/(U_r/D),$$

$$\omega^* = \omega/(U_r/D), \qquad \Pi^* = \Pi_{r\theta}/(2\eta_rU_r/D), \qquad Q^* = Q_r/(\lambda_r\Delta/DT_r),$$

$$N_i^* = N_i/(2\eta_rU_r/D), (i=1, 2), \qquad Q_\theta^* = Q_\theta/(\lambda_r\Delta/DT_r),$$

$$\rho^* = \rho/\rho_r, \qquad \chi^* = \chi D$$

where

$$\gamma = (r/2)[d(u_\theta/r)/dr], \qquad \omega = (2r)^{-1}[d(ru_\theta)/dr],$$

$$\chi = d\ln T/dr,$$

and where u_θ is the tangential component of the velocity vector u, $\hat{C}_p$ the specific heat per unit mass, $\Pi_{r\theta}$ the shear stress (i.e., the rθ component of stress tensor Π), N_i the primary and secondary normal stress differences and Q_r the radial component of heat flux Q. We will specify more explicitly the reference set of variables later when we discuss numerical analysis. It is convenient to introduce the following dimensionless numbers well known in fluid dynamics:

Mach No.: $\qquad M = U_r/(\gamma_0 RT_r)^{1/2}$,

$\gamma_0 = \hat{C}_p/\hat{C}_v$; R = gas constant per unit mass,

Reynolds No.: $\qquad Re = \rho_rU_rD/\eta_r$,

Eckert No.: $\qquad E = U_r^2/\hat{C}_p\Delta, [\hat{C}_p = \hat{C}_p(T_r)]$,

Prandtl No.: $\qquad Pr = \hat{C}_p\eta_rT_r/\lambda_r$,

Knudsen No.: $\qquad Kn = \ell/D, (\ell = \text{mean free path})$

Since these dimensionless numbers appear in a product form, we will find it convenient to define the following composite reduced numbers:

$\delta \equiv (2\gamma_0/\pi)^{1/2}$ MKn and $\varepsilon \equiv A/4T_r EPr$. Note that $Re = (\pi\gamma_0/2)^{1/2}$ M/Kn. The generalized hydrodynamic equations include in general the normal stresses as well. Therefore, to be rigorous it would be necessary to include them, but since we ignore them, the present theory should be regarded as a model theory.

A. General Set of Equations for Cylindrical Couette Flow

With the reduced variables defined, the reduced generalized hydrodynamic equations at the steady state are as follows:

$$\gamma_0 M^2 \frac{\rho^* u^{*2}}{\xi} = \frac{d}{d\xi}[p^* + \frac{2}{3}\delta(N_2^* - N_1^*)] - 2\delta N_1^*/\xi \tag{1}$$

$$\frac{d}{d\xi}(\xi^2 \Pi^*) = 0 \tag{2}$$

$$\frac{1}{\xi}\frac{d}{d\xi}(\xi Q^*) + 4PrE\gamma^* \Pi^* = 0 \tag{3}$$

$$\Pi^* q = -\eta^* \gamma^* - \frac{2}{3}\delta(\eta^* \gamma^*/p^*)(2N_1^* + N_2^*) \tag{4}$$

$$N_1^* q = 4\delta(\eta^* \gamma^*/p^*)\Pi^* \tag{5}$$

$$N_2^* q = -4\delta(\eta^* \gamma^*/p^*)\Pi^* \tag{6}$$

$$Q^* q = -\lambda^* \chi^* - (\delta/Pr)(\lambda^*/\alpha^*)[(2\gamma^* + \omega^*)Q_\theta^*$$
$$+ \frac{4}{3}Pr(h^* \chi^* - E\frac{u^{*2}}{\xi})(N_2^* - N_1^*)] \tag{7}$$

$$Q_\theta^* q = -(\delta/Pr)(\lambda^*/\alpha^*)[2Pr(h^* \chi^* - E\frac{u^{*2}}{\xi})\Pi^* - \omega^* Q^*] \tag{8}$$

$q = \sinh\kappa/\kappa$

$\kappa = \delta\kappa^*$

$$\equiv \delta(\pi^{3/2}/\gamma_0)^{1/2}(T^{*1/4}/\eta^{*1/2} p^*)[\Pi^{*2} + \frac{1}{3}(N_1^{*2} + N_2^{*2} + N_1^* N_2^*)$$
$$+ \varepsilon(\eta^*/\lambda^*)(Q^{*2} + Q_\theta^{*2})]^{1/2} \tag{9}$$

Since parameter δ is proportional to Kn, the terms containing δ will become increasingly important as Kn increases. The δ or Kn dependence of Π^*, N_1^*, N_2^*, etc., is important for understanding the mode by which Eqs. (1–8) approach Euler's ideal fluid dynamic equations. This aspect is investigated in a series of works of which the present paper is a part; see refs.10 and 14. Here we will study a particular case of Eqs.(1–8) in order to reduce the mathematical complexity. In the limit of $\delta \to 0$ the set of Eqs. (1–8) can be shown to reduce to the NSF equations. This limit is indeed attained by the numerical result obtained for the density profile as shown in Fig. 1.

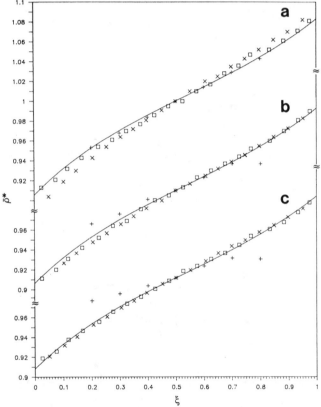

Fig. 1 Reduced density vs reduced distance.
(a) Kn = 0.0544 and p_0 = 0.05 mm Hg. Symbols: + = experimental value by Alofs and Springer(1971); □ = simulation value by Nanbu(1984); x = Navier–Stokes theory with slip boundary conditions — = the present theory. $\bar{\rho}^*$ = $\rho(\xi)/\rho(\text{midstream})$. The same symbols are used in Figs. 1b–2c. (b) Kn = 0.1046 and p_0 = 0.026 mm Hg.
(c) Kn = 0.1388 and p_0 = 0.00196 mm Hg.

The Chapman–Enskog viscosity and thermal conductivity in the NSF equations are independent of the gas density. However, it is known that transport coefficients are density dependent even in the low–density regime. There is evidence that they also depend on thermodynamic gradients. To reflect these facts in the theory, it is necessary to use effective transport coefficients that have such features incorporated into them. This means that we abandon the Chapman–Enskog first approximation formulas for the transport coefficients. This aim can be achieved if we modify the constitutive equations for stress and heat flux so that the viscosity and thermal conductivity depend on density and thermodynamic gradients. Attempts have been made in refs. 15 and 16 to include the higher order Chapman–Enskog solutions, i.e., the Burnett solutions, in the transport coefficients. However, in such approaches there are some difficulties, since the presence of higher order derivatives of velocity, temperature and pressure requires additional boundary conditions that are not known. Moreover, the constitutive equations are not certain to satisfy the second law of thermodynamics because some Burnett terms make the fluxes inconsistent with the H theorem. The generalized hydrodynamic equations used here are free from such difficulties mentioned above.

B. Approximate Set of Equations Used

In the present paper we use the following generalized hydrodynamic equations obtained from Eqs. (1–8) by setting equal to zero:

$$\gamma_0 M^2 \frac{\rho^* u^{*2}}{\xi} = \frac{dp^*}{d\xi} \tag{10}$$

$$\frac{d}{d\xi}(\xi^2 \Pi^*) = 0 \tag{11}$$

$$\frac{1}{\xi}\frac{d}{d\xi}(\xi Q^*) + 4\mathrm{Pr}\mathrm{E}\gamma^* \Pi^* = 0 \tag{12}$$

$$\Pi^* q = -\eta^* \gamma^* \tag{13}$$

$$Q^* q = -\lambda^* \chi^* \tag{14}$$

$$q = \sinh\kappa/\kappa, \quad (\kappa = \delta\kappa^*) \tag{15}$$

$$\kappa^* = (\pi^{3/2}/\gamma_0)^{1/2}(T^{*1/4}/\eta^{*1/2} p^*)[\Pi^{*2} + \epsilon(\eta^*/\lambda^*)Q^{*2}]^{1/2} \tag{16}$$

The constitutive Eqs. (13) and (14) are simply the non–Newtonian stress equation and the non–Fourier heat flux equation which reduce to the Newtonian and Fourier constitutive equations as the magnitude of the stress and heat flux decreases, that is, if q = 1. We have studied a set equivalent to Eqs. (10–16) in the case of plane Couette flow in ref. 17. These equations, and also Eqs. (1–8), not only do not require more boundary conditions than those required for the Navier–Stokes and Fourier equations but also are fully in conformation to the thermodynamic laws. We have investigated the utility of this set of equations and the present paper reports on some aspects not communicated in ref. 10. Since the density profiles are the quantities that can be compared with experiment, we repeat the comparison here to establish the utility of the theory in a self–contained manner. It is also relevant to a theme of this symposium.

The entropy production for the process described by Eqs. (1–8) or Eqs. (10–14) is given by the formula

$$\sigma_{ent} = k_B (n\sigma)^2 (2k_B T/m')^{1/2} \kappa \sinh\kappa \qquad (17)$$

where n is the number density, σ is the size parameter of the molecule, m' is the reduced mass, and κ is given by Eq. (9) or (16) according as Eqs. (1–8) or Eqs. (10–14) is taken for calculating flow properties. We remark that if $\sinh\kappa$ is approximated with κ, then Eqs. (17) becomes the usual entropy production (Rayleigh–Onsager dissipation function) appearing in linear irreversible thermodynamics.

III. Boundary Conditions and Linear Transport Coefficients.

The system of differential equations [Eqs. (10–14)] requires five boundary conditions. Four boundary conditions are provided by the boundary values for velocity and temperature. Lin and Street take the boundary condition for pressure to be the mean pressure in ref. 16 on cylindrical Couette flow of a gas. Such a boundary condition seems inappropriate, especially when the pressure is expected to change with position. Since the density at the wall can be used as a boundary condition but there is no boundary value known for it, it cannot be exploited. Given this situation, we make use of the fact that the mass density must be conserved in time. Thus, if ρ_o is the initial mass density, then it must be equal to the mean value of the steady–state density $\rho(r)$:

$$\rho_o = [2\pi(R_o^2 - R_i^2)]^{-1} 2\pi \int_{R_i}^{R_o} dr \, r\rho(r) \equiv <\rho(r)>$$

This can be used as a condition for the remaining integration constant. In reduced form it may be written as

$$\langle \rho^*(\xi) \rangle = 1 \qquad (18a)$$

For the velocity and temperature boundary conditions we take

$$u^* = 0; \quad T^* = T_o/T_r \qquad \text{at } \xi = R_o/D \qquad (18b)$$

$$u^* = 1; \quad T^* = T_i/T_r \qquad \text{at } \xi = R_i/D \qquad (18c)$$

These are no–slip boundary conditions. In this series of work we take the viewpoint that the generalized hydrodynamic equations with nonlinear transport coefficients (or nonlinear constitutive equations) can account for flow properties even if no–slip boundary conditions are used for velocity and temperature. This viewpoint is supported by the numerical results presented below. Therefore, what seems to be essential for accounting for flow properties of rarefied gases is nonlinear transport coefficients, or nonlinear constitutive equations for the stress tensor and heat flux, rather than the slip boundary conditions. The slip boundary conditions contain accommodation coefficients which act in practice as empirical parameters. In the present theory there does not appear such parameters. Therefore, the present theory is simpler than the slip boundary condition theory.

There now remains the specification of linear transport coefficients η_o and λ_o. For these quantities we take the forms provided by Ashurst and Hoover[18] and Eu(1979b). For lack of space we do not list them here, but refer the reader to refs. 10 and 14.

IV. Numerical Results

Eqs. (10–14) are numerically integrated subject to the boundary conditions [Eqs. (18a–18c)] presented earlier. We have used a combination of the sixth–order Runge–Kutta method and the shooting method.

To compare our reults with those obtained experimentally, we have to use the same values for the dimensionless parameters involved. However, there is some ambiguity in the way some of the parameters were defined and in the choice of some reference quantities in the papers by Alofs and Springer(1971). The wall temperatures were initially identical, but their steady–state values were found to be different by 8 deg. We therefore define our Mach number on the basis of the mean value of the two steady–state wall temperatures. This value of temperature is then our reference temperature. The Mach number thus calculated turns out to be comparable to that used by

Alofs and Springer(1971); i.e., 0.9908 (ours) compared to 0.9917 (theirs). The aspect ratio A and the Eckert and Prandtl numbers are set equal to the experimental conditions, i.e., $A = 2/3$, $E = 25.115$, and $Pr = 0.666$. Note that the specification of the Eckert number fixes that of the temperature ratio T_i/T_o. The Chapman–Enskog transport coefficients are those for a Lennard–Jones gas, whereas the ones used in the experiment are based on the Maxwellian model for the intermolecular force. For this reason the Kn values in the present investigation are slightly higher than the values quoted in the papers by Alofs and Springer(1971) and by Nanbu(1984), although they are both based on the same value of the initial density. In the experiment by Alofs and Springer the initial chamber pressures (p_o) varied from 0.050 to 0.0020 mm Hg with the corresponding Knudsen number ranging from 0.0426 to 1.065. When converted to the present choice of reference variables, the Knudsen numbers range from 0.0544 to 1.3962.

The density profiles for different initial chamber pressures are shown in Figs. 1a–2c where the + symbols represent the experimental data(Alofs and Singer, 1971), the squares the simulation results by Nanbu(1984), the x symbols the values obtained by the Navier–Stokes theory with slip boundary conditions, and the solid curve the profile by the present theory. They are of particular interest since density measurements were mainly the object of attention in the experiment. The corresponding velocity, temperature, and pressure distributions were also calculated, but since they have already been reported in ref. 10, we do not present them here. As indicated in Figs. 1a–1c, for Kn = 0.0544, 0.1388 and 0.3201 the present profiles are virtually identical with the simulation results, and those by the NSF theory with slip conditions. However, the results by all three methods begin to show a qualitative difference from the experimental results as the Knudsen number increases, although absolute numerical differences from the experimental values are invariably less than a few percent at most. From Kn = 0.5184 and up, the simulation profiles show a noticeable minimum in density in the vicinity of the inner cylinder wall whereas the results by the present theory, i.e., generalized hydrodynamic Eqs. (10–14), show only a rather weak minimum at the Knudsen number in question. However, such a minimum appears noticeably at 1.362 and at higher Knudsen numbers of 9.459 and 212.8; see ref. 10 for such a minimum. Because of limitations on measurements, the experiment is not able to determine the presence of such a minimum. In any case, the numerical differences between the two sets do not exceed 5–6% at most. Since the simulation method itself is by no means exact, it is reasonable to state that the present nonlinear hydrodynamic equations with stick boundary conditions are able to reproduce the density profiles comparable with the simulation results and those by the NSF theory with slip boundary conditions in the range of Knudsen numbers so far studied. Although the present theory does not have accommodation coefficients, it is not only as accurate as the usual

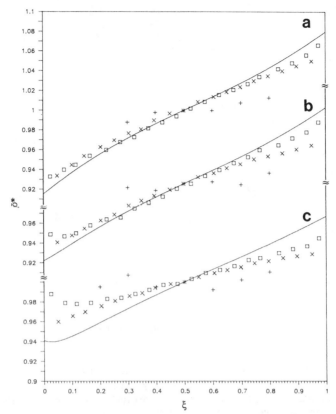

Fig. 2 Reduced density vs reduced distance.
(a) Kn = 0.3201 and p_o = 0.0085 mm Hg.
(b) Kn = 0.5184 and p_o = 0.00525 mm Hg.
(c) Kn = 1.362 and p_o = 0.009 mm Hg.

hydrodynamic theory employing slip boundary conditions, but is also simpler than the latter.

For lack of space we do not present the velocity, temperature, and pressure profiles, but they do not exhibit slips as the corresponding profiles obtained by Nanbu(1984) do even at the lowest Knudsen number studied; see ref. 10. We have instead observed a gradual formation of boundary layer near the inner boundary as Kn increases. In the present theory the boundary layers are due to the nonlinear terms in the constitutive equations for the stress tensor and heat flux. Since there is no experimental measurement available to sort out these conflicting numerical results regarding slips in velocity and temperature, we are not able to draw a definite conclusion at the present time. According to the calculations that include normal stresses, which are reported in ref.14, the lack of slip in the present

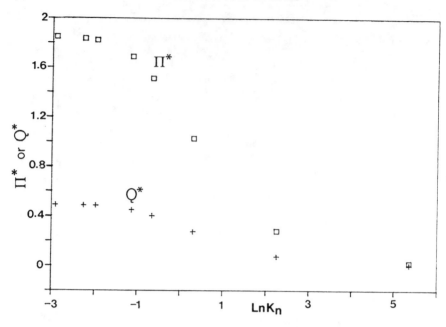

Fig. 3   Reduced shear stress and heat flux vs lnKn.

results is due in part to the neglect of normal stresses. In any case, a thin boundary layer may be easily mistaken for a slip within experimental errors, and we will have to leave the question unresolved for now and reserve it for future study.

The classical hydrodynamic theory does not predict a shear stress and heat flux which vary with Kn since η_o and λ_o are independent of density if their Chapman–Enskog formulas are used. The present generalized hydrodynamic equations, on the other hand, predict the effective transport coefficients varying with Kn. To see their Kn dependence we have first plotted the reduced shear stress and heat flux against lnKn in Fig. 3. As shown, they practically vanish by the time lnKn has reached 5. Thus we see that the generalized hydrodynamic equations approach the ideal fluid dynamic equations as the gas rarefies, and this is what is expected. We stress that in the Navier–Stokes theory the shear stress and heat flux do not change with Kn unless the boundary conditions depend on Kn as in the slip boundary condition theory.

The effective transport coefficients may be defined by $\eta_e^* = -\Pi^*/\gamma^*$ and $\lambda_e^* = -Q^*/\chi^*$. Figure 4 presents the reduced effective shear viscosity and heat conductivity as a function of position. The Kn dependence of Π^* and Q^* in Fig. 3 is basically due to q appearing in the constitutive Eqs. (13) and (14). This is the factor that makes Eqs. (10–14) different from the Navier–Stokes and Fourier equations.

Figure 5 presents the reduced entropy production $\Sigma = \kappa \sinh\kappa / Kn^2$ accompanying the flow process in hand. Since it is sufficient to compute Σ at a point in ξ in view of the fact that the shear stress and heat flux have simple distributions in ξ, we have computed Σ at the inner wall. The squares indicate the values of the reduced entropy production so calculated from Π^* and Q^* values by using Eq. (17), and the broken line indicates the estimated value for the entropy production in the Navier–Stokes–Fourier theory: $\Sigma = \kappa^2 / Kn^2$ which is independent of Kn. Thus, the classical theory predicts an unrealistic value for Σ in the high Kn regime. This situation is remedied in the case of the Navier–Stokes theory with slip boundary conditions by a modification of the boundary conditions. If there is a slip in velocity and temperature, the values of Π^* and Q^* at the wall will be accordingly reduced, and a diminished entropy production will result. A similar reduction in Σ is achieved in the present theory because of the nonlinear transport processes as described by Eqs. (13) and (14). This result therefore lends further support of our proposition that the

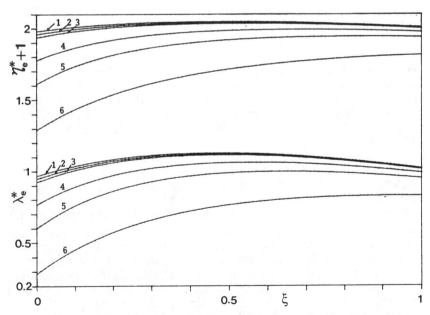

Fig. 4 Reduced Effective shear viscosity and thermal conductivity vs ξ. The values of Kn for curves 1, 2, ..., 6 are 0.0544, 0.1046, 0.1388, 0.3201, 0.5184 and 1.362, respectively. Note that the effective transport coefficients are not constant over the interval.

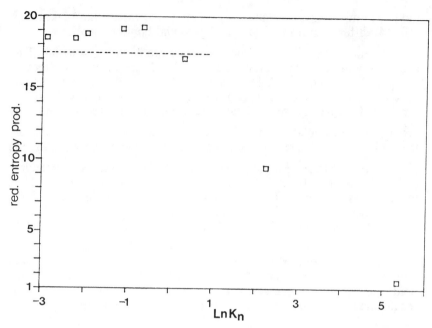

Fig. 5   Reduced entropy production vs lnKn. The broken line is the reduced entropy production for the linear (Navier–Stokes) theory.

generalized hydrodynamic equations with no–slip boundary conditions can be as effective as a slip boundary condition theory in accounting for flow properties. Thus, we see that the decreasing tendency of the stress and heat flux with increasing Kn is nature's way of reducing the entropy production as the gas rarefies and thus becomes less viscous and heat conducting.

The present theory employs a set of approximate Eqs. (10–14) that neglect the normal stresses and the tangential heat flux. Yet it yields density profiles comparable to those by the MCDS method, and the NSF theory with slip conditions, whereas three sets of density profiles display some significant discrepancies with the experimental data. There is then a question of experimental errors that one might invoke to reconcile the discrepancies, and, of course, the experimental errors cannot be ignored. However, if we can safely assume that the experiment was sufficiently accurate and reliable, there is an important question raised by the comparison of the results by the MCDS method and the present theory. That is, although the decoupling approximation is used, the Monte Carlo method solves the Boltzmann equation and the results are believed to be accurate. However, the Monte Carlo density profiles are basically in agreement with the density profiles by the approximate Eqs. (10–14) and also by the Navier–Stokes theory with slip boundary conditions, as we have already shown. Therefore it is possible to draw the conclusion that either the approximate Eqs. (10–14) are sufficiently accurate or the Monte Carlo method is not as accurate as it is believed to be. In view of the discrepancies with the experimental data, it is likely that the latter possibility is more probable, and it may therefore be necessary to further improve the MCDS method. In this regard we would like to mention that the GH Eqs. (1–8) not only improve the density profiles qualitatively but also show some new features unseen in gas dynamics, according to our recent study(1989). Therefore, the approximate set [Eqs. (10–14)] does not give completely accurate results, although it can be quite useful for calculating flow properties as is shown here.

In conclusion, we have shown that the generalized hydrodynamic Eqs. (10–14) can yield density profiles in agreement with the Monte Carlo direct simulation method, even if no–slip boundary conditions are used. It is also shown that the set [Eqs. (10–14)] approaches asymptotically the ideal fluid dynamic equations as Kn increases. The entropy production is shown to vanish as Kn increases to a large value.

References

[1] Eu, B. C., "A Modified Moment Method and Irreversible Thermodynamics", Journal of Chem. Phys., vol. 73, 2958–2969, 1980.

[2] Eu, B. C. "On the Extended Gibbs Relations and Nonlinear Irreversible Thermodynamics", Journal of Chem. Phys., vol 74, 2998–3005, 1981.

[3] Eu, B. C., "Nonlinear Transport Processes in Gases", Journal of Chem. Phys., vol. 74, 3006–3015, 1981.

[4] Eu, B. C., "Kinetic Theory of Nonlinear Transport Processes in Dilute Ionized Gases subject to an Electromagnetic Field", Journal of Chem. Phys., vol. 82, 4283–4302, 1985.

[5] Eu, B. C., "On the Modified Moment Method and Irreversible Thermodynamics", Journal of Chem. Phys., vol. 85, 1592–1602, 1986.

[6] Eu, B. C., Kinetic Theory of Dense Fluids. I.", Annals of Physics (N.Y.), vol. 118, 187–229, 1979.

[7] Eu, B. C., "The Modified Moment Method, Irreversible Thermodynamics, and Nonlinear Viscosity of a Dense Fluid", Journal of Chem. Phys., vol. 74, 6362–6367,1981.

[8] Eu, B. C., "Kinetic Theory of Dense Fluids Subject to an External Field", Journal of Chem. Phys., vol. 1220–1237, 1987.

[9] Eu, B. C., "Connection between Extended Thermodynamics and Transport Theory", J. Non–Equilibrium Thermodynamics, vol. 11, 211–239, 1986.

[10] Khayat, R. E. and Eu, B. C., "Nonlinear Transport Processes and Fluid Dynamics: Cylindrical Couette Flow of Lennard–Jones Fluids", Physical Review A, vol. 38, 2492–2507, 1988.

[11] Alofs, D. J. and Springer, G. S., "Cylindrical Couette Flow Experiments in the Transition Regime", Physics of Fluids, vol. 14, 298–305, 1971.

[12] Nanbu, K., "Analysis of Cylindrical Couette Flows by Use of the Direct Simulation Method", Physics of Fluids, vol. 27, 2632–2635, 1984.

[13] Bird, G. A., Molecular Gas Dynamics, Oxford, London, 1976.

[14] Khayat, R. E. and Eu, B. C., "Generalized Hydrodynamics, Normal Stress Effects and Velocity Slips in the Cylindrical Couette Flow of Lennard–Jones Fluids, Physical Review A, vol. 39, January, 1989.

[15] Wang Chang, C. S. and Uhlenbeck, G. E., in Studies in Statistical Mechanics, North–Holland, Amsterdam, the Netherlands, 1970, vol. V, J. de Boer and G. E. Uhlenbeck, eds.

[16] Lin, T. C. and Street, R. E., "Effect of Variable Viscosity and Thermal Conductivity on High Speed Slip Flow between Concentric Cylinders", National Advisory Committee for Aeronautics Report No. 1175, U. S. Government Printing Office, Washington, D. C., 1954

[17] Bhattacharya, D. K. and Eu, B. C., "Nonlinear Transport Processes and Fluid Dynamics: Effects of Thermoviscous Coupling and Nonlinear Transport Coefficients on Plane Couette Flow of Lennard–Jones Fluids", Physical Review A, vol. 35, 821–836, 1987.

[18] Ashurst, W. T. and Hoover, W. G., "Dense–Fluid Shear Viscosity via Nonequilibrium Molecular Dynamics", Physical Review A, vol. 11, 658–678, 1975.

[19] Eu, B. C., "Kinetic Theory of Dense Fluids. III. Density Dependence of Transport Coefficients", Annals of Physics (N.Y.), vol. 120, 423–455, 1979.

Chapter 5. Flowfields

Direct Simulation of AFE Forebody and Wake Flow with Thermal Radiation

James N. Moss* and Joseph M. Price[†]
NASA Langley Research Center, Hampton, Virginia

Abstract

Calculated results for the flowfield structure and surface quantities are presented for an axisymmetric representation of an aeroassist flight experiment vehicle. The direct simulation Monte Carlo method is used to perform the calculations, since the flow is highly nonequilibrium about the vehicle during both the compression and expansion phases. The body configuration is an elliptically blunt nose followed by a skirt with a circular radius and an afterbody. Freestream conditions correspond to a single point along the entry trajectory at an altitude of 90 km and a velocity of 9.9 km/s. The calculations account for nonequilibrium in the translational and internal modes, dissociation, ionization, and thermal radiation. The degree of dissociation is large, but the maximum ionization is only about 2% by mole fraction. The blunt forebody flow experiences a high degree of thermal nonequilibrium in which the translational temperature is generally greater than the internal temperature. However, as the flow expands about the aerobrake skirt and afterbody, the internal temperature is generally greater than the translational temperature. Furthermore, the calculated results clearly show mass separation effects in the wake with a preferential increase in the concentration of the light (atomic) species relative to their values at the corner expansion on the aerobrake skirt. Forebody heating is dominated by the convective component, however, the stagnation-point radiative heating under the assumption of no absorption is about 12% of the convective value. Afterbody heating is very small compared with forebody values.

Copyright © 1989 by the American Institute of Aeronautics and Astronautics, Inc. No copyright is asserted in the United States under Title 17, U.S. Code. The U.S. Government has a royalty-free license to exercise all rights under the copyright claimed herein for governmental purposes. All other rights are reserved by the copyright owner.
  *Research Engineer, Space Systems Division.
  †Mathematician, Space Systems Division.

Nomenclature

a	=	major axis of ellipsoidal nose
b	=	minor axis of ellipsoidal nose
p	=	surface pressure
Q	=	radiative emission
q_c	=	surface convective heat flux
q_r	=	surface radiative heat flux
R_C	=	corner (skirt) radius of curvature
R_N	=	stagnation radius of curvature
s	=	coordinate along body surface
U_∞	=	freestream velocity
u	=	velocity component tangent to body surface
v	=	velocity component normal to body surface
X_i	=	mole fraction of species i
x	=	coordinate measured along body centerline
y	=	coordinate measured normal to body centerline
η	=	coordinate normal to body surface
λ	=	wavelength
ρ	=	density
τ	=	shear stress

<u>Subscripts</u>

i	=	ith species
w	=	wall value
∞	=	freestream value

Introduction

The potential economic benefit of a reusable aeroassisted orbital transfer vehicle (AOTV) over its all-propulsive counterpart is such that AOTV's are being actively studied[1-3] as a class of vehicles for providing transporation between low-Earth orbit and various locations within the inner solar system. Results of the studies show that the preferred vehicle concept for these missions is one that has a low ballistic coefficient and features a large, blunt, lightweight aerobrake. On return from high-Earth orbit or the moon, the vehicles will enter the Earth's atmosphere with a velocity of approximately 10 km/s, fly a roll-modulated trajectory with a perigee of 75-100 km, skip back out of the atmosphere, and rendezvous with a space station after having achieved the velocity decrement required for capture into low-Earth orbit.

Both the velocity and altitude for the atmospheric pass are sufficiently high to produce a highly nonequilibrium flow environment where high-temperature and low-density gas effects will significantly impact the aerodynamic and

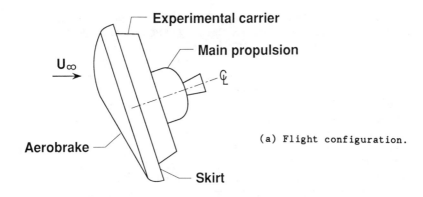

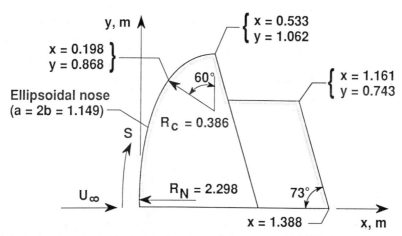

Fig. 1 Aeroassist Flight Experiment vehicle.

thermal loads. Since such an environment cannot be simulated in ground-based test facilities, the ultimate design of the AOTV's will rely heavily on numerical calculations.

In order to advance the technology for the AOTV vehicles, NASA has undertaken the Aeroassist Flight Experiment (AFE).[3] The AFE is a subscale vehicle (Fig. 1a) that will be launched from the Space Shuttle, fly a representative aeroassist trajectory, and be recovered by the Shuttle. Flight measurements will provide an opportunity for clarifying issues associated with a radiating nonequilibrium flowfield where rarefaction effects will be present for a significant portion of the atmospheric encounter.

Two numerical simulation approaches are being developed and applied to the problem of calculating the flow about AFE and AOTV vehicles. One is the continuum approach[4-7] and the other is the particle approach as implemented through the direct simulation Monte Carlo (DSMC) method.[8-13] In the present paper, the DSMC method is applied to a single point along an AFE entry trajectory corresponding to an altitude of 90 km and a velocity of 9.9 km/s. An axisymmetric representation of the AFE vehicle is used to reduce the computational effort. The upper rather than the lower portion of the AFE vehicle is considered so that the most severe heating can be calculated for the carrier panel (Fig. 1), which is shadowed by the aerobrake. The present study is an extension of the work reported in Ref. 12, in that the AFE vehicle afterbody is used rather than an arbitrary afterbody. Furthermore, the cell grid and time steps used in the present study are much smaller than those used in Ref. 12. Results for radiative and convective heating are presented along both the blunt forebody and the afterbody carrier panel. Also, details concerning the flowfield structure for an eleven-species dissociation and ionizing gas mixture are presented.

DSMC Method

The DSMC method involves the simultaneous computation of the trajectories of some thousands of simulated molecules in simulated physical space. The time parameter in the simulation may be identified with real time, and the flow is always calculated as an unsteady flow. The initial conditions do not depend on a prediction of the flowfield but can be specified in terms of states, such as a uniform flow or a vacuum, that permit exact specification. Any steady flow is the large time state of the unsteady flow. There are no iterative procedures and no stability or convergence problems. A computational grid is required only in physical space, rather than phase space, and then only for the choice of collision pairs and the sampling of flow properties. The boundary conditions are specified in terms of the behavior of individual molecules, rather than the molecular distribution function, and all procedures may be specified such that the computation time is linearly dependent on the number of molecules. Advantage may be taken of flow symmetries to reduce the number of dimensions of the grid and the number of position coordinates that need to be stored for each molecule, but the collisions are always calculated as three-dimensional (3-D) phenomena.

Gas Model

This section presents a brief summary of the models used to describe molecular collisions, internal energy, chemical reaction, and thermal radiation. The gas models for reacting air included 11 chemical species, 41 chemical reactions, 35 electronic states or groups of states, and 26 bound-bound radiative transitions for molecular-band and atomic events. For a more detailed description along with tabulated data, see Refs. 10 and 12.

The variable hard sphere (VHS) model[14] was used for the intermolecular collisions. This is the simplest model that satisfies the basic requirement to model both the coefficient of viscosity and the temperature dependence of this coefficient. The viscosity coefficient was assumed to be proportional to the 0.7 power of temperature, and the molecular diameters at a reference temperature of 288 K were 0.396, 0.407, 0.300, 0.300, and 0.400 nm for O_2, N_2, O, N, and NO, respectively. Molecular diameters for the ions were assumed to be the same as for the five corresponding neutral species. For the electron, the effective elastic diameter generally is assumed to be less than that of the atoms and molecules. A reference diameter of 0.1 nm was chosen for the present study.

For the rotational and vibrational energy modes, the Borgnakke-Larsen[15] model is used. The essential feature of this model is that a fraction of the collisions is regarded as completely inelastic, and for these, new values of the translational and internal energies are sampled from the distributions of these quantities that are appropriate to an equilibrium gas. The remainder of the molecular collisions are regarded as elastic. The fraction of inelastic collisions can be chosen to match the real-gas relaxation rate. For this study, constant relaxation collision numbers of 5 and 50 were used for the rotational and vibrational modes, respectively. The effective number of degrees of freedom in the partially excited vibrational states is calculated from the harmonic oscillator theory.

The procedures for the nonequilibrium chemical reactions are extensions of the elementary collision theory of chemical physics. The binary reaction rate is obtained as the product of the collision rate for collisions with energy in excess of the activation energy and the probability of reaction or steric factor. A form of the collision theory[16] that is consistent with the VHS model has been used to convert these temperature-dependent rate coefficients of continuum theory into collisional energy-dependent steric factors. The reactive cross section is the product of the steric factor and the elastic cross section. The chemical reactions considered in this study consisted of 41 reactions for 11 species, and the data are listed in Ref. 10.

Accompanying the partial ionization of a gas are electronic excitation and thermal radiation. Radiation from bound-bound transitions between electronic states can be significant in 10-km/s flows. The procedures used for calculating the population of electronic states are analogous to the Borgnakke-Larsen model that has proved successful for the rotational and vibrational degrees of freedom. For a specified fraction of the collisions, the electronic states are sampled from the equilibrium distribution appropriate to the effective temperature based on the sum of the relative translational energy and the electronic energy of the molecules in the collision. (Note that this temperature is defined on the basis of the relative energy for the single collision pair.) The specified fraction is related to the ratio of the cross section for electronic state excitation to the elastic cross section. Separate fractions are specified for collisions of each species with neutrals, ions, and electrons. (See Ref. 10 for a tabulation of the fraction of collisions that leads to electronic excitation and the rationale for the selection of the fractions.) Unlike the procedures for the rotational and vibrational modes in which each molecule is assigned a single energy or state, each molecule is assigned a distribution over all of the available electronic states. This overcomes the computational problems associated with radiation from sparsely populated states.

The molecular-band system is the same as that employed by Park and Menees[17] and involves the electronic states of molecular oxygen, neutral and ionized nitrogen, and nitric oxide. Radiation from six molecular-band transitions is included in the simulation with a specified mean time to spontaneous emission for each radiating state. The actual time to emission is exponentially distributed about this mean time.

Because the number of electronic states and radiative transitions is large for the atomic species, both the electronic states and radiative transitions are combined to form a manageable number of groups. Eight groups of electronic states are used for both atomic oxygen and atomic nitrogen.[10] The radiative transitions are divided into seven groups for atomic oxygen and 13 groups for atomic nitrogen. (Reference 10 tabulates specific data concerning the electronic state and radiative transitions.)

The radiation absorption model currently implemented in the DSMC simulation is one in which the photon will be absorbed only by atoms or molecules in the end state of the transition that produced the photon. If the number density of absorbing molecules in n_a and the absorption cross section is σ_a, the probability of absorption of a photon traversing a length $\Delta \ell$ while moving through a cell is

$$\Delta \ell \ n_a \sigma_a$$

Each time a radiation even occurs, the orientation is chosen such that all directions are equally possible, and the trajectory of the photon is followed until it is absorbed in the flow, is absorbed at the surface, or exits from the flow. In the present application, absorption is assumed to be zero.

Conditions for Calculations

The freestream conditions for an altitude of 90 km are as follows: velocity = 9.89 km/s, temperature = 188 K, and density = 3.418×10^{-6} kg/m^3. The freestream mole fractions were 0.209, 0.787, and 0.003 for oxygen, nitrogen, and atomic oxygen, respectively. The surface temperature is assumed constant at 1000 K on the forebody of the aerobrake and constant at 300 K for all surfaces in the shadow of the aerobrake. Also, the surface is assumed to be diffuse with full thermal accommodation and to promote recombination of atoms, ions, and electrons. Recombination probabilities appropriate for the Shuttle thermal protection tiles were imposed for atom recombination, and the ions and electrons are assumed to be fully recombined at the surface.

The AFE vehicle is composed of three basic components: aerobrake, carrier vehicle, and main propulsion unit (Fig. 1a). The aerobrake is a blunt elliptic cone raked off

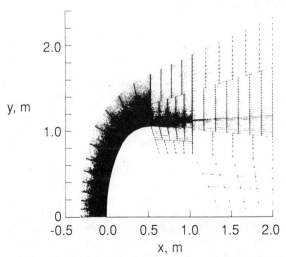

Fig. 2 Computational domain and cell distribution.

at the base and fitted with a skirt-type afterbody. Therefore, the vehicle is a three-dimensional configuration and has an effective diameter of 4.25 m. The present calculations are for an axisymmetric representation of the upper portion of the AFE to examine the maximum heating to the carrier and to reduce the computational effort, although general 3-D DSMC codes[13] currently are being applied to the actual AFE configuration at higher altitude conditions. For the axisymmetric calculations, the geometry shown in Fig. 1b was used.

The computational domain for the DSMC simulation is shown in Fig. 2, where the outermost boundary is displaced sufficiently from the vehicle so as to capture the flowfield disturbance generated by the vehicle. The outflow boundary downstream of the vehicle is treated as a vacuum boundary. The computational domain in subdivided into 23 regions, and each region is further subdivided into computational cells. (Cells are quadrilaterals.) The computational time step and the physical scaling relation (relates the number of simulated molecules to the actual number of physical molecules) are prescribed independently for each region to achieve an adequate number of modeled molecules per cell (order of 10) and to ensure that the computational time step be less than the molecular mean transit time in each cell. The scaling relation and the time step were constrained to preserve flux continuity at the region boundaries. In addition, the cell size was selected such that the dimension normal to the body was less than the local mean free path length (minimum cell dimensions ranged from 2.5×10^{-5} to 0.20 m). A total of 2561 cells was used in the simulation, and the number of simulated molecules was 63,000.

Results and Discussion

The flowfield structure and surface quantities resulting from a DSMC simulation are highlighted to demonstrate the change in these quantities as the flow expands from the stagnation region of the aerobrake to the skirt section and finally to the carrier panel, which is shadowed by the aerobrake. The first section reviews existing calculations for the 90-km case contrasting the continuum and DSMC results. This is followed by a review of the present calculation for the flowfield structure by indicating the extent of thermal and chemical nonequilibrium and the region within the flowfield where radiation emission dominates. Finally, the surface quantities in terms of heating rate, pressure, and shear are presented.

Existing Results for the 90-km Case

The AFE entry condition at 90 km has been examined in several recent studies.[5,6,11,12] Results of Ref. 11 for a five-species dissociating air model show that the local Knudsen number, where the characteristic length is based on the density gradient normal to the body, is greater than 0.1 in the region near the surface and throughout the shock-wave region. Recall that the Navier-Stokes (NS) equations are valid[18] only as long as the Chapman-Enskog theory for the transport properties is valid and that this condition is satisfied if the <u>local</u> Knudsen number <u>based on macroscropic gradients</u> is small compared with unity (about 0.1 or less). This condition suggests that the modeling of the flow at 90 km, even along the stagnation streamline, using the NS equations is deficient for this problem. Evidence of this is suggested in Ref. 5, where stagnation profile quantities obtained using viscous shock-layer (VSL) and NS (shock-capturing) calculations are compared with the DSMC results of Refs. 11 and 12. The comparisons show major differences in density and composition profiles. For example, near the center of the shock wave, the density calculated with the NS model (Ref. 5, Fig. 10a) is about a factor of 2 greater than the DSMC results, and these differences persist throughout most of the flow domain. The fact that the NS value for the density is higher indicates that the ratio of specific heats should be smaller. However, the amount of dissociation is less than the DSMC calculation. The problems with this NS calculation are its inability to describe the shock-wave structure under hypersonic conditions,[19] and also the inconsistency in obtaining the thermodynamic properties for the individual chemical species from curve fits based on the assumption of local thermodynamic equilibrium. Since the population of internal states (rotional, vibrational, and electronic) in a Mach 36 shock wave is quite different from that based on the calculated temperatures of Ref. 5, the ratio of specific heats would be too low and, hence, the density too high. As for the surface heating, the NS value was 58% of the DSMC value for the 90-km entry condition.

Another important comparsion for AFE conditions can be made by comparing the VSL[5] and DSMC[12] stagnation-point convective heating-rate values. The stagnation streamline solutions reported in Ref. 12 were mistakenly reported as being made for a finite catalytic surface. They were actually made for a noncatalytic surface, and this error is carried over in the comparisons of Ref. 5. Once the correction is made, we find that the DSMC and VSL calculations agree the best at perigee conditions (VSL low by about 8%)

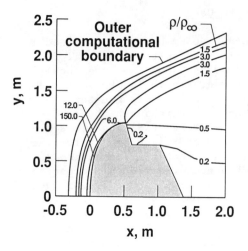

Fig. 3 Density ratio contours.

and experience increasing differences with increasing altitude (VSL low by about 40% at 90 km). Qualitatively, this is the type of behavior one would expect to be caused by the low-density and high-temperature effects giving rise to a highly nonequilibrium flow.

Flowfield Structure

Figure 3 displays selected contours of the local density, expressed as a ratio to the freestream value, for the forebody and near-wake region. The density variation is in excess of three orders of magnitude. In the stagnation region, the compression combined with a relatively cool wall produces a maximum density that is 155 times the freestream value. Because of the expansion about the elliptically blunt nose, the density adjacent to the surface decreases to 112 times the freestream in the last cell on the nose, and then with the vary rapid expansion on the skirt, to a value of 9.5 in the last cell on the skirt. Along the carrier panel, the density is only a small fraction of the freestream value.

Variation of flow properities in the body normal (η) direction is presented in Figs. 4-6 (density and velocity in Fig. 4, translational and internal temperature in Fig. 5, and species mole fractions in Fig. 6), where each figure presents data at three body stations: part a along the stagnation streamline, part b along the last column of cells above the skirt section, and part c above the carrier panel at an x location of 0.8 m.

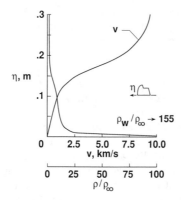

(a) Stagnation streamline.

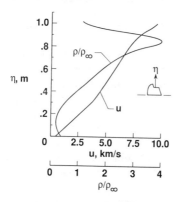

(b) Skirt corner.

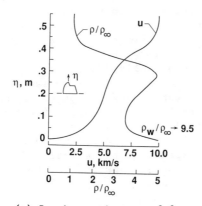

(c) Carrier region, x = 0.8 m.

Fig. 4 Density and velocity profiles.

Along the stagnation streamline, the various profiles show no evidence of a distinct shock wave, only a gradual merging of the shock wave and shock layer. The translational temperature (Fig. 5a) peaks at about 40,000 K, whereas the internal temperature (defined in terms of the rotational and vibrational energies and the number of internal degrees of freedom) has a peak value of about 17,000 K. Thermal nonequilibrium is evident throughout the flowfield. As the flow expands about the skirt (Fig. 5b) and onto the carrier panel (Fig. 5c), the translational temperature decreases by almost a factor of 4, whereas the peak internal temperature decreases to about 60% of its peak stagnation value. Over both the skirt and carrier panel, a substantial portion of the flowfield has a higher internal temperature than the corresponding translational value. Even though the density and temperature decrease substantially as the flow expands onto the skirt (Fig. 4b and 5b), the gas remains highly dissociated as is indicated in Fig. 6, where the mole fraction profiles for only the electron and neutral species are shown. The electron concentration is 2% or less. For the region immediately above the surface, the gas composition consists primarily of three species: atomic and molecular nitrogen and atomic oxygen. The dominant species is atomic nitrogen with a mole fraction value of approximately 0.6. As the gas expands from the skirt corner (Fig 6b) over the carrier panel, mass separation effects are evident where concentration of heavy species (N_2) decreases in relation to the light (atomic species).

Figure 7 shows the radiative emission profiles along body normals at the same three body locations. The peak value at each of the three body stations occurs near the location where the internal temperature is a maximum. In contrast to the results of Ref. 6, the maximum total radiation emission occurs along the stagnation streamline and then decreases substantially downstream of the stagnation region. In fact, the variation is so large that a tangent slab approximation (an approximation often used in calculating the surface radiative heating, but not used in the present calculation) would probably introduce significant errors, particularly on the skirt and carrier panel. Beyond the skirt corner expansion, the emission decreases rapidly with increasing distance downstream of the corner. Also, the emission is confined to a region that is radially above the corner.

Surface Distributions

Figures 8 and 9 present the calculated radiative and convective heating distributions, respectively, along the

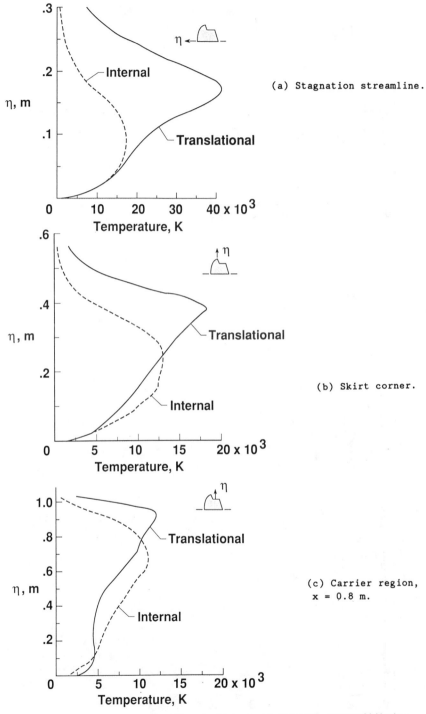

(a) Stagnation streamline.

(b) Skirt corner.

(c) Carrier region, x = 0.8 m.

Fig. 5 Extent of thermal nonequilibrium.

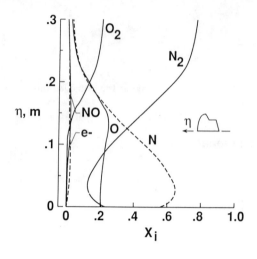

(a) Stagnation streamline.

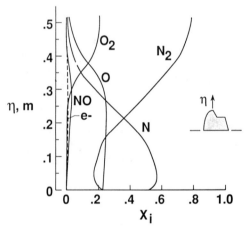

(b) Skirt corner.

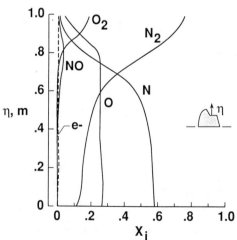

(c) Carrier region, x = 0.8 m.

Fig. 6 Mole fraction profiles of electron and neutral species.

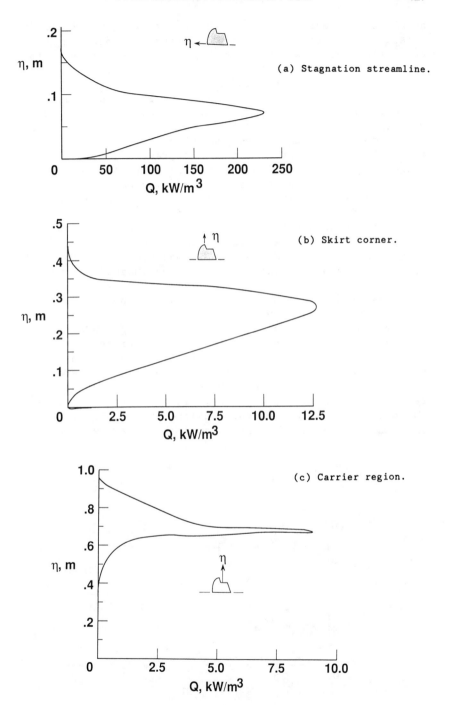

Fig. 7 Radiation emission profile.

aerobrake and carrier panel. In concert with the radiative emission, the surface radiative heating is a maximum in the stagnation region and decreases along the aerobrake. The radiative heat flux to the skirt is only 10% of the stagnation value. For the carrier panel, the radiative flux is about 1% of the stagnation value. For the aerobrake, Fig. 8 presents both the total radiative flux distribution and that resulting from bound-bound transitions where the wavelength is greater than 0.2 μ. The present calculation with no absorption shows that a significant fraction (62% in the stagnation region and 24% on the skirt) of the aerobrake heating originates at the shorter wavelengths. If radiation absorption were included, then the contribution of the shorter wavelength radiation would decrease, since the gas is almost transparent to the longer wavelength radiation, but there is significant absorption of the ultraviolet radiation. This effect has been demonstrated previously[10] with the absorption model currently implemented in the DSMC method.

The present calculation for radiative heating differs from the preliminary axisymmetric results reported in Ref. 12 in several respects. First, the magnitude of the stagnation radiative heating is about two times that reported in Ref. 12. Also, the present calculation shows a significant decrease in heating with increasing distance from the stagnation region, whereas the previous values remained approximately constant. Three differences between the previous and current calculations exist. First, the axisymmetric calculation reported in Ref. 12 used a constant

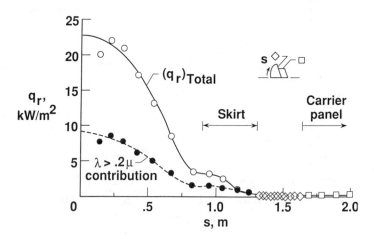

Fig. 8 Radiative heating distribution.

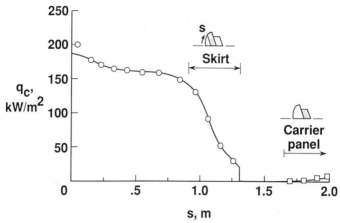

Fig. 9 Convective heating distribution.

absorption cross section of 5.0×10^{-20} m^2. The present calculation used a value of zero. Inclusion of absorption causes a larger reduction in the stagnation region than on the skirt. The second and third differences were that the current calculation used a much smaller cell size and time step. Of more importance for the radiation calculation is the time step, because the time step was about two orders of magnitude smaller, such that it was comparable to the smallest mean radiation lifetime.

In the stagnation region, for example, most of the radiation to the surface is caused by atomic nitrogen (67% from atomic nitrogen, 26% from molecular bands, and 7% from atomic oxygen). In the region where the radiation emission is a maximum, atomic nitrogen has a mole fraction of about 0.55.

Figure 9 presents the convective heating distribution on the aerobrake and carrier panel. The calculation shows that the convective heating decreases gradually from a maximum at the stagnation point on the elliptical nose and then decreases significantly as the flow accelerates about the circular skirt. This trend is consistent with previous DSMC calculations[11] but differs from those reported in Ref. 6, where the maximum heating occurred on the elliptic nose but downstream of the stagnation point. The magnitude of the heating on the carrier panel is small, increasing from about 1% of the stagnation-point value at the carrier-aerobrake juncture to about 5% at the most downstream location.

Concluding Remarks

Through the use of the DSMC method, flowfield structure and surface quantities about an axisymmetric representation of the AFE vehicle have been calculated for entry conditions at 90 km. Results of the calculation which include the effects of ionization and thermal radiation show the following:

(1) Convective heating is dominant, yet thermal radiation is significant.
(2) For zero absorption, a significant portion of the surface radiative flux is due to ultraviolet radiation.
(3) The carrier panel heating rates are small with values ranging from 1 to 5% of the corresponding stagnation-point value.
(4) The flow is highly nonequilibrium with mass separation effects evident in the wake region.

References

[1] *Pioneering the Space Frontier: The Report of the National Commission on Space*, Bantam, New York, 1986.

[2] Walberg, G. D., "Aeroassisted Orbit Transfer-Window Opens on Missions," *Astronautics & Aeronautics*, Vol. 21, Nov. 1983, pp. 36-43.

[3] Walberg, G. D., Siemers, P. M., Calloway, R. L., and Jones, J. J., "The Aeroassist Flight Experiment," IAF Paper 87-197, Oct. 1987.

[4] Gnoffo, P. A. and Green, F. A., "A Computational Study of the Flowfield Surrounding the Aeroassist Flight Experiment Vehicle," AIAA Paper 87-1575, June 1987.

[5] Gupta, R. N. and Simmonds, A. L., "Stagnation Flowfield Analysis for an Aeroassist Flight Experiment Vehicle," AIAA Paper 88-2613, June 1988.

[6] Candler, G. and Park, C., "The Computation of Radiation from Nonequilibrium Hypersonic Flows," AIAA Paper 88-2678, June 1988.

[7] Li, C. and Wey, T. C., "Numerical Simulation of Hypersonic Flow Over an Aeroassist Flight Experiment Vehicle," AIAA Paper 88-2675, June 1988.

[8] Bird, G. A., "Low-Density Aerothermodynamics," *Progress in Astronautics and Aeronautics: Thermophysical Aspects of Re-Entry Flows*, Vol. 103, edited by J. N. Moss and C. D. Scott, AIAA, New York, 1986, pp. 3-24.

[9] Bird, G. A., "Direct Simulation of Typical AOTV Entry Flows," AIAA Paper 86-1310, June 1986.

[10] Bird, G. A. "Nonequilibrium Radiation During Re-Entry at 10 km/s," AIAA Paper 87-1543, June 1987.

[11]Dogra, V. K., Moss, J. N., and Simmonds, A. L. "Direct Simulation of Aerothermal Loads for an Aeroassist Flight Experiment Vechicle," AIAA Paper 87-1546, June 1987.

[12]Moss, J. N., Bird, G. A., and Dogra, V. K., "Nonequilibrium Thermal Radiation for an Aeroassist Flight Experiment Vehicle," AIAA Paper 88-0081, Jan. 1988.

[13]Celenligil, M. C., Moss, J. N., and Blanchard, R. C., "Three-Dimensional Flow Simulation About the AFE Vehicle in the Transitional Regime," AIAA Paper 89-0245, Jan. 1989.

[14]Bird, G. A., "Monte-Carlo Simulation in an Engineering Context," Progress in Astronautics and Aeronautics: Rarefied Gas Dynamics, Vol. 74, Pt. 1, edited by Sam S. Fisher, AIAA, New York, 1981, pp. 239-255.

[15]Borgnakke, C. and Larsen, P. S., "Statistical Collision Model for Monte Carlo Simulation of Polyatomic Gas Mixtures," Journal of Computational Physics, Vol. 18, Aug. 1975, pp. 405-420.

[16]Bird, G. A. "Simulation of Multi-Dimensional and Chemically Reacting Flows," Rarefield Gas Dynamics, edited by R. Campargue, CEA, Paris, 1979, pp. 365-388.

[17]Park, C., "Calculation of Nonequililbrium Radiation in the Flight Regimes of Aeroassisted Orbital Transfer Vehicles," Progress in Astronautics and Aeronautics, edited by H. F. Nelson, Vol. 96 of 1985, pp. 395-418.

[18]Bird, G. A., "Direct Simulation of Gas Flows at the Molecular Level," Communications in Applied Numerical Methods, Vol. 4, March-April 1988, pp. 165-172.

[19]Fiscko, K. A. and Chapman, D. R., "Hypersonic Shock Structure with Burnett Terms in the Viscous Stress and Heat Flux," AIAA Paper 88-2733, June 1988.

Direct Monte Carlo Simulations of Hypersonic Flows Past Blunt Bodies

W. Wetzel* and H. Oertel†

DFVLR/AVA, Göttingen, Federal Republic of Germany

Abstract

This paper reports on the first steps toward the gaskinetic simulation of hypersonic flows past the nose region of re-entry vehicles. For the validation of the direct simulation Monte Carlo (DSMC) method and the molecular dynamics (MD) method concerning the conservation of angular momentum, an analytical solution of the Navier-Stokes equations, the Oseen vortex decay, was chosen. The results obtained with both methods correlate well with the analytical solution. The angular momentum is conserved exactly in the MD method; however, in the DSMC method the box sizes must be small enough to conserve the angular momentum approximately because of the statistical assumptions. In this paper, the developed two- and three-dimensional DSMC-codes have been verified for the hypersonic flow past a circular cylinder and a sphere, respectively. The shock stand-off distance and the pressure distribution along the body surface were compared with appropriate experimental results. In both cases, the agreement is very good. As an application to more realistic configurations, the gaskinetic flow past the nose of a typical spacecraft is discussed for different Knudsen numbers.

Introduction

Because of the development or planning of new spacecrafts such as Hermes,[1] Hotol,[2] Sänger,[3] or future hyper-

---

Copyright © 1989 by the American Institute of Aeronautics and Astronautics, Inc. All rights reserved.
* Dipl.-Ing., Institute for Theoretical Fluid Mechanics.
† Professor, Institute for Theoretical Fluid Mechanics; currently at Institute for Fluid Mechanics, Technical University of Braunschweig, FRG.

sonic airplanes,[4] there is renewed interest in the understanding of hypersonic flows. It is known that the development of the U.S. Space Shuttle mainly took place in wind tunnels or by means of other experimental investigations. The experiences gained in free flight compared with the experimental and theoretical design are described in the literature.[5] These experiences show that quite a few flow phenomena lack basic understanding and require further studies. In contrast to the past design of the Space Shuttle, the Hermes vehicle is planned to be designed with the help of advanced numerical methods in order to reduce the expensive and time-consuming experimental investigations, especially in the rarefied flow regime of the re-entry trajectory.

We are interested in the low-density region in the upper atmosphere, where numerical gaskinetic procedures are available to simulate hypersonic flows. The simulation of high speed and low density results in the deviation of air from a perfect gas behavior because of the excitation of molecular rotation, vibration, and dissociation. At high altitudes and therefore low density, the rate of collisions between the particles is not high enough so that all these processes are at nonequilibrium state. For the simulation of such complicate flow phenomena, models and assumptions are necessary that must be checked separately.

Therefore, in this paper the numerical gaskinetic codes were first applied to perfect gas and simple configurations, where real gas effects do not occur. A further presented study treats the flowfield simulation past a typical spacecraft nose under re-entry conditions. The simulation of gas viscosity and the excitation of internal modes have been approximated with gaskinetic models that are often used in engineering context.[6]

Gaskinetic Procedures

In the transitional flow regime between continuum mechanics and free molecular flow, two gaskinetic simulation procedures, the molecular dynamics (MD)[11] simulation procedure and the direct simulation Monte Carlo (DSMC)[14] method, have been investigated[9] and applied.[16]

In the MD simulation procedure the calculation starts with randomly positioning a given number of particles within the chosen control volume. The initial velocities of the particles are related to the physical problem; they can, for example, have a Maxwell distribution superimposed by a given macroscopic velocity. After the definition of the initial conditions, the particles move with their individual velocities. The change of the motion depends only on the chosen particle potential and the conditions at the boundaries of the computational domain. For the hard sphere potential, which has been assumed for all present MD results, there is no interaction between particles unless they collide with each other. Therefore, the time step must not be chosen larger than the time span between two subsequent collisions. Thus, in principle, for the estimation of the smallest time step all possible collisions need to be checked. Then all molecules are moved according to this smallest time step, and just the single collision is then considered based on the governing equations for fully elastic collisions.

Macroscopic quantities can be calculated at any time by sampling the properties of the particles. The advantage of this procedure is that no grid is required for the calculation, and it can therefore be applied to complex geometries. Although modifications to the MD method will speed up the computations considerably, it remains a very time-consuming method.

The DSMC method is similar to the MD procedure in that the paths of the model particles are determined, and the macroscopic quantities are obtained by sampling the microscopic quantities. The main difference between both methods exists in the treatment of the collisions where statistical assumptions are employed in the DSMC method. Because this results in less computational effort, the DSMC method is preferred. A detailed desciption of the DSMC method is given, for example, by Bird.[14]

Diatomic Simulation Models

The gas viscosity in the DSMC calculations was simulated by using the variable hard sphere (VHS) collision model of Bird.[6] In this model the temperature dependency of the viscosity is considered by a variable cross section that is inversely proportional to the relative collision energy between the colliding particles.

The internal energies of molecules have been considered with the phenomenological model introduced by Borgnakke and Larsen.[7] The essential feature of this model is that a part of the collisions is treated as completely inelastic. The probability for inelastic collisions, where energy transfer between translational and rotational energies may occur, is directly related to the rotation relaxation time. For the temperature range in our calculations, the rotational energy is always fully excited. At higher temperatures the vibrational degrees of freedom become excited.[12] To predict the degree of excitation at a particular temperature, the harmonic oscillator model has been used. The knowledge of the vibrational degrees of freedom enables the consideration of the translational-rotational and vibrational energy exchange in a modified version[13] of the Borgnakke-Larsen model. The proportion of such inelastic collisions is adjusted to the vibration relaxation time.

Validation of the Gaskinetic Procedures

For the validation of the DSMC method and the MD procedure the decay of Oseen's vortex was chosen. This flow allows us to check the sensitivity of the DSMC method with respect to the conservation of angular momentum. The advantage of this test case is that there is an analytical solution of the Navier-Stokes equations to compare with in the transitional regime.

The control volume for the simulation of the vortex decay is sketched in Fig. 1. The characteristic length scale L of the problem is chosen to be the distance between the vortex center and initial location of the maximum of the velocity. At the boundary of the control volume, the model particles are reflected specular, which is physically incorrect. This way, however, the total amount of angular momentum within the control volume must remain constant. Thus, the conservation properties of the considered method can be checked directly. Knowing that the boundary condition is critical, care is taken that the influence of that

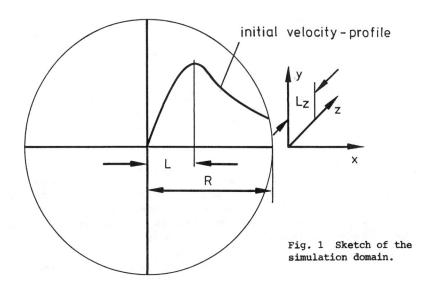

Fig. 1 Sketch of the simulation domain.

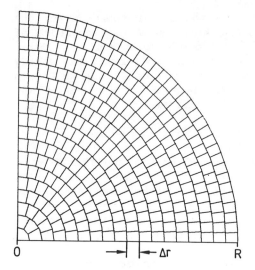

Fig. 2 Monte Carlo cell discretization.

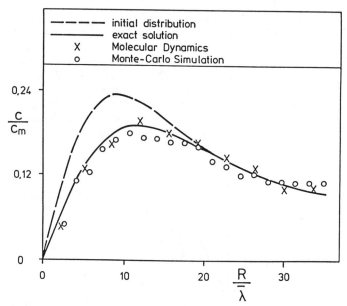

Fig. 3 Velocity distribution for Kn = 0.1.

condition does not penetrate too far in the direction of the vortex center. This extension of the disturbance can be limited by choosing the simulation time approximately. The interaction between the particles is considered with the hard sphere collision model.

In the case of the DSMC method the circular computational domain has been divided into cells as shown in Fig. 2.

All boxes have the same size, and the radial extent of each box is 0.8 mean free paths. The total number of particles is roughly 80,000.

Figure 3 shows for Kn = 0.1 a comparison of results for the MD and the DSMC method with the analytical solution in terms of velocity distribution. The velocity c is related to the most probable thermal speed c_m. In the case of the MD method 10,000 particles are employed for the simulation. The ratio between mean particle distance and particle diameter is 2.5. Eight independent runs have been made, and the corresponding results have been averaged. The agreement between the predicted and the analytical solution is good for both methods. As expected, only near the boundary is some systematic disagreement recognized because of the chosen boundary condition.

After having investigated these two gaskinetic techniques, only the DSMC was used in the following because of its superior computational efficiency.

Verification of the Two- and Three-Dimensional DSMC Method

For the verification of the developed two-dimensional (2D) DSMC code, the adiabatic hypersonic flow past a circular cylinder is simulated, the fluid being argon. The calculations have been carried out with the VHS collision model for Kn = 0.1 and 0.3 and a freestream Mach number of 5.48. Diffuse reflection with complete thermal accommodation served as body surface boundary conditions. Particles are generated at the upstream boundary producing constant freestream conditions. The particles that leave the control volume will be neglected. The stationary state is achieved, if the total number of particles does not change with time. After the stationary state is reached, time averages have been made to reduce the fluctuations in the simulation results. Similar investigations were carried out in 1974.[15] Therein, however, the inverse 12-th-power-law potential was used to simulate the gas flow of argon. Figure 4 shows the velocity distribution along the stagnation streamline in comparison with the simulation results of Crawford and Vogenitz.[15] The shock standoff distance is referenced to the radius R. An additional comparison is the continuum flow approximation, which has been plotted as well into Fig. 4. One recognizes the shock wave building up in front of the body with reduced Kn, with the shock thickness decreasing toward the continuum flow limit. The agreement between our DSMC simulation results and the literature results[15] is good.

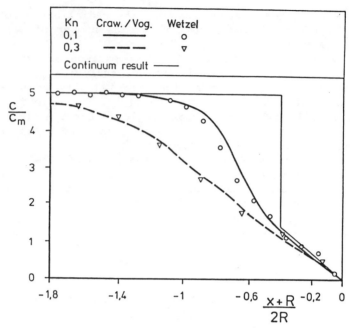

Fig. 4 Velocity profiles along the stagnation streamline. $M_\infty = 5.48$, adiabatic wall.

As another test case the DSMC method was applied to the hypersonic flow past an isothermal cylinder. The simulation of nitrogen molecules is carried out with the VHS collision model for Kn = 0.1, M_∞ = 22.4, and a wall temperature of 293°K. The internal energy exchange of the diatomic gas has been considered with the Borgnakke-Larsen[7] model. This test case was experimentally investigated by Koppenwallner.[9] In Fig. 5 the pressure distribution along the body surface is shown in the angle range of 0-90 deg.. The comparison between the measurements and the simulation results is very good. As expected, the Newtonian theory, which is also plotted in Fig. 5, deviates considerably from the experimental results for larger angles, because the tangential momentum component dominates over the normal component for larger angles.

For the verification of the three-dimensional (3D) DSMC code, the supersonic flow past a sphere was simulated. The sphere serves as test body for the 3D code as the circular cylinder for the 2D code.

In Fig. 6 a comparison of the calculation results with electron beam measurements by Russell[8] is shown for the density distribution along the stagnation streamline. The

MONTE CARLO SIMULATIONS OF HYPERSONIC FLOWS 439

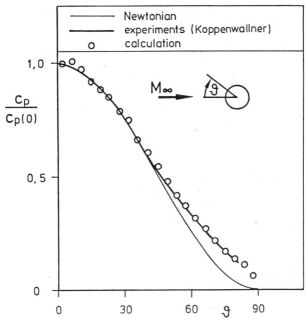

Fig. 5 Pressure distribution along the body surface. $M_\infty = 22.4$, $Kn = 0.1$, $T_w = 293°K$.

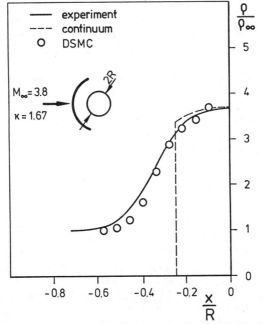

Fig. 6 Comparison of experimental and theoretical density profiles on the stagnation line of a sphere. $Kn = 0.06$, $M_\infty = 3.8$.

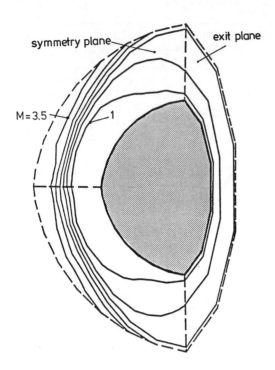

Fig. 7 Lines of constant Mach numbers in a perspective view.

flowfield conditions are $M_\infty = 3.8$, $Kn = \bar{\lambda}/R = 0.06$. At the body surface the adiabatic wall temperature was fixed. The predicted 3D DSMC results are in a good agreement with the experimentally estimated density distribution ahead of a sphere.

Figure 7 shows the Mach-isolines in the calculated 3D flowfield. The sonic line that separates the subsonic and supersonic region looks different in comparison to the continuum mechanic case. The calculated subsonic region is extended around the front part of the sphere and points out a low-density effect.

Application of the Validated Codes for a Typical Spacecraft Nose

In the following paragraphs the 2D flow simulation past the nose of a realistic spacecraft is discussed for different Knudsen numbers in the symmetry plane. The flow conditions are chosen from realistic re-entry parameters, and the boundary conditions are similar to the conditions given earlier. In Fig. 8 the body-fitted grid that we used for the flow simulation at $Kn = 1.5$ is shown. The cell sizes in the normal direction to the body are one mean free path of

MONTE CARLO SIMULATIONS OF HYPERSONIC FLOWS 441

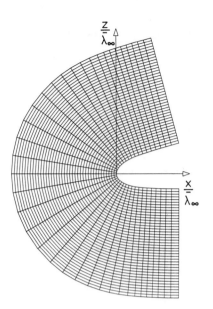

Fig. 8 Body-fitted cell discretization.

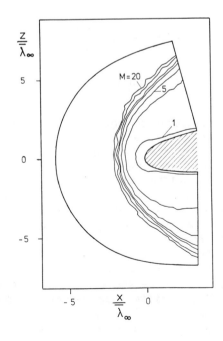

Fig. 9 Lines of constant Mach numbers.

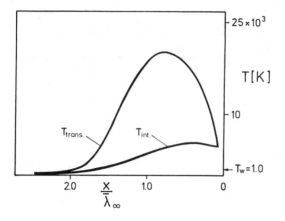

Fig. 10 Temperature distribution along the stagnation streamline.

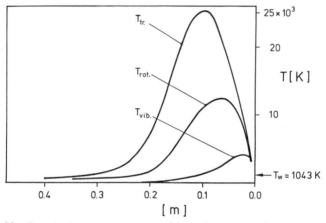

Fig. 11 Temperature components along the stagnation streamline.

the undisturbed gas flow. This mean free path corresponds to an altitude of about 110 km, and the characteristic length is the average nose radius in the symmetry plane. As a first step the gas viscosity of nitrogen molecules has been simulated with the VHS collision model.

Figure 9 shows lines of constant Mach numbers in the symmetry plane of the flow past the typical spacecraft nose. A shock layer in front of the nose is seen. The thickness is about two mean free paths, which corresponds to about 1.5 m. In Fig. 10 the temperature components along the stagnation streamline are plotted. The translation temperature increases across the shock wave to about 20,000°K. The large difference between translational temperature and internal temperature points out a strong nonequilibrium state of the flow near the stagnation point. This difference between the temperature components decreases at lower altitudes, since the collision rate increases because of the higher density.

A further simulation was carried out at the lower Knudsen number of 0.08, which corresponds to an altitude of about 95 km. Again the temperature components along the stagnation streamline are drawn (see Fig. 11). At this

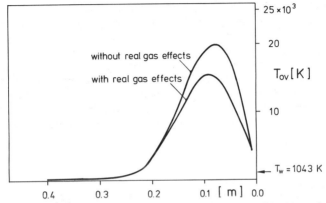

Fig. 12 Overall temperature distribution. $M_\infty \cong 24$, Kn = 0.08.

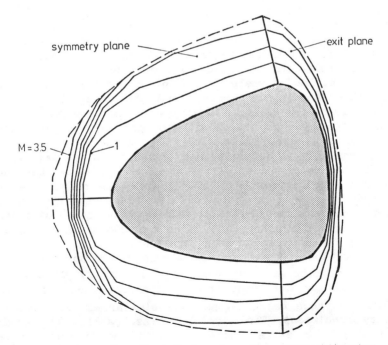

Fig. 13 Lines of constant Mach numbers in a perspective view.

lower altitude the vibration degrees of freedom of the diatomic molecules are excited. Also, the difference between translational and rotational temperature becomes smaller, because of the larger collision rate as was mentioned earlier.

Figure 12 shows the overall temperature difference along the stagnation line, which is due to the excitation of the vibration modes. The overall temperature is defined as the weighted mean of the translational and internal temperature. It can be recognized that, for the chosen parameters, the real gas effect causes a decrease of the temperature maximum of roughly 20%.

The last plot (see Fig. 13) presents a preliminary result for the supersonic flow past the 3D nose of a typical re-entry vehicle. The flow conditions are similar to the conditions of the sphere. To recognize the flow behavior in the symmetry and exit plane, the Mach-isolines are plotted in a perspective view.

Conclusion

This paper describes results of the first step toward the gaskinetic simulation of ralistic hypersonic flowfields past blunt bodies. The validation of the MD procedure and the DSMC method in a vortex flow was carried out. It was shown that the DSMC is as good as the other method even with respect to the conservation of angular momentum in vortical flows, if the parameters of the method are chosen appropriately. For the simulation of hypersonic flowfields the DSMC method is preferred to the more accurate MD method because of the larger computational efficiency.

At first, the 2D and 3D DSMC codes were applied to simulate the hypersonic/supersonic flow past a circular cylinder and a sphere. These simple bodies were investigated experimentally in detail by many authors and are well suited for the verification of the numerical codes because, due the low temperatures ($T \lesssim 1500°K$), real gas models are unnecessary. The comparison of the pressure distribution along the cylinder surface as well as the density distribution ahead of a sphere are in good agreement with the appropriate measurements. As an application to more realistic configurations, the hypersonic flow past the nose of a typical spacecraft was discussed, in the symmetry plane of the flow, for two different Knudsen numbers. The flow conditions in these calculations were similar to the real re-entry conditions. The simulation results show the strong nonequilibrium state across the shock wave in front of the body, which becomes smaller at lower altitudes since the density and therefore the collision rate increase. The preliminary result of a fully 3D simulation was presented. The next step will be the numerical investigation concerning real gas effects in fully 3D hypersonic flows.

References

[1] Langereux, P., "Europe Aims for Space Independence," *Aerospace America*, Vol. 10, Special Rept., Feb. 1986.

[2] DeMeis, R., "HOTOL: The Other Aerospaceplane," *Aerospace America*, July 1986, pp. 10-12.

[3] Vogels, H. A., "Von der Mikrosystemtechnik zum Raumtransporter Sänger II," *Luft- und Raumfahrt*, Vol. 7, No. 4-86, Oct. 1986, pp. 99-108.

[4] Williams, R. M., "National Aero-Space Plane-Technology for America's Future," *Aerospace America*, Nov. 1986, pp. 18-22.

[5] Arrington, J. P. and Jones, J. J. (eds.), "Shuttle Performance: Lessons Learned," NASA CP-2283, 1983.

[6] Bird, G. A., "Monte-Carlo Simulation in an Engineering Context," *Progress in Astronautics and Aeronautics: Rarefied Gas Dynamics*, Vol. 74, AIAA, New York, 1981, pp. 239-255.

[7] Borgnakke, C. and Larsen, P. S., "Statistical Collision Model for Monte-Carlo-Simulation of Polyatomic Gas Mixtures," *Journal of Computational Physics*, Vol. 18, Aug. 1975, pp. 405-420.

[8] Russell, D. A., "Density Disturbance ahead of a Sphere in Rarefied Supersonic Flow," *Physics of Fluids*, Vol. 11, Aug. 1968, pp. 1679-1685.

[9] Koppenwallner, G., "Experimentelle Untersuchung der Druckverteilung und des Widerstands von querangeströmten Kreiszylindern bei hypersonischen Machzahlen im Bereich von Kontinuums- bis freier Molekularströmung," *Zeitschrift fuer Flugwissenschaften*, Vol. 17, 1969, pp. 321-332.

[10] Wetzel, W. and Oertel, H., "Gas-Kinetical Simulation of Vortical Flows," *Acta Mechanica*, Vol. 70, 1987, pp. 127-143.

[11] Meiburg, E., "Comparison of the Molecular Dynamics Method and the Direct Simulation Monte Carlo Technique for Flows around Simple Geometries," *Physics of Fluids*, Vol. 29, Oct. 1986, pp. 3107-3113.

[12] Vincenti, W. G. and Kruger, C. H., *Introduction to Physical Gas Dynamics*, Wiley, New York, 1965.

[13] Bird, G. A., "Direct Molecular Simulation of a Dissociating Diatomic Gas," *Journal of Computational Physics*, Vol. 25, Dec. 1977, pp. 353-365.

[14]Bird, G. A., *Molecular Gas Dynamics*, Clarendon, Oxford, UK, 1976.

[15]Crawford, D. R. and Vogenitz, F. W., "Monte-Carlo Calculations for the Shock Layer Structure on an Adiabatic Cylinder in Rarefied Supersonic Flow," *Rarefied Gas Dynamics*, Deutsche Forschungs- und Versuchsanstalt für Luft- und Raumfahrt Press, Porz-Wahn, FRG, 1974.

[16]Riedelbauch, S., Wetzel, W., Kordulla, W. and Oertel, H., Jr., "On the Numerical Simulation of Three-Dimensional Hypersonic Flow," AGARD CP-428, 1987.

Direct Simulation of Three-Dimensional Flow About the AFE Vehicle at High Altitudes

M. Cevdet Celenligil*
Vigyan Research Associates, Inc., Hampton, Virginia

James N. Moss[†]
NASA Langley Research Center, Hampton, Virginia
and
Graeme A. Bird[‡]
University of Sydney, Sydney, Australia

Abstract

Three-dimensional hypersonic rarefied flow about the Aeroassist Flight Experiment vehicle has been studied using the direct simulation Monte Carlo technique. Results are presented for the transitional flow regime encountered between altitudes of 120 and 200 km with an entry velocity of 9.92 km/s. In the simulations, a five-species reacting real-gas model, which accounts for rotational and vibrational internal energies, is used. The results show that the transitional effects are significant even at 200-km altitude and influence the overall vehicle aerodynamics. For the cases considered, surface pressures, convective heating, flowfield structure, and aerodynamic coefficient variations with rarefaction effects are presented.

Introduction

Over the past 20 years, increasing attention has been given to aeroassisted flights in which a space vehicle is decelerated and/or maneuvered through a planet's atmosphere, by making use of aerodynamic forces rather than an all-propulsive system. The objective of these efforts

---

Copyright © 1989 by the American Institute of Aeronautics and Astronautics, Inc. No copyright is asserted in the United States under Title 17, U.S. Code. The U.S. Government has a royalty-free license to exercise all rights under the copyright claimed herein for governmental purposes. All other rights are reserved by the copyright owner.
   * Research Engineer.
   [†] Research Engineer, Space Systems Division.
   [‡] Professor.

has been to reduce the weight of the propellants carried by the vehicle and to increase the payload. Presently, attention has been focused on applying this idea to reusable aeroassisted space transfer vehicles (ASTV) that would transfer payloads between two orbits in space. Recently, NASA initiated the Aeroassist Flight Experiment (AFE) in order to provide flight data to be used in the design of an ASTV. In this experiment, a subscale model vehicle will be deployed from the Space Shuttle and accelerated by a solid rocket motor to enter the Earth's atmosphere down to an altitude of about 75 km. The vehicle will then exit the atmosphere to be retrieved by the Shuttle. Aerothermal loads will be recorded constantly by the AFE vehicle during this representative flight, which is scheduled for the mid-1990s.[1]

Today, estimation of the aerothermal loads for the design of an ASTV and the AFE vehicle primarily depends on numerical simulations because ground-based facilities fail to simulate the highly nonequilibrium flows about these vehicles. After the AFE experiment, the recorded flight data will be used to validate these simulations.

Analysis

At high altitudes, the mean free path of molecules is comparable with the size of space vehicles, thereby invalidating the use of the Navier-Stokes equations because the continuum assumptions are not applicable. For these flows, the direct simulation techniques appear to be the most feasible methods from the computational point of view.

In this study, the direct simulation Monte Carlo (DSMC) method as developed by Bird[2] is used. This method simulates transitional flows by following the motion of thousands of representative molecules in the flowfield. The molecular motion is computed for small time intervals. Intermolecular collisions appropriate to these time intervals are uncoupled from the molecular motion and are calculated by choosing collision partners on a probabilistic basis. In this study, the collisions are modeled by the variable hard sphere (VHS) model,[3] as is the case with most contemporary DSMC applications. The VHS model treats the molecules as hard spheres, as far as the scattering is concerned, but the collision cross section depends on the relative speed of the colliding partners. A reacting real-gas model with five chemical species (O_2, N_2, O, N, and NO) is used for these computations. Ionization is neglected on the expectation that it will not be significant for the present conditions.

In the three-dimensional (3-D) DSMC applications, the computational domain is divided into a network of deformed hexahedral cells, and each cell is further divided into tetrahedral subcells. The collision partners are selected from the same tetrahedral subcells (rather than from the same cells) to improve the results with regard to the conservation of angular momentum.[4]

Computational Results

In this study, flow about the AFE vehicle has been investigated. The shape of the AFE vehicle is that proposed by the NASA Johnson Space Center (see Fig. 1). The vehicle basically consists of a forebody aerobrake, a hexagonal experimental carrier, and a solid rocket motor, which is ejected prior to atmospheric entry. The axis of the solid rocket motor is inclined at a 17-deg angle with respect to the incoming freestream.

The angle of incidence remains constant for the present solutions. The wind vector was assumed to be parallel to the x axis (Fig. 1), which corresponds to zero incidence for the present coordinate system. (This is equivalent to an angle of incidence of 17 deg for the AFE body axes system in which the x coordinate is coincident with the solid rocket motor axis.)

The forebody of the AFE vehicle is generated by a 60-deg elliptic cone raked at a 73-deg angle. Its nose is blunted in the shape of an ellipsoid, and the aft skirt is appended such that at each meridional plane it is tangent

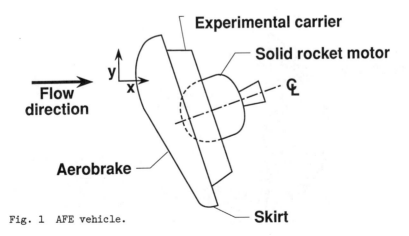

Fig. 1 AFE vehicle.

to the cone by a circle of fixed radius. The base length of the vehicle is 4.25 m. Cheatwood et al.[5] describe mathematically the shape of the AFE forebody and have written a computer program to generate the forebody surface coordinates. Their program has been incorporated in these computations to describe the forebody surface grid. The coordinate system used in this study is located at the tip of the elliptic cone such that the x axis becomes parallel to the freestream velocity, and the x-y plane is located on the symmetry plane, as shown in Fig. 1. According to this system, the geometric stagnation point of the AFE vehicle lies at $x = 0.3$ m, $y = 0$, and $z = 0$.

In general, calculations for high altitudes need to be performed on large computational domains where the cells can be quite large because the molecular mean free path, and, consequently, the disturbance field in front of the body, is large. For example, the computational domain in front of the body starts at $x = -4.7$ m for the 120-km-altitude case, at $x = -10.3$ m for the 140-km-altitude case, and at $x = -18$ m for the 150-km- and higher-altitude cases. The 130-km-altitude computations show that the afterbody portion does not affect the vehicle aerodynamics; hence, for the 140-km and higher altitudes, only the forebody domain is used. In each case, only half of the geometry is considered because of the plane of symmetry.

Figure 2 shows the computational grid used for the 120-km-altitude case. In this figure, both the cells and subcells are shown on the outer freestream boundary. However, on the plane of symmetry, only the cell structure is drawn for clarity. For all of the cases considered, the cells and subcells are matched exactly with the neighboring ones throughout the computational domain. For the computa-

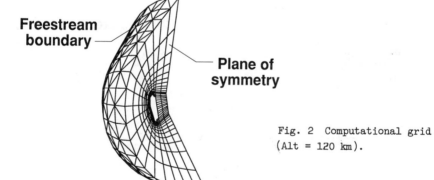

Fig. 2 Computational grid (Alt = 120 km).

tions where both the forebody and afterbody are considered (i.e., the 120- and 130-km-altitude cases), a total of 1,820 cells is used, and these flows are simulated using about 55,000 molecules. The forebody domain is divided into 600 cells, and the 140-km- and higher-altitude cases are simulated using about 20,000 molecules. In all cases, the AFE forebody surface consists of 100 triangles, or, in other words, 50 distorted rectangles with 10 divisions in the longitudinal direction (see Fig. 2) and five divisions in the spanwise direction. Recently, computations of this study have been repeated using a finer surface grid with twice as many cells both in longitudinal and spanwise directions, and the results have been confirmed that they are grid independent.

At the freestream boundaries, atmospheric conditions appropriate for the particular altitude are used. Table 1 lists the atmospheric conditions for altitudes between 120 and 200 km. The uniform forebody temperature is based on free-molecule radiative equilibrium heat transfer to the stagnation point. This corresponds to forebody wall temperatures of 230, 300, 370, 425, 500, 650 K, for the altitudes of 200, 170, 150, 140, 130, 120 km, respectively. The AFE vehicle is insulated with very low conductivity materials. The afterbody wall temperature is, therefore, assumed to be at a uniform 300 K for all cases. At the body surface, full thermal accommodation with diffuse reflection is assumed for the gas-surface interaction.

Once steady state was achieved in the simulations, computations were continued for an additional 8,000 time steps, and samples were taken every other time step to arrive at the time-averaged results. Examples of the time-averaged results presented in Figs. 3-12 are based on these 4,000 samples. Figures 3-6 show pressure and heat-transfer-rate distributions on the forebody surface. Both

Table 1 Freestream conditions

Altitude, km	Density, kg/m^3 x 10^9	Temperature, K	Mole fractions			Mean free path,
			X_{O_2}	X_{N_2}	X_O	
200	0.3	1026	0.031	0.455	0.514	197.00
170	0.9	892	0.044	0.548	0.408	75.61
150	2.1	733	0.055	0.616	0.330	30.93
140	3.9	625	0.062	0.652	0.286	16.92
130	8.2	500	0.071	0.691	0.238	7.72
120	22.6	368	0.085	0.733	0.183	2.69

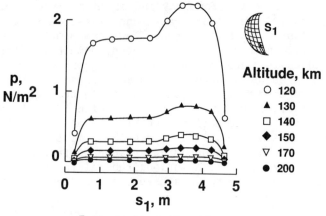

a) Longitudinal direction.

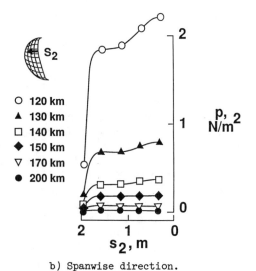

b) Spanwise direction.

Fig. 3 Surface pressure distributions.

pressure and surface heating have their maxima slightly below the geometric stagnation point, are fairly uniformly distributed along the forebody surface (this was one of the design aims), and quickly decrease along the skirt.

Figures 7-9 show various flowfield contours in the x-y plane for the 120-km case. In these figures, the vehicle is half a cell size larger in all directions. The distortion in the picture is greatest in the afterbody, where rather large cells are used. Clearly, the flowfield

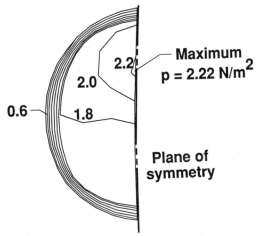

Fig. 4 Surface pressure contours (Alt = 120 km).

density reaches its maximum at the body surface, whereas the maximum translational (kinetic) temperature occurs somewhere away from the surface in the upstream direction. On the other hand, the internal temperature shows a rather complicated picture and is discussed in the next section.

Figures 10-12 present the aerodynamic coefficients (C_D = drag coefficient, C_L = lift coefficient, L/D = lift-to-drag ratio) at selected altitudes for an angle of incidence of 0 deg (using the present coordinate system shown in Fig. 1). The forces are normalized with respect to $1/2 \, \rho_\infty \, U_\infty^2 \, A_{ref}$ where ρ_∞ and U_∞ are the freestream density and velocity, respectively; and A_{ref}, the reference area, equals 14.1 m^2. These figures also show the calculated free-molecule and modified Newtonian results, along with experimental wind-tunnel data. These experiments were conducted in the NASA Langley Research Center Mach 10 air and Mach 6 CF_4 (Freon) wind tunnels using high-fidelity models.[6] Clearly, the DSMC results approach the free-molecule limit very slowly at higher altitudes, and even at an altitude of 200 km, the flow is not completely collisionless. Prior to this study, it was generally acknowledged that free-molecule flow existed for the AFE vehicle for altitudes near 150 km, but this study shows that the transitional effects are significant at these altitudes and influence the overall aerodynamic coefficients. Figures 10-12 also contain the results of the Lockheed bridging formula, which empirically connects the

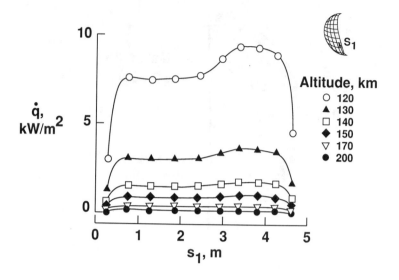

a) Longitudinal direction.

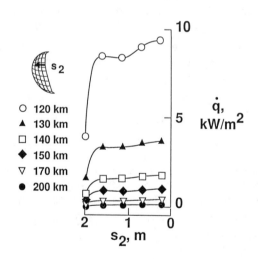

b) Spanwise direction.

Fig. 5　Surface heat-transfer-rate distributions.

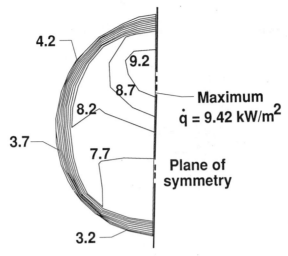

Fig. 6 Surface heat-transfer-rate contours (Alt = 120 km).

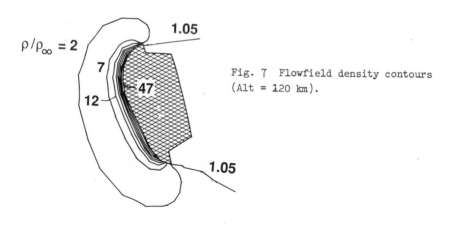

Fig. 7 Flowfield density contours (Alt = 120 km).

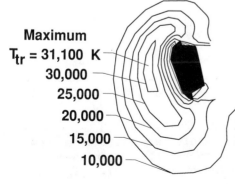

Fig. 8 Flowfield translational temperature contours (Alt = 120 km).

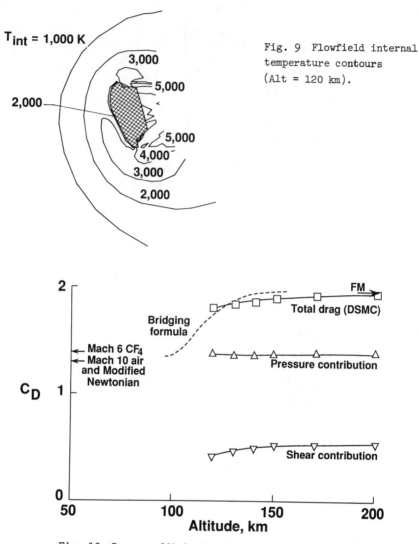

Fig. 9 Flowfield internal temperature contours (Alt = 120 km).

Fig. 10 Drag coefficient variation with altitude.

axial and normal aerodynamic force coefficients between the continuum and free-molecule limits. This is accomplished with a sine-square function by assuming continuum flow at a Knudsen number $Kn_\infty = 0.01$ and free-molecule flow at $Kn_\infty = 10$, which correspond to altitudes of 90 and 150 km, respectively. The bridging-formula results are plotted to show the general trend even though its results are erroneous for the conditions considered in the present study. Figures 10

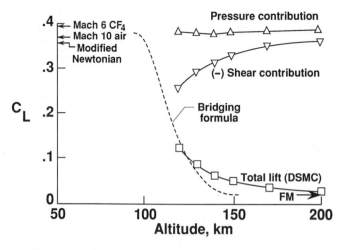

Fig. 11 Lift coefficient variation with altitude.

and 11 also show effects of the pressure and shear forces on the total drag and lift coefficients. Clearly, the pressure contributions are fairly constant at all altitudes and are approximately equal to the measured continuum wind-tunnel values. However, the shear contribution increases with altitude and, in effect, increases the drag and reduces the lift. The pressure coefficient at the stagnation point is 1.994 for the 120-km case and 2.033 for the 200-km case. The corresponding limiting value for the continuum regime is 1.839, and the free-molecule result is 2.051.

One of the design aims of the AFE was to have the vehicle trim at 0-deg angle of attack. At this attitude, therefore, the pitching moment should be zero. The computed pitching-moment coefficients with respect to the center of gravity are less than 10^{-2}, and unlike the drag and lift, are sensitive to the surface grid chosen. Considering the relative coarseness of the grid and the scatter due to the small sample, it is more meaningful to report the equation of the line along which the resultant force acts. For the 120-km case, this line is $y = 0.07x - 1.031$, according to the coordinate system shown in Fig. 1, and varies smoothly with altitude to $y = 0.013x - 0.984$ for the 200-km case.

Discussion

The computations in this study have been performed using a general 3-D DSMC computer program written by

G. A. Bird. This program has already been applied to various wedge flow problems by the present authors.[7] Through advancements in computer technology, 3-D DSMC applications recently have become possible and various cylinder/cone problems have been studied by other groups as well.[8] However, this is the first time that a general 3-D DSMC code has been used to describe the flow about a full-scale space vehicle.

In this study, the 3-D transitional flow about the actual AFE vehicle was investigated. This problem was studied previously using an axisymmetric DSMC code between altitudes of 90 and 130 km.[9] Because higher-altitude flows are, in general, easier to compute, initial 3-D efforts concentrated on the 130-km case, first without and then with the afterbody. It should be noted that the 3-D shape of the AFE vehicle is exactly generated without any approximations. In the computations, only the forebody and the experimental carrier were included because the solid rocket motor is ejected during entry near 130 km. Above this altitude the afterbody has no effect on the aerodynamics.

In analyzing these results, one must remember that these flows are in a highly nonequilibrium state and that these macroscopic properties are obtained by averaging over various classes of molecules whose distribution functions are separated widely in velocity space. Therefore, it is not possible to interpret properly these kinetic tempera-

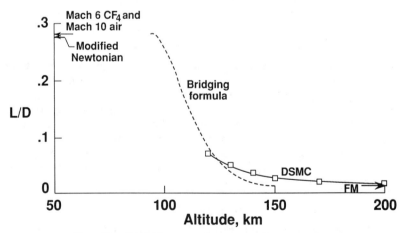

Fig. 12 Lift/drag variation with altitude.

ture contours entirely from the viewpoint of classical thermodynamics.

The complicated internal energy distribution shown in Fig. 9 is the result of the three distinct classes of molecules present in the flowfield. First, at these altitudes, there are quite a few freestream molecules with low internal energies crossing the entire domain without any collisions. Second, there is the effect of intermolecular collisions. Although there are very few collisions, the internal energies of the molecules that have undergone a collision are altered drastically because of the energy exchange with the high translational energies of the gas molecules. Third, there is a class of molecules that has been reflected from the body surface. These molecules have low internal energies because the surface temperature is low, and it is assumed that these molecules are reflected with full thermal accommodation. The relative abundance of these three classes of molecules varies with location in the flowfield, and their mixing produces a highly nonequilibrium flow. Figure 9 shows the effects of this on the macroscopic flowfield properties. As the flow approaches the body along the stagnation streamline, the average internal energy first increases as a result of intermolecular collisions and then decreases because of the increase in the concentration of surface-reflected molecules. As the molecules move around the body, they have more opportunity to have collisions, and since the relative abundance of the collided molecules increases, so does the average internal energy. In the wake region, the molecular number density is extremely small due to the "shadow effect" of the forebody. Consequently, this region has basically one class of molecules; they are the ones that have undergone a collision. Because this class of molecules has high internal energies, the average internal energies in the wake region tend to be higher.

As the altitude increases to 200 km, the internal temperature picture changes entirely. This is not surprising, because at this altitude the freestream temperature is higher, the body surface temperature is cooler, and, more importantly, the number of intermolecular collisions is much lower. In other words, the flowfield has basically two classes of molecules, namely, freestream and surface-reflected molecules. With full thermal accommodation, the surface-reflected molecules have low internal energies due to the cool surface temperature, and naturally, their concentra- tion is high near the body. Consequently, the average internal energy in the forebody region continuously decreases as the flow approaches the body.

For 10-km/s flows under highly rarefied conditions, the assumption of full thermal accommodation is almost certainly incorrect. As stated before, for most cases considered in this study, there are quite a few freestream molecules that hit the body surface with very large translational energies. According to the full thermal accommodation assumption used in this study, all of the energy of the molecules is converted to surface heating, and the molecules are reflected with low internal energy that is characteristic of the surface temperature. However, this assumption is probably unrealistic, and it is quite likely that some of the translational energy appears in excited internal modes during surface reflection. Actually, this phenomenon already has been observed on the Shuttle Orbiter.[10] Unfortunately, the mechanism underlying this complicated interaction and the nature of the energy exchange are not known. These questions need to be clarified in future experiments. Until then, the internal temperature results of this study should be interpreted with reservation. At lower altitudes, this phenomenon is less significant because the molecules that hit the surface have less translational energy due to intermolecular collisions. Even though the nature of the gas-surface interaction could significantly alter the internal temperatures, it should not affect the translational temperatures and the aerodynamic coefficients because the translational energy is most likely fully accommodated to the surface.

Summary and Conclusions

This study applies the direct simulation Monte Carlo technique to assess the 3-D flow about the Aeroassist Flight Experiment (AFE) vehicle. For altitudes of 120 to 200 km, the results show the following:

(1) The flow approaches the free-molecule limit very gradually at higher altitudes, and even at 200 km, the flow is not completely collisionless.

(2) Maximum surface pressure and heating occur slightly below the geometric stagnation point of the vehicle.

(3) At altitudes of 130 km and above, the effect of the AFE afterbody on the vehicle aerodynamics is negligible.

(4) Dissociation is negligible.

For the cases investigated, radiation is negligible and is not considered. At these altitudes, the aerothermal loads on the vehicle are small, and the peak loads and heating occur at much lower altitudes. This should not overshadow the significance of the present study because these results are the first 3-D solutions to the AFE flow problem in the rarefied regime, and they show that the transitional effects are significant even up to an altitude of 200 km. Results such as these are essential for the proper interpretation of the high-altitude aerodynamic measurements.

Acknowledgments

The authors wish to express their thanks to Robert Blanchard and Jim Jones for their many suggestions, and to Neil Cheatwood for the use of his computer program in generating the forebody surface grid.

References

[1] Walberg, G. D., Siemers, P. M., III, Calloway, R. L., and Jones, J. J., "The Aeroassist Flight Experiment," 38th International Astronautical Federation Congress, Paper IAF 87-197, Oct. 1987.

[2] Bird, G. A., Molecular Gas Dynamics, Clarendon Press, Oxford, 1976.

[3] Bird, G. A., "Monte-Carlo Simulation in an Engineering Context," AIAA Progress in Astronautics and Aeronautics: Rarefied Gas Dynamics, Vol. 74, Pt. 1, edited by Sam S. Fisher, AIAA, New York, 1981, pp. 239-255.

[4] Bird, G. A., "Direct Simulation of Gas Flows at the Molecular Level," Communications in Applied Numerical Methods, Vol. 4, 1988, pp. 165-172.

[5] Cheatwood, M. F., DeJarnette, F. R., and Hamilton, H. H., II, "Geometrical Description for a Proposed Aeroassist Flight Experiment Vehicle," NASA TM-87714, July 1986.

[6] Wells, W. L., "Wind Tunnel Preflight Test Program for Aeroassist Flight Experiment," AIAA Paper 87-2367-CP, Aug. 1987.

[7] Celenligil, M. C., Bird, G. A., and Moss J. N., "Direct Simulation of Three-Dimensional Hypersonic Flow About Intersecting Blunt Wedges," AIAA Paper 88-0463, Jan. 1988.

[8] Riedelbauch, S., Wetzel, W., Kordulla, W., and Oertel, H., Jr., "On the Numerical Simulation of Three-Dimensional Hypersonic Flow," AGARD Conference Preprint No. 428, Bristol, U.K., Apr. 1987.

[9] Dogra, V. K., Moss, J. N., and Simmonds, A. L., "Direct Simulation of Aerothermal Loads for an Aeroassist Flight Experiment Vehicle," AIAA Paper 87-1546, June 1987.

[10] Mende, S. B., "Experimental Measurement of Shuttle Glow," AIAA Paper 84-0550, Jan. 1984.

Knudsen-Layer Properties for a Conical Afterbody in Rarefied Hypersonic Flow

G. T. Chrusciel* and L. A. Pool†
Lockheed Missiles and Space Company, Inc., Sunnyvale, California

Abstract

Nonequilibrium flow properties in the Knudsen layer have been obtained from a direct simulation Monte Carlo code; profiles of heat transfer, viscous stress, normal stress, mean molecular speed, and collision frequency are provided for flow over a cold-wall afterbody of a sphere-cone configuration. Freestream conditions are for hypersonic transitional rarefied flow at a Knudsen number of 0.24. Slip effects on tangential velocity and temperature obtained from mean flow profiles closely replicate the nonequilibrium behavior of the translational-to-rotational temperature ratio. Breakdown of the continuum description for heat transfer and viscous stress using Monte Carlo flow properties is demonstrated. Implications to slip modeling are provided.

Nomenclature

$\underline{C}$ = total molecular velocity
$\overline{C'}$ = mean molecular speed
$\overline{C'}_{eq}$ = $2(2R\,T_{av}/\pi)^{1/2}$
C_p = specific heat at constant pressure
C_f = skin-friction coefficient, = $2\tau_t/\rho_\infty U_\infty^2$
d_{eff} = effective molecular diameter
H = enthalpy
k = coefficient of thermal conductivity
M = Mach number
n = number density
$\underline{P}$ = scalar pressure, = $\rho(\overline{C^2})/3$
$\overline{P}$ = nondimensional normal stress or pressure, = $P_n/\rho_\infty U_\infty^2$

Copyright © 1988 by the American Institute of Aeronautics and Astronautics, Inc. All rights reserved.
*Staff Engineer, Aerothermodynamics Department.
†Research Specialist, Aerothermodynamics Department.

P_{ij} = pressure tensor, = $\rho \overline{C_i C_j}$
Pr = Prandtl number
$\dot{q}$ = heat-transfer rate, = $(\rho C_i C^2/2) + n C_i \epsilon_{int}$
R = gas constant
Re = freestream Reynolds number, = $\rho_\infty U_\infty R_N / \mu_\infty$
R_N = nose radius
S = surface distance from stagnation point
S_T = Stanton number, = $\dot{q}_n / (\rho_\infty U_\infty)(H_{o,\infty} - H_w)$
S_r = speed ratio
T = temperature, = T_{av} for Monte Carlo results
T^r = T/T_{∞}
T_{av} = $(3 T_{tr} + 2 T_{rot})/5$
U = freestream velocity
$\overline{U}$ = $u/U \cos \theta_s$
u = velocity tangent to body surface
Y = distance normal to body surface
γ = ratio of specific heats
δ_{ij} = Kronecker delta: = 1 for i=j; = 0 for i ≠ j
ϵ_{int} = internal energy
θ_s = body surface angle relative to velocity vector
λ = molecular mean free path, = $\overline{C'}/\nu$
μ = viscosity
ν = collision frequency
ν_{eq} = $(\rho/\rho_\infty) n_\infty 2\pi d_{eff} (4 R T_{av}/\pi)^{1/2}$
ρ = mass density
τ = shear stress
τ_{ij} = viscous stress, = $-(\rho \overline{C_i C_j} - P \delta_{ij})$

Subscripts

eq = equilibrium value
i,j = components along the x,y, and z coordinate axes
n = evaluated normal to surface
rot = rotational energy mode
s = Knudsen-layer properties near the surface
t = evaluated tangent to surface
tr = translational energy mode
w = wall conditions
o,∞ = stagnation freestream values
∞ = freestream value

Superscript

− = mean value

Background

Previous studies[1] employing a kinetic-theory Monte Carlo (MC) code for rarefied transitional flow over a cold spherical nose at hypersonic speeds provided detailed resolution of flow quantities within the Knudsen layer for $\lambda_\infty/R_N = 0.24$. Slip velocity, temperature jump, and delineation of thermodynamic nonequilibrium regions differed significantly from currently available slip models used with continuum-flow viscous shock-layer codes. Breakdown of the continuum description for heat transfer and shear stress in the Knudsen layer was demonstrated, using the MC temperature and velocity profiles together with a variable hard sphere (VHS) molecular model for transport properties of the gas. These results clearly indicate the deficiency of present modeling of rarefied flow effects in continuum codes assuming the gas to be perturbed slightly from thermodynamic equilibrium.

Subsequent MC calculations were performed for the flow downstream of the nose in a region where rarefied effects should be more pronounced. The purpose of this study was to derive detailed flow characteristics in the Knudsen layer over an afterbody that could serve as the basis for evaluating existing kinetic models.

Approach

The MC kinetic code[2] developed by McGregor and Bird was used to provide details of flowfield properties, and new

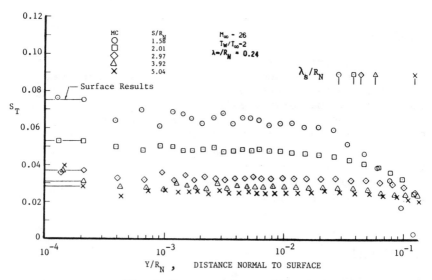

Fig. 1 Heat-transfer profiles.

sampling routines were added to provide terms required for
determination of the viscous stress coefficient (C_f), normal
stress coefficient ($\bar{P}$), and heat-transfer coefficient (S_t).
Radial-cell thickness and spacing were selected to provide
high resolution of the flow gradients in the shock layer, the
outer disturbance layer, and the Knudsen layer. Knudsen-
layer definition was provided by approximately eight cells in
the stagnation region, and 19 cells aft of the nose-cone
junction. Cell structure consisted of a 60-axial by 50-
radial mesh, with 30 axial cells along the conical afterbody.

A modified weighting factor[3] option was employed that
inserts or deletes molecules to compensate for the large
ratios of cell volumes and number density distribution
throughout the flowfield. Previous MC solutions[4] were used
to estimate the number density distribution. Discussion of
the expressions used are provided in the initial study of
this flowfield.[1] Minima of five simulated molecules were
obtained in the Knudsen layer with this method; maxima of
20 molecules were realized in the outer flow region. These
numbers were sufficient to provide a realistic collision
frequency.

The Borgnakke-Larsen[5] molecular collision model is used
to determine the energy exchange in collisions; a VHS molecular model described by Bird[6] determines the molecular properties of monatomic and diatomic gases. Results obtained in
this study are for hard sphere modeling for a perfect diatomic gas ($\gamma = 1.4$), diffuse reflection with thermal accom-

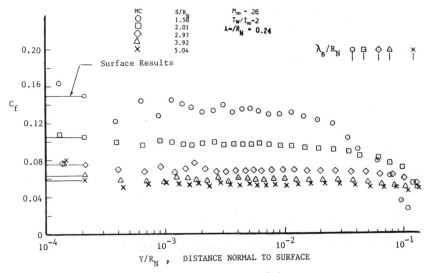

Fig. 2 Skin-friction profiles.

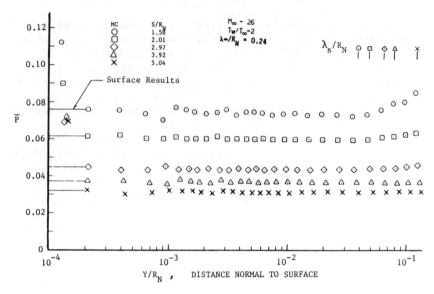

Fig. 3 Normal-stress profiles.

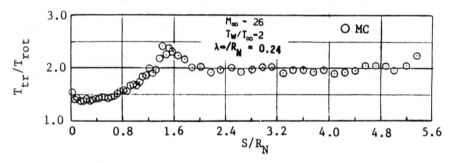

Fig. 4 Near-surface temperature ratio distribution.

modation for molecules striking the body surface, and zero angle of attack (axisymmetric flow).

Continuum solutions of the parabolized Navier-Stokes equations, using the three dimensional viscous shock layer (VSL3D) flow code,[7] were obtained[4] for boundary conditions simulating 1) rarefied flow effects of a viscous shock over the nose and afterbody ("shock slip"), and 2) surface temperature jump and velocity slip over the nose ("body slip"). VSL3D predictions are compared with MC results and an evaluation of first-order terms in the Chapman-Enskog[8] formulation for heat transfer and viscous stress using the MC data, i.e.,

$$\dot{q} = -k\left(\frac{\partial T}{\partial Y}\right)_s$$

$$\tau = \mu\left(\frac{\partial U}{\partial Y}\right)_s$$

where $k = \mu\, Cp/Pr$ and μ are determined from the VHS molecular model.[6] $Pr = 0.71$ was utilized for this analysis, consistent with assumed properties for the VSL3D solutions.

Results

The geometry is a spherically blunted nose with a conical afterbody segment ($\theta_s = 7.0$ deg, $S/R_N = 6$); freestream conditions are $\lambda_\infty/R_N = 0.24$; $Sr = 21.8$; Mach number = 26; wall-to-freestream temperature ratio of 2.0 ($\bar{T}_w = 0.0148$); Re = 197.

Knudsen-Layer Profiles

Smooth heat-transfer (S_t) profiles for cells normal to the surface were obtained at all locations aft of the sphere-cone juncture (Fig. 1). Similar trends were observed in the viscous stress (C_f) shown in Fig. 2 and normal stress ($\bar{P}$) profiles (Fig.3) in the Knudsen layer calculated from the molecular velocity terms at the final simulated-flow time. (Total molecular velocity terms were used in formulating P, τ, and $\dot{q}$ to include the effects of finite mass average velocity.) Of importance to rarefied flow modeling was the relatively constant values of properties within the Knudsen layer demonstrated in the results. With the exception of $\bar{P}$ in the first cell above the surface, flowfield S_t, C_f, and $\bar{P}$ results were in good agreement with corresponding surface values. It should be noted that outside the Knudsen layer, S_t and C_f results over the afterbody decreased rapidly with radial distance above the surface, whereas $\bar{P}$ indicated a gradual increase. These results correspond to flowfield gradients in the shock layer arising from the spherical nose and the cold wall.

The ratio of translational (T_{tr}) to rotational temperature (T_{rot}) is a direct indication of thermodynamic nonequilibrium; however, with the exception of results in the first cell near the surface, the temperature ratio profiles are essentially constant in the Knudsen layer at any given axial location over the afterbody. Consequently, only the near-surface distribution of temperature ratio, representative of the Knudsen layer, is presented in Fig. 4. Thermodynamic nonequilibrium effects are illustrated in the Knudsen profiles shown in Fig. 5 for the collision frequency (ν) and Fig. 6 for the mean thermal speed ($\bar{C'}$). Equilibrium values for and C' were determined from Bird[6] for hard sphere molecules using the MC T_{av} and number density. A large change in

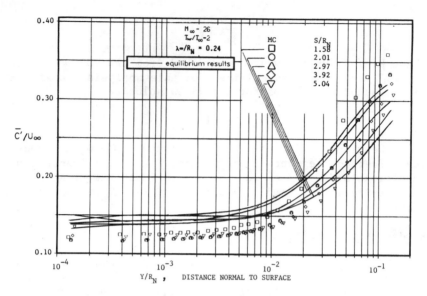

Fig. 5 Collision frequency profiles.

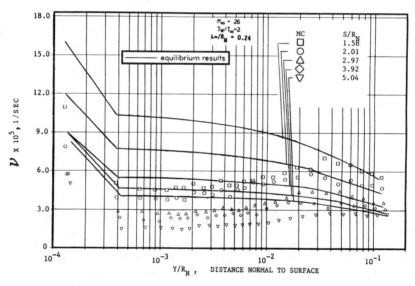

Fig. 6 Mean thermal speed profiles.

for the first cell above the surface corresponds to the rapid change in density as the flow approaches accommodation to the relatively cold wall. Similar behavior was observed in the normal stress profiles (Fig. 3) and temperature ratios. Near the stagnation region, ν approached equilibrium conditions[1]; however, the low-density conditions over the afterbody produced significant translational nonequilibrium temperature results (Fig. 4) with attendant departure of ν from equilibrium near the surface. $\overline{C'}$ profiles are similar to equilibrium characteristics; this result is consistent with the hard sphere molecular model, where the collision cross section is constant, providing the mean thermal speed proportional to the mean temperature. Equilibrium values are approached for ν and $\overline{C'}$ near the edge of the Knudsen layer.

Mean Flow Profiles and Slip Effects

Flowfield profiles for the mean temperature ($\overline{T}$) and tangential velocities ($\overline{U}$) are compared in Figs. 7 and 8 with VSL3D predictions[4] using shock-slip boundary conditions. Significant differences between the present MC $\overline{T}$ results and the continuum predictions (Fig. 7) occur in the Knudsen layer; temperature jump is demonstrated by the difference in the wall value ($\overline{T}_w$ = 0.015) and MC $\overline{T}$ near the surface. Slip velocity is apparent as the mean of the MC mass average tangential velocity ($\overline{U}$) profiles near the surface (Fig. 8). The small magnitudes of temperature jump and slip velocity are relatively constant over the afterbody, contrasting the expected trend of increasing values with decreasing density (increasing λ) as the flow expands over the afterbody. Mean $\overline{U}$ flow profiles toward the Knudsen-layer edge are in reasonable agreement with VSL3D results for no-slip boundary conditions, particularly at the more aft locations. Mean MC $\overline{T}$ profiles, however, diverge from the VSL3D results near the Knudsen-layer edge on the afterbody (Fig. 7); this contrasts the reasonable continuum comparisons found for flow on the forward portion of the spherical nose,[1] demonstrating the

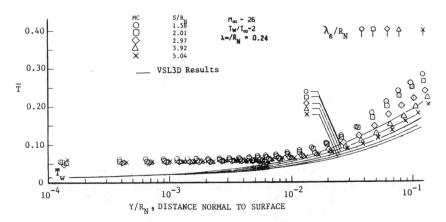

Fig. 7 Temperature profiles.

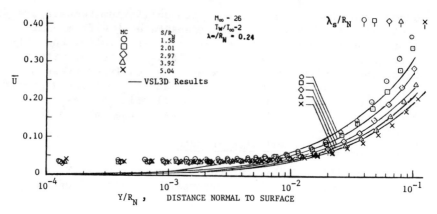

Fig. 8 Tangential velocity profiles.

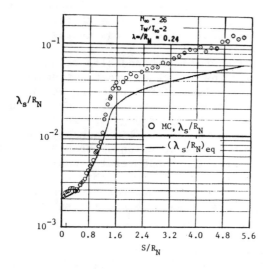

Fig. 9 Near-surface mean free path.

more pronounced thermodynamic nonequilibrium effects on the afterbody.

The distribution of Knudsen-layer thickness (λ_s/R_N) near the surface together with equilibrium results are shown in Fig. 9. The entire body is shown to demonstrate an approximate factor of 50 in magnitude and more pronounced nonequilibrium effect on λ_s between the stagnation and aft end of the afterbody. Note that near the rear of the body, the local value of mean free path approaches 50% of the freestream, which contrasts the small magnitude of λ_s in the stagnation region. The distribution of near-surface tangential gas velocity ($\bar{U}_s$, slip velocity), and mean temperature ($\bar{T}_s$), determined from MC profiles are presented along the

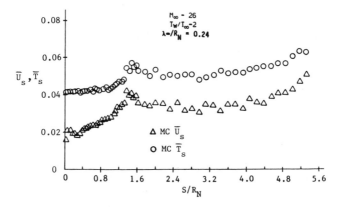

Fig. 10 Near-surface slip velocity and temperature distributions.

entire body surface in Fig. 10. The similar behavior of $\bar{U}_s$ and $\bar{T}_s$ with the temperature ratio (provided in Fig. 4) illustrates the thermodynamic nonequilibrium influence on slip properties. VSL3D results using body-slip boundary conditions were not determined over the afterbody as a consequence of the poor comparison found over the spherical nose.[1]

Surface Properties

Heat transfer and skin-friction coefficient distributions over the entire body surface are compared with previous results[4] in Figs. 11 and 12, respectively. Although heat-transfer and skin-friction coefficient distributions for VSL3D predictions[4] using shock slip compared well with MC results over the nose, the continuum theory demonstrated overpredictions approaching 100% on the afterbody. Evaluation of S_t and C_f from the Chapman-Enskog (C-E) theory, using smoothed MC profiles and VHS molecular models for transport properties of the gas, provided significant underpredictions of S_t and C_f over the nose; this trend continues for C_f on the afterbody. However, continuum C-E S_t results are overpredicted along the afterbody. These anomalous results demonstrate the breakdown of continuum theory in the Knudsen layer for rarefied transitional flow. Failure of the C-E theory is attributed to the non-negligible contribution of the heat transfer and viscous stress to the velocity distribution function.

MC pressure ($\bar{P}$) distributions shown in Fig. 13 are in reasonable agreement with VSL3D results[4] for shock slip and previous MC results; the continuum theory slightly overpredicts the afterbody pressures. Note that the S_t, C_f, and $\bar{P}$ results near the stagnation point from the present MC code

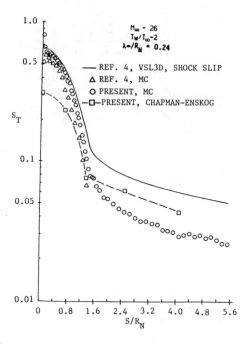

Fig. 11 Surface heat-transfer distribution.

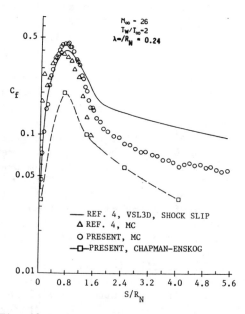

Fig. 12 Surface skin-friction distribution.

are not considered valid as a consequence of the distortion of collision statistics related to excess duplication of molecules in those regions.[1]

Slip Modeling Implications

Current slip models are based on assumptions that the flow within the Knudsen layer can be represented as a slight departure from thermodynamic equilibrium, with a velocity distribution function approximated as a small perturbation from an equilibrium (Maxwellian) distribution.[9] The Chapman-Enskog theory is typically employed for the velocity distribution function, thus establishing a relationship among the molecular transport properties, thermal velocities, and gradients of mean flow properties to provide heat transfer and shear stress. Additional assumptions for the slip analysis applied to a perfect gas require that the following properties are constants in the Knudsen layer: mass; momentum; energy; and that the velocity distribution function is uniform.

Knudsen-layer results from the current MC simulations on the afterbody and the previous studies on the spherical nose[1] at hypersonic transitional rarefied flow ($\lambda_\infty/R_N = 0.24$) for a cold wall ($\bar{T}_w = 0.015$) provide the following indications:

1) Essentially constant S_t, C_f, and $\bar{P}$ demonstrate constant momentum and energy fluxes.

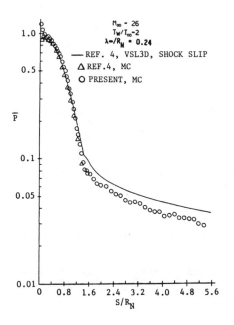

Fig. 13 Surface pressure distribution.

2) Normal stress ($\bar{P}$) agrees with wall pressure; consequently, no pressure slip effect is evident. This result is contrary to the existing VSL3D body-slip model, which provides a factor of 2 reduction in surface pressure at the stagnation point at this flow condition.

3) Slip velocity and temperature jump over the spherical nose are poorly predicted using the slip model in the VSL3D continuum code for the present case and for more rarefied conditions^{10} (λ_∞/R_N = 0.75).

4) Collision frequencies ($\bar{\nu}$) are reasonably well represented by an equilibrium description over the forward portion of the spherical nose (S/R_N < 1.0); differences between the MC and equilibrium ν results can exceed a factor of 2 in portions of the Knudsen layer on the afterbody. Gasdynamic models such as the well-known Bhatnagar-Gross-Krook (BGK) approach, which assumes that ν is proportional to density and dependent on temperature (local equilibrium) but not on molecular velocity, may not provide realistic results for slip modeling.

5) The first-order Chapman-Enskog theory fails to predict S_t and C_f using the MC profiles and VHS molecular model for gas properties over the entire configuration. This implies that the small-perturbation approach to the velocity distribution function is not valid at these energetic flow conditions.

Acknowledgment

The assistance of R. D. McGregor of TRW in formulating the Monte Carlo sampling routines and determining flow properties used in this study is greatly appreciated.

References

[1] Chrusciel, G. T. and Pool, L. A., "Details of the Knudsen Layer for a Cold Wall Sphere" *Rarefied Gas Dynamics*, edited by Vinicio Boffi and Carlo Cercignani, Vol. I, B.G. Teubner, Stuttgart, 1986, pp. 226-235.

[2] McGregor, R. D., "Users Manual for the TRW 'MCSASC' Computer Program for the Aerodynamic Analysis of Blunt Body Reentry Vehicles Using the Direct Simulation Monte Carlo Method," TRW Defense and Space Systems Group, July 1981.

[3] Bird, G. A., *Rarefied Gas Dynamics*, Oxford Univ. Press, London, 1976, pp. 159-160.

[4] Chrusciel, G. T., Lewis, C. H., and Sugimura, T., "Slip Effects in Hypersonic Rarefied Flow," *Progress in Astronautics and Aeronautics: Rarefied Gas Dynamics*, edited by Sam S. Fisher, Vol. 74, Pt. II, AIAA, New York, 1981, pp. 1040-1054.

[5] Borgnakke, C. and Larsen, P. S., "Statistical Collision Model for Monte Carlo Simulation of Polyatomic Gas Mixtures," Journal of Computational Physics, Vol. 18, 1975, pp. 405-420.

[6] Bird, G. A., "Monte Carlo Simulation in an Engineering Context," Progress in Astronautics and Aeronautics: Rarefied Gas Dynamics, edited by Sam S. Fisher, Vol. 74, Pt. I, AIAA, New York, 1981, pp. 239-255.

[7] Lewis, C. H., Gogineni, P. R., and Murray, A. L., "Comparisons of Viscous Shock-Layer Predictions with Experiment for Blunted Cones in Rarefied Transitional Regime," Eleventh International Symposium on Rarefied Gas-Dynamics, Cannes, France, Vol. I, 1979.

[8] Chapman, S. and Cowling, T. G., The Mathematical Theory of Non-Uniform Gases, 2nd ed., Cambridge Univ. Press, London, 1952.

[9] Gupta, R. N., Scott, C. D., and Moss, J. D., "Slip Boundary Equations for Multicomponent Nonequilibrium Airflow," NASA TP-2452, Nov. 1985.

[10] Chrusciel, G. T. and Pool, L. A., "Knudsen Layer Characteristics for a Highly Cooled Blunt Body in Hyersonic Rarefied Flow," AIAA Paper 83-1424, June 1983.

Approximate Calculation of Rarefied Aerodynamic Characteristics of Convex Axisymmetric Configurations

Y. Xie* and Z. Tang†
China Aerodynamics Research and Development Centre, Mianyang, Sichuan, People's Republic of China

Abstract

This paper presents an engineering calculation for rarefied aerodynamic characteristics of convex axisymmetric configurations. Within the framework of the local method for rarefied gasdynamics, the calculation is completed by means of experimental data obtained from the low-density hypersonic wind tunnel (LDHWT) at China Aerodynamics Research and Development Centre (CARDC). The present paper gives the approach formulas for local force coefficients involving the terms of s^{-2}, the correlation functions having a uniform form, and considers the difference between the molecular models for flight and experimental conditions. Although the measured data for longitudinal moment coefficients have not been used in the process to determine correlative functions, the present calculation curves are in good agreement with relevant measured data, not only for axial and normal force coefficients but also for longitudinal moment coefficients.

Introduction

It is well known that the common approach for the different experimental equipment of rarefied gasdynamics is to make the experimental models very

---

Copyright©1989 by the American Institute of Aeronautics and Astronautics, Inc. All rights reserved.
*Senior Research Engineer.
†Associate Research Engineer.

small in order to get large Knudsen numbers. This requirement makes it very difficult to measure rarefied aerodynamic characteristics for complex bodies. How can we predict these characteristics for complex bodies on the basis of rarefied aerodynamic characteristics measured for simple bodies at experimental conditions or for existing complex bodies at high-altitude flight conditions? This paper is just part of the program to resolve the problem.

The present paper describes the calculation procedure of the rarefied aerodynamic longitudinal characteristics for two convex bodies at angles of attack that are varied from 0 to 180 deg. The calculation has been completed by means of experimental data obtained from LDHWT at CARDC and the framework of the local method. The calculation results are in good agreement with measured data.

Simple Description of the Calculation Method

The local method is based on the model of local interaction. It was assumed that every surface element of a body would be independent, i.e., momentum and energy transfer between it and rarefied gas flow were not affected by its neighboring elements.[1,2]
In this paper, we consider that
If $\theta > \pi/2$, then $C_p = C_\tau = 0$.
If $\theta \leq \pi/2$, then, we suppose following approach formulas:

$$C_p = P_x \cos\theta + P_{y*} \sin\theta$$
$$C_\tau = P_x \sin\theta - P_{y*} \cos\theta \qquad (1)$$

$$P_x = f_1(\theta, s, \lambda_i)$$
$$= \cos\theta\,(\lambda_1 + \lambda_2 \cos 2\theta) + \frac{\sin^2\theta}{2\pi^{\frac{1}{2}}s}(\lambda_1 - \lambda_2)$$
$$+ (1/s^2)\cos\theta\,[0.25(\lambda_1 + 3\lambda_2 - \lambda_3) - (2\lambda_2 - \lambda_3)\cos 2\theta]$$

$$P_{y*} = f_2(\theta, s, \lambda_i)$$
$$= \sin\theta\,\{\lambda_3 \cos^2\theta - (\lambda_1 - \lambda_2)\frac{\cos\theta}{2\pi^{\frac{1}{2}}s}$$
$$+ (1/s^2)[1/4(\lambda_1 - \lambda_2 + \lambda_3) + (3\lambda_2 - 3/2\lambda_3)\cos^2\theta]\} \qquad (2)$$

where

$$\theta = \arccos(-\vec{n}_0 \cdot \vec{x}_0)$$

$$S = V_\infty / (2RT_\infty)^{\frac{1}{2}}$$

Furthermore,

$$\vec{F} = 0.5 \rho_\infty V_\infty^2 \int_{A*} (-C_p \vec{n}_0 + C_\tau \vec{\tau}_0) dA$$

$$\vec{M} = 0.5 \rho_\infty V_\infty^2 \int_{A*} \vec{r} \times (-C_p \vec{n}_0 + C_\tau \vec{\tau}_0) dA \qquad (3)$$

where $A*$ is the wetted surface area of the body, λ_i ($i = 1, 2, 3$) are global parameters, $\vec{x}_0$ is $\vec{V}_\infty / V_\infty$, $\vec{n}_0$ and $\vec{\tau}_0$ are the unit normal vector and the unit tangent vector to the surface element of the body, respectively. $\vec{\chi}$ and $\vec{y}_*$ are shown in Fig. 1.

Determination of Global Parameters

Suppose that a uniform form of correlative functions for global parameters is as follows:

$$\lambda_i = \lambda_i^C + (\lambda_i^F - \lambda_i^C)[1 + \text{erf}(X_i / 2^{\frac{1}{2}})] / 2$$

$$X_i = (\log_{10} Kn_* + \alpha_i) / \sigma_i \qquad (4)$$

where λ_i^F is the free molecular flow limit of λ_i. If the molecular model, the temperature on surfaces

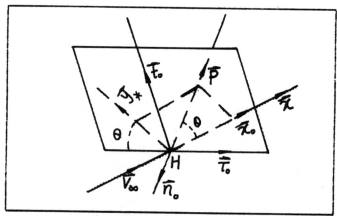

Fig. 1 Local coodinate system to the surface element of the body.

and the condition of coming flow are given in advance, the local force coefficients are as follows, according to the free molecule flow theory.

$$P_x^F = f_3(\theta, s, T_w/T_\infty), \quad P_{y*}^F = f_4(\theta, s, T_w/T_\infty)$$

Define

$$J_x^F = \int_0^{\frac{\pi}{2}} [f_1(\theta, s, \lambda_i^F) - f_3(\theta, s, T_w/T_\infty)]^2 \, d\theta$$

$$J_{y*}^F = \int_0^{\frac{\pi}{2}} [f_2(\theta, s, \lambda_i^F) - f_4(\theta, s, T_w/T_\infty)]^2 \, d\theta$$

and from

$$\frac{\partial J_x^F}{\partial \lambda_1^F} = 0, \quad \frac{\partial J_x^F}{\partial \lambda_2^F} = 0, \quad \frac{\partial J_{y*}^F}{\partial \lambda_3^F} = 0, \quad \lambda_i^F$$

($i = 1, 2, 3$) can be determined.

λ_i^c is the another limit of λ_i, which is obtained by means of modified Newtonian theory.

If subscript J represents the Jth experimental condition in LDHWT and if Kn_{*J} is the Knudsen number for the Jth condition, the measured data of drag and lift coefficients, which vary with the angle of attack, can be expressed as

$$C_{xe} = f_5(Kn_{*J}, \alpha), \quad C_{ye} = f_6(Kn_{*J}, \alpha)$$

From Eqs.(1)-(3), furthermore, the approach formulas of drag and lift coefficients can be derived as the following forms

$$C_x = f_7(\lambda_{iJ}, s)$$
$$C_y = f_8(\lambda_{iJ}, s)$$

Define

$$J_{xs} = \int_0^{\alpha*} [f_7(\lambda_{iJ}, s) - f_5(Kn_{*J}, \alpha)]^2 \, d\alpha$$

$$J_y = \int_0^{\alpha*} [f_8(\lambda_{iJ}, s) - f_6(Kn_{*J}, \alpha)]^2 \, d\alpha$$

where α_* is a characteristic angle of attack, and from

$$\frac{\partial J_{xs}}{\partial \lambda_{1J}} = 0 \quad , \quad \frac{\partial J_{xs}}{\partial \lambda_{2J}} = 0 \quad , \quad \frac{\partial J_y}{\partial \lambda_{3J}} = 0$$

therefore, λ_{iJ} (i = 1, 2, 3) can be found out.

If

$$A_J = arc\{erf[2(\lambda_{iJ} - \lambda_i^C)/(\lambda_i^F - \lambda_i^C) - 1]\}$$

$$B_J = -\log_{10} Kn_{*J}$$

$$A_* = arc\{erf[2(\lambda_{i*} - \lambda_i^C)/(\lambda_i^F - \lambda_i^C) - 1]\}$$

$$B_* = -\log_{10} Kn_{**}$$

we have

$$\alpha_{iJ} = A_J(B_* - B_J)/(A_J - A_*) + B_J$$

$$\sigma_{iJ} = (B_* - B_J)/[2^{\frac{1}{2}}(A_J - A_*)]$$

$$\alpha_i = (1/n) \sum_{J=1}^{n} \alpha_{iJ}$$

$$\sigma_i = (1/n) \sum_{J=1}^{n} \alpha_{iJ}$$

where λ_{i*} and Kn_{**} are the values of λ_i and Kn_* at the relevant experimental condition, respectively. Arc[erf( )] is the antifunction value for error function, and it can easily be calculated on a computer.

Definition of Knudsen Number

The molecular model is approximate to the Maxwellian molecule for the experimental conditions in LDHWT but, on the other hand, the molecular model is approximate to the hard sphere molecule for the real flight conditions at high altitude. In order easily to use Eq.(4) in calculating for real flight conditions at high altitude,[3] the Knudsen number is defined as

$$Kn_* = \frac{(\gamma \pi)^{\frac{1}{2}}}{4} M_\infty [1+(\frac{\gamma-1}{\gamma})^{\frac{1}{2}}(\frac{T_w}{T_0})^{\frac{1}{2}}] Kn$$

$$kn = (\frac{\pi \gamma}{2})^{\frac{1}{2}} \frac{M_\infty}{Re_{\infty,L^+}} \frac{1}{\Phi(s_1)\pi^{\frac{1}{2}}}$$

$$\Phi(s_1)\pi^{\frac{1}{2}} = 5\pi 4^{(3-S_1)/(S_1-1)} / \Gamma(\frac{4s_1-6}{s_1-1}), s_1 = 5$$

$$L^+ = \frac{1}{\alpha_*} \int_0^{\alpha_*} A_*^{\frac{1}{2}} d\alpha, \quad A_* = \int_{\cos\theta \geq 0} dA$$

where γ is the ratio of specific heats, L^+ characteristic length, and $\Gamma(\ )$ the gamma function; M_∞ and Re_{∞,L^+} are Mach number and Reynolds number calculated at the coming flow condition for the experiments, respectively. For the calculation at real flight conditions, we have the following relation between Kn_* and Mach number and Reynolds number

$$Kn_* = (\frac{\pi \gamma}{2})^{\frac{1}{2}} [M/Re_{L^+}]_f$$

where subscript f represents the freeflow condition in the flight.

Results and Discussion

Figures 2 and 3 show the variation of rarefied aerodynamic longitudinal characteristics calculated for L-1 and F-2 configurations at angles of attack, respectively. L-1 configuration is made up of a blunt-nose and three-segmental cone, and F-2 configuration is a slender cone with a slightly blunt nose. The comparison between the present calculation curves and corresponding experimental points is shown in these figures, too. Although the measured data for longitudinal moment coefficients have not been used in the process to determine correlative functions of global parameters, the present calculation curves are in good agreement with relevant measured data not only for axial and normal force coefficients but also for longitudinal moment coefficients. Although

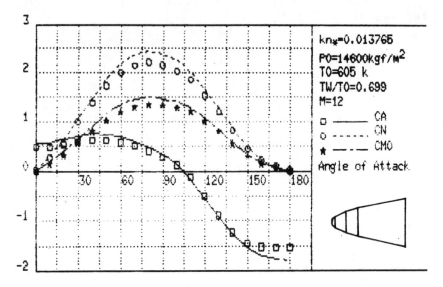

Fig. 2 Variation of aerodynamic coefficients on L-1 with angle of attack, comparison of experiment with calculation.

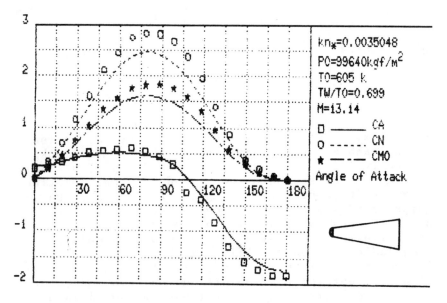

Fig. 3 Variation of aerodynamic coefficients on F-2 with angle of attack, comparison of experiment with calculation.

there are some differences between calculated results and experimental data, it is possible further to improve the present calculation by analyzing the differences.

References

[1] Alexeeva, E. V. and Barantsev, R.G., <u>The Local Method of Aerodynamical Calculation in Rarefied Gas</u>, Leningrad State Univ., Leningrad, USSR, 1976, (in Russian).

[2] Bunimovich, A. I. and Dubinskii, A. V., "Local method in Rarefied Gas Dynamics," <u>Proceedings of the Thirteenth International Symposium on Rarefied Gas Dynamics</u>, Vol. 1, edited by O. M. Belotserkovskii, Plenum Press, New York and London, 1982, pp. 431-438.

[3] Kogan, M. N., <u>Rarefied Gas Dynamics</u>, Translation edited by L. Trilling, Plenum Press, New York, 1969, pp. 479-491.

Procedure for Estimating Aerodynamics of Three-Dimensional Bodies in Transitional Flow

J. Leith Potter*
Vanderbilt University, Nashville, Tennessee

Abstract

Based on considerations of fluid dynamic simulation appropriate to hypersonic, viscous flow over blunt-nosed lifting bodies, a method was presented earlier by the author for estimating drag coefficients in the transitional-flow regime. The extension of the same method to prediction of lift coefficients is presented in this paper. Correlation of available experimental data by a simulation parameter appropriate for this purpose is the basis of the procedure described. Under low-density, hypersonic flow conditions, the overriding importance of projected and wetted areas, rather than details of body configuration, support this approach for finding aerodynamic forces. The ease of application of the method makes it useful for preliminary studies that involve a wide variety of three-dimensional vehicle configurations or a range of angles of attack of a given vehicle. It is also easy to examine the influence of different assumptions regarding gas/surface interaction parameters.

Nomenclature

C_D = drag coefficient
C_e = $(\mu_e/\mu_\infty)(T_\infty/T_e)$
C_L = lift coefficient
D = drag
H = enthalpy
L = lift
M = Mach number
PFA = projected frontal area
PPA = projected planform area
P_n = similarity parameter, Eqs. (2), (5)
Re = Reynolds number
Re_o = $Re_\infty(\mu_\infty/\mu_o)$

This paper is declared a work of the U. S. Government and therefore is in the public domain.
*Research Professor of Mechanical Engineering.

s   = streamwise wetted length
s*   = characteristic length, Eq. (3)
T   = temperature
U   = velocity
$\overline{V}_e$ = $M_e(C_e/Re_e)^{1/2}$
WA = wetted area
μ = absolute viscosity
ν = kinematic viscosity
ω = exponent in viscosity-temperature relation

Subscripts

e   = edge of boundary layer
fm   = free molecular flow
i   = inviscid flow
o   = stagnation region
w   = wall
∞ = freestream

Introduction

Many experimental and theoretical investigations of the drag of simple shapes in hypersonic, rarefied flow have been reported. Very few reports on lift and static stability have appeared. During this time it has been recognized that even the best laboratory facilities available for this work did not meet all of the desired simulation requirements and that computational techniques had not been developed to the level necessary for dealing with complex shapes having unknown gas/surface interaction, slip flow, and nonequilibrium gas processes. It is remarkable that adequate re-entry vehicles have been designed on the basis of this earlier research. That continues the history of aeronautics, which also is a record of successful but not generally optimal aircraft produced despite obvious deficiencies in the theoretical and laboratory base for the work.

The low air densities that characterize the transitional flow flight regime lead to low aerodynamic forces and heating. For many vehicles, this makes precision in aerodynamic predictions of lesser importance. Errors in aerodynamic coefficients lead to only small errors in absolute magnitudes of forces or heating rates. This had made many aerospace designers indifferent to rarefied flow phenomena despite the rather spectacular changes in aerodynamic coefficients that often occur in the transitional regime. However, aerospace vehicles of increased aerodynamic and structural sophistication, capable of maneuvering through use of aerodynamic controls and dependent upon much more accurate predictions of fluid dynamics for safe and efficient operation are now of greater concern. Lift, lift-to-drag ratio, and stability have assumed increased importance.

The prediction of aerodynamic, or more generally gasdynamic forces under hypervelocity, rarefied flow conditions is based upon approximate theoretical methods or correlations of the relatively small collection of relevant experimental data. At present, and probably for several more years, the two avenues to follow are the use of experimental data in conjunction with carefully chosen simulation and scaling parameters and the direct simulation Monte Carlo (DSMC) computational technique. The latter is a powerful tool, but it requires certain arbitrary inputs; the necessity to assume a gas/surface interaction model is perhaps the main point in this case. On the other hand, the shortcomings of wind tunnels and other laboratory devices in regard to duplication of in-flight real gas and gas/surface effects are well known and do not need repeating here. Complex configurations, with interfering flows associated with different components of the vehicle usually are no problem in wind tunnels. However, inability to fully duplicate the full-scale vehicle's flight environment, possibly including the gas/surface interaction, raises various questions. Thus, the data must be scaled or extrapolated with care, to say the least.

This paper addresses the problem of predicting lift and lift-to-drag ratio of aerospace craft in transitional flow. The approach used earlier by this author[1] in scaling drag coefficients is extended for the present purpose. That may be described as the correlation of normalized aerodynamic coefficients with a simulation parameter which is designed to account for the principal flow phenomena that cause the coefficients to vary. The derivation of this simulation parameter is fully described in Ref. 1, so only its extension to include lift coefficients is covered here.

Lift in Hypersonic Transitional Flow

In marked contrast to the amount of data now available on drag of bodies in transitional flow, very few measurements of lift have been published. Several reports from the Arnold Engineering Development Center in the 1960's (e.g., Refs. 2 and 3) contain wind tunnel data on lift, drag, and pitching moment in the midst of the transitional regime. Space Shuttle re-entry data[4,5] are now available and are, of course, extremely valuable because no wind tunnels can fully match the flow conditions of hypersonic re-entry. No theoretical/numerical computations of lift coefficients for bodies throughout the transitional regime were previously known to the writer, therefore it was necessary to work mainly with the experimental data of Refs. 2-5. However, it is noted that relevant results from the DSMC method are presented in this volume by Celenligil, Moss, and Bird.

The Parameters

The simulation parameter described in Ref. 1 was devised as a hybrid of $\overline{V}_e$ and Re_0 with the purpose of obtaining a parameter suitable for scaling viscous, hypersonic flow effects on typical hypersonic re-entry configurations, either lifting or non-lifting. These were visualized as blunted, cold-wall bodies at varying angles of attack. The characteristic length in the parameter was taken to be a modified wetted length in the streamwise direction. The modification was made to obtain a length more representative of the overall body. For example, a simple wetted length or body diameter tells nothing about fineness ratio or angle of attack of the body. Yet it is well known that both projected area and wetted area are significant in regard to forces on bodies in transitional flow. Therefore, a "shape factor" was applied to the wetted length to yield the characteristic length. This approach is suggested by the empirical data showing that projected and wetted areas override detailed body shape variations in determining overall forces in hypersonic, transitional flows.

To briefly summarize, in Ref. 1, it is shown that the normalized form of drag coefficient,

$$\overline{C}_D = (C_D - C_{Di})/(C_{Dfm} - C_{Di}) \tag{1}$$

of a variety of practical vehicle shapes is correlated with a simulation parameter defined as follows:

$$Pn_D = [(U/\nu)_\infty s^* (H_\infty/H^*)^\omega]^{1/2} \tag{2}$$

where

$$s^* = s (PFA/WA)^{1/2} \tag{3}$$

$$H_\infty/H^* = H_\infty/(0.2H_0 + 0.5H_W) \tag{4}$$

and $\omega \approx 0.63$ for high-temperature conditions in nonreacting air or N_2.

The basis of the parameter Pn_D is presented in Ref. 1, where it is shown that, for hypersonic, blunt-body rarefied flow,

$$\overline{V}_e \propto Re_0^{-1/2} = [Re_\infty (H_\infty/H_0)^\omega]^{-1/2}$$

When H_0 is replaced by a reference enthalpy H^* to better represent local flow at points not in the stagnation region and a shape factor is included to modify the wetted length, s, the bracketed right-hand term above becomes Pn. Thus, Pn may be regarded as a derivative of $\overline{V}_e$ or Re_0 with modified reference enthalpy and length. Note that PFA was denoted as CSA in Ref. 1. Again drawing on empirical knowledge and with emphasis on overall forces in flows of low density, a nonreacting gas is assumed. The alteration proposed here for the accommodation of both drag and lift by the same basic parameter is to now define the pair,

$$Pn_D = \text{drag parameter in Eq. (2)} \tag{5a}$$

and

$$Pn_L = Pn_D (PPA/PFA)^{1/4}, \text{ the lift parameter} \tag{5b}$$

For correlating lift coefficients, the shape factor is based upon projected planform area (PPA), while projected frontal area (PFA) is retained for correlating drag coefficients. As a result, Pn_D incorporates a reference length appropriate to drag, while Pn_L involves a reference length more suitable for lift

The counterpart of Eq. (1) in the case of lift is

$$\overline{C}_L = (C_L - C_{Li})/(C_{Lfm} - C_{Li}) \qquad (6)$$

It will be noticed that the form of Eqs. (1) and (6) encourages scatter in the correlation of experimental data when C_D or C_L is close to the corresponding value for near-inviscid flow, i.e., at large Pn_D or Pn_L. That deficiency is not as serious as it may seem at first, because the designer often will have both experimental and theoretical results available for the higher Reynolds number, continuum, viscous-flow part of the transitional regime for which the scatter may be troublesome.

There are other features of the normalized coefficients that must be pointed out. These concern the specific bases of the inviscid and free molecular flow coefficients used in this paper. It should be noted that either the modified Newtonian theory or an experimental high Reynolds number C_{Li} has been used. In the first instance, C_{Li} has been calculated on the basis of a windward surface pressure coefficient,

$$C_p = K \cos^2 \eta \qquad (7)$$

where, following Lees,[6]

$$K = (\gamma + 3)/(\gamma + 1) \qquad (8)$$

for the stagnation regions of blunt bodies with detached shock waves at large Mach numbers. The conventional assumption, $C_p = 0$, is made for leeward surfaces. The user of the correlations must not substitute an experimental value for the inviscid term unless it is a high Reynolds number result, say Pn greater than 600. A more sophisticated and accurate calculation for obtaining C_{Di} or C_{Li} may be used if consistency is maintained.

The values of C_{Dfm} or C_{Lfm} needed for the correlation were based upon the gas/surface interaction model of fully accommodated, completely diffuse re-emission. This is the most common model assumed, even through it is well known that some degree of specular remission and less-than-complete accommodation are more realistic assumptions. Blanchard[5] has deduced from NASA Space Shuttle flight data that approximately 90% energy accommodation and on the order of 10% specular reflection would lead to the best agreement of conventional free molecular theory[7] and flight results for lift-to-drag ratio. It is worth noting that these small changes in gas/surface interaction parameters correspond to very large percentage changes in L/D. It is easy to study the potential effect of gas/surface interaction simply by recalculating the free molecule force coefficients for various assumed interaction models. In any event, the basis for $\overline{C}_D$ and $\overline{C}_L$ must be kept in mind and consistency in the use of inviscid and free molecular coefficients is essential.

Correlation of Data

Not only is the amount of data on lift in transitional flow small, but accuracy is also a concern because of the low forces encountered in rarefied flows. Nevertheless, several sets of data are available.[2-5] These have been normalized according to the procedure described and the results are plotted in Fig. 1 along with all of the data of Ref. 1 and additional drag data for slender, slightly blunted cones from Ref. 8. In Fig. 1, $\overline{C}$ represents either $\overline{C}_D$ or $\overline{C}_L$ and P_n represents the corresponding P_{nD} or P_{nL}. This method of presentation facilitates a comparison of the lift data with the much larger number of drag points and leads to the conclusion that one analytical curve seems adequate for fitting both lift and drag data. A simple form for such a curve is

$$\overline{C}_L = [2.6/(2.6 + P_{nL}^{1.6})]^{0.5} \tag{9a}$$

or

$$\overline{C}_D = [2.6/(2.6 + P_{nD}^{1.6})]^{0.5} \tag{9b}$$

There is evidence that drag coefficients of very slender, sharp-nosed bodies at low angles of attack may exceed the free molecular level in the near-free-molecular or first-collision regime. Therefore, the part of the curve at very low Pn's is uncertain in regard to its suitability for bodies fitting that description.

Considering the limited data, uncertainty in those data, and the inherent limitations on a correlation of the type presented, an anticipated uncertainty in C_D or C_L of approximately ±10% may be cautiously proposed at this time. In view of the apparently good agreement with full-scale flight data

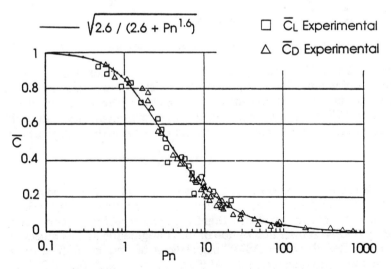

Fig. 1 Normalized lift and drag coefficients as functions of Pn

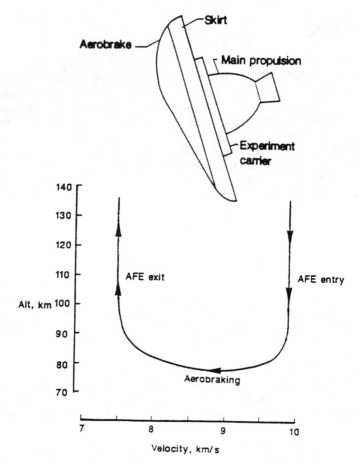

Fig. 2 Aeroassist flight experiment (AFE) vehicle shape and trajectory (from Ref. 9).

demonstrated in Ref. 1 and Fig. 1, it is suggested that Eqs. (9a) and (9b) represent useful tools for preliminary design studies wherein numerous variants may be screened and for other problems where wind tunnel experiments or CFD solutions are not feasible. The present method also could be used for quickly determining if results from new methods seem plausible.

Application to AFE Spacecraft

There is current interest in spacecraft which will utilize aerodynamic lift to transfer payloads between different orbits. To obtain flight data on one vehicle design, an aeroassist flight experiment (AFE) is being considered by NASA. A brief description of the configuration and its trajectory are given in Ref. 9, where DSMC calculations of drag coefficient of an axisymmetric approximation to the AFE are also presented. Figure 2, from Ref. 9, shows the elevation or side view of the AFE and the trajectory.

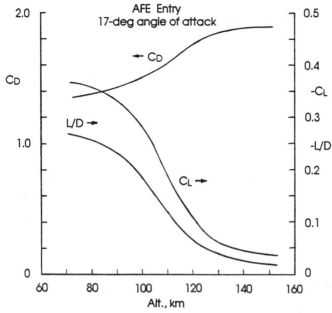

Fig. 3 Lift and drag coefficients predicted for the AFE during re-entry.

Lift and Drag

With the use of the trajectory information in Ref. 9 and supplemental information from Robert C. Blanchard of NASA Langley Research Center, the values of P_{nD} and P_{nL} have been computed. Corresponding values for $\overline{C}_D$ and $\overline{C}_L$ were obtained from Eqs. (9a) and (9b). Experimental data for C_{Di} and C_{Li} at high Reynolds number and computed C_{Dfm} and C_{Lfm} for fully accommodated, diffuse reemission were used. The results for C_D, C_L, and L/D appear in Fig. 3.

Base Flow

Although it is not an issue in this study of lift and drag, the neglect of base pressure in calculating aerodynamics of hypersonic bodies should be reviewed when moments are analyzed. If the stability of an asymmetric shape in low-density hypersonic flow is marginal, the possibility of non-uniform base pressure distribution and the resultant moment contribution may be a factor to consider. Major radial non-uniformity of base pressures on cones in hypersonic flows at low Reynolds numbers has been reported.[10]

Real gas processes are also likely to be important in determining base pressures and related quantities when there is a chemically reacting flow involving strong compression followed by rapid expansion. Bluff bodies such as the AFE fit this description.

Acknowledgments

This work was performed as a part of programs supported by NASA Research Grants NAG-1-549 and NAG-1-921, with J. N. Moss and R. C. Blanchard the respective NASA technical officers.

References

[1]Potter, J. L.,"Transitional, Hypervelocity Aerodynamic Simulation and Scaling," *Progress in Astronautics and Aeronautics: Thermophysical Aspects of Re-Entry Flows*, Vol. 103, edited by J. N. Moss and C. D. Scott, AIAA, New York, 1986, pp. 79-96.

[2]Boylan, D. E. and Potter, J.L., "Aerodynamics of Typical Lifting Bodies Under Conditions Simulating Very High Altitudes," *AIAA Journal,* Vol. 5, Feb. 1967, pp. 226-230.

[3]Boylan, D. E., "Aerodynamics of Blunt Heat Shields at Simulated High Altitudes," AEDC-TR-67-126, July 1967.

[4]Blanchard, R. C. and Buck, G. M., "Rarefied-Flow Aerodynamics and Thermosphere Structure from Shuttle Flight Measurements," *Journal of Spacecraft and Rockets*, Vol. 23, Jan. - Feb. 1986, pp. 18-24.

[5]Blanchard, R. C., "Rarefied Flow Lift-to-Drag Measurements of the Shuttle Orbiter," ICAS Paper 86-2.10.2, presented at 15th Congress of International Council of Aeronautical Sciences, London, Sept. 1986.

[6]Lees, L., "Hypersonic Flow," *Proceeding of Fifth International Aeronautical Conference*, Institute of the Aeronautical Sciences, June 1955, pp. 241-276.

[7]Schaaf, S. A. and Chambre, P. L., "Flow of Rarefied Gases," *Fundamentals of Gas Dynamics*, edited by H. W. Emmons, Princeton University Press, Princeton, NJ, 1958, pp. 687-738.

[8]Lyons, W. C., Jr., Brady, J. J., and Levensteins, Z. J., "Hypersonic Drag, Stability, and Wake Data for Cones and Spheres," *AIAA Journal*, Vol. 2, Nov. 1964, pp. 1948-1956.

[9]Dogra, V. K., Moss, J.N., and Simmonds, A. L., "Direct Simulation of Aerothermal Loads for an Aeroassist Flight Experiment Vehicle," AIAA Paper 87-1546, June 1987.

[10]Cassanto, J. M., "Radial Base-Pressure Gradients in Laminar Flow," *AIAA Journal*, Vol. 5, Dec. 1967, pp. 2278-2279.

Drag and Lift Measurements on Inclined Cones Using a Magnetic Suspension and Balance

R. W. Smith* and R. G. Lord†
Oxford University, Oxford; England, United Kingdom

Abstract

Drag and lift forces on a sharp, 10-deg cone in hypersonic, rarefied flow at angles of incidence up to 10 deg have been measured using magnetic suspension and balance techniques. The measurements are therefore free from errors resulting from sting interference. The flow Mach number was approximately 6, and freestream Knudsen numbers, based on cone base diameter, varied between 0.008 and 0.02. Flow gradients were believed to be negligible. The results have been compared with calculated limiting values for inviscid continuum flow and with the experimental results of other workers.

Introduction

A number of studies of the drag experienced by bodies of simple shapes in rarefied hypersonic flow have been carried out in the Oxford University low density wind tunnel using magnetic suspension and balance techniques. The most recent of these[1,2] have been concerned with the drag on slender cones at zero incidence and, in particular, with the effects of varying wall-to-freestream temperature ratio and bluntness. The present magnetic suspension and balance system was designed specifically for this purpose. The aim of the present investigation was to extend these studies to the case of nonzero incidence and, if possible, to the measurement of lift force and pitching moment as well as drag. The model wall temperature was not varied in these studies, and the results presented all refer to the adiabatic wall case.

Copyright © 1989 by the American Institute of Aeronautics and Astronautics, Inc. All rights reserved.
*Research Assistant, Department of Engineering Science.
†University Lecturer, Department of Engineering Science.

The main potential advantages of magnetic suspension and balance techniques are the absence of sting interference effects, which are particularly large and troublesome in low Reynolds number flows, and the high accuracy to which very small aerodynamic forces can be measured by monitoring the currents in the control coils. This allows measurements to be made under much more rarefied flow conditions than is possible with a conventional mechanical balance. Despite these advantages, only one other group, the University of Virginia at Charlottesville, has employed magnetic suspension and balance techniques to obtain drag and lift forces on cones in rarefied hypersonic flow.[3] The data from this group are restricted to a single cone geometry (9-deg semivertex angle) and the adiabatic wall case; thus, there is considerable scope for further study. The Virginia group had one advantage in that it was able to rotate the flow axis with respect to the axes of the magnetic suspension system, but this is not possible with the Oxford system. It is therefore necessary to calibrate the system for each separate angle of incidence, as well as for each model.

Experimental Details

Full details of the Oxford magnetic suspension and balance system can be found elsewhere,[1,2] and only a brief description is included here. The layout of the magnet coils is shown in Fig. 1. The sum of the currents in the front and rear drag coils is kept constant to provide a constant axial magnetizing field at the model position. The difference between the front and rear drag coil currents produces an axial gradient of the axial field component without changing value of the axial field at the model position, and this exerts a forward thrust on the model. The pitch coils are designed to produce a uniform vertical field, and hence a pitching moment, and the lift coils superimpose an axial gradient on this field component to produce lift. The model position detector (Fig. 2) consists of an array of phototransistors and sends signals proportional to vertical and horizontal displacement and pitch to the appropriate control circuits. With the model at zero incidence, the three control circuits are completely decoupled and operate independently. At nonzero incidence, however, coupling between the circuits is introduced for two reasons. First, the position detector, which has to rotate with the model, still sends signals proportional to displacements along, and normal to, its own axis to the drag and lift coils, respectively. Second, the magnetic forces on the model now depend to some extent on the vertical gradients of the field components also. A final compli-

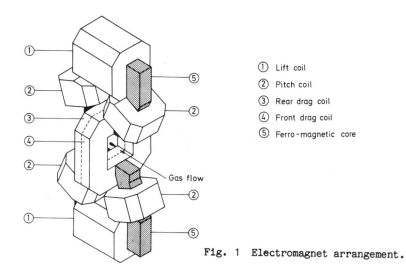

Fig. 1 Electromagnet arrangement.

① Lift coil
② Pitch coil
③ Rear drag coil
④ Front drag coil
⑤ Ferro-magnetic core

cation is that the magnetization of the model depends on the pitch coil current as well as the sum of the currents in the two drag coils, although this problem could possibly be overcome by using a permanently magnetized model.

These coupling effects have been investigated experimentally, and it has been found that the coupling is linear (i.e., the coupling affects the intercepts but not the slopes of the calibration curves). The actual calibration curves themselves are also fairly linear, so that calibration becomes a matter of determining the coefficients of a 3 x 3 matrix that converts lift and drag and pitch coil currents into some system of forces equivalent to the resultant aerodynamic force on the model. For simplicity of calibration, we have chosen to represent the resultant force by a single drag force acting through the center of the base of the cone and two lift forces, one through the centre of the base and one through the nose. No measurements of pitching moment will be presented because it has been found that the pitching moment about the center of mass of the cone is small; in every case, the line of action of the resultant aerodynamic force on the cone passes through a point on the axis of the cone not more than 2 mm from the center of mass.

The magnetic suspension and balance system is installed in the Oxford University low density wind tunnel, also described in detail in Refs. 1 and 2. The measurements presented here were obtained in flows at Mach numbers of approximately 6, produced by a contoured nozzle, using stagnation pressures from 15 to 60 Torr and stagnation

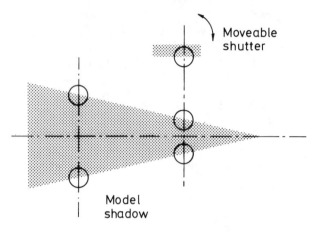

Fig. 2 Phototransistor arrangement.

temperatures of 293 and 373 K. Flow gradients in the neighborhood of the model were small and had no significant effect on the results. The model used was sharp-nosed and had a semivertex angle of 10 deg and a base diameter of 10 mm.

Results and Discussion

Drag and lift coefficients are plotted as functions of freestream Knudsen number based on the model base diameter, the values of fluid viscosity being those calculated for a Lennard-Jones intermolecular potential using data from Hirschfelder et al.[4]

The results for air and nitrogen at angles of incidence of 3.4 and 7.25 deg are shown in Figs. 3 and 4. In Fig. 3, drag data for zero incidence obtained by Dahlen[2] for the adiabatic wall case are included for comparison. Not surprisingly, no significant difference is observable between the two gases. The angle of incidence, always less than the cone angle, appears to have no detectable effect on the drag coefficient at any Knudsen number in the range covered. The lift coefficient, however, increases roughly in proportion to the angle of incidence.

The axial and normal force coefficients on a slender cone at small angles of incidence in inviscid supersonic flow have been considered by Stone[4] and, fortunately, the case of a 10-deg cone is one of the three for which detailed calculations are presented. According to these calc-

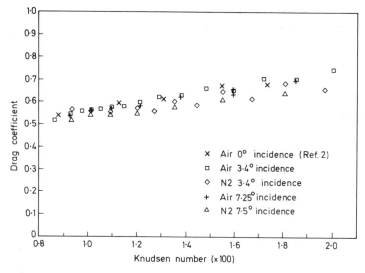

Fig. 3 Drag coefficients for air and nitrogen at 0-, 3.4-, and 7.25-deg incidence.

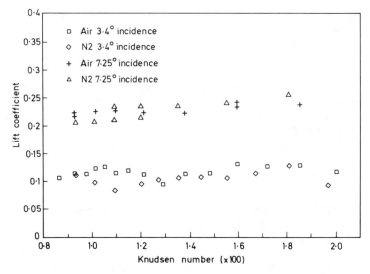

Fig. 4 Lift coefficients for air and nitrogen at 3.4- and 7.25-deg incidence.

ulations, the normal force coefficient on a 10-deg. cone at Mach 6 should be proportional to the angle of incidence, being 0.245 at 7.25 deg, which is about the limit to which the approximation might be expected to hold. Because the drag coefficient is constant, to the same degree of approx-

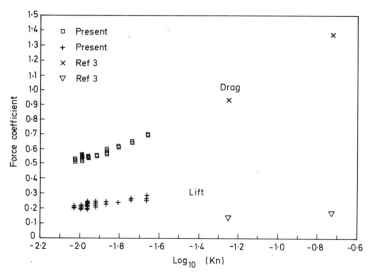

Fig. 5 Comparison of data at 7.25-deg incidence with data of Bharathan and Fisher.[3]

imation, over this range of incidence, it is reasonable to take this value as a lift coefficient rather than a normal force coefficient. As can be seen from Fig. 3, the present results appear to approach this limiting value reasonably closely at the lowest Knudsen numbers. It should be mentioned, however, that the calculated force coefficient is that due to the pressure distribution on the curved surface of the cone only, skin friction and the force on the base being neglected.

Comparison with Results of Other Workers

Figure 5 shows a comparison between the present results at 7.25-deg incidence and those of Bharathan and Fisher[3] for a 9-deg cone in nitrogen at a Mach number of 8.2. In order to make the comparison, it was necessary to interpolate between their data points, to convert from normal and axial force coefficients to lift and drag coefficients, and to plot them against a freestream Knudsen number rather than a Reynolds number based on freestream velocity, fluid viscosity at the wall temperature, and model length. The drag coefficients are seen to be in extremely good agreement, but the present lift coefficients are significantly higher than those calculated from Bharathan and Fisher's data.[3] The reason for this is not known.

Further Work

Attempts are currently being made to extend the measurements to greater angles of incidence and to other cone geometries.

Acknowledgment

This work has been carried out with the support of the Procurement Executive, Ministry of Defence.

References

[1] Haslam-Jones, T. F., "Measurements of the Drag of Slender Cones in Hypersonic Flow at Low Reynolds Numbers using a Magnetic Suspension and Balance," Oxford University Engineering Laboratory, Oxford, England, U.K., Rept. 1235/78, March 1978.

[2] Dahlen, G. A., "Cone Drag in the Transition from Continuum to Free Molecular Flow," Oxford University Engineering Laboratory, Oxford, England, U.K., Rept. 1538/84, March 1984.

[3] Bharathan, D. and Fisher, S. S., "Measured Axial and Normal Force Coefficients for 9-Degree Cones in Rarefied Hypersonic Flow," AIAA Paper 73-154, Jan. 1973.

[4] Hirschfelder, J. O., Curtiss, C. F., and Bird, R. B., Molecular Theory of Gases and Liquids, Wiley, New York, 1954.

[5] Stone, A. H., "On Supersonic Flow past a Slightly Yawing Cone," Journal of Mathematics and Physics, Vol. 27, Dec. 1948, pp.67-81.

Three-Dimensional Hypersonic Flow Around a Disk with Angle of Attack

K. Nanbu,* S. Igarashi,† and Y. Watanabe†
Tohoku University, Sendai, Japan
and
H. Legge ‡ and G. Koppenwallner§
DFVLR, Göttingen, Federal Republic of Germany

Abstract

Three-dimensional rarefied gas flows around a disk in a hypersonic stream are analyzed using the direct simulation Monte Carlo method. The collision process in this method is treated by the modified Nanbu method. Accepting the assumption of diffuse reflection and the model of hard sphere, no ambiguous assumptions are employed. A skew-coordinates system is introduced to make it easy to reset molecular indexing. The numerical calculations are performed for the disk with angles of attack of α = 45, 60, 75, and 90 deg. The rarefaction effects on the flowfield and the drag, lift, and heat-transfer coefficients are clarified for Kn = 0.1-20. A particular examination is made when α = 90 deg, i.e., the flow is axisymmetric; the wall temperature effect on the drag and heat transfer is examined, the drag being in good agreement with the experimental data for argon by Legge. The recovery temperature vs Knudsen number is also presented.

Introduction

One of the most challenging problems in rarefied gasdynamics is a rarefied flow around the aeroassisted orbital

---

Copyright © 1989 by the American Institute of Aeronautics and Astronautics,Inc. All rights reserved.
* Professor, Institute of High Speed Mechanics.
† Research Associate, Institute of High Speed Mechanics.
‡ Research Scientist, Institute for Experimental Fluid Mechanics.
§ Division Chief, Institute for Experimental Fluid Mechanics.

transfer vehicle. It is a three-dimensional flow of gas
mixture with translational, rotational, vibrational, chemical, and radiative nonequilibria. Only the direct simulation Monte Carlo (DSMC) method, together with a newborn
supercomputer, may make it possible to compute such a complicated flow. Dogra et al.[1] calculated an axisymmetric
flow along this line by use of the DSMC method. The problem of adopting the DSMC method is that one must have re-
course to many assumptions or models that have not yet been
verified sufficiently by experimental or theoretical studies. Our position on the problem is that such assumptions
or models must be checked one by one for flows that are
simple enough to be treated by both the DSMC method and experiment.

A fully three-dimensional hypersonic flow of monatomic
gas around a disk with angle of attack is considered. The
purpose of the present work is to examine the effect of the
Knudsen number on the flowfield around and the aerodynamic
characteristics of the lifting disk. The collision process
in the DSMC method is treated by the modified Nanbu method.[2,3] The calculations are carried out without using any
empirical techniques such as the weighting factor method.
The assumption of diffuse reflection and the hard sphere
model are all that are employed here. It is hoped that the
detailed results presented herein will become a standard
for checking the validity of various methods to calculate
three-dimensional rarefied flows.

Only when the disk is oriented perpendicular to the
flow direction, advantage may be taken of flow axisymmetry.
This simple case was investigated experimentally by Legge[4]
and theoretically by Hermina,[5] who used the Bird method[6] together with the assumption of the weighting factor. A network of cells is devised herein as to be applicable to fully
three-dimensional flows. That is, a skew-coordinates system is introduced, which makes it easy to reset molecular
indexing. This system of coordinates is a natural extension
of a rectangular-coordinates system, the use of which is
limited to axisymmetric flows. In the next section, the
simulation calculation procedure, based on the skew-coordinates system is described. In the last section, we discuss
the results obtained for axisymmetric as well as fully
three-dimensional flows.

Numerical Simulation Procedure

We consider a thin disk of radius r_D immersed in a
hypersonic stream. The disk has the angle of attack α to
the direction of the freestream with velocity U_∞. First,

we introduce the skew-coordinates system (x,y,z) shown in Fig. 1; the origin 0 is at the center of the disk, the z axis is oriented to the upstream direction, the disk is on the xy plane, the y axis is taken so that the yz plane is the plane of flow symmetry, the angle between the y and z axes is α, and the x axis is perpendicular to both the y and z axes. Next, we consider a rectangular-coordinates system (x',y',z') auxiliary to the skew system; the x' and z' axes coincide with the x and z axes, respectively. The relations between the coordinates are

$$x = x', \quad y = y'/\sin\alpha, \quad z = z' - y' \cot\alpha \quad (1)$$

The domain of calculation is the inside of the elliptic cylinder shown in Fig. 1; each of the slant end surfaces is parallel to the disk and has a circular form. Let the radius of this circle be R and the length of the cylinder be (a+b). Note that the form of the A-A cross section in Fig. 1 is an ellipse with the major axis of 2R and the minor axis of 2 R sinα. The cell network is based on the skew coordinates as follows: first, the calculation domain is divided into N_z wafers with the skew thickness Δz; next, each wafer is subdivided by circles and radial lines, as shown in Fig. 2. The radius r_k of the kth circle is given by

$$r_k = k\Delta r, \quad k = 1, 2, \ldots, N_r \quad (2)$$

where r_1 (= Δr) is the radius of the innermost circle. The annular region $r_{k-1} < r < r_k$ (k = 2,3,...,N_r) is divided

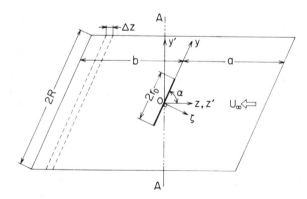

Fig. 1 Domain of calculation in the plane of flow symmetry.

into N_k subregions, where

$$N_k = 3k, \quad k = 2,3,\ldots,N_r \tag{3}$$

It is to be noted that the innermost circle is not divided by radial lines. This network of cells is reasonable since the maximum distance L between two molecules in a cell is of the same order for all cells. In fact, the distance L is

$$L = 2\Delta r \quad \text{for } k = 1$$

$$= 2r_k \sin(\Delta\phi/2) \text{ for } k = 2,3,\ldots,N_r$$

where $\Delta\phi = 2\pi/N_k$. Then we have

$$L/2\Delta r = k \sin(\pi/3k) < \pi/3 \ (\approx 1.05) \tag{4}$$

This means that $L/2\Delta r$ is between 1 and 1.05 for all cells. Next, we consider the cell volume. The volume V_1 of the innermost cell is $\pi(\Delta r)^2 \Delta z \sin\alpha$, and the volume V_k of each of N_k cells in the annular region $r_{k-1} < r < r_k$ is given by

$$V_k/V_1 = (2k-1)/3k, \quad k = 2,3,\ldots,N_r \tag{5}$$

This ratio is between 1/2 and 2/3. The total number of cells, N_c, is

$$N_c = [(3/2)N_r(N_r+1)-2] N_z \tag{6}$$

Our choice of the sizes of the cell and the calculation domain is as follows:

$$R = 3r_D, \quad a = (2-6)r_D, \quad b = (2-6)r_D,$$

$$\Delta r = 0.2r_D, \quad \Delta z = (0.1-0.2)r_D$$

In a typical case, N_c is 21,480.

The simulation starts by putting molecules in the half-domain upstream of the disk. These molecules are given the velocity of the freestream. They have no peculiar velocity since the Mach number of the freestream is assumed to be infinitely large. As time goes on, fresh molecules come in across the upstream boundary ($z = a$). Molecules that go out of the domain of calculation are eliminated. This is reasonable for hypersonic flow. Molecules that strike the disk are reflected diffusely with complete ther-

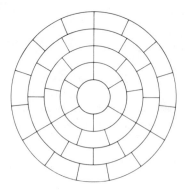

Fig. 2 Network of cells on the xy plane.

mal accommodation to the wall temperature T_w. The intermolecular collision is simulated by the modified Nanbu method,[2,3] whose computing task is proportional to the number of molecules in a cell, not to its square.

Before the start of calculation, all of the variables are made dimensionless by dividing the length by r_D, velocity by U_∞, and time by r_D/U_∞. Then it is immediately seen that the parameters governing the present problem are the Knudsen number Kn $(= \lambda_\infty/2r_D)$; the wall-to-stagnation temperature ratio T_w/T_0; and the angle of attack α; where λ_∞ $(= 1/\sqrt{2}\pi d^2 n_\infty)$ is the mean free path at the freestream density n_∞, d being the molecular diameter. In addition to the values of Kn, T_w/T_0, and α, we have to give the values of the dimensionless time step $\Delta \hat{t}$ $(= U_\infty \Delta t/r_D)$ and the number N_0 of simulated molecules in a far-upstream standard cell with volume V_1. In a typical case, our choice is $N_0 = 30$ and $\Delta \hat{t} = 0.01$. The time step is always chosen in such a way that the principle of uncoupling and the Courant condition are satisfied.

Results and Discussion

In order to clarify the rarefaction effects on the flowfield and the drag, lift, and heat-transfer coefficients, the numerical calculations are performed for the following Knudsen numbers: Kn = 0.1, 0.2, 0.5, 1, 2, 5, 10, and 20. The angles of attack are α = 45, 60, 75, and 90 deg. The wall-to-stagnation temperature ratio T_w/T_0 is fixed at 1. In the case of α = 90 deg, however, the wall temperature is varied to find the recovery temperature and to see the effect of the cold wall (T_w/T_0 = 0.5).

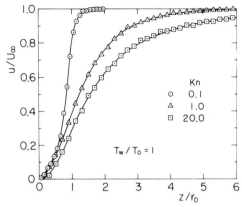

Fig. 3 Velocity profiles along the stagnation streamline for α = 90 deg and $T_w/T_0 = 1$.

Axisymmetrical Flow (α = 90 deg)

Figures 3-11 show the results for α = 90 deg. Figure 3 is the velocity profiles along the z axis, i.e., the stagnation streamline. As the Knudsen number decreases, the velocity gradient becomes very large, which means the formation of a shock front. Since the wall temperature has little effect on the profiles, the result for the cold-wall case is omitted. Figure 4 shows the density profiles along the stagnation streamline. In the case of $T_w/T_0 = 1$, the density ratio at the surface of the disk is roughly equal to that behind the normal shock, which is 4 in the hypersonic limit. In the cold-wall case, however, a large increse in density occurs near the disk. As will be shown later, the density profile along the stagnation streamline for the free molecular flow is given by

$$\frac{\rho}{\rho_\infty} = 1 + \left(\frac{\pi\gamma}{\gamma-1}\frac{T_0}{T_w}\right)^{1/2}\left[1 - \frac{\hat{z}}{(1+\hat{z}^2)^{1/2}}\right] \qquad (7)$$

where $\hat{z} = z/r_D$. The present solution for Kn = 20 agrees exactly with Eq. (7). Figure 5 shows the temperature profiles along the stagnation streamline. It can be seen that a larger temperature jump occurs for a greater Knudsen number. In the case of Kn = 0.1, the temperature profile for $T_w/T_0 = 0.5$ beyond the peak agrees with that for $T_w/T_0 = 1$; that is, only the gas between the disk and the location of maximum temperature is cooled by the cold wall. Figures 6 and 7 show the density and temperature profiles on the

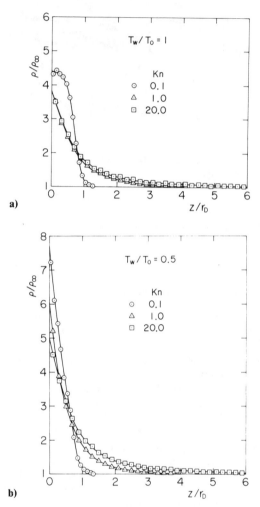

Fig. 4 Density profiles along the stagnation streamline for $\alpha = 90$ deg: a) $T_w/T_0 = 1$; b) $T_w/T_0 = 0.5$.

front surface of the disk as a function of the radial distance, respectively. The temperature is uniform on the disk because the flow is almost stagnant there. Both the density and temperature suddenly lower at the edge of the disk due to gas expansion. The streamlines for Kn = 0.1 and 1 are compared in Fig. 8 in the cold-wall case. For the smaller Knudsen number, there is a flow toward the backside of the disk due to the higher frequency of molecular collisions.

The aerodynamic characteristics, i.e., the drag and heat-transfer coefficients and the recovery temperature,

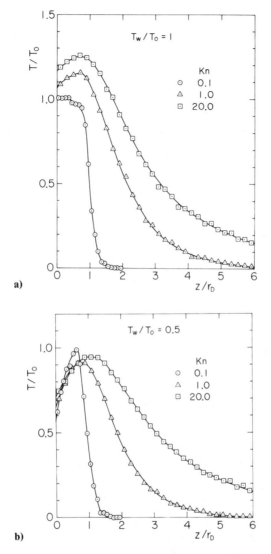

Fig. 5 Temperature profiles along the stagnation streamline for $\alpha = 90$ deg: a) $T_w/T_0 = 1$; b) $T_w/T_0 = 0.5$.

are shown in Figs. 9, 10, and 11. The DSMC data are connected with the smooth curves to guide our eyes. The drag coefficient for $T_w = T_0$ is in good agreement with the experimental data (circular symbols) for agron by Legge.[4] Here the drag coefficient C_D is defined by

$$C_D = D/(\tfrac{1}{2}\rho_\infty U_\infty^2 A) \qquad (8)$$

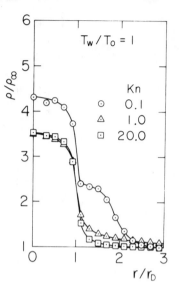

Fig. 6 Density profiles on the disk surface for $\alpha = 90$ deg and $T_w/T_0 = 1$.

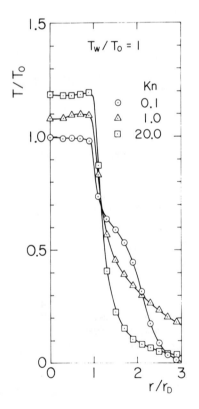

Fig. 7 Temperature profiles on the disk surface for $\alpha = 90$ deg and $T_w/T_0 = 1$.

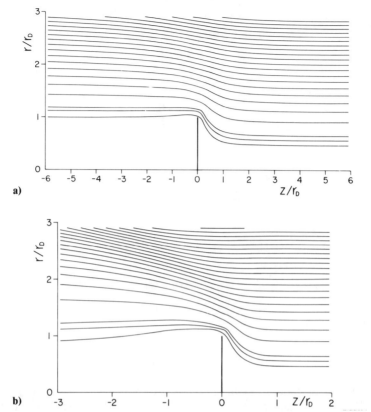

Fig. 8 Streamlines for α = 90 deg and T_w/T_0 = 0.5: a) Kn = 1; b) Kn = 0.1.

where D is the drag and A (= $\pi r_D^2 \sin\alpha$) is the projected area. We now give the definitions and formulas for an arbitrary angle of attack which will be used later. In the limit of the free molecular flow, we have

$$C_D = 2 + \left[\frac{\pi(\gamma-1)}{\gamma}\frac{T_w}{T_0}\right]^{1/2} \sin\alpha, \quad Kn \gg 1 \qquad (9a)$$

with γ (= 5/3) the ratio of specific heats. In the continuum limit, we have, for α = 90 deg,

$$C_D = (\gamma+3)/(\gamma+1), \quad Kn \ll 1 \qquad (9b)$$

from the modified Newtonian theory.[7] The C_D values obtained for large and small Kn tend to these two limits. Figure

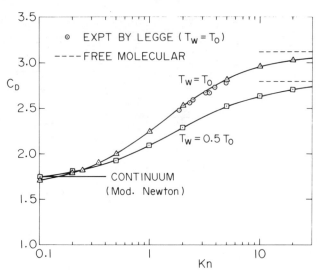

Fig. 9 Drag coefficient for $\alpha = 90$ deg.

10 shows the heat-transfer coefficient C_H defined by

$$C_H = \dot{Q}/(\tfrac{1}{2}\rho_\infty U_\infty^3 A) \qquad (10)$$

with $\dot{Q}$ the net heat transferred per unit time from gas to the disk. In the case of the free molecular flow, we have[8]

$$C_H = 1 - \frac{\gamma+1}{2\gamma}\frac{T_w}{T_0}, \qquad Kn \gg 1 \qquad (11)$$

It is to be noted that the coefficient C_H in Eq. (11) is independent of the angle of attack. We can see from Fig. 10 that the present data tend to the limit of Eq. (11). In Fig. 11, we show the recovery temperature T_r, which is found by changing the wall temperature until the heat flux $\dot{Q}$ vanishes. The temperature T_r decreases with the Knudsen number and tends rather abruptly to a limiting value near $Kn = 0.2$. A similar behavior is seen in the data of C_H for $T_w = T_0$. For the free molecular flow, by setting $C_H = 0$, we derive from Eq. (11),

$$T_r/T_0 = 2\gamma/(\gamma+1) \qquad (12)$$

Of course, this T_r is independent of the angle of attack α. It can be seen that the obtained recovery temperature tends to the value of 1.25 given by Eq. (12).

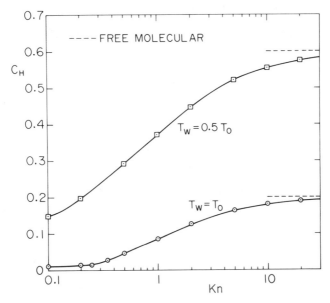

Fig. 10 Heat-transfer coefficient for $\alpha = 90$ deg.

Three-Dimensional Flow ($\alpha \neq 90$ deg)

We now show the results for lifting flows. The density contours and velocity vectors in the symmetrical plane (yz plane) are shown for $Kn = 0.1$ in Figs. 12 and 13, respectively. Figure 12 shows that the compression region above the leading edge becomes narrower as the angle of attack decreases. It can be seen from Fig. 13 that a downward flow becomes strong and the stagnation streamline moves toward the leading edge as the angle of attack decreases. In the case of $\alpha = 45$ deg, the stagnation point is close to the leading edge. Note that, although there is a flow into the backside of the disk, the corresponding mass flux is negligibly small because of low density. This can be also seen from Fig. 8b. In order to see the flow patterns for high Knudsen numbers, the velocity field superimposed on the density contours is shown for $Kn = 1$ in Fig. 14. Comparison of the velocity fields in Figs. 13 and 14 shows that the domain of calculation should have been greater for $Kn = 0.1$, especially for smaller α, because there is an appreciable downward flow even at the lower boundary of the calculation domain. In the case of $Kn = 1$, the density contours are approximately symmetrical with respect to the normal ζ in Fig. 1. This fact can be

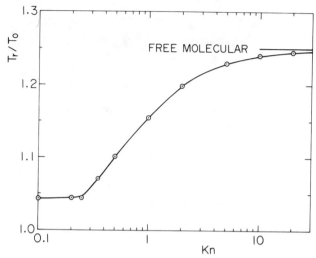

Fig. 11 Recovery temperature for α = 90 deg.

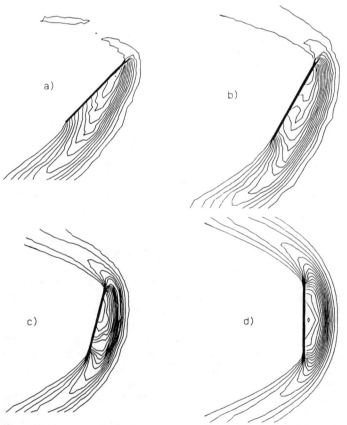

Fig. 12 Density contours for Kn = 0.1 and T_w/T_0 = 1: a) α = 45 deg; b) α = 60 deg; c) α = 75 deg; d) α = 90 deg.

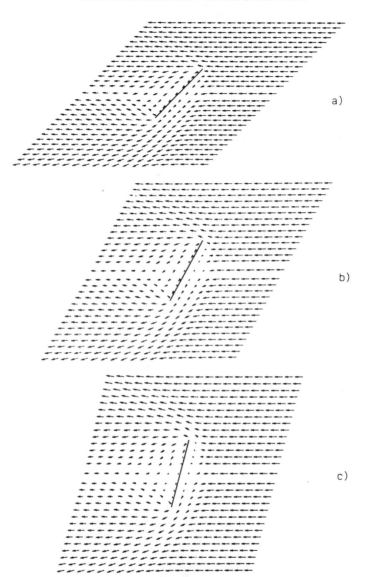

Fig. 13 Velocity field on the plane of symmetry for Kn = 0.1 and $T_w/T_0 = 1$: a) $\alpha = 45$ deg; b) $\alpha = 60$ deg; c) $\alpha = 75$ deg.

verified analytically for the free molecular flow as follows. We take account of the fact that the molecules in a volume element located upstream of the disk are the sum of those in the uniform flow and those coming from the whole front surface of the disk after reflection. Somewhat lengthy algebra yields

$$\frac{\rho}{\rho_\infty} = 1 + \left(\frac{\pi\gamma}{\gamma-1} \frac{T_0}{T_W}\right)^{1/2} \sin\alpha \left[\frac{\hat{\zeta}}{(\hat{\zeta}^2+\hat{y}^2)^{1/2}} - \frac{\hat{\zeta}}{\pi}\int_0^{\pi/2}\left(\frac{1}{R_D^+} + \frac{1}{R_D^-}\right)d\phi\right]$$

(13)

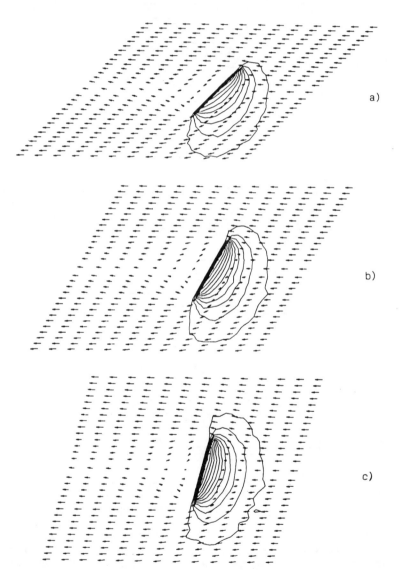

Fig. 14 Density contours and velocity field on the plane of symmetry for Kn = 1 and T_W/T_0 = 1: a) α = 45 deg; b) α = 60 deg; c) α = 75 deg.

where
$$R_D^{\pm} = (\hat{\zeta}^2+\hat{y}^2+1\pm 2\hat{y}\sin\phi)^{1/2}$$

and $\hat{y} = y/r_D$. This equation gives the density field in the $y\zeta$ plane. Equation (13) can be expressed by the incomplete elliptic integral. It can be seen from Eq. (13) that ρ is an even function of $\hat{y}$, i.e., the density field is symmetrical with respect to the normal ζ. If we put $\hat{y} = 0$ and $\alpha = \pi/2$ (hence, $\hat{\zeta} = \hat{z}$) in Eq. (13), it reduces to Eq. (7). In deriving Eq. (13), we have used the hypersonic approximation and the continuity condition of the number flux.[8] Note that increment $(\rho-\rho_\infty)$ is inversely proportional to the square root of the wall temperature.

Figure 15 shows the drag coefficient for various angles of attack. The coefficient increases with the angle of attack although the projected area is used in its definition. The DSMC data for larger Kn tend to the free molecular limits of Eq. (9a). Figure 16 shows the lift coefficient C_L defined by

$$C_L = L/(\tfrac{1}{2}\rho_\infty U_\infty^2 A) \qquad (14)$$

where L is the lift and A is the projected area as before. The coefficient C_L depends weakly on Kn. In the range of Kn > 1, C_L is almost equal to that for the free molecular flow given by

$$C_L = \left[\frac{\pi(\gamma-1)}{\gamma}\frac{T_w}{T_0}\right]^{1/2}\cos\alpha, \qquad Kn \gg 1 \qquad (15)$$

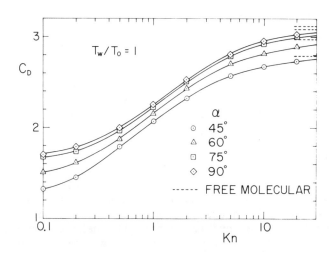

Fig. 15 Drag coefficient.

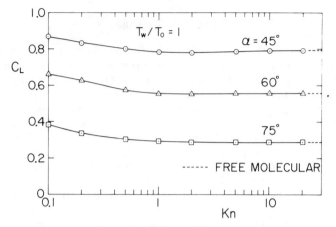

Fig. 16 Lift coefficient.

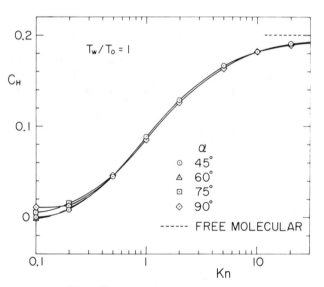

Fig. 17 Heat-transfer coefficient.

Figure 17 shows the heat-transfer coefficient for various angles of attack. In the regime of Kn > 0.5, the data for various angles of attack are indistinguishable, so that only the data for α = 45 and 90 deg are plotted. It is seen that not only in the free molecular limit but also in a wide range of the Knudsen number, the heat-transfer coefficient hardly depends on the angle of attack. From this fact and Eq. (12), it is most probable that the recovery temperatures for the angles of attack of 45, 60, and 75 deg are given by Fig. 11, which is the result for α = 90 deg.

Acknowledgment

The numerical calculations were carried out by using the supercomputer SX-1 at the Computing Center of Tohoku University.

References

[1]Dogra, V. K., Moss, J. N., and Simmonds, A. L., "Direct Simulation of Aerothermal Loads for an Aeroassist Flight Experiment Vehicle," AIAA Paper 87-1546, June 1987.

[2]Nanbu, K., "Theoretical Basis of the Direct Simulation Monte Carlo Method," Proceedings of the 15th International Symposium on Rarefied Gas Dynamics, B. G. Teubner, Stuttgart, Vol. 1, 1986, pp. 369-383.

[3]Illner, R. and Neunzert, H., "On Simulation Methods for the Boltzmann Equation," Transport Theory and Statistical Physics, Vol. 16, Nos. 2 & 3, 1987, pp. 141-154.

[4]Legge, H., "Wiederstandsmessungen an parallel angeströmten Platten und senkrecht angeströmten Scheiben in Überschall-freistrahlen grosser Verdünnung," Deutsche Luft- und Raumfahrt Forschungsbericht, No. 73-17, 1973.

[5]Hermina, W. L., "Monte Carlo Simulation of Transitional Flow Around Simple Shaped Bodies," Proceedings of the 15th International Symposium on Rarefied Gas Dynamics, B. G. Teubner, Stuttgart, Vol. 1, 1986, pp. 452-460.

[6]Bird, G. A., Molecular Gas Dynamics, Clarendon Press, Oxford, 1976, pp. 118-132.

[7]Koppenwallner, G. and Legge, H., "Drag of Bodies in Rarefied Hypersonic Flow," Thermophysical Aspects of Re-Entry Flows, Vol. 103 edited by J. N. Moss and C. D. Scott, AIAA, New York, 1986, pp. 44-59.

[8]Hayes, W. D. and Probstein, R. F., Hypersonic Flow Theory, Academic, New York, 1959, pp. 395-406.

Direct Simulation Monte Carlo Method of Shock Reflection on a Wedge

F. Seiler*
*Deutsch-Französisches Forschungsinstitut Saint-Louis (ISL),
Saint-Louis, France*
H. Oertel†
DFVLR-AVA, Göttingen, Federal Republic of Germany
and
B. Schmidt‡
University of Karlsruhe, Karlsruhe, Federal Republic of Germany

Abstract

The direct simulation Monte Carlo method was used to investigate numerically the pseudosteady oblique shock-wave reflection by a low densities. In this numerical study, the real-gas flow is simulated at the molecular level by a large number of model particles following their positions in phase space. In a two-dimensional computational model, the incident shock wave is generated by a piston, which is suddenly set in motion. Then it moves into the flowfield at constant velocity. The incident shock wave, simulated in a model gas of about 10,000 particles, propagates down into the gas, and after some time arrives at a wedge. The reflected shock develops by oblique shock reflection. The calculations are carried out for a shock Mach number of 3 and a monatomic gas. The wedge angles have been set at from 30 to 70-deg. The simulation results give information about the time-dependent regular and Mach reflection flowfield development, beginning at the wedge leading edge and switching into a quasistationary shock reflection arrangement. Regular reflection occurs just at the wedge's edge, whereas for the Mach reflection formation the reflection point has to travel some mean free path along the wedge's surface. It also can be seen that the influence of the boundary layers shifts the transition condition to higher incident shock angles. As expected, smaller reflected

---

Copyright © 1989 by ISL. Published by the American Institute of Aeronautics and Astronautics, Inc. with permission.

* Scientist, Department for hypersonic gasdynamics
† Head of the institute, fluid mechanics, Institut für Theoretische Strömungsmechanik
‡ Professor, fluid mechanics, Institut für Strömungslehre und Strömungsmaschinen

shock angles are produced by the boundary-layer effect. Comparison of experimentally and theoretically determined density distribution and the subsequent flow pattern at the same point in time show a shortcoming of our simulation calculations as a consequence of 1) choosing the cell sizes larger than the mean free path and 2) the small number of particles per cell due to the limited computer capacity. A qualitative agreement can be ascertained with the introduction of a factor for adjusting the length scale of the simulation results. The reflected shock angle and the transition to Mach reflection seem unaffected by these simplifications toward the cell size and particle number.

Introduction

Phenomenon

When a planar moving shock wave hits a reflecting wedge, as shown schematically in Fig. 1, the incident shock wave (I) is reflected at the wedge surface, and a reflected shock wave (R) can be observed. Two fundamental types of shock reflection appear: regular reflection (RR) and Mach reflection (MR). The Mach reflection can be subdivised into single-Mach (SMR), complex-Mach (CMR), and double-Mach reflections (DMR). The so-called irregular or Mach-type reflection of an oblique shock wave was described for the first time by Mach[1] in the last century. Since then, despite many theoretical and experimental efforts, its behavior has not been fully understood. The available theoretical descriptions of this phenomenon are in agreement with experiments only for narrow ranges of parameters. For weak shocks, for example, the agreement is usually poor.

The present paper aims to improve the understanding of the nonstationary shock reflection at a wedge with new numerical results obtained using the direct simulation Monte Carlo (DSMC) method[2] in comparison with similar experimental data published by Walenta.[3] Monte Carlo calculations for the steady shock reflection already were done by Auld and Bird,[4] emphasizing more the macroscopic aspect of the flow than the microscopic one.

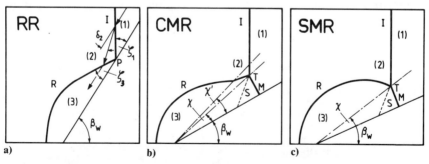

Fig. 1 Illustration of a) regular reflection (RR), b) complex-Mach reflexion (CMR), and c) single-Mach reflection (SMR).

Shock-Wave Reflection

With a Galilean transformation of the unsteady shock reflection by a wedge with respect to the moving shock reflection point P, the flowfield can be considered pseudostationary. In this coordinate system, the wedge surface and the gas in front of the shock wave (1) move at speed u_1. The oblique incident shock wave (I) deflects the flow toward the wall by an angle δ_2. Since the wall is impermeable, the flow must be deflected once more so as to return parallel to the wall by a new reflected shock (R). As shown by several authors,[5] the shock reflection under stationary and quasistationary conditions can be described using the same theoretical approach. The process of shock reflection is best illustrated by using the "shock polar" diagram, in which the pressure p is plotted against the deflection angle δ as shown in Fig. 2. The conditions upstream of the incident shock are represented by point 1 with $p = p_1$ and $\delta = 0$. The region downstream of the incident shock (2) is denoted by point 2 with $\delta = \delta_2$. As the wedge angle β_w decreases, i.e., the shock angle ζ_1 increases, δ_2 increases until the reflected shock reaches the maximum flow deflection angle. This is the conventional criterion for the onset of Mach reflection and is sometimes called the "detachment condition." Hornung et al.[6] advanced another criterion that differs slightly from the detachment condition. They suggest that regular reflection terminates when the flow behind the reflected shock becomes sonic with respect to the reflection point P (sonic criterion). Mach reflection becomes possible where point 3 just equals the pressure achieved from 1 through a normal shock. The smallest deflection angle δ_2, at which Mach reflection is theoretically possible, is denoted usually as the von Neumann criterion.

Boundary-Layer Influence

Regular reflection is theoretically not possible for shock angles $\zeta_1 > \zeta_{1,d}$. Some results[6,7] indicate that the transition occurs for the pseudostationary flow conditions at significantly higher angles $\zeta_{1,tr} > \zeta_{1,d}$ up to several degrees. Taking into account viscosity and heat conduction, the flowfield shown in Fig. 3b can explain the modified transition conditions. The reason for this discrepancy can be seen as an

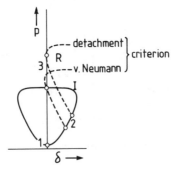

Fig. 2 p, δ map.

effect of the boundary layer behind the incident shock wave, as described by Hornung and Taylor[7] and Hornung.[8] The boundary layer generates a negative displacement thickness $-\delta^*$ and acts as a mass sink. Therefore, the flow deflection passing the reflected shock is smaller than in the inviscid case. The p, δ map of Fig. 3a illustrates this boundary-layer effect. The flow develops a component toward the wall, causing the flow to turn additionally by a small angle ε. It is evident that this behavior results in a transition shock angle $\zeta_{1,tr}$ shifted to larger values by an amount of approximately ε. Consequently, the wedge angle $\beta_w = 90\text{-deg} - \zeta_1$ at which transition occurs moves to smaller values of about $\beta_w - \varepsilon$. A further consequence of the boundary-layer influence results in a smaller reflected shock angle ζ_3, not always being confirmed by the experimental results.[6]

Direct Simulation Monte Carlo Technique

Theoretical Background

To get a better insight into this gasdynamic problem, the flowfield under consideration was investigated numerically by the direct simulation Monte Carlo technique developed by Bird,[2] which was found by many applications to be a successful gas-kinetic procedure for theoretical modeling of rarefied gas flows, although recently it had been observed[9] that the DSMC method cannot conserve exactly the angular momentum. It depends on the number of particles per cell. However, this fact seems to be an important issue if vortical flows are investigated.[9-11] Therefore, the vortex-free oblique shock reflection will not be influenced, at least by the loss of angular momentum.

The DSMC technique is a statistical one, describing the behavior of a gas flow on the molecular level. In this gas-kinetic theory, the distribution function for the molecular velocities provides a statistical description of the gas pattern. The basic equation to be solved is the Boltzmann equation. If vorticity moments are not present, the Monte Carlo calculation results will be a good solution of the Boltzmann equation.[12] However, if vortex flows are investigated, it is necessary to conserve the angular momentum for the DSMC method, too. This can be done by optimizing the cell size and the number of particles.

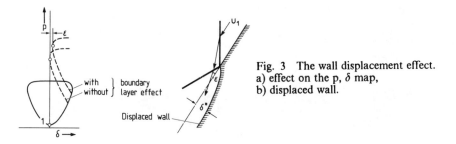

Fig. 3 The wall displacement effect.
a) effect on the p, δ map,
b) displaced wall.

In the DSMC numerical study, the real wedge flow is simulated by using a large number of model gas particles, following their positions within a network of cells according to the velocity components. In order to simulate the shock reflection from the wedge, a computational model was developed, as shown in principle in Fig. 4. The incident shock wave is generated by a piston that moves with constant velocity into the flowfield from the left. The shock wave propagates down the gas and impinges on the wedge. At the wedge, the particles on the one hand are reflected specularly to suppress viscosity and heat conduction effects, and on the other hand are reflected diffusely to model a boundary-layer development at the wedge surface. At the other boundaries, specular reflection is used. A detailed description of the theoretical model is given by Seiler.[13] In addition, it should be noticed that the hard sphere model was used as an intermolecular potential model. Moreover, the fluctuations of the macroscopic quantities, extracted by sampling and averaging over appropriate molecular quantities in each cell, are reduced by smoothing the computational results.

Calculation Parameters

The calculations presented in Figs. 5 and 6 were carried out at the computing center of DFVLR in Oberpfaffenhofen on an IBM 3081 computer. In the simulation procedure, all of the lengths are normalized by the mean free path λ_1 of the real-gas particles in front of the incident shock wave (condition 1). The initial number of simulation particles within the whole flowfield was 10,290 for all runs. The cell size in the ξ direction ($\xi = x/\lambda_1$) was about $20 \lambda_1$, being kept approximately constant during the ongoing simulation procedure. The cell size in the η direction ($\eta = y/\lambda_1$) has been enlarged with increasing distance from the surface, beginning with a width of $20 \lambda_1$. The dimensions of the model space were set to $\xi = 1600$ and $\eta = 10,000$. Since the cell size was chosen larger than the mean free path, a smearing effect of the shock-wave profiles has been taken into account, because in this study the macroscopic effects of the oblique shock reflection are investigated and not the microscopic flow structure. For this reason, it was tolerated that the shock width was set by the cell dimensions rather than by the

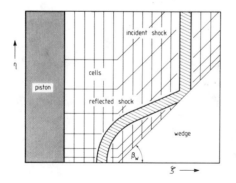

Fig. 4 Computational model.

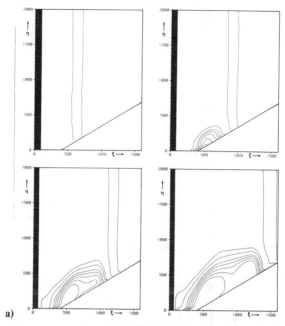

Fig. 5a Density contours from the DSMC calculations, $M_s = 3$, 30-deg wedge, contour values without boundary layer: $\rho_n = 0.1, 0.6, 1.1, 1.2, 1.3, 1.35, 1.4, 1.45, 1.5$.

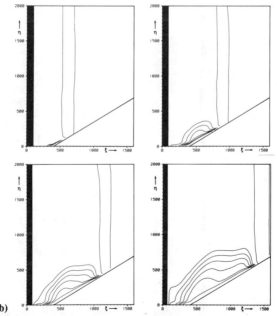

Fig. 5b Density contours from the DSMC calculations, $M_s = 3$, 30-deg wedge, contour values with boundary layer: $\rho_n = 0.1, 0.72, 1.1, 1.2, 1.3, 1.4, 1.5, 1.7$.

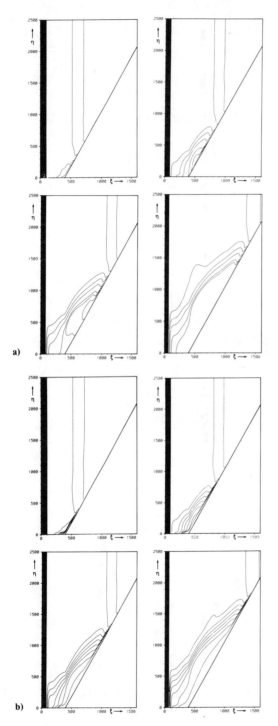

Fig. 6a Density contours from the DSMC calculations, $M_s = 3$, 60-deg wedge, contour values without boundary layer: $\rho_n = 0.1$, 0.72, 1.2, 1.8, 2.1, 2.2, 2.3.

Fig. 6b Density contours from the DSMC calculations, $M_s = 3$, 60-deg wedge, contour values with boundary layer: $\rho_n = 0.1$, 0.72, 1.1, 1.3, 1.5, 1.7, 1.9, 2.1, 2.3.

mean free path. The main interest of these calculations was to find out if the DSMC method in the presence of the discrepancies mentioned earlier still provides a realistic picture of the oblique shock reflection on a wedge.

The shock reflection was calculated for an incident shock Mach number of 3 in a perfect monatomic gas. The wedge angles have been set at 30 and 60-deg. The time counter in the simulation calculation was continued until the incident shock wave moved about 1) 1500 λ_1 in the case of the 30-deg wedge and 2) 2000 λ_1 for the 60-deg wedge on the wedge surface downstream of the leading edge.

Results

Reflection Types

The results of the Monte Carlo calculations are presented for the 30-deg wedge in Fig. 5 and for the 60-deg wedge in Fig. 6 by means of lines of constant density $\rho_n = (\rho - \rho_1)/(\rho_2 - \rho_1)$. The curves of equal density are shown for subsequent positions of the incident shock wave. The simulation calculations give information about the formation of the oblique shock reflection beginning at the leading edge of the wedge up to a quasistationary shock reflection condition. The differences between the diffuse and the specular reflections at the surface can be seen clearly. Inside the boundary layer, the density increases strongly toward the wall. The maximal density values at the wedge surface are the following: 1) 30-deg specular: $\rho_n = 1.7$, 2) 30-deg diffuse: $\rho_n = 4.7$, 3) 60-deg specular: $\rho_n = 2.55$, and 4) 60-deg diffuse: $\rho_n = 8.7$. For a 30-deg wedge angle, a well-established Mach-type reflection can be recognized. This wedge angle is below the detachment criterion value of $\beta_{w,d} = 54.2$-deg, at which Mach reflection should appear. Above this wedge angle, e.g., for the 60-deg wedge, regular reflection occurs as predicted by the detachment criterion. It seems that, for the Mach reflection (Fig. 5) as well as for the regular reflexion (Fig. 6), a length of several mean free paths λ_1 is required. This has been found experimentally by Walenta,[3] too. The comparison between the experimental results of Walenta,[3] which are given in Fig. 7, and the results of Fig. 5, shows obvious differences in regard to the ξ length scale. However, there appears to be a reasonable qualitative similarity between the shape of the constant density curves obtained experimentally at small distances from the wedge's leading edge and those from the Monte Carlo calculations further downstream (Fig. 5b). Also, the experimentally gathered kink in the density contours of Fig. 7, as seen for CMR, can be described by the boundary-layer-affected results of Fig. 5b. Nevertheless, the suspicion is that the effective mean free path in the Monte-Carlo calculation is really controlled by the size of the cells, chosen here to be much larger than the mean free path. The reason for this effect will be found in the collision procedure of the DSMC technique: the collision pairs are selected randomly and, for not a very large number of cell particles, a particle may undergo only one collision in a cell and then move to another one. As a consequence,

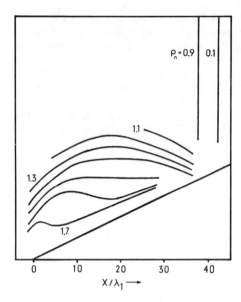

Fig. 7 Experiment Walenta,[3] shock Mach number $M_s = 2.75$, wedge angle $\beta_w = 25$-deg.

the cell size acts as a "fictitious mean free path." For experimental comparison, the length scales of the DSMC results of Figs. 5 and 6 have to be reduced by a factor taking into account the dimension of the cell width (about 20 λ_1). Then a qualitative similarity also with regard to the ξ length scale is evident.

Further Monte Carlo calculations[12] have been done on a Bourroughs 7700 computer at the University of Karlsruhe with much smaller cell sizes (ξ and η directions about $2\lambda_1$) with wedge angles varying from 30 to 70-deg in 10-deg steps for the same shock Mach number, $M_s = 3$. Some results of these calculations are presented in Fig. 8, showing equal density curves for the wedge angles $\eta_w = 30$ to 60-deg. Without boundary-layer formation (Fig. 8a) for wedge angles $\beta_w = 70$ and 60-deg the density contours show a well-developed RR. On the contours for wedge angles of 50 and 40-deg the Mach stem can be seen clearly. At both angles, there appears a growing region of subsonic flow viewed from the frame of reference of the reflection point, which is a characteristic sign for MR. In addition, a kink in the outer density contours is established, as can be seen for CMR. Finally, for $\beta_w = 30$-deg, pure SMR is present. A clear Mach stem is visible, and the triple point has moved significantly off from the wall.

The density results obtained with diffuse reflection at the wedge surface are given in Fig. 8b. RR appears for wedge angles $\beta_w = 70$, 60, and 50-deg. That means that RR exists at a wedge angle of 50-deg, where MR should be observed. As discussed earlier, the wall-displacement effect shifts the transition point to smaller wedge angles β_w in agreement with the expected boundary-layer effect. For $\beta_w = 40$-deg, the reflection type cannot be determined exactly, because the density field contains elements of both RR as well as MR.

DSMC METHOD OF SHOCK REFLECTION 527

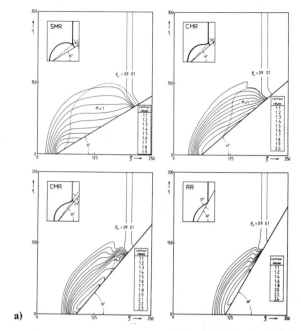

Fig. 8a Lines of constant density without boundary-layer influence.

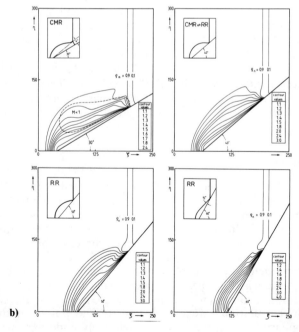

Fig. 8b Lines of constant density with boundary-layer influence.

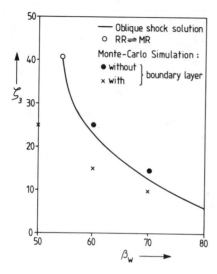

Fig. 9 Reflected shock angle ζ_3 vs. wedge angle β_w, DSMC data from Fig. 8.

Regarding the density contours for β_w = 30-deg, the kink mentioned earlier is visible again, therefore, CMR can be assumed.

Comparing corresponding wedge angles of Fig. 5 resp. 6 with those of Fig. 8, it is apparent that the Monte Carlo calculations, made with different cell widths, lead to qualitatively similar results. Approximate quantitative agreement can be obtained by reducing the ξ and η length scales by the factor given from the cell sizes used (Figs. 5 and 6: factor of about 20; Fig. 8: factor of about 2). Then there also appears a reasonable similarity of both Figs. 5b and 8b (β_w = 30-deg) with the shapes of the constant density curves of Fig. 7, which are obtained experimentally by Walenta[3] (β_w = 25-deg). These facts support the requirement of DSMC calculations with many more model particles moving inside the cells with cell dimensions smaller than the mean free path.

Displacement effects

In Fig. 9, the reflected shock angle ζ_3 is plotted against the wedge angle β_w. The solid line represents the solution of the oblique shock equations for the two shock configurations. It can be seen that the simulation Monte Carlo results without boundary-layer effects are slightly above the two shock solutions, whereas the values with displacement are lower, as theoretically expected.

The dependence of the triple-point trajectory angle χ on the wedge angle β_w is plotted in Fig. 10. The calculated solid line as well as the experimental points are given by Ben-Dor and Glass.[14] The results of the Monte Carlo calculations are also included. The points in the cases without boundary-layer influence are in relatively good agreement with the data of Ben-Dor and Glass. The simulation done with the

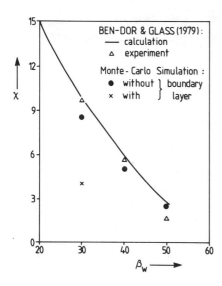

Fig. 10 Triple-point trajectory χ vs. wedge angle β_w, DSMC data from Fig. 8.

boundary-layer considerations mentioned earlier shows much lower values of χ.

The Monte Carlo simulated reflection-type results are compared in Fig. 11 with the nonstationary shock-wave reflection regions indicated by Ben-Dor and Glass[14] for monatomic gases in a M_s, β'_w map and a M_s, β_w map. The effective wedge angle β'_w equals $\beta_w + \chi$ in the domains of SMR, CMR, and DMR, and β_w in the domain of RR. The points deduced from the calculations executed without displacement effects are well within the predicted regions, whereas those with wall displacement show a slight disagreement. These points are shifted to smaller effective wedge angles, i.e., the transition to Mach reflection occurs at higher incident shock angles ζ_1, which is in agreement with the considerations made about the displacement influence on the transition condition.

Remarks

The pseudostationary oblique shock-wave reflection at a wedge has been investigated using the direct simulation Monte Carlo method. Calculations have been carried out for monatomic perfect gases at a shock Mach number of 3. Viscosity and heat conduction were taken into account by diffuse reflection of the model particles at the wedge surface. Wedge angles ranging from 30 to 70-deg were considered. Following the detachment criterion, transition from regular (RR) to Mach reflection (MR) occurs at the wedge angle $\beta_w = 54.2$-deg. The inviscid calculations are in agreement with the predicted transition to MR. The boundary-layer influence produces a smaller transition wedge angle, resulting in a higher shock transition angle ζ_1. This result agrees with the negative wall-displacement effect and is supported by

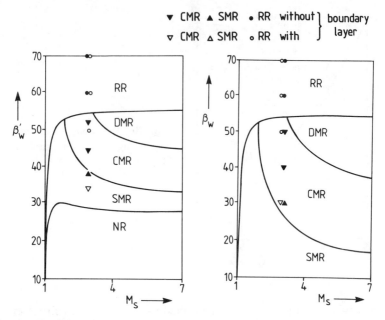

Fig. 11 Domains of pseudo-stationary shock wave reflections from Ben-Dor and Glass,[14] DSMC data obtained from Fig. 8.

experimental results.[7] As theoretically predicted, the calculated reflected shock angles ζ_3 with displacement effect are below the inviscid calculations.

The main aspect of this paper is a more macroscopic description of the influence of the boundary layer on 1) the transition to MR, 2) the reflected shock angles ζ_3, and 3) the growth of the length of the Mach stem, i.e., the triple-point trajectory angle χ. Therefore, it was accepted that the shock thickness is affected by the cell sizes, being chosen larger than the mean free path λ_1. Moreover, the calculations have been carried out with a cell size of about 20 λ_1. The influence of the cell size is demonstrated by using calculations with a cell width of about 2 λ_1. These results are more in agreement with experimental data.[3] Both DSMC results have to be reduced by a factor 20, resp. 2, indicating that the cell size dominates the mean free path. For an efficient application of the DSMC method, it is necessary to use a powerful computer for eliminating the influence of cell size and particle number by optimizing the grid division ($< 1 \lambda_1$) and the number of particles per cell (> 10). Then, as the recent DSMC results of Wetzel and Oertel[11] demonstrate, the loss of angular momentum also can be diminished to that point, where the angular momentum will be computed almost exactly. Further calculations should be done by giving preference to the mentioned cell and particle dimensioning to support the given oblique shock reflection interpretation of the presented results gained with the direct simulation Monte Carlo method.

References

[1] Mach, E., Akademie der Wissenschaften, Wien, 77, 1878, pp. 1228.

[2] Bird, G. A., Molecular Gas Dynamics, Clarendon Press, Oxford, 1976.

[3] Walenta, Z. A., "Mach Reflexion of a Moving Plane Shock Wave Under Rarefied Flow Conditions," Proceedings of the 16th International Symposium on Shock Tubes and Waves, Edited by H. Grönig, VCH Verlagsgesellschaft mbH, Weinheim, 1988.

[4] Auld, D. J. and Bird, G. A., "The Transition from Regular to Mach Reflexion," Proceedings of the 9th Fluid and Plasma Dynamics Conference, San Diego, CA, 1976.

[5] Sternberg, J., "Triple-Shock-Wave Intersections," Physics of Fluids Vol. 2, No. 2, 1959, pp. 179-206.

[6] Hornung, H. G., Oertel, H., and Sandeman, R., "Transition to Mach Reflexion of Shock Waves in Steady and Pseudosteady Flow With and Without Relaxation," Journal of Fluid Mechanics, Vol. 90, Pt. 3, 1979, pp. 541-560.

[7] Hornung, H. G. and Taylor, J. R., "Transition to Mach Reflexion of Shock Waves. Part 1. The Effect of Viscosity in the Pseudosteady Case," Department of Physics, Australian National University.

[8] Hornung, H., "Regular and Mach Reflection of Shock Waves," Annual Review of Fluid Mechanics, Vol. 18, No. 33, 1985, p. 58.

[9] Meiburg, E., "Direct Simulation Techniques for the Boltzmann Equation," DFVLR-FB 85-13, 1985.

[10] Meiburg, E., "Comparison of the Molecular Dynamics Method and the Direct Simulation Technique for Flows Around Simple Geometries," Physics of Fluids, Vol. 29, 1986, pp. 3107-3113.

[11] Wetzel, W. and Oertel, H., "Gas-Kinetical Simulation of Vortical Flows," Acta Mechanica, Vol. 70, 1987, pp. 127-143.

[12] Bird, G. A., "Direct Simulation and the Boltzmann Equation," Physics of Fluids Vol. 13, No. 11, 1970.

[13] Seiler, F., "Pseudo-Stationary Mach Reflexion of Shock Waves", Proceedings of the 15th International Symposium on Shock Tubes and Waves, Edited by D. Bershader and R. Hanson, Stanford University Press, Stanford, CA, 1986.

[14] Ben-Dor, G. and Glass, I. I., "Domains and Boundaries of Non-stationary Oblique Shock-Wave Reflexions. 2. Monatomic Gas," Journal of Fluid Mechanics, Vol. 96, Pt. 4, 1980, pp. 735-756.

Interference Effects on the Hypersonic, Rarefied Flow About a Flat Plate

Richard G. Wilmoth*
NASA Langley Research Center, Hampton, Virginia

Abstract

The Direct Simulation Monte Carlo method is used to study the interference effects under hypersonic, rarefied flow conditions of nearby surfaces on a flat plate. Calculations focus on shock-boundary-layer and shock-lip interactions in hypersonic inlets. Results are presented for two cases: (1) a flat plate with different leading-edge shapes located over a flat lower wall and (2) a blunt-edged flat plate over a 5-deg wedge. The problems simulated correspond to a typical entry flight condition of 7.5 km/s at altitudes of 75-90 km. The results show increases in predicted local heating rates for shock-boundary-layer and shock-lip interactions that are qualitatively similar to those observed experimentally at much higher densities.

Nomenclature

C_H	= heat-transfer coefficient, $2q/\rho_\infty U_\infty^3$
C_p	= pressure coefficient, $2p/\rho_\infty U_\infty^2$
Kn_∞	= freestream Knudsen number, λ_∞/R_N
ℓ	= characteristic dimension
M_∞	= freestream Mach number
p	= pressure
q	= heat flux
R_N	= leading-edge nose radius
Re_∞	= Reynolds number, $\rho_\infty U_\infty \ell/\mu_\infty$
T	= temperature
U_∞	= freestream velocity

Copyright © 1989 by the American Institute of Aeronautics and Astronautics, Inc. No copyright is asserted in the United States under Title 17, U. S. Code. The U. S. Government has a royalty-free license to exercise all rights under the copyright claimed herein for governmental purposes. All other rights are reserved by the copyright owner.

*Aerospace Engineer, Aerothermodynamics Branch, Space Systems Division.

X_i	= mole fraction of species i
x	= coordinate measured from leading edge of flat plate
λ_∞	= freestream mean free path
μ_∞	= freestream viscosity
ρ_∞	= freestream density
θ	= angular position on circular leading edge

Subscripts

o	= stagnation value
is	= isolated value

Introduction

Viscous-inviscid interactions occurring in high-Mach-number flight represent one of the most severe problems associated with hypersonic cruise and entry vehicles. The strongest interactions are usually associated with some form of shock impingement such as that occurring on the cowl lip or wall boundary layers of hypersonic inlets as illustrated in Fig. 1. These interactions can result in high local heating which may lead to component failures. An excellent review of the variety of problems that can occur along with examples of some actual failures is given by Korkegi.[1]

The most severe conditions for viscous-inviscid interaction problems are usually associated with the higher densities of the continuum flow regime. However, similar interactions are expected to occur at the lower densities associated with initial entry or suborbital maneuver conditions (typically above 70 km). Although the strength of the interactions may be expected to be weaker in these transitional flows, localized problems may still occur. Furthermore, even when the overall flowfield is essentially continuum, the analysis of the problem may require modifications to account for local transitional or rarefied flow

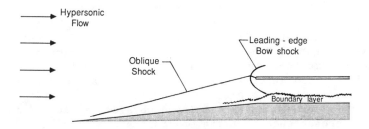

Fig. 1 Typical shock interactions in hypersonic inlets.

effects.[2,3] The use of a purely continuum-based analysis typically leads to an overprediction of the local heating rates in such cases.

The Direct Simulation Monte Carlo (DSMC) method of Bird[4,5] has been used extensively to study a variety of transitional and rarefied flow problems in hypersonic flows. For the most part, these studies have focused on modeling the physics[4-9] or on calculations of complex flows about simple geometries.[10-15] Studies of more complex geometries have been limited, since they require the more difficult three-dimensional codes[16] or require prohibitively large computer storage and CPU time to resolve the flow around complex shapes even for two-dimensional problems. However, with the current availability of supercomputers and with recent improvements to existing DSMC computer codes[17,18] that provide easier setup for complex geometries, multi-surface interaction problems can be more readily studied.

The purpose of the present paper is to present initial results of a study of some of the basic interactions shown in Fig. 1 for hypersonic inlets. Although calculations are presented for specific geometries (see Fig. 2), the results are mainly intended to demonstrate the nature of the interactions at low densities and to show both the similarities and the differences in comparison to analogous interactions in continuum flows. A flat plate was used to simulate an inlet cowl, since extensive studies have been made of the flow over isolated flat plates and the expected flow behavior is reasonably well understood.[12,13,18-22]

The results in this paper focus on some of the basic interference effects between two parallel flat plates and on the interference caused by an oblique shock impinging on the leading edge of the plate. The calculations were performed at nominal entry conditions of 7.5 km/s at an altitude of 90 km for the parallel flat plates and over an altitude range of 75-90 km for the shock impingement calculations.

Direct Simulation Monte Carlo Method

The DSMC method of Bird has been described extensively in the literature.[4,5] The method consists of simulating a gas by some thousands of molecules whose positions, velocities, and internal states are stored and modified with time as the molecules move, collide, and undergo boundary interactions in simulated physical space. A time step is chosen such that the movement and collision processes can be assumed to be uncoupled over the duration

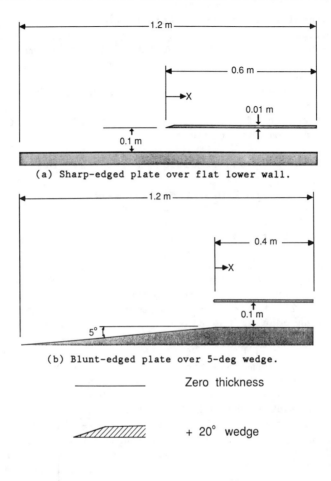

Fig. 2 Geometries used for the calculations.

of the step. The calculations are started with an initial state such as a vacuum or uniform equilibrium flow. The simulation time is identified with physical time in the real flow, and all calculations are treated as unsteady. When boundary conditions appropriate to steady flow are imposed, the solution is the asymptotic limit of the unsteady flow.

A physical space computational cell network is required only to facilitate the choice of collision pairs and the sampling of flow properties. The cells may be divided into subcells for selection of collision pairs to better simulate the nearest neighbor collisions in regions of high vorticity. The collisions are simulated using the variable hard sphere (VHS) molecular model with modifications to account for rotational and vibrational states in diatomic or polyatomic gases.[5]

Geometries and Conditions for the Calculations

Two basic two-dimensional geometries, shown in Fig. 2, were studied. Figure 2a shows a sharp-edged flat plate parallel to a flat lower wall. This geometry was used to study the interactions between the lower wall and the flat plate with different boundary conditions imposed on the lower wall. The effects of the leading-edge shapes shown in Fig. 2c were also computed. A set of calculations was also made with the lower wall representing a 5-deg wedge ramp (Fig. 2b) to produce an oblique shock that impinged on the leading edge of the plate. For these calculations, only the blunt (circular) leading edge was used.

Calculations for the flat lower wall were performed only for the 90-km condition using the freestream values given in Table 1. Calculations for the 5-deg wedge lower wall were made at all four altitudes given in Table 1. The freestream Mach number was 27.2 in all cases. The species mole fractions at altitudes below 90 km were held constant at values appropriate to 90 km using the tables of Ref. 23. Thus the chemical species simulation is realistic only at the 90-km condition. Some of the flow parameters, expressed in terms of the leading-edge nose radius, are given in Table 2.

Table 1 Freestream conditions

Altitude, km	Density, kg/m^3	U_∞, km/s	T_∞, K	Mole fractions[a]			Re_∞/ℓ, m^{-1}	λ_∞, m
				X_{O_2}	X_{N_2}	X_O		
90	3.4x10^{-6}	7.5	188	0.209	0.788	0.003	2.0x10^3	2.4x10^{-2}
85	8.2x10^{-6}	7.5	188	0.209	0.788	0.003	4.9x10^3	9.9x10^{-3}
80	1.8x10^{-5}	7.5	188	0.209	0.788	0.003	1.0x10^4	4.4x10^{-3}
75	4.0x10^{-5}	7.5	188	0.209	0.788	0.003	4.3x10^4	2.2x10^{-3}

[a]Held constant at values appropriate for 90 km.

Table 2 Parameters based on nose radius of blunt-edged flat plate
$R_N = 0.005$ m

Altitude, km	Re_∞	Kn_∞
90	10	4.74
85	24	1.98
80	52	0.88
75	109	0.41

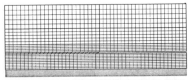
SHARP-EDGED PLATE OVER FLAT LOWER WALL

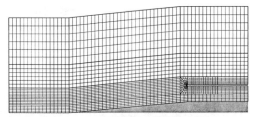
BLUNT-EDGED PLATE OVER 5 DEGREE WEDGE

Fig. 3 Typical grids for DSMC calculations.

The computational domains and typical cell grids for the two geometries are shown in Fig. 3. The heavy solid lines represent region boundaries within which the time step and the molecule weighting factor (relating the number of computational molecules to the number of physical molecules) are constant. For the circular leading-edge cases, cells were clustered around the leading edge. The cell size was reduced at the lower altitudes in order to minimize the flow gradients across the cell in a direction normal to the leading-edge surface.

For all calculations, the wall temperature of the flat plate was assumed to be constant at 1000°K, and gas-surface collisions were assumed to be diffuse with full thermal accommodation. The same wall conditions were generally used for the lower wall except for a few cases with the flat lower wall in which the boundary was treated as specular or as a freestream boundary.

Results and Discussion

Isolated Flat Plate Predictions

To establish the undisturbed nature of the isolated flat plate flow, calculations were made at the 90-km conditions with each of the leading-edge shapes and with the lower wall treated as a freestream boundary. The pressure and heat-transfer distributions along the upper surface of the plate for these cases are shown in Fig. 4. At 90 km, the length of the plate corresponds to about 25 λ_∞, and the results are typical of transitional flow behavior.[13,24] For the -20-deg wedge, the upper surface is flat, and the results are nearly identical to the zero-thickness case. The slight differences between the -20-deg and zero-thickness results are attributed to the fact that molecules striking the lower surface of the leading edge have a fairly high probability of reaching the upper surface through gas-gas collisions at this Knudsen number. With the +20-deg wedge and circular leading edges, the pressure and heat-transfer coefficients are highest near the leading edge and decrease downstream, approaching the zero-thickness results near the end of the plate. Although the +20-deg wedge shows higher pressures and heat transfer in Fig. 4 than the circular leading edge, results are shown only for the flat portion of the plate downstream of the leading edge. The peak stagnation values on the leading edge itself are actually higher for the circular leading edge. In all cases, the pressure and heat-transfer coefficients are much higher than the calculated free molecular values of 0.003 for C_p and 0.01 for C_H at these conditions. The trends shown in Fig. 4 are qualitatively similar to those expected in continuum flows.

Interactions with Flat Lower Wall

The pressure and heat-transfer distributions along the lower surface of the plate with the blunt leading edge are shown in Fig. 5 for different lower wall boundary conditions. The freestream lower boundary condition represents the isolated or undisturbed flat plate. A specular lower boundary is equivalent to a plane of symmetry, and the geometry for this calculation is analogous to the entrance of a two-dimensional channel whose height is twice the distance between the plate and the symmetry plane. The results show a gradual rise in the pressure and heat transfer due to the growth of the boundary layer along the lower surface of the (upper) flat plate.

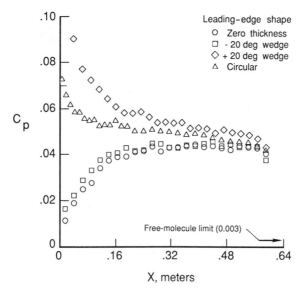

(a) Pressure distributions.

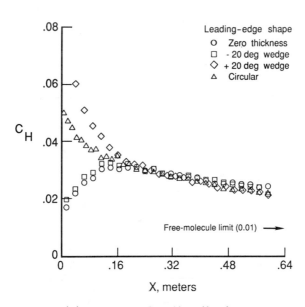

(b) Heat-transfer distributions.

Fig. 4 Effect of leading-edge shape on pressure and heat transfer along upper surface of isolated flat plate (altitude = 90 km).

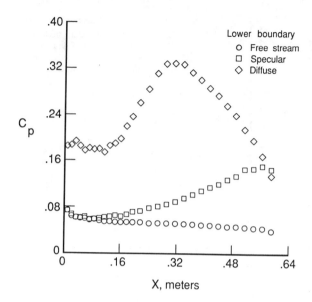

(a) Pressure distributions.

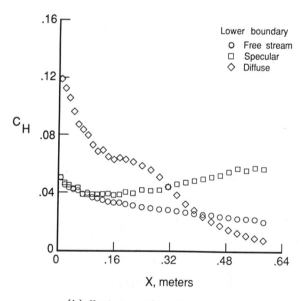

(b) Heat-transfer distributions.

Fig. 5 Effect of lower wall boundary on pressure and heat transfer along lower surface of flat plate (altitude = 90 km).

With a diffuse lower wall, the geometry becomes similar to an inlet with a flat ramp, and the results show a more complex behavior. The pressures are much higher than the undisturbed predictions, and the distribution shows a steep rise starting at about 0.16 m (approximately 7 λ_∞). The pressure reaches a maximum at about 0.32 m and then decreases continuously to the exit of the channel. The heat-transfer distribution exhibits a maximum near the leading edge and decreases continuously downstream. However, there does appear to be an inflection in the heat-transfer distribution near the location of the pressure rise.

The previously described behavior is caused by the interaction of a diffuse bow shock generated by the leading edge, which strikes and is reflected from the lower wall. This reflection is evident in the pressure distribution along the specular lower wall shown in Fig. 6. Significant differences in the location of the reflected bow shock are seen for different leading-edge shapes. These differences are probably more pronounced at the 90-km conditions than might be expected at lower altitudes because of the large curvature and thickness of the shock, which is roughly the same order of magnitude as the channel height.

For the diffuse lower wall (see Fig. 7), the differences in the interaction for different leading-edge

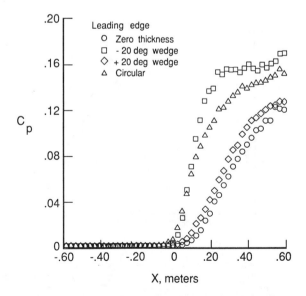

Fig. 6 Interaction of leading-edge shock with a specular lower wall boundary (altitude = 90 km).

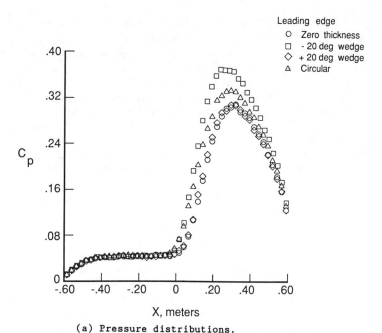

(a) Pressure distributions.

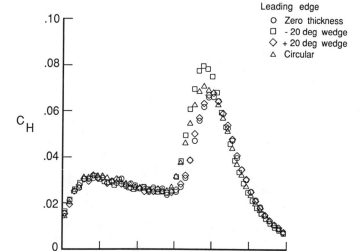

(b) Heat transfer distributions.

Fig. 7 Interaction of leading-edge shock with a diffuse lower wall boundary (altitude = 90 km).

shapes are not as pronounced. Also, the relative strength of the interaction (in terms of the ratio of pressures across the interaction region) is reduced by the presence of the lower wall boundary layer (although the overall pressures are higher due to boundary-layer displacement effects). The results shown here are qualitatively similar to the experimental data of Kaufman and Johnson[25] for weak shocks incident on laminar boundary layers. Although the experiments were conducted at much higher densities ($Re_\infty \approx 10^6$/m) and at a lower Mach number ($M_\infty=8$), a reasonable comparison can be made based on the peak pressure rise. For the results in Fig. 7, the ratio of peak to upstream pressures is about 8-10, whereas the ratio of the peak to upstream heating rates is about 2.5-3.0. Based on the correlation of heating-rate amplification to peak-pressure amplification given in Ref. 25, the heating-rate amplification at the same pressure amplification would be roughly 15-20 at the Reynolds numbers of the experiment. The weaker interaction for the present calculations is perhaps not too surprising, since the shock is much more diffuse at the lower densities.

The pressure and heat-transfer distributions downstream of the shock interaction on the diffuse lower wall shows significant streamwise gradients extending to the exit of the internal region. Since the boundary layers are quite thick relative to the channel height, a significant portion of the internal flow is subsonic. Therefore, it is quite likely that the location of the exit and the boundary conditions imposed on the exit plane may have a significant influence on the shock-boundary-layer interaction. Because of this possible influence, the postshock flow quantities may not be representative of those for isolated shock reflections from laminar boundary layers.

Interaction with 5-deg Wedge Lower Wall

For the calculations with the 5-deg wedge lower wall, the main interest in this study is on the interaction of the oblique shock generated by the wedge with the leading edge of the flat plate. Therefore, the following results focus on the surface properties on the blunt lip of the plate.

The pressure and heat-transfer distributions around the leading edge of an isolated flat plate are shown in Fig. 8 for several altitudes. The stagnation pressure coefficient shows a slight decrease with decreasing altitude, whereas the stagnation heat-transfer coefficient decreases somewhat more rapidly. (It should be noted that

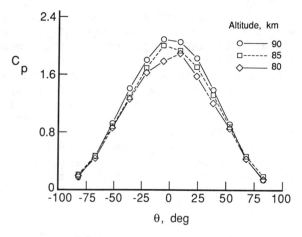

(a) Pressure distributions.

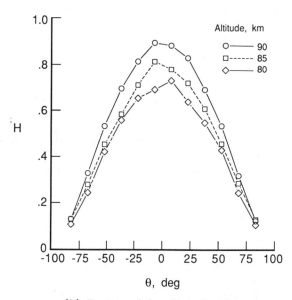

(b) Heat transfer distributions.

Fig. 8 Isolated blunt leading edge.

although C_H decreases with decreasing altitude, the absolute heating rate actually increases.) The results shown here are in reasonable agreement with the predictions of Cuda and Moss,[10] which were obtained under similar freestream conditions but with a larger nose radius. A correlation of the stagnation heat-transfer predictions for the present calculations with those of Cuda and Moss is presented later in this paper.

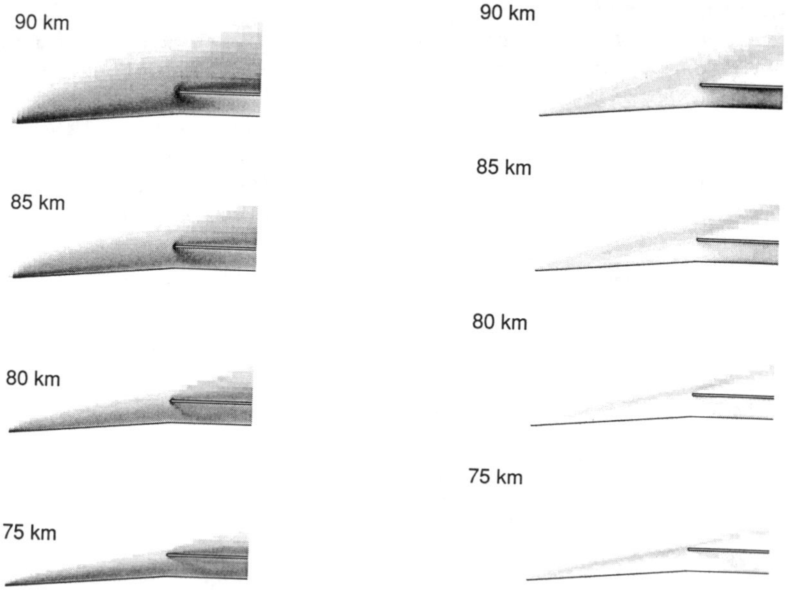

Fig. 9 Gray-shaded contours of flowfield for 5-deg wedge lower wall (darker shades denote higher values).

(a) Temperature.   (b) Density.

As a result of the viscous shock layer caused by the 5-deg wedge, the leading edge of the flat plate is subjected to higher temperatures and densities. The qualitative features of the flowfield are demonstrated by the gray-shaded contours shown in Fig. 9, where the darker regions represent higher temperatures and densities. The variation in the extent of the overall thermal environment with altitude is clearly shown by the temperature contours, whereas the thickness and location of the oblique wedge shock are reasonably evident in the density contours. At 90 km, the distance from the start of the 5-deg wedge to the leading edge of the flat plate corresponds to about 50 λ_∞. At 75 km, this distance corresponds to about 800 λ_∞. At 90 km, the shock is very thick relative to the leading-edge radius, and the peak density in the undisturbed shock is about twice the freestream value. At 75 km, the shock is more clearly defined with a peak density that is about four times the freestream density.

It should be noted that the gradients in flow quantities across the shock are probably not adequately resolved at the lower altitudes because of the relatively coarse grid used for sampling. Although an attempt was made to accurately simulate the collisions in the shock by increasing the number of subcells at the lower altitudes, the average flow properties are always determined from the total sample within a cell. Since the gray contours are produced by filling each cell with a level-of-gray halftone shading that is proportional to the flow quantity, the shock may appear to be artificially smeared in these contours. However, the cell size used near the leading-edge surface was adequate to give reasonably

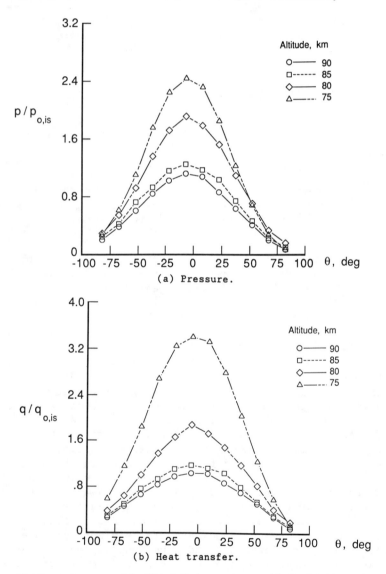

Fig. 10 Interference effects of 5-deg wedge on blunt leading edge.

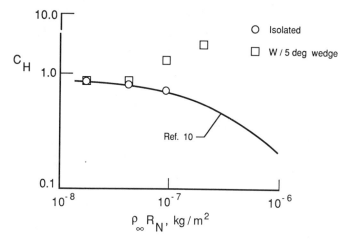

Fig. 11 Stagnation heat transfer on blunt leading edge.

accurate predictions of surface heating as determined by a cell-size sensitivity study.

The pressure and heating-rate distributions around the plate leading edge with the 5-deg wedge shock are shown in Fig. 10. Both quantities have been nondimensionalized by the peak stagnation values for the isolated leading edge at the corresponding altitude. Therefore, peak values greater than 1 indicate an amplification of the pressure or heating rate resulting from impingement of the shock. At 90 and 85 km, there is little effect of the shock, whereas the 80- and 75-km results show significant amplification of both pressure and heating rate in the stagnation region.

A comparison of the heat-transfer coefficients for the 5-deg wedge with those for the isolated leading edge is given in Fig. 11 as a function of the product of the freestream density and the leading-edge radius. The solid line is taken from the results of Cuda and Moss,[10] and the present results correlate well for isolated blunt leading edges. The impingement of the wedge shock causes the heat-transfer coefficient to increase significantly with increasing density. Since the actual heating rate is proportional to the product of C_H and the freestream density, the increase in heating is quite significant.

Shock impingement effects on circular leading edges have been studied extensively for hypersonic flows at much higher densities than the present study. The works of Edney,[26] Johnson and Kaufman,[27] and more recently Wieting and Holden[28] describe in detail the nature of the interaction at continuum flow conditions. Although a direct comparison of the flow details from the present DSMC

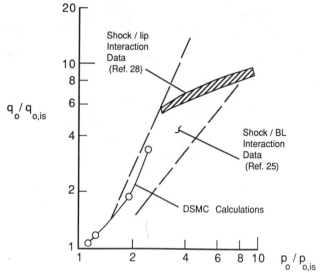

Fig. 12 Correlation of peak heating amplification with pressure amplification.

calculations with those given in these references is inappropriate, a qualitative comparison of the correlation of heating-rate amplification with pressure amplification may be appropriate. Using such a correlation, a comparison of the DSMC results with the shock-boundary-layer interaction data of Kaufman and Johnson[25] and the shock-lip interaction data of Wieting and Holden[28] is given in Fig. 12. Although no direct conclusions should be drawn from this comparison, the interaction effects predicted by the DSMC method do show similar trends to those for shock interactions with laminar boundary layers. It also appears that the DSMC results adequately represent the shock-lip interactions that might be expected at the lower densities. Further DSMC calculations at the higher densitites and/or experimental measurements at the lower densities are needed.

Concluding Remarks

This initial DSMC study of interference effects on the hypersonic, rarefied flow about flat plates has focused on viscous-inviscid interactions relevant to inlet-type geometries. The results have been interpreted from somewhat of a continuum viewpoint. However, for the flight conditions presented, the distinction between viscous and inviscid regions is difficult at best. Therefore, the DSMC results need to be examined in more detail to better understand the physics involved in viscous-inviscid interactions at low densities.

The geometries studied and the boundary conditions used do not necessarily represent realistic hypersonic vehicle problems. This is particularly true for the internal flow simulations where the mass flow rate depends strongly on the internal geometry and exit boundary conditions. The physical scale of the geometry is also important, and calculations for much higher Reynolds numbers would be useful. However, the DSMC method is not currently practical for very large vehicles at the high Reynolds numbers. Therefore, methods that can combine the computational capabilities of the DSMC method with more conventional continuum methods are needed.

In spite of the limitations mentioned, the results of this initial study have demonstrated the usefulness of the DSMC method for the analysis of complex interaction problems such as those present in hypersonic inlets. The results show qualitative trends that appear to be similar to those given by available experimental data.

Acknowledgments

The author wishes to express his appreciation to G. A. Bird, J. N. Moss, and E. V. Zoby for many helpful discussions.

References

[1]Korkegi, R. H., "Survey of Viscous Interactions Associated with High Mach Number Flight," AIAA Journal, Vol. 9, May 1971, pp. 771-784.

[2]Moss, J. N., "Direct Simulation of Hypersonic Transitional Flow," Rarefied Gas Dynamics, Vol. 1, edited by V. Boffi and C. Cercignani, Teubner, Stuttgart, FRG, 1986, pp. 384-399.

[3]Gupta, R. N. and Simmonds, A. L., "Hypersonic Low-Density Solutions of the Navier-Stokes Equations with Chemical Nonequilibrium and Multicomponent Surface Slip," AIAA Paper 86-1349, June 1986.

[4]Bird, G. A., Molecular Gas Dynamics, Clarendon, Oxford, U.K., 1976.

[5]Bird, G. A., "Monte-Carlo Simulation in an Engineering Context," Progress in Astronautics and Aeronautics: Rarefied Gas Dynamics, Vol. 74, Part I, edited by Sam S. Fisher, AIAA, New York, 1981, pp. 239-255.

[6]Bird, G. A., "Direct Simulation of Typical AOTV Entry Flows," AIAA Paper 86-1310, June 1986.

[7]Bird, G. A., "Nonequilibrium Radiation During Re-Entry at 10 km/s," AIAA Paper 87-1543, June 1987.

[8]Bird, G. A., "Direct Simulation of Gas Flows at the Molecular Level," *Communications in Applied Numerical Methods*, Vol. 4, No. 2, March/Apr. 1988, pp. 165-172.

[9]Bird, G. A., "Thermal and Pressure Diffusion Effects in High Altitude Flows," AIAA Paper 88-2732, July 1988.

[10]Cuda, V., Jr., and Moss, J. N., "Direct Simulation of Hypersonic Flows Over Blunt Slender Bodies," AIAA Paper 86-1348, June 1986.

[11]Dogra, V. K., Moss, J. N., and Simmonds, A. L., "Direct Simulation of Aerothermal Loads for an Aeroassist Flight Experiment Vehicle," AIAA Paper 87-1546, June 1987.

[12]Harvey, J. K., "Direct Simulation Monte Carlo Method and Comparison with Experiment," *Progress in Astronautics and Aeronautics: Thermophysical Aspects of Re-Entry Flows*, Vol. 103, edited by J. N. Moss and C. D. Scott, AIAA, New York, 1986, pp. 25-43.

[13]Hermina, W. L., "Monte Carlo Simulation of Rarefied Flow Along a Flat Plate," AIAA Paper 87-1547, June 1987.

[14]Moss, J. N. and Bird, G. A., "Monte Carlo Simulations in Support of the Shuttle Upper Atmospheric Mass Spectrometer Experiment," AIAA Paper 85-0968, June 1985.

[15]Moss, J. N, Cuda, V., Jr., and Simmonds, A. L., "Nonequilibrium Effects for Hypersonic Transitional Flows," AIAA Paper 87-0404, Jan. 1987.

[16]Celenligil, M. C., Bird, G. A., and Moss, J. N., "Direct Simulation of Three-Dimensional Hypersonic Flow About Intersecting Blunt Wedges," AIAA Paper 88-0463, Jan. 1988.

[17]Bird, G. A., Private Communication, 1987.

[18]Olynick, D. P., Moss, J. N., and Hassan, H. A., "Grid Generation and Adaptation for the Direct Simulation Monte Carlo Method," AIAA Paper 88-2734, July 1988.

[19]Becker, M. and Boylan, D. E., "Experimental Flow Field Investigations Near the Sharp Leading Edge of a Cooled Flat Plate in a Hypervelocity, Low-Density Flow," *Rarefied Gas Dynamics*, Vol. II, edited by C. L. Brundin, Academic, 1967, pp. 993-1014.

[20]Becker, M., Robben, F., and Cattolica, R., "Velocity Distribution Functions Near the Leading Edge of a Flat Plate," *AIAA Journal*, Vol. 12, Sept. 1974, pp. 1247-1253.

[21]Hurlbut, F. C., "Sensitivity of Hypersonic Flow Over a Flat Plate to Wall/Gas Interaction Models Using DSMC," AIAA Paper 87-1545, June 1987.

[22]Nagamatsu, H. T. and Sheer, R. E., Jr., "Heat Transfer on a Flat Plate in Continuum to Rarefied Hypersonic Flows at Mach 19.2

and 25.4," Progress in Astronautics and Aeronautics: Thermophysical Aspects of Re-Entry Flows, Vol. 103, edited by J. N. Moss and C. D. Scott, AIAA, New York, 1986, pp. 60-78.

[23] U. S. Standard Atmosphere, U.S. Government Printing Office, Washington, 1976.

[24] Moss, J. N., Unpublished Results, 1987.

[25] Kaufman, L. G. II, and Johnson, C. B., "Weak Incident Shock Interactions with Mach 8 Laminar Boundary Layers," NASA TN D-7835, Dec. 1974.

[26] Edney, B. E., "Effects of Shock Impingement on the Heat Transfer Around Blunt Bodies," AIAA Journal, Vol. 6., Jan. 1968, pp. 15-21.

[27] Johnson, C. B. and Kaufman, L. G. II, "Interference Heating from Interactions of Shock Waves with Turbulent Boundary Layers at Mach 6," NASA TN D-7649, 1974.

[28] Wieting, A. R. and Holden, M. S., "Experimental Study of Shock Wave Interference Heating on a Cylindrical Leading Edge at Mach 6 and 8," AIAA Paper 87-1511, June 1987.

Numerical Simulation of Supersonic Rarefied Gas Flows Past a Flat Plate: Effects of the Gas-Surface Interaction Model on the Flowfield

Carlo Cercignani* and Aldo Frezzotti†
Politecnico di Milano, Milano, Italy

Abstract

The supersonic flow of a rarefied gas past a flat plate at zero angle of attack is analyzed by solving numerically the Boltzmann equation. The effects of the gas-surface interaction on the flowfield are studied assuming that molecules are re-emitted from the plate according to the Cercignani-Lampis model. After a preliminary analysis of the sensitivity of the flowfield to changes of the model coefficients, the numerical results are compared with the experimental data obtained by Becker et al. ("Velocity distribution function near the leading edge of a flat plate." <u>AIAA</u> Journal 12, 1247-1253).

I. Introduction

It is well known that gas-surface interaction plays a fundamental role in determining both the forces and the thermal loads on spacecrafts during the early stages of re-entry into the Earth's atmosphere. In this work the effects of the gas-surface interaction on the supersonic flow of a rarefied gas past a flat plate are analyzed in view of the applications to hypersonic flight. This model problem has been selected because both experimental data[1,2] and previous numerical simulations[3,4] are available for a comparison. In an interesting paper Hurlbut[3] presented an analysis of the influence of the gas-surface interaction model on the flowfield
mentioned above. Bird's direct simulation method in

Copyright © 1989 by the American Institute of Aeronatuics and Astronautics, Inc. All rights reserved.
  * Professor, Department of Mathematics.
  † Assistant Professor, Department of Mathematics.

combination with various gas-surface interaction models (mainly the Maxwell model and the Nocilla-Hurlbut-Sherman model) was used to calculate the flow variables. Hurlbut's calculations exhibit good agreement with the experimental results close to the leading edge but deviate as the trailing edge is approached. Our work differs from Hurlbut's mainly in that the Cercignani-Lampis model[5] is used to describe the gas-surface interaction. The numerical technique to solve the Boltzmann equation is also different, being based on a finite difference discretization of the Boltzmann equation in combination with the Monte Carlo evaluation of the collision integral[4].

An experimental validation of the Cercignani-Lampis model was already attempted by Bellomo et al.[6] According to these authors, the model does not reproduce the experimental drag and the heat flow on a metal plate placed in a supersonic nearly free molecular flow. In our opinion the interpretation of the experimental results given in Ref. 6 is not completely correct for the following reasons:

1) One tries to fit the curves giving the drag coefficient C_D and the heat-transfer coefficient C_H as a function of the plate temperature with the corresponding expressions derived from the Cercignani-Lampis model with temperature-independent α_n and α_t. This is not correct because the coefficients could depend on the plate temperature, as the measurements themselves seem to indicate. A simple calculation based on the experimental C_D at 90 deg of incidence shows that although the ratio of the plate temperature to the stagnation temperature ranges from 1 to 2.5, the accommodation coefficient of the normal momentum varies from 0.84 to 1.98.

2) The measurements were limited to high angles of incidence (45-90 deg) for which the aerodynamic coefficients C_D are practically insensitive to the model coefficients.

A firm conclusion that can be drawn from the experimental data is that the recovery temperature does not depend on the angle of attack; therefore the accommodation coefficients of the tangential and normal kinetic energy must be equal[7]. As pointed out in Refs. (2-3) the measurements by Becker et al.[3] and the use of any numerical tool to solve the Boltzmann equation provide a further opportunity to test gas-surface interaction models.

II. Description of the Problem

The steady two-dimensional Boltzmann equation

$$\xi_x \frac{\partial F}{\partial x} + \xi_y \frac{\partial F}{\partial y} = \left(\frac{\partial F}{\partial t}\right)_{Coll} \qquad (1)$$

has been solved numerically to simulate the flow of a rarefied monatomic gas past a flat plate at zero incidence. During a preliminary sensitivity analysis it was assumed that far from the plate the distribution function is a Maxwellian having prescribed number density N_∞, temperature T_∞, and bulk velocity U_∞ parallel to the plate surface. The preceding condition at infinity was later modified to match the experimental conditions described in Ref.2 as closely as possible. The boundary condition for the distribution function $F(\underline{\xi})$ on the plate surface is given as

$$(\underline{\xi} \cdot \hat{n}) F(\underline{\xi}) = \int_{\underline{\xi}' \circ \hat{n} < 0} R(\underline{\xi},\underline{\xi}') F(\underline{\xi}') |\underline{\xi}' \circ \hat{n}| \, d\underline{\xi}', \quad \underline{\xi} \cdot \hat{n} > 0 \qquad (2)$$

As was anticipated in the previous section the scattering kernel $R(\underline{\xi},\underline{\xi}')$ was assumed to have the form

$$R(\xi,\xi'|\alpha_n,\alpha_t) = \frac{[\alpha_n \alpha_t (2-\alpha_t)]^{-1}}{2\pi(RT_w)^2} \xi_n \exp\left\{-\frac{\xi^2 + (1-\alpha_n)\xi'^2}{2RT_w \alpha_n}\right.$$

$$\left. - \frac{1}{\alpha_t(2-\alpha_t)} \frac{[\xi-(1-\alpha_n)\xi']^2}{2RT_w}\right\} \times I_0\left(\frac{\sqrt{1-\alpha_n}}{\alpha_n RT_w}\xi_n\xi'_n\right)$$

$$\alpha_n \in [0,1]$$
$$\alpha_t \in [0,2] \qquad (3)$$

where

$$I_0(x) = \frac{1}{2\pi}\int_0^{2\pi} e^{x\cos\phi} \, d\phi \qquad (4)$$

It is easily shown[8] that the coefficients α_n and α_t are the accommodation coefficients of the "normal part" of the kinetic energy $\xi_n^2/2$ and of the tangential momentum ξ_t, respectively. The scattering kernel (3) reduces to Maxwell model with $\alpha=1$ if $\alpha_n=1$ and $\alpha_t=1$. Specular reflection is obtained by setting $\alpha_n=\alpha_t=0$, and the choice $\alpha_n \to 0$ and $\alpha_t \to 2$ gives $R(\xi,\xi')=\delta(\xi+\xi')$.

Although our attention was concentrated on the preceding model, the Maxwell model was also used to obtain a number of reference solutions.

The steady two-dimensional Boltzmann equation Eq.(1) is solved numerically by means of a simple numerical technique.[4] A rectangular region enclosing the plate is divided into a number of rectangular cells of variable size. The distribution function is defined at the cell centers and is assumed to be constant within each cell. A three-dimensional grid in the velocity space is associated with each spatial cell. The velocity grid is organized in levels of growing resolution. First, a coarse uniform grid is generated, then each cell of this grid is allowed to be a cell or a finer subgrid. This technique was very effective in reducing the required amount of computer memory in the case of problems in which the macroscopic variables can suffer large changes across the flowfield. A first-order conservative upwind scheme is used to represent the streaming term, and the collision term is evaluated according to the BGK model[9] or the Boltzmann collision integral. In the latter case the fivefold collision integral is evaluated by a Monte Carlo technique with standard methods for variance reduction. The resulting difference equation is solved by iteration. A rigid sphere gas was assumed in the case of the Monte Carlo simulations and the collision frequency associated with the BGK model was evaluated from helium viscosity.

III. Results

Two distinct series of numerical tests have been performed. The first group of tests was aimed at studying the sensitivity of the flowfield to changes of the coefficients of the Cercignani-Lampis model. For computational convenience Mach numbers were selected in the range of 4-5 and Knudsen numbers in the range of 0.1-0.5. The numerical results lead to the following conclusion.

The flowfield is extremely sensitive to changes of the coefficient α_t, which affects the entire flowfield. The flowfield is much less sensitive to variations of α_n in the range of temperature ratios and Mach number considered (see Fig. 1). A change of α_n affects only the heat flow. A

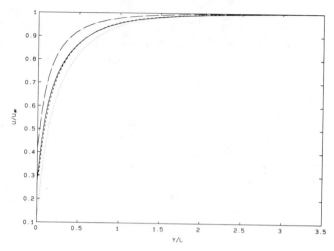

Fig. 1 $M_\infty=5$, Kn=0.5, $T_w/T_\infty=1$, velocity field vs distance from the plate at x=L/2. Solid line $\alpha_t=1$, $\alpha_n=1$; dashed line $\alpha_t=1$, $\alpha_n=0.8$; dotted line $\alpha_t=1.2$, $\alpha_n=1.$; dashed-dotted line $\alpha_t=0.8$, $\alpha_n=1$.

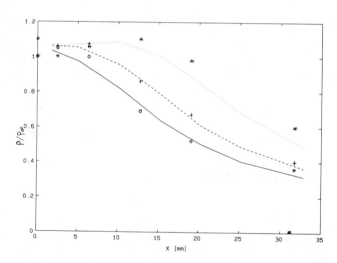

Fig. 2 Maxwell model $\alpha=1$. Normalized density field ρ/ρ_∞ vs experimental data.

comparison of Cercignani-Lampis model predictions with those of Maxwell model was also made. Strictly speaking, a comparison of the two models for equal values of the accommodation coefficients is not possible because, if we set $\alpha=\alpha_t$ to have the same drag, then the predictions of the heat flow will be different for the two models. The choice $\alpha_t(2-\alpha_t)=\alpha_n=\alpha$ would make the heat flow values equal in free molecular flow, but different values for the drag would result. A second series of numerical tests has then been performed in order to compare the calculation with the experimental results reported in Ref. 2. In the experiment a model plate having a length of 50.8 mm was placed along the centerline of an orifice jet. The distance of the model from the orifice was such that the following conditions resulted at the leading edge: Mach number M_∞ = 8.9, gas temperature T_∞ = 10.7° K, mean free path λ_∞ = 1.286 mm. The plate temperature was kept constant at 290°K.

The Knudsen number based on model length was approximately equal to 1/40 and the wall temperature-to-gas temperature ratio T_w/T_∞ was equal to 27.1.

The distribution function was measured at a number of stations along the plate for three values of the distance from the plate: y=1.25, 2.6, 5.1 mm. The corresponding experimental points have been denoted by circles, plus signs, and asterisks in the figures.

Most of the calculations have been performed using the BGK model to evaluate the collision term because the Monte Carlo technique described earlier did not give good results. Some of the macroscopic quantities exhibited oscillations that did not damp in a reasonable number of iterations. The reason for the bad performance of this technique is probably the use of the same set of random velocities to calculate the collision integral in a number of cells of the spatial grid. This procedure is usually very effective in reducing the computational time in problems in which the parameter T_w/T_∞ is not too high. In the case under examination it is not possible to tune the random number generator to produce a single set of collisional velocity that is suitable for both the high-and low-temperature region of the flowfield. It possible to generate a set for each cell or for a smaller group of nearby cells, but the increase in computational time would put the problem beyond the capabilities of our Micro-Vax II. The iteration procedure starts usually from the free molecular flow solution. Sixty iterations are usually needed to obtain a fully converged flowfield.

In a first series of calculations it has been assumed that the plate was immersed in a uniform stream flowing parallel to the plate itself. However, since the model was only five diameters apart from the sonic orifice, an effect of the jet divergence on the entire flowfield is to be expected. Accordingly, the calculations were performed assuming that the plate was placed inside an axisymmetric jet originating from the adiabatic expansion of a monatomic gas from a sonic orifice. The flowfield far from the jet has been calculated from the well-known Ashkenas-Sherman formulas [10]. Slip velocities 20% higher than those originating from the uniform stream were in fact calculated. It should be noticed that the interaction between the axisymmetric jet and the plate was not treated

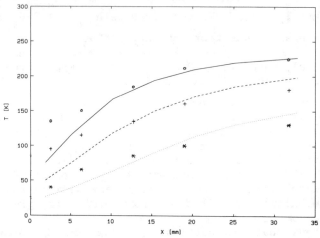

Fig. 3 Maxwell model α=1. Temperure field vs experimental data.

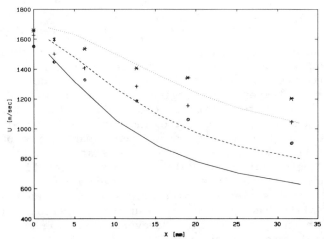

Fig. 4 Maxwell model α=1. Axial velocity field vs experimental data.

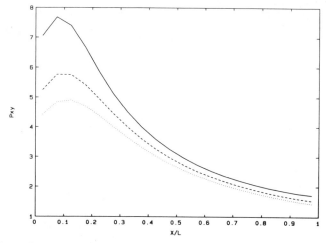

Fig. 5 Pxy along the plate surface as a function of α.

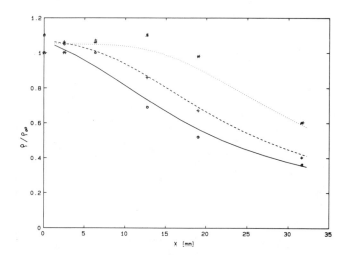

Fig. 6 Maxwell model α=0.7. Normalized density field ρ/ρ_∞ vs experimental data.

correctly because the resulting fully three-dimensional flow was calculated as a two-dimensional one. The Maxwell model was first used to obtain a first group of baseline calculations. The results obtained from Maxwell model are summarized in Figs. 2-9. The values of the Pxy component of the stress tensor was normalized to $\rho_\infty RT_\infty$, and the E_y component of energy flux was normalized to $1/2\, \rho_\infty (RT_\infty)^{3/2}$.

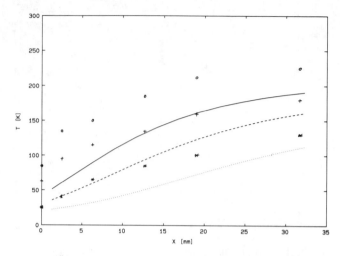

Fig. 7 Maxwell model α=0.7. Temperure field vs experimental data.

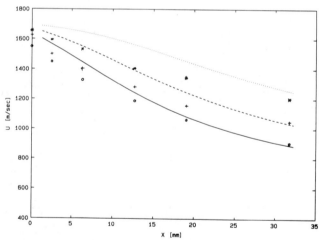

Fig. 8 Maxwell model α=0.7. Axial velocity field vs experimental data.

It is immediately seen that the flowfield is very sensitive to variations of the accommodation coefficient α. The comparison with the experimental data shows that the calculated density and temperature field are rather close to the experimental points in the case of complete accommodation (see Figs. 2 and 3). The experimental temperature exhibits a sharper rise close to the leading edge than the calculated one. However the theoretical

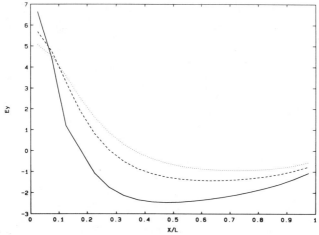

Fig. 9 Ey along the plate surface as a function of α. Solid line $\alpha=1$; dashed line $\alpha=0.8$; dotted line $\alpha=0.7$.

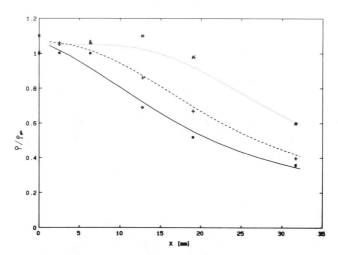

Fig. 10 Cercignani-Lampis model, $\alpha_t=0.7$, $\alpha_n=1$. Normalized density field ρ/ρ_∞ vs experimental data.

velocity field is far from the experimental one. The calculated velocity close to the plate drops much faster than the experimental profile (see Fig. 4). If the accommodation coefficient is reduced to 0.7, the calculated velocity field meets the three experimental points at the last station on the plate, but the overall agreement is not

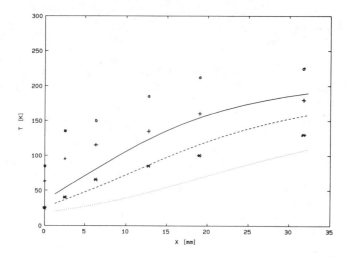

Fig. 11 Cercignani-Lampis model, $\alpha_t=0.7$, $\alpha_n=1$. Temperature field vs experimental data.

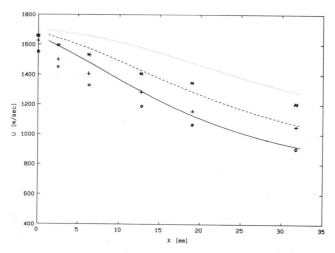

Fig. 12 Cercignani-Lampis model, $\alpha_t=0.7$, $\alpha_n=1$. Axial velocity field vs experimental data.

improved. Furthermore, the increase in the slip velocity has been obtained at the expense of a worse temperature field (see Figs. 6-8). The situation does not change appreciably when the Cercignani-Lampis model is used in place of Maxwell's to describe gas-surface interaction (see Figs. 10-12). As is clearly seen from Figs. 13-16 the

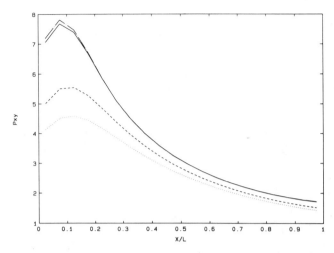

Fig. 13 Cercignani-Lampis model. Pxy along the plate. Solid line $\alpha_t = \alpha_n = 1$; dashed line $\alpha_t = 0.8$, $\alpha_n = 1$; dotted line $\alpha_t = 0.7$, $\alpha_n = 1$; dashed-dotted line $\alpha_t = 1$, $\alpha_n = 0.75$.

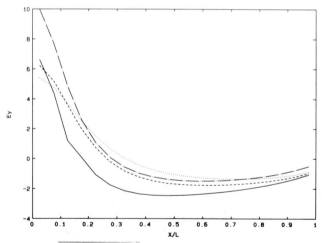

Fig. 14 Cercignani-Lampis model. Ey along the plate. Solid line $\alpha_t = \alpha_n = 1$; dashed line $\alpha_t = 0.8$, $\alpha_n = 1$; dotted line $\alpha_t = 0.7$, $\alpha_n = 1$; dashed-dotted line $\alpha_t = 1$, $\alpha_n = 0.75$.

density, temperature and velocity field are only very slightly affected by variations of the coefficient α_n which have an appreciable effect on the energy flux only (See Fig. 14). Therefore the number of parameters that can be used to fit the experimental data is again one as in

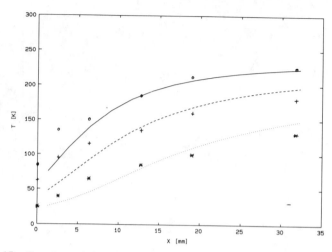

Fig. 15 Cercignani-Lampis model $\alpha_t=1$, $\alpha_n=0.75$. Temperature field vs experimental data.

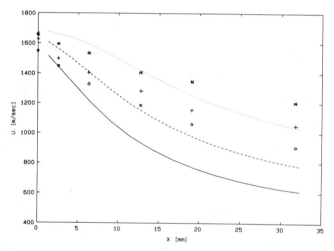

Fig. 16 Cercignani-Lampis model $\alpha_t=1$, $\alpha_n=0.75$. Axial velocity field vs experimental data.

Maxwell model. If the results obtained from the two models are compared in the case $\alpha_t=\alpha$, $\alpha_n=1$, it is seen that the density, velocity and temperature fields practically coincide, the only difference being found in the energy fluxes, as expected.

IV. Conclusions

The examination of the numerical results leads to the conclusion that neither the Maxwell model nor the Cercignani-Lampis model can reproduce the whole set of experimental results. Whereas temperature and density are well reproduced by the Maxwell model with complete accommodation, the velocity field is not. Any attempt to reproduce the correct velocity profiles led to temperatures lower than the right ones. Unfortunately, the coefficient α_n seems to have no effect on the measured fields, affecting mainly the heat flow, which was not measured. Apparently a gas-surface interaction model is needed that describes perfect accommodation as far as temperatures are concerned but gives higher slip velocities. A possible source of uncertainty in the results described earlier comes from the use of the BGK model, but it should be noted that the numerical tests at lower Mach number gave results reasonably close to those obtained from the Monte Carlo calculation. The results were also close to those obtained by Hurlbut[3] in the case of complete accommodation if the same grid size is used. Calculations with a modified Monte Carlo procedure are under way to check this point.

References

[1] Boettcher, R.D., Koppenwallner, G., and Legge, H., "Flat Plate Skin Friction in the Range between Hypersonic Continuum and Free Molecular Flow, " *Proceedings of the Tenth Symposium (International) on Rarefied Gas Dynamics*, Academic Press, New York, 1976, pp. 349-359.

[2] Becker, M., Robben, F., and Cattolica, R., "Velocity Distribution Functions near the Leading Edge of a Flat Plate, " *AIAA Journal*, Vol. 12, September 1974, pp. 1247-1253.

[3] Hurlbut, F.C., "Sensitivity of Hypersonic Flow over a Flat Plate to Wall/Gas Interaction Models Using DSMC, " AIAA Paper 87-1545, June 1987.

[4] Tcheremissine, F.G., "Numerical Methods for the Direct Solution of the Kinetic Boltzmann Equation, " *U.S.S.R. Comput. Math. Math. Phys.*, Vol. 25, 1985, pp. 156-166.

[5] Cercignani, C., Lampis, M., "Kinetic Models for Gas-Surface Interactions, " *Journal of Stat. Phys.*, Vol. 1, 1971, pp. 101-114.

[6] Bellomo, N., Dankert, C., Legge, H., "Drag, Heat Flux, and Recovery Factor Measurments in Free Molecular Hypersonic Flow and Gas-Surface Interaction Analysis, " *Proceedings of the Thirteenth Symposium (International) on Rarefied Gas Dynamics*, Plenum Press, New York, 1985, pp. 421-430.

[7] Cercignani, C., and Lampis, M., "On the Recovery Factor in Freemolecular Flow, " *Journal of Appl. Math. and Phys.(ZAMP)*, Vol.27, 1976, pp. 733-738.

[8] Lampis, M., "Some Applications of a Model for Gas-Surface Interaction, "*Proceedings of the Eighth Symposium (International) on Rarefied Gas Dynamics*, Academic Press, New York, 1974, pp. 369-380.

[9] Cercignani, C., *Mathematical Methods in Kinetic Theory*, Plenum Press, New York, 1969.

[10] Ashkenas, H., and Sherman, F., "The Structure and Utilization of Supersonic Free Jets in Low Density Wind Tunnel, " *Proceedings of the Fifth Symposium (International) on Rarefied Gas Dynamics*, Academic Press, New York, 1966, pp. 84-105.

Rarefield Flow Past a Flat Plate at Incidence

Virendra K. Dogra*
ViRA, Inc., Hampton, Virginia
and
James N. Moss† and Joseph M. Price‡
NASA Langley Research Center, Hampton, Virginia

Abstract

Results of a numerical study using the direct simulation Monte Carlo (DSMC) method are presented for the transitional flow about a flat plate at 40-deg incidence. The plate has zero thickness and a length of 1 m. The flow conditions simulated are those experienced by the Shuttle Orbiter during re-entry at 7.5 km/s. The range of freestream conditions are such that the freestream Knudsen number values are between 0.02 and 8.4, i.e., conditions that encompass most of the transitional flow regime. The DSMC simulations show that transitional effects are evident when compared to free molecule results for all cases considered. The calculated results clearly demonstrate the necessity of having a means of identifying the effects of transitional flow when making aerodynamic flight measurements as are currently being made with the Space Shuttle Orbiter vehicles. Previous flight data analyses have relied exclusively on adjustments in the gas-surface interaction models without accounting for the transitional effect, which can be comparable in magnitude. The present calculations show that the transitional effect at 175 km would increase the Space Shuttle lift-to-drag ratio by 90% over the free molecular value.

This paper is declared a work of the U.S. Government and therefore is in the public domain.
 * Research Engineer.
 † Research Engineer, Space Systems Division.
 ‡ Mathematician, Space Systems Division.

Nomenclature

C_D	=	drag coefficient, $2D/\rho_\infty U_\infty^2 \ell$
C_f	=	skin friction coefficient, $2F/\rho_\infty U_\infty^2 \ell$
C_H	=	heat-transfer coefficient, $2q/\rho_\infty U_\infty^3$
C_L	=	lift coefficient, $2L/\rho_\infty U_\infty^2 \ell$
C_p	=	pressure coefficient, $2p/\rho_\infty U_\infty^2$
D	=	drag force
F	=	skin friction
Kn_∞	=	freestream Knudsen number, λ_∞/ℓ
L	=	lift force
ℓ	=	length of the flat plate
$\bar{M}$	=	molecular weight of air
p	=	pressure
q	=	heat flux
R	=	universal gas constant, $R = 8.3143$ J/mole-°K
S_∞	=	freestream speed ratio, $U_\infty(\bar{M}/2RT_\infty)^{1/2}$
T	=	thermodynamic temperature
T_∞	=	freestream temperature
T_w	=	surface temperature
U_∞	=	freestream velocity
u	=	velocity component tangent to the plate surface
v	=	velocity component normal to the plate surface
x	=	coordinate measured along the plate surface
y	=	coordinate measured normal to the plate surface
α	=	angle of incidence
ε	=	fraction of specular reflection
λ_∞	=	freestream mean free path
ρ_∞	=	freestream density
ρ	=	density

Introduction

The development and application of future hypersonic space vehicles require accurate predictions of aerothermal loads during re-entry. A portion of the re-entry for these vehicles will take place in the transitional flow regime where nonequilibrium effects become important in establishing the thermal and aerodynamic response of these vehicles. In order to simplify the computational requirements, the lift-drag characteristics for vehicles such as the Space Shuttle Orbiter are often approximated[1,2] with a flat plate at incidence for the free molecular regime. For the transitional flow regime, empirical approximations are normally used to predict the aerodynamic forces for such vehicles. Examples of such applications are given in

Refs. 1, 2, and 3 for the Space Shuttle Orbiter. Therefore, accurate numerical studies on basic configurations such as a plate at incidence can provide useful information and physical insight concerning the nature of transitional flows.

A flat plate as a basic aerodynamic surface has remained the focus of the transitional flow research by several authors (Refs. 4-8). But most of these investigations are for zero incidence and do not cover the range of flow parameters of interest. There are very few experimental and theoretical investigations of the flat plate at incidence (Refs. 9 and 10). Even these do not cover the large incidence and high-speed ratio characteristics of hypersonic flight. Furthermore, they are limited to the continuum flow regime.

In the present paper the direct simulation Monte Carlo (DSMC) method of Bird[11,12] with the variable hard sphere molecular model is used to simulate the transitional flow past a flat plate at 40-deg incidence. The flow conditions and incidence angle simulated are those that the Space Shuttle Orbiter experiences during re-entry. The DSMC is the only suitable numerical technique to simulate the transitional flow regime accurately because it allows the direct implementation of nonequilibrium flow models. The various nonequilibrium phenomena are the main characteristics of the transitional flow regime, and the DSMC method simulates the flow physics adequately. The present calculations show that the transitional effect for a diffuse surface would increase the Space Shuttle lift-to-drag ratio by 90% over the free molecular value at approximately 175 km altitude.

Computational Approach

The DSMC method[11,12] models the real gas by some thousands of simulated molecules in a computer. The position coordinates, velocity components, and internal state of each molecule are stored in the computer and are modified with time as the molecules are concurrently followed through representative collisions and boundary interactions in simulated physical space. The time parameter in the simulation may be identified with physical time in the real flow, and all calculations are unsteady. When the boundary conditions are such that the flow is steady, then the solution is the asymptotic limit of unsteady flow. The computation is always started from an initial state that permits an exact specification such as a vacuum or uniform equilibrium flow. Consequently, the method does not require an initial approximation to the flowfield and does not involve any iterative procedures. A computational cell network is required in physical space only, and then only to facilitate the choice of potential collision pairs and the sampling of the macroscopic flow properties. Furthermore, advantage may be taken of flow symmetries to reduce the dimensions of the cell network and the number of position coordinates that need to be stored for each molecule, but the collisions are always treated as three-dimensional phenomena. The boundary conditions are specified in terms of the behavior of the individual molecules rather than the distribution functions. All procedures may be specified in such a manner that the computational time is directly proportional to the number of simulated molecules.

Conditions for Calculations

The freestream conditions considered in the present study are for an altitude range of 90-130 km. For the 1-m flat plate with zero thickness, the corresponding freestream Knudsen numbers are 0.023-8.439. During the Shuttle's re-entry trajectory, the same freestream Knudsen numbers based on the mean aerodynamic chord of 12 m correspond approximately to an altitude range of 100-175 km. The freestream velocity, incidence angle, and wall temperature are assumed constant at 7.5 km/s, 40 deg, and 1000°K, respectively, i.e., conditions experienced by the Space Shuttle during re-entry. These conditions are summarized in Table 1, where the freestream values are

Table 1 Freestream Conditions

Altitude, km	$\bar{M}$, g/mole	Kn_∞	U_∞ km/s	ρ_∞ kg/m^3	T_∞, °K	S_∞
90	28.810	0.023	7.5	3.418×10^{-6}	188	23.1
100	28.257	0.137	7.5	5.640×10^{-7}	194.3	21.6
120	26.159	3.146	7.5	2.269×10^{-8}	367.8	15.7
130	25.441	8.439	7.5	8.220×10^{-9}	499.7	13.3

Length of the flat plate, ℓ=1m, and angle of incidence, α = 40 deg.

those given by Jacchia[13] for an exospheric temperature of 1,200°K.

The surface of the plate is assumed to be diffused with full thermal accommodation and to promote recombination of the oxygen and nitrogen atoms. Recombination probabilities appropriate for the Shuttle thermal protection tiles are imposed. The oxygen and nitrogen recombination probabilities are 0.0049 and 0.0077, respectively.

Since the computational requirements increase significantly with increasing freestream density, computations are performed only in the transitional flow regime.

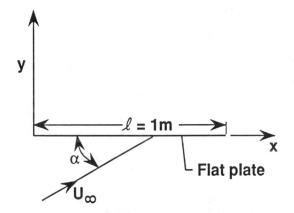

Fig. 1 Flat plate configuration at incidence.

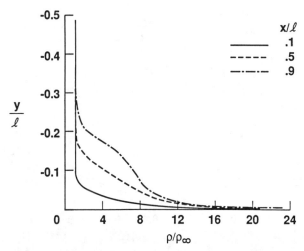

Fig. 2 Density profiles normal to the plate surface; Kn_∞ = 0.023 and α = 40 deg.

Results and Discussion

Since the flat plate at incidence is a basic element of lifting surfaces, it is often used for numerical and experimental studies in hypersonic flow research. Furthermore, the configuration is also helpful in understanding the re-entry of space vehicles such as the Space Shuttle. Therefore, attention is focused on the flow structure, surface quantities, and aerodynamic characteristics resulting from low-density flow about a flat plate at 40-deg angle of attack (Fig. 1).

Flowfield Structure

Figures 2-7 present calculated flowfield quanities on the compression side (lower surface) of a flat plate at two freestream Knudsen numbers (0.023 and 8.439). Results at three different locations along the surface are presented. Two regions of interest are those near the plate surface and the shock-wave. Near the surface, a large increase in density occurs that is characteristic of a high-velocity flow about a cold wall. These results clearly demonstrate that the shock wave is fully merged with the viscous layer for Knudsen number values of 0.023 and 8.439. The density and velocity profiles for a Knudsen number value of 0.023 (Figs. 2 and 3) show that the shock wave thickness increases gradually along the surface of the plate.

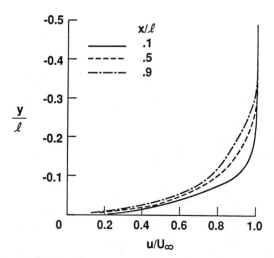

Fig. 3 Tangential velocity profiles normal to the plate surface; Kn_∞ = 0.023 and α = 40 deg.

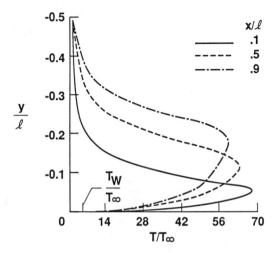

Fig. 4 Temperature profiles normal to the plate surface; $Kn_\infty = 0.023$ and $\alpha = 40$ deg.

However, for a freestream Knudsen number of 8.439 (Figs. 5 and 6), the extent of the flowfield disturbances remains almost constant along the surface due to the large degree of rarefaction at this Knudsen number.

The overall nondimensional kinetic temperature T/T_∞, shown in Figs. 4 and 7 for the freestream Knudsen number value of 0.023 and 8.439, respectively, is defined for a nonequilibrium gas as the weighted mean of the translational and internal temperatures. It can be seen from these figures that overall kinetic temperature rise in the shock wave precedes the density rise. The initial rise in temperature is due to the bimodal velocity distribution: the molecular sample consists of mostly undisturbed freestream molecules with just a few molecules that have been affected by the shock. The large velocity separation between these two classes of molecules give rise to the early temperature increase.

The temperature rise in the shock wave is comparatively large near the leading edge and then gradually decreases toward the trailing edge (Figs. 4 and 7). This obviously shows that the strength of shock wave decreases along the surface of the flat plate. The difference in peak shock-wave temperatures at different locations is less for the higher Knudsen number condition. The shock wave is very diffuse for the 8.439 Knudsen number condition due to the large freestream mean free path. The temperature and velocity profiles also show a significant temperature jump and velocity slip at the surface for both

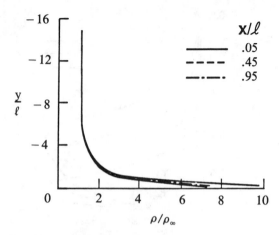

Fig. 5 Density profiles normal to the plate surface; $Kn_\infty = 8.439$ and $\alpha = 40$ deg.

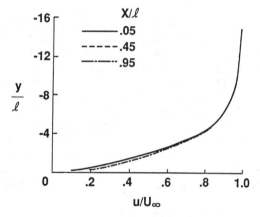

Fig. 6 Tangential velocity profiles normal to the plate surface; $Kn_\infty = 8.439$ and $\alpha = 40$ deg.

values of Knudsen number. Consequently, the flowfield exhibits the effects of rarefaction for all the cases considered.

Surface Quantities

The surface pressure, skin friction, and heat-transfer coefficients are presented in Figs. 8, 9, and 10, respectively, for the lower surface of the flat plate at 40-deg incidence. The effects of rarefaction are shown by comparing the results for different freestream Knudsen

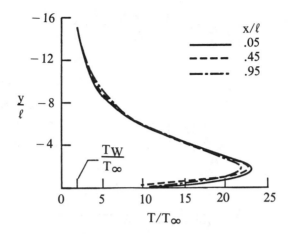

Fig. 7 Temperature profiles normal to the plate surface; Kn_∞ = 8.439 and α = 40 deg.

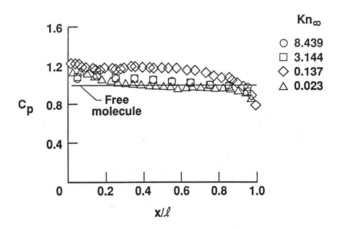

Fig. 8 Effect of rarefaction on the compression side surface pressure coefficient; α = 40 deg.

numbers. The variation of the pressure coefficient with rarefaction is moderate, provided the gas-surface interaction is diffuse, as assumed in the present calculations. The calculated pressure coefficient is greater than the free molecule value for large freestream Knudsen numbers. In contrast, the skin friction and heat-transfer coefficients (Figs. 9 and 10) are very sensitive to rarefaction effects and approach the free molecule value with increasing rarefaction.

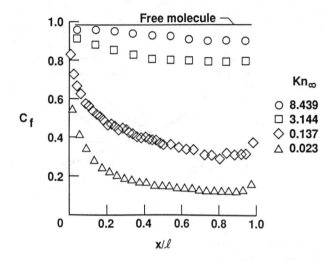

Fig. 9 Effect of rarefaction on the compression side surface skin friction coefficient; α = 40 deg.

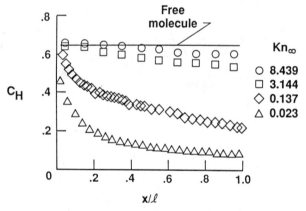

Fig. 10 Effect of rarefaction on the compression side surface heat transfer coefficient; α = 40 deg.

Aerodynamic Characteristics

Figures 11 and 12 present the drag and lift coefficients as a function of freestream Knudsen number for the flat plate at 40-deg incidence. These results show the expected variation in the transitional flow regime; that is, the drag coefficient increases and the lift coefficient decreases substantially with increasing rarefaction and both approach the free molecule limit. The change in the

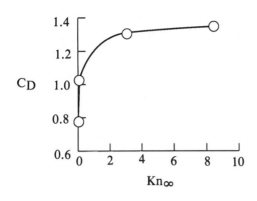

Fig. 11 Drag coefficient vs Kn; α = 40 deg.

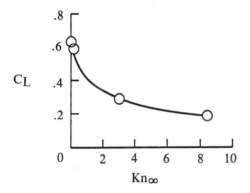

Fig. 12 Lift coefficient vs Kn; α = 40 deg.

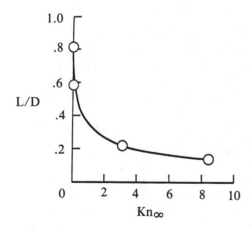

Fig. 13 Lift-to-drag ratio vs Kn; α = 40 deg.

drag and lift coefficients is due primarily to an increase in the skin friction rather than to a change in the pressure coefficient. Figure 13 presents the lift-to-drag ratio as a function of freestream Knudsen number where the trend is the same as the lift coefficient data, which

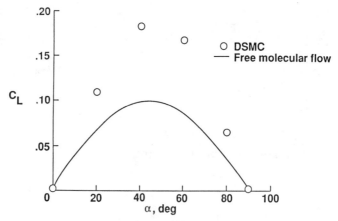

Fig. 14 Lift coefficient vs incidence angle for $Kn_\infty = 8.439$.

experiences a significant decrease with increasing rarefaction.

The effect of angle of incidence variation on the aerodynamic coefficients is demonstrated in Figs. 14-16 for the 8.4 Knudsen number case. The DSMC results are compared with those obtained using free molecule expressions for drag, lift, and lift-to-drag ratio. Figure 14 shows that the lift coefficient increases with angle of attack, reaches a maximum value at 45 deg, then decreases with further increase in angle of attack. The DSMC values are higher than the free molecule values, indicating that transitional effects are evident even at this highly rarefied condition. Figure 15 shows that the drag coefficient agrees well with the free molecule calculations for small incidence. However, at higher incidence the DSMC values are slightly lower than free molecule values due to the overprediction of skin friction by the free molecule method. Figure 16 presents the corresponding lift-to-drag ratio comparison and shows the same trend of the transitional effects as in the lift coefficient data.

These results have important implications for the interpretation of the flight measurements used to deduce aerodynamic coefficients under rarefied conditions. As early as 1985, it was recognized (Ref. 14) that transitional effects rather than specular reflection might be influencing the interpretation of the flight measurements; however, no calculations were available to establish the fact. At altitudes of 160 km and above, the conventional procedure[1-3] has been to interpret the flight measurements using the free molecule flow calculations. Such procedures are used to establish what fraction of the gas-surface

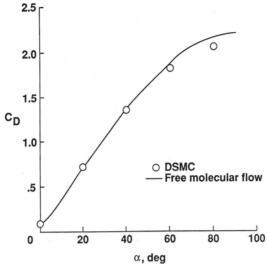

Fig. 15 Drag coefficient vs incidence angle for $Kn_\infty = 8.439$.

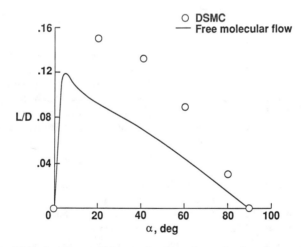

Fig. 16 Lift-to-drag ratio vs incidence angle for $Kn_\infty = 8.439$.

interaction is specular. But it can be seen from the present calculations that the transitional effects persist even at very high altitudes (160 km and above). This is clearly demonstrated in Fig. 16, where the transitional effect increases the lift-to-drag ratio by 90% for a flat plate at 40-deg incidence. The freestream Knudsen number for this condition is 8.4, which corresponds to Shuttle

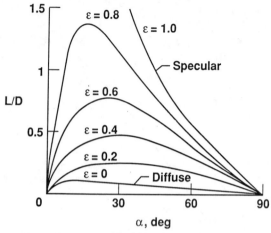

Fig. 17 Flat plate free molecule lift-to-drag ratio vs incidence angle; $S_\infty = 13.1$ and $T_w/T_\infty = 2.0$.

conditions at approximately 175 km. This transitional effect is quite large, and if not properly interpreted could be mistaken for a contribution due to the specular reflection. For example, the free molecule results for lift-to-drag ratio as a function of angle of incidence for different fractions of specular reflection are shown in Fig. 17. As the fraction of specular reflection increases, the lift-to-drag ratio also increases for a given incidence angle. Since these two separate effects both produce increased lift-to-drag ratio, interpretation of flight measurements must account for the transitional effects.

Conclusion

Results obtained with the DSMC method for hypersonic flow past a flat plate at incidence show the extent of the transitional flow effects on the aerodynamic characteristics. These effects are very significant, even for large freestream Knudsen numbers. Thus, the interpretation of aerodynamic flight data for space vehicles such as the Space Shuttle Orbiter must be done in concert with calculations that describe the transitional effects. Failure to account for this effect could significantly distort the interpretation of the gas-surface interactions under highly rarefied conditions.

References

[1]Blanchard, R. C., "Rarefied Flow Lift-to-Drag Measurements of the Shuttle Orbiter," 15th Congress of International Council of

Aeronautical Sciences, ICAS Paper 86-2.10.2, London, England, U.K., Sept. 7-12, 1986.

[2]Blanchard, R. C., Hendrix, M. K., Fox, J. C., Thomas, D. J., and Nicholson, J. Y., "Orbital Acceleration Research Experiment," *Journal of Spacecraft and Rockets*, Vol. 24, Nov./Dec. 1987, pp. 504-511.

[3]Aerodynamic Design Data Book, Volum I, Orbiter Vehicle 102, SD72-SH-0060, Volume 1M, November 1980, Space Division, Rockwell International.

[4]Vogenitz, F. V., Broadwell, J. E., and Bird, G. A., "Leading Edge Flow by Monte Carlo Direct Simulation Technique," *AIAA Journal*, Vol. 8, March 1971, pp. 304-310.

[5]Vogenitz, F. V., and Takata, G. Y., "Rarefied Gas Flow about Cones and Flat Plates by Monte Carlo Simulation," *AIAA Journal*, Vol. 9, Jan. 1971, pp. 91-100.

[6]Shakhore, E. M., "Rarefied Gas Flow Over a Plate Parallel to the Stream," *USSR Fluid Dynamics*, 1973, pp. 119-126.

[7]Beottcher, R. D., Koppenwallner, G., and Legge, H., "Flat Plate Skin Friction in the Range between Hypersonic Continuum and Free Molecular Flow," *Proceedings of the 10th International Symposium on Rarefied Gas Dynamics*, Aspen, CO, Vol. 1, 1976, pp. 348-359.

[8]Hermina, W. L., "Monte Carlo Simulation of Rarefied Flow Along a Flat Plate," AIAA Paper 87-1547, June 1987.

[9]Allegre, J., and Bisch, C., "Angle of Attack and Leading Edge Effects on the Flow about a Flat Plate at Mach Number 18," *AIAA Journal*, Vol. 6, May 1968, pp. 848-852.

[10]Gai, S. L., "An Experimental Study of the Flow Past a Flat Plate at Incidence in Supersonic Low Density Stream," *Rarefied Gas Dynamics*, Vol. 1, edited by H. Oguchi, University of Tokyo Press, 1984, pp. 257-264.

[11]Bird, G. A., "Monte-Carlo Simulation in an Engineering context," *AIAA Progress in Astronautics and Aeronautics: Rarefied Gas Dynamics*, Vol. 74, Part 1, edited by S. S. Fisher, 1981, pp. 239-255.

[12]Bird, G. A., *Molecular Gas Dynamics*, Clarendon, Oxford, England, U.K., 1976.

[13]Jacchia, L. C., "Thermospheric Temperature Density, and Composition: New Models," *Research in Space Science*, SAO Special Report No. 375, March 1977.

[14]Blanchard, R. C., and Rutherford, J. F., "Shuttle Orbiter High Resolution Accelerometer Package Experiment: Preliminary Flight Results," *Journal of Spacecraft and Rockets*, Vol. 22, July/Aug. 1985, pp. 474-480.

Monte Carlo Simulation of Flow into Channel with Sharp Leading Edge

Michiru Yasuhara,* Yoshiaki Nakamura,† and Junichi Tanaka‡
Nagoya University, Nagoya, Japan

Abstract

The direct simulation Monte Carlo method is applied to some simple configurations: the flat plate at zero incidence, the channel, and the channel with obstacle. In the flat plate, comparison between Bird's and Nanbu's methods shows that these methods give almost the same results. Moreover, the effect of time step on the simulation is studied. Consequently, one-fifth of the mean free time is selected as a time step. In the channel flow, the two interactions of shock wave are observed: the interaction of two shock waves emanated at the leading edge, and, thereafter, their hitting on the boundary layer. As a result, the flow patterns in the rarefied flow are similar to those in the continuous flow, though they are very diffuse.

I. Introduction

Recently, much attention is paid to the high-speed vehicle at very high altitude. The air at such high altitude must be treated as rarefied gas. Hence, the analysis of rarefied gas is becoming more significant. In rarefied gasdynamics, the Boltzmann equation is mainly used as a governing equation. This equation is very complicated, even in comparatively simplified form. For this reason, the analytical solution cannot be acquired, except for restricted cases. Thus numerical approach is expected to solve the Boltzmann equation. Although this approach to the Boltzmann equation requires much computing time and memory, the progress of computer hardware has made it practical.

Copyright © 1989 by the American Institute of Aeronautics and Astronautics, Inc. All rights reserved.
* Professor, Department of Aeronautical Engineering.
† Associate Professor, Dept. of Aeronautical Engrg.
‡ Graduate Student, Dept. of Aeronautical Engrg.

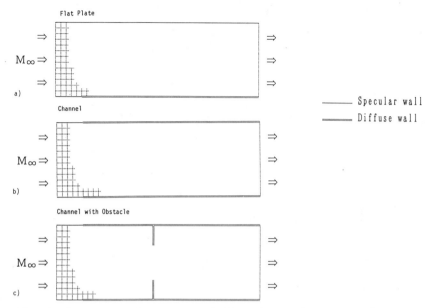

Fig. 1 Computational regions: a) flat plate; b) channel; c) channel with obstacle.

The most expected numerical method is the direct simulation Monte Carlo (DSMC) method. There are some variations in the DSMC method and, in the present study, Bird's and Nanbu's methods are used. Apart from their theoretical basis, one purpose of this study is to compare these two methods to one another by their simulation results.

Another purpose is to search the phenomena in the rarefied-gas flow, which is quite different from the continuous flow. We simulate the flow around simple configurations -- flat plate, channel, and channel with obstacle -- to observe phenomena such as the boundary layer, the shock wave, and the interactions between them.

First, the flat plate at zero incidence is calculated by both Bird's and Nanbu's methods. With these results, comparison between those methods is presented. Moreover, the effect of the various time steps is examined in Nanbu's method. Then we observe particularly the development of the boundary layer and the shock wave emanating from the leading edge. Next, we proceed to the channel flow to see the collision of oblique shock waves and the reflection of shock waves incident to the boundary layer.

Section 2 describes the outline of the selected problems and the simulation conditions. The results are shown in Sec. 3. Finally, the concluding remarks are made in Sec. 4.

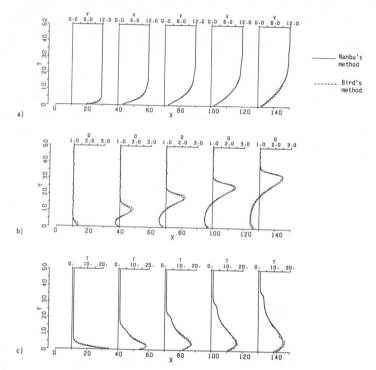

Fig. 2 Comparison of properties along the Y axis between Bird's and Nanbu's method (flat plate): a) velocity; b) density; c) temperature.

II. Simulation Conditions

The DSMC method is applied to three problems: a flat plate at zero incidence, channel flow, and a channel flow with obstacle. To the flat plate, both Bird's and Nanbu's methods are applied. To the channels, only Nanbu's method is applied.

The configurations of the computational regions are shown in Fig. 1. The diffuse wall is used as the actual wall and is completely diffuse (accommodation coefficient = 1.0). The specular wall before the diffuse wall prevents the disturbance affecting the upstream boundary. The computational regions are divided into cells whose size is all over the computational region. Each cell has 100 particles in the upstream uniform flow.

For the case of the flat plate, the computational region extends over $150\lambda_\infty \times 50\lambda_\infty$ in the parallel and normal directions to the upstream uniform flow, respectively. As a

SIMULATION OF CHANNEL FLOW

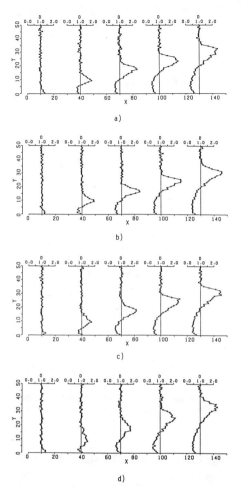

Fig. 3 Effect of Δt on density distributions (flat plate): a) Δt = Tf/20; b) Δt = Tf/10; c) Δt = Tf/5; d) Δt = Tf/2.5.

result, the Knudsen number for the plate length is 0.0067. For the channel flow, we simulate two cases: $150\lambda_\infty \times 20\lambda_\infty$ and $150\lambda_\infty \times 40\lambda_\infty$. These flow corresponds to Kn = 0.05 and 0.025, respectively, where the width of the channel is taken as the reference length.

Through the present simulation, we use the hard sphere molecule model. In the upstream uniform flow, the Mach number is set to 12.9, and the wall temperature is fixed as Tw = $5.4 T_\infty$ = $0.095 To$.

The following boundary conditions are employed in this study. At the upstream boundary, properties of the incident molecules are calculated using the Maxwell distribution. At

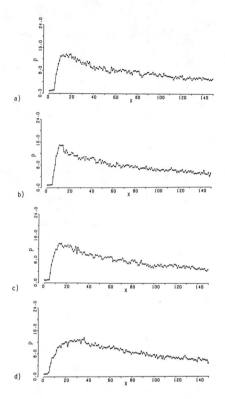

Fig. 4 Effect of Δt on plate pressure distributions (flat plate):
a) Δt = Tf/20; b) Δt = Tf/10; c) Δt = Tf/5; d) Δt = Tf/2.5

the downstream boundary, the following two different boundary conditions are tried:
(1) The same method as used by Vogenitz et al. (1970),[2]
(2) No particle from outside.
As a calculation procedure, the data are averaged for some time interval to reduce the fluctuation after the flow reaches the steady state.

III. Results

Flat Plate

First, the comparison between Bird's and Nanbu's methods is presented. Figure 2 presents the velocity, density, and temperature distributions along the Y axis at several X positions. The values shown in the figures are the ratio to the freestream values. The fluctuation seen in the lines of Bird's method is due to the short averaging time, not to the method. Figure 2 shows that the difference between Bird's

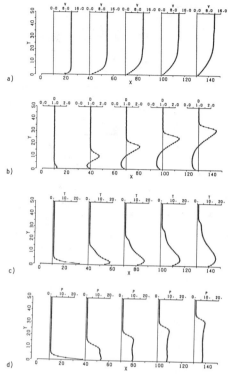

Fig. 5 Distributions of properties along the Y axis (flat plate): a) velocity; b) density; c) temperature; d) pressure.

and Nanbu's profiles is very slight. This is natural because both methods give the solution of the Boltzmann equation to some extent. Consequently, Nanbu's method is able to be applied to the rarefied-gas problems as Bird's is. In our program codes, the computing times by Bird's and Nanbu's methods was 70 and 40 min, respectively, with VP200. Therefore, we use the Nanbu's method through the following simulations.

Then the dependence of the Nanbu's method on time step is examined. Four values for time step are taken: Δt = Tf/20, Tf/10, Tf/5, and Tf/2.5, where Tf denotes the mean free time. This result is presented in Figs. 3 and 4, where the density and plate pressure are presented. The first three time steps show almost the same results, and the last one is apparently different from the others. In particular, the distribution of the plate pressure for Δt=Tf/2.5 is dull near the leading edge in contrast to the other Δt. Thus we determined the time step as Tf/5.

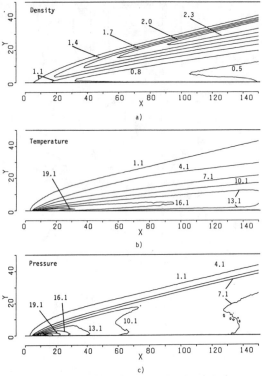

Fig. 6 Contours of properties (flat plate): a) density; b) temperature; c) pressure.

Figures 5 to 7 present the results for the flat plate. Figure 5 show the velocity, density, temperature, and pressure distributions, respectively. The velocity distributions indicate that the boundary layer develops and that the slip velocity at the wall is very large near the leading edge. The steep rises in the pressure distributions, where the flowfield pressure is calculated using the equation of state, clearly show the location of the shock wave. The density and temperature also increase at the shock wave. In the temperature distributions, a rather cold region appears near the wall by the cooling effect due to the fixed low value of the wall temperature. The temperature distributions also indicate the development of the thermal boundary layer.

Figure 6 presents the contours for density, temperature, and pressure. These contours clearly show that the shock wave emanates from the leading edge of the plate and that the thermal boundary layer develops along the plate. In the temperature contour, the line 1.1 is getting separated

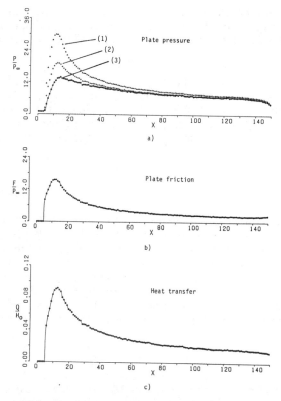

Fig. 7 Properties on the plate: a) pressure (line 1 shows the flowfield pressure at the wall calculated by the equation of state with the mean temperature and line 2 with the temperature in the Y direction; line 3 shows the pressure on the plate); b) friction; c) heat transfer.

from other lines 4.1, 7.1, etc., at about X=100. This indicates that the shock wave begins to separate from the boundary layer.

Figure 7 shows the plate pressure, plate friction, and heat transfer. Figure 7a indicates the distributions of the plate pressure and the flowfield pressure at the wall calculated by the equation of state. In this nonequilibrium flow, the temperature depends on the direction. Hence, the flowfield pressure calculated by the equation of state also depends on the direction. The mean pressure and the pressure in the Y direction are shown in Fig. 7a. It should be noted that there is a large difference between them near the leading edge. This is due to the strong nonequilibrium and large slip velocity in that area. As a result, the equation of state cannot be used around the leading edge. However, such difference is rapidly reduced toward the downstream. Hence, we use the equation of state to get the approximate value of pressure. Figures 7b and 7c have peak points near the leading edge because of the large slip velocity.

Channel Flow

First, the channel flow with the width of $40\lambda_\infty$, Kn = 0.025, is presented. Figure 8 show the contours of density, temperature, and pressure, respectively. The shock waves are generated from the leading edges of the two plates as those in the flat plate are. These shock waves interact at about X=70. The location of shock waves cannot be determined from the contours of density and temperature. However, the pressure contour presents the clear shock wave location. Two incident shock waves cross, making the shape of X, although they are quite thick and diffuse. For the oblique shock waves after interaction, it should be noted that there is almost no temperature jump across it, but a rapid density rise. After interaction, the shock waves separate, incident to boundary layer, and are reflected, though the pattern is obscure. In short, the same flow pattern as in the continuous flow can be seen. This is due to the comparatively low Knudsen number. Figure 9 presents the distributions of the

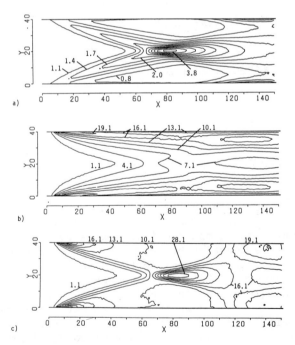

Fig. 8 Contours in channel flow; Kn = 0.025: a) density; b) temperature; c) pressure.

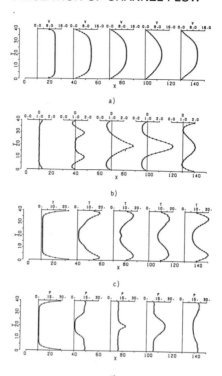

Fig. 9 Distributions of properties along the Y axis (channel flow; Kn = 0.025): a) velocity; b) density; c) temperature; d) pressure.

velocity, density, temperature, and pressure, respectively. The pressure distributions clearly show the characteristics of this flow. Figure 10 shows the plate pressure, plate friction, and heat transfer, respectively. The upstream half of each profile is the same as the profile for the flat plate. These properties, in particular the plate pressure, increase around the point at which the shock waves are reflected by the boundary layer.

Moreover we examine the effect of the downstream boundary condition. Figure 11 presents the comparison between the two types of prescribed boundary conditions for Kn = 0.025 and 0.05. For the case of Kn=0.025, these conditions give almost the same results contrary to our expectation.

Then we doubled the Knudsen number of the flow by setting the width of the channel to $20\lambda_\infty$. This means that the flow corresponds to a more rarefied region. The result is shown in Figs. 12 and 13. Figure 12 presents the density, temperature, and pressure contours, respectively. The upstream half-regions of the contours are almost the same as

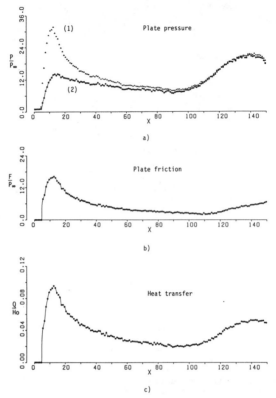

Fig. 10 Properties on the plate (channel flow; Kn = 0.025): a) pressure (line 1 shows the flowfield pressure at the wall calculated by the equation of state with mean temperature; line 2 shows pressure on the wall); b) friction; c) heat transfer.

the previous case, Kn=0.025. However, a characteristic feature appears in the downstream part. The density becomes remarkably large near the wall, and the temperature also shows a high value in the middle region where the growing thermal boundary layers merge. As a result, the pressure has a high value and is almost constant along the Y axis. Figure 13 presents the velocity, density, temperature, and pressure distributions, respectively. The deceleration of the flow velocity can be seen in the velocity distributions.

The two downstream boundary conditions mentioned earlier are also applied to this narrow channel flow. The result is shown in Fig. 11. An appreciable difference in the density contour is seen near the downstream boundary. Nevertheless, the flows in the other region are not affected.

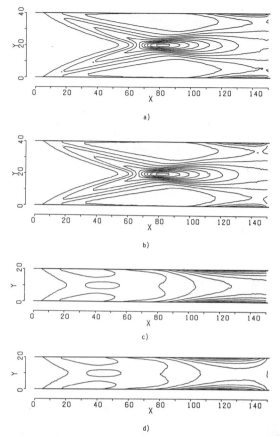

Fig. 11 Comparison of density contours between two different downstream boundary conditions in channel flow: a) the boundary condition used by Vogenitz et al.(Ref. 2), Kn = 0.025; b) no particles from downstream, Kn = 0.025; c) the condition by Vogenitz et al., Kn = 0.05; d) no particles from downstream, Kn = 0.05.

Channel with Obstacle

We demonstrate the channel flow with obstacle plates in Fig. 14, where density, temperature, and pressure contours are indicated, respectively. The thermal boundary layers are raised into the flow by the obstacles. Consequently, the shock waves and thermal boundary layers concentrate at the same place. The pressure therefore becomes remarkably high there. However, the temperature is not so much affected. After that, the high-temperature parts separate and move toward the channel walls, which produces rather low-temperature region around the center in the downstream part.

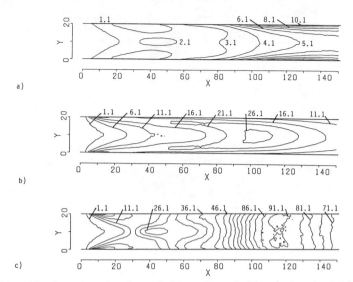

Fig. 12 Contours in channel flow; Kn = 0.05: a) density; b) temperature; c) pressure.

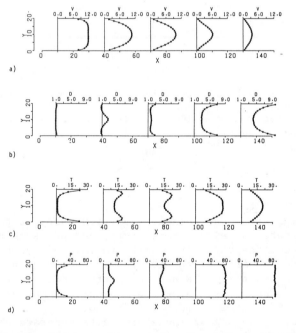

Fig. 13 Distributions of properties along the Y axis (channel flow ; Kn = 0.05): a) velocity; b) density; c) temperature; d) pressure.

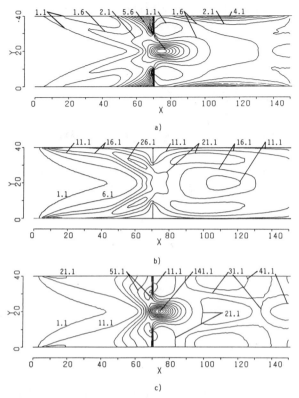

Fig. 14 Contours in channel with obstacle: a) density; b) temperature; c) pressure.

IV. Concluding Remarks

We showed several results simulated by the DSMC method. This method was applied to the three configurations: the flat plate, the channel, and the channel with obstacles.

First we compared Nanbu's method with Bird's. There was almost no difference between them; however, Nanbu's method was more efficient in the computational time for our program code. Hence, we chose Nanbu's method.

The shock wave was clearly seen to be generated from the sharp leading edge of the plate as in the previous results.

The channel flow presented several interesting results. In particular, the interaction between the two oblique shock waves and the interaction of the oblique shock wave with the boundary layer were well simulated although they showed a very diffuse flow pattern compared with the continuous flow. In the present research, the relatively low Knudsen number was chosen, so that the simulated results included the characteristics of the continuous flow.

References

[1] Bird, G.A., <u>Molecular Gas Dynamics</u>, Oxford University Press, Oxford, England, U.K. 1976.

[2] Vogenitz, F.W., Broadwell, J.E., and Bird, G.A., "Leading Edge Flow by the Monte Carlo Technique," <u>AIAA Journal</u>, Vol. 8, March 1970, pp. 504-510.

[3] Nanbu, K., "Theoretical basis of the Direct Simulation Monte Carlo Method," <u>Rarefied Gas Dynamics</u>, Vol. 1, 1986, pp. 369-383.

Structure of Incipient Triple Point at the Transition from Regular Reflection to Mach Reflection

B. Schmidt*
Universitaet Karlsruhe, Karlsruhe, Federal Republic of Germany

Nomenclature

a	= speed of sound
E	= distance between leading edge and interferometer
e	= distance between the two interferometer beams
M_s	= shock Mach number
l_v	= introduced second length due to the viscosity
M_∞	= shock Mach number parallel to wedge surface
MR	= Mach reflection
p_1	= initial pressure
RR	= regular reflection
u_∞	= shock wave velocity parallel to the wedge surface
u_s	= shock wave velocity upstream of the wedge
α	= halfangle of the wedge
$-\delta^*$	= displacement of the boundary layer
ε	= additional turning angle due to viscosity
$\bar{\lambda}_1$	= mean free path upstream of the shock wave
μ	= viscosity of the fluid
ρ	= density
χ	= angle between wedge surface and triple point trajectory

Abstract

After some historical remarks concerning the investigation of the shock-wave reflection from an inclined flat or curved wall, more recent and unsolved problems are dealt with. The aim of our mainly experimental effort in connection with the shock reflection is to answer the question: "Where does the triple-point come into existence?" or "What is the location of the incipient triple-point?" The present paper reports about

---

Copyright © 1989 by the American Institute of Aeronautics and Astronautics, Inc. All rights reserved.

*Professor, Insistut für Störmungslehre und Störmungsmachinen.

measurements of the triple-point trajectory for strong incoming shock waves ($5 \leq Ms \leq 8.5$), three different angles of inclination (α = 25, 45 and 65 deg.) and a pressure range of $6.66 \leq p_1 \leq 26.66$ N/m^2 ($0.05 \leq p_1 \leq 0.2$ Torr) or this expressed in the mean free path $\bar{\lambda}_1$ in argon: $0.25 \leq \bar{\lambda}_1 \leq 1.0$ mm. The report concerns work in progress.

Introduction

More than 100 years ago the physicist Ernst Mach[1] reported the results of an extensive investigation concerning shock wave reflection from flat and curved walls. After Mach, the interest in shock reflection problems faded away. The problems stayed untouched for about 65 years, before it became of more interest again. Von Neumann's[2] theoretical work was a publication that inspired new interest and, in the last 20 - 25 years, a number of differences between theoretical and experimental data found their explanation. I will not go more in details, but I would like to mention the review paper about "Regular and Mach Reflection of Shock Waves" by Hornung[3]. This paper describes well the problems, the background of solutions and has a long list of the more important papers, published in the last 20 - 25 years.

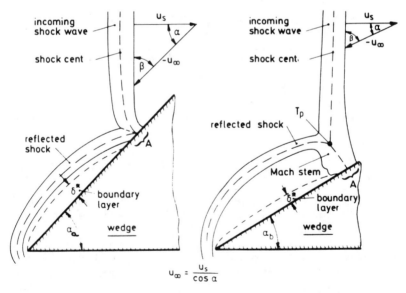

a) Regular reflection (RR).   b) Mach reflection (MR).

Fig. 1   Two main types of reflection ($u_\infty = u_s/\cos \alpha$).

Problem

Without viscosity and heat transfer in the flow, the shock reflection process in a shock tube can be treated as pseudosteady; i.e., the flow is self-similar. The point or region A in Fig´s.1a and b becomes stationary by superimposing $-u_\infty$ onto the flow. For Mach reflection this means that the length of the Mach stem grows proportional to the distance of the foot point A from the leading edge. The trajectory of the triple-point is straight, passes through the leading edge, and has a constant angle χ to the wedge surface.

Introducing viscosity means the introduction of a second length into the problem and the flow cannot be any longer self-similar or pseudosteady (see Hornung[3]). The length introduced can be taken as being proportional to the viscosity in the form $l_\mu \sim \mu/(\rho u_s)$. This length is also proportional the mean free path $\bar{\lambda}_1$ ($\bar{\lambda}_1 \sim \mu_1/(\rho_1 u_s) \sim \mu_1/(\rho_1 a_1)$). (Index 1 means the state in region 1 upstream of the incoming shock wave). $\bar{\lambda}_1$ should be the scaling length for viscous flow at low density. However we shall see later on that this is generally not true.

Theoretical Background and Numerical Approach

As mentioned earlier, the introduction of viscosity into the flow destroys the self-similarity and the flow is no longer pseudosteady. A second length comes into the problem, and the Mach stem does not grow proportional to the distance between A and the leading edge of the wedge. The viscosity causes shear stress as soon as there is a velocity difference between the gas close to the wall and the wall. The shear layer, growing at the wall within the incoming shock wave, is the head of the boundary layer downstream of the shock wave. The shear flow and the boundary layer develop a negative displacement $(-\delta^*)$. With viscosity the transition from regular reflection (RR) to Mach reflection (MR) does not take place at the theoretical deflection angle, calculated for inviscid flow. The reason is the action of the boundary layer. This action can be modelled (see Fig's 2a-c) (Hornung[3]). In Fig.2a, the additional turn of the flow by the angle ε takes care of $-\delta^*$; Fig.2b demonstrates the action of suction by perforation of the wall. The action of suction can be calculated by a suitable distribution of sinks at the wall; Fig.2c shows a model that compares the amount of mass passing through the Mach stem in the inviscid case with the pseudosteady triple-point Tps and T_μ in the viscous case. T_μ is closer to the wall because a considerable part of the mass passing the Mach stem is sucked into the boundary layer. For the viscous flow the Mach stem becomes shorter than that one for the inviscid flow. Calculating the displacement of the laminar

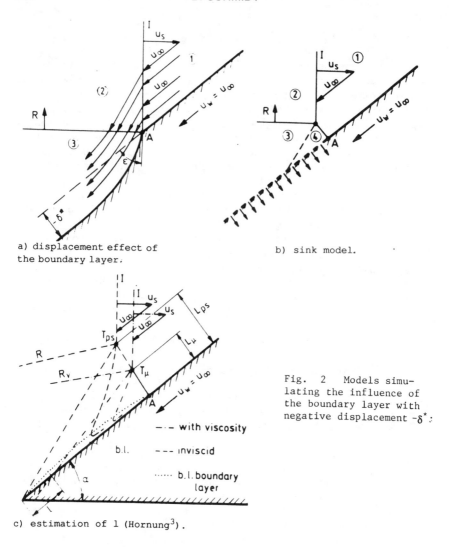

Fig. 2 Models simulating the influence of the boundary layer with negative displacement $-\delta^*$:

a) displacement effect of the boundary layer.

b) sink model.

c) estimation of l (Hornung[3]).

boundary layer makes it possible to get reasonable results for the triple-point trajectory to cut the wedge surface somewhat downstream from the leading edge. The trajectory for the inviscid case is straight and passes through the leading edge.

No analytical solution to this problem is known, but numerical approaches have been performed with sucess. One possibility is the direct simulation Monte Carlo method as developed by Bird[4]. The results of this method are reasonable if the subdivision of the flowfield is done with cells that have an extension less than a mean free path close to the wall and may become larger with dwindling gradients in the flow with

STRUCTURE OF INCIPIENT TRIPLE POINT

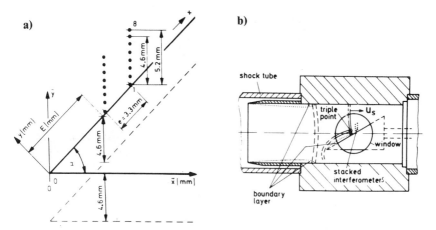

Fig. 3   Arrangement of the 8 stacked interferometers at the wedge surface: a) doubling the range by lowering the shock tube (here = 4,6mm); b) test section.

growing distance from the wall. Besides this the number of gas particles should not be less than 10 - 15 particles. Numerical results, using Birds method, have been calculated by Seiler et al[5]. Results will be shown on the present RGD meeting.

Experimental Setup

The measurements have been done in test sections with cross sections of 90 x 90 mm and 57 x 126 mm. The test sections are attached at the downstream end to the 150 mm i.d. shock tube of our Institute. A sharp edged "cookie cutter" cuts out the central part of the incoming shock wave (see Fig. 3). The cookie cutter has been placed far enough upstream of the place of measurement that the waves generated at the cookie cutter are attenuated sufficiently. To obtain an impression about the whole flow field, that is visible through the test section windows (90 mm diameter), an optical "field method" was used. To get pictures of the fast moving shock waves at low pressure, a double-exposure differential interferometer has been set up. Two sparks, which are arranged in some distance off to the left and the right of the optical center line, generate flashes a fraction of a microsecond long, and they are triggered with a fraction of a microsecond delay. The two light beams passing the test section are complementary polarized. On the photo, only moving density anomalies, such as the shock wave, are visible against the uniform gray background.

The method produces acceptable pictures down to an initial pressure p_1 in the shock tube of p_1 = 200 N/m^2 (1.5 Torr). The results for two initial pressures p_1 = 200 N/m^2 and p_1 = 400 N/m^2 are shown in Fig's. 4a, b. Extending the triple-point

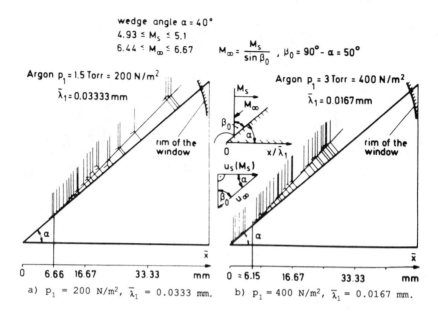

Fig. 4 Results of the double-exposure method for $\alpha = 40$ deg, $M_s \cong 5$:

trajectory up to the leading edge, it does not reach the leading edge but cuts the wedge surface some distance before. This distance, l, is for 200 and 400 N/m² almost the same in millimeters. Normalizing the results with $\bar{\lambda}_1$ shows, that $\bar{\lambda}_1$ is not the right characteristic length in this case. All results, gained with this method show the same behaviour concerning the triple point trajectory. The resolving power of the double exposure differential interferometer is not good enough for the structure in the region close to the triple-point or the real location of the transition from RR to MR to be seen.

Changing over to low density methods of observation of the transition from RR to MR or the incipient triple-point, we expected to see the same. The prolonged triple-point trajectory cuts the wedge surface at a certain distance l from the leading edge. Walenta[6] used an electron beam densitometer to measure the density distribution in the vicinity of the reflection of the incoming shock wave from an inclined flat wall (wedge) for different combinations of the parameters α, M_s and p_1. Fig. 5 shows a typical result for a weak shock wave with $M_s = 1.42$, $\alpha = 25$ deg. in argon with $\bar{\lambda}_1$ about 0.5 mm ($p_1 = 13.3$ N/m²). Here, too, the backward prolonged trajectory of the triple-point cuts into the wedge not far from the leading edge (here l = 69 mm).

For low density measurements an eight beam laser differential interferometer was set up. The signals of the eight interferometers are recorded with four transient recorders

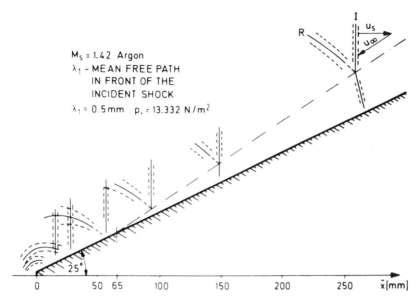

Fig. 5 Electron-beam density measurements[6], argon, M_s = 1.42, p_1 = 13,33 N/m², $\bar{\lambda}_1$ = 0.5 mm.

(Biomation 8100). To have a control over the quality of the recorded signals, the contents of the recorder memory were displayed on an oscilloscope screen and photographed. The transfer of the recorded data to the computer (PDP 11/24) is performed by punched paper tape. The data are reduced and processed by the computer and are finally plotted in a suitable form.

The interferometer beams are aligned to the wedge surface (see Fig. 3a). The eight interferometers are stacked with a vertical distance of about 0.76 mm (mean free path in argon at 13.33 N/m² (0.1 Torr) is $\bar{\lambda}_1$ = 0.5 mm).

Experimental Results

The data reduction starts with a plot of the raw data of the eight interferometers, recorded with the stack of interferometers at a certain distance E (mm) from the leading edge. Synchronized in time to each other, these are plotted on a graph with the axes Volt vs. time. If the flow is locally treated as being pseudosteady, $x = u_\infty \cdot t$ (mm) can be used which is much easier to reduce than nonsteady flow data. The next step is to mark the leading edges of the shock waves, the center and the trailing edges, too, with a suitable scale; a drawing as seen in Fig. 6 is generated. Here, in this plot, the location of the triple-

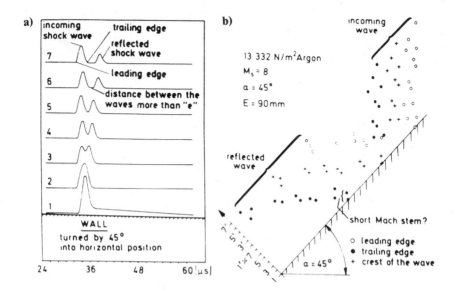

Fig. 6   Reduction of the measured traces (wall turned by 45 deg.):
a) signals from interferometers 1 - 7; b) reflection of the incoming wave
and the reflected wave, argon, $M_s \cong 8$, $p_1 = 13.33$ N/m², E = 90 mm.

point becomes visible. Furthermore, the point of separation between the trailing edge of the incoming wave to the leading edge of the reflected wave by a distance e (mm) becomes visible. (here: e = 3.3 mm). The location of these two points, the triple point and the separation by e point are transfered to a plot, where there points are displayed as a function of the distance E (mm) from the leading edge. Fig. 7 shows the result for α = 45 deg., ($6.66 \leq p_1 \leq 26.664$ N/m², $5.5 \leq Ms \leq 8.5$, argon). The triple-point seems to come into existence at the leading edge or at least very close to it. To clear this point, detailed measurements in the region close to the leading edge are in progress. In the first 10 mm the trajectory of the triple-point gains a distance of about 1 mm from the wall and then stays constant with the same distance up to about E = 120 mm. Above E = 120 mm the triple-point distance from the wall becomes larger. The trajectory of the point where the distance between the trailing edge of the incoming shock wave and the leading edge of the reflected wave becomes equal to e, is also displayed. This trajectory runs with almost constant distance to the triple-point trajectory, except for the region close to the leading edge. Here the mean free path λ_1 is not the scaling length. A change in p_1 by a factor of 4 ($6.66 \leq p_1 \leq 26.664$ N/m²) gives no systematic change in the results, except for the thickness of the shock waves. As is well known, the thickness scales with λ_1.

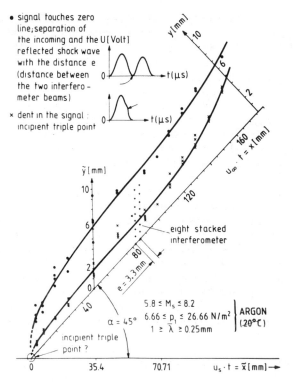

Fig. 7   Triple-point trajectory and separation by the distance e trajectory for $5.8 < M_s < 8.2$, $6.66 < p_1 < 26.66$ N/m², $1 > \bar{\lambda}_1 > 0.5$ mm, $\alpha = 45$ deg, argon.

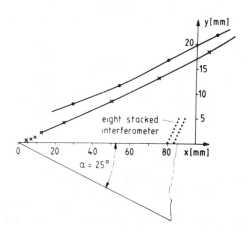

Fig. 8   As Fig. 7, but $\alpha = 25$ deg.

The triple-point trajectory shows at low density some unexpected features. For α = 25 degrees and the other parameters kept constant, the development of the triple-point and the distance e point trajectory, are not parallel to the wedge surface (see Fig. 8) but the distance begins at E = 5 mm to increase. Here the region close to the leading edge, also needs more detailed measurements. These are planned. For α = 65 deg. the number of experimental points is too limited for the development of the triple-point trajectory and the distance e (mm) separation trajectory to be described. Much more detailed experiments are necessary. The same measurements should be done at least for some shock Mach numbers in the weak shock range with $M_s \leq 2.25$.

Conclusions

The main result of this study is that more experimental results and numerical solutions are needed, to get a clear picture of the flow and the development of the wave pattern in the flow. As soon as the incoming shock touches the leading edge of the wedge, the reflected wave begins to develop and a boundary layer starts to grow already inside and at the wall of the incoming shock wave. The boundary layer develops a negative displacement $-\delta^*$. This means, the boundary layer acts like a distribution of sinks along the wall.

In continuum flow the experiments show, that the transition from RR to MR takes place on the wedge somewhat downstream of the leading edge. In low density flow and at wedges with α = 25 and 45 deg., the triple-point comes into existence near the leading edge. It may be that in the case of continuum flow the resolution is not sufficient to show the real place of transition. Another possibility may be the fact that the continuum result cannot be backed by a low density result.

This paper reports about work in progress, and there are now more questions than answers.

Acknowledgment

This work is supported by the Deutsche Forschungsgemeinschaft (German Research Association), Bonn-Bad Godesberg, Federal Republic of Germany.

References

[1] Mach, E., ``Über den Verlauf von Funkenwellen in der Ebene und im Raum'', Sitzungsberichte der Akademie der Wissenschaften in Vienna, Austria, Mathematisch Naturwissenschaftliche Kl. Abt. 2A 78, 1878, pp 819 - 838

[2]Neumann, J. von, ``Oblique Reflection of Shocks'', U.S. Department of the Navy, Washington, D.C. Explosion Research Report 12, 1943; see also: Collected Works, Vol 6 pp 238 - 299, Pergamon, Oxford 1963, U.K.

[3]Hornung, H., ``Regular and Mach Reflection of Shock Waves'', Annual Review of Fluid Mechanics 18, 1986, pp 33 - 58.

[4]Bird, G. A., Molecular Gas Dynamics, Clarendon, Oxford, U.K., 1976.

[5]Seiler, F., Oertel, H., and Schmidt, B., ``Direct Simulation Monte Carlo of Shock Reflection on a Wedge with Boundary Layer'', Poster Session, 16th International Symposium on Rarefied Gas Dynamics, Pasadena, CA. 1988.

[6]Walenta, Z. A., ``Formation of the Mach Type Reflection of Shock Waves'', Archives of Mechanics, Vol. 35, No.2, pp 187 - 196, 1983.

Author Index

Abe, T. 258
Aoki, K. 70, 297
Azmy, Y. Y. 15
Babovsky, H. 85
Bird, G. A. 211, 447
Boffi, V. C. 48
Boyd, I. D. 245
Broadwell, J. E. 100, 155
Cabannes, H. 148
Caflisch, R. E. 3
Celenligil, M. C. 447
Cercignani, C. 552
Chapman, D. R. 374
Chen, Y.-K. 359
Chrusciel, G. T. 462
Cornille, H. 131
Dogra, V. K. 567
Erwin, D. A. 271
Eu, B. C. 396
Fiscko, K. A. 374
Frezzotti, A. 552
Furlani, T. R. 227
Gaveau, B. 61
Gaveau, M.-A. 61
Goldstein, D. 100
Greber, I. 194
Greenberg, W. 29
Gropengiesser, F. 85
Howell, J. R. 359
Igarashi, S. 500
Ivanov, M. S. 171
Kawashima, S. 148
Khayat, R. E. 396
Koppenwallner, G. 500
Lee, K. D. 337
Legge, H. 500
Longo, E. 118
Lord, R. G. 493
Lordi, J. A. 227
Mausbach, P. 182

Monaco, R. 118
Moss, J. N. 413, 447, 567
Nadiga, B. T. 155
Nakamura, Y. 582
Nanbu, K. 500
Neunzert, H. 85
Oertel, H. 432, 518
Ohwada, T. 70
Pham-Van-Diep, G. C. 271
Polewczak, J. 39
Pool, L. A. 462
Potter, J. L. 484
Price, J. M. 413, 567
Protopopescu, V. 15
Rogasinsky, S. V. 171
Rudyak, V. Ya. 171
Schmidt, B. 284, 518, 597
Seiler, F. 284, 518
Shimada, T. 258
Smith, R. W. 493
Sone, Y. 70
Spiga, G. 48
Stark, J. P. W. 245
Struckmeier, J. 85
Sturtevant, B. 100, 155
Tan, Z. 359
Tanaka, J. 582
Tang, Z. 476
Tcheremissine, F. G. 343
van der Mee, C. V. M. 29
Varghese, P. L. 359
Wachman, H. 194
Watanabe, Y. 500
Wetzel, W. 432
Wiesen, B. 85
Wilmoth, R. G. 532
Xie, Y. 476
Yasuhara, M. 582
Yen, S. M. 323, 337

PROGRESS IN ASTRONAUTICS AND AERONAUTICS SERIES VOLUMES

VOLUME TITLE/EDITORS

*1. **Solid Propellant Rocket Research** (1960)
Martin Summerfield
Princeton University

*2. **Liquid Rockets and Propellants** (1960)
Loren E. Bollinger
The Ohio State University
Martin Goldsmith
The Rand Corporation
Alexis W. Lemmon Jr.
Battelle Memorial Institute

*3. **Energy Conversion for Space Power** (1961)
Nathan W. Snyder
Institute for Defense Analyses

*4. **Space Power Systems** (1961)
Nathan W. Snyder
Institute for Defense Analyses

*5. **Electrostatic Propulsion** (1961)
David B. Langmuir
Space Technology Laboratories, Inc.
Ernst Stuhlinger
NASA George C. Marshall Space Flight Center
J.M. Sellen Jr.
Space Technology Laboratories, Inc.

*6. **Detonation and Two-Phase Flow** (1962)
S.S. Penner
California Institute of Technology
F.A. Williams
Harvard University

*7. **Hypersonic Flow Research** (1962)
Frederick R. Riddell
AVCO Corporation

*8. **Guidance and Control** (1962)
Robert E. Roberson
Consultant
James S. Farrior
Lockheed Missiles and Space Company

*9. **Electric Propulsion Development** (1963)
Ernst Stuhlinger
NASA George C. Marshall Space Flight Center

*10. **Technology of Lunar Exploration** (1963)
Clifford I. Cummings
Harold R. Lawrence
Jet Propulsion Laboratory

*11. **Power Systems for Space Flight** (1963)
Morris A. Zipkin
Russell N. Edwards
General Electric Company

12. **Ionization in High-Temperature Gases** (1963)
Kurt E. Shuler, Editor
National Bureau of Standards
John B. Fenn, Associate Editor
Princeton University

*13. **Guidance and Control—II** (1964)
Robert C. Langford
General Precision Inc.
Charles J. Mundo
Institute of Naval Studies

*14. **Celestial Mechanics and Astrodynamics** (1964)
Victor G. Szebehely
Yale University Observatory

*15. **Heterogeneous Combustion** (1964)
Hans G. Wolfhard
Institute for Defense Analyses
Irvin Glassman
Princeton University
Leon Green Jr.
Air Force Systems Command

16. **Space Power Systems Engineering** (1966)
George C. Szego
Institute for Defense Analyses
J. Edward Taylor
TRW Inc.

17. **Methods in Astrodynamics and Celestial Mechanics** (1966)
Raynor L. Duncombe
U.S. Naval Observatory
Victor G. Szebehely
Yale University Observatory

18. **Thermophysics and Temperature Control of Spacecraft and Entry Vehicles** (1966)
Gerhard B. Heller
NASA George C. Marshall Space Flight Center

---

*Out of print.

*19. **Communication Satellite Systems Technology** (1966)
Richard B. Marsten
Radio Corporation of America

20. **Thermophysics of Spacecraft and Planetary Bodies: Radiation Properties of Solids and the Electromagnetic Radiation Environment in Space** (1967)
Gerhard B. Heller
NASA George C. Marshall Space Flight Center

21. **Thermal Design Principles of Spacecraft and Entry Bodies** (1969)
Jerry T. Bevans
TRW Systems

22. **Stratospheric Circulation** (1969)
Willis L. Webb
Atmospheric Sciences Laboratory, White Sands, and University of Texas at El Paso

23. **Thermophysics: Applications to Thermal Design of Spacecraft** (1970)
Jerry T. Bevans
TRW Systems

24. **Heat Transfer and Spacecraft Thermal Control** (1971)
John W. Lucas
Jet Propulsion Laboratory

25. **Communication Satellites for the 70's: Technology** (1971)
Nathaniel E. Feldman
The Rand Corporation
Charles M. Kelly
The Aerospace Corporation

26. **Communication Satellites for the 70's: Systems** (1971)
Nathaniel E. Feldman
The Rand Corporation
Charles M. Kelly
The Aerospace Corporation

27. **Thermospheric Circulation** (1972)
Willis L. Webb
Atmospheric Sciences Laboratory, White Sands, and University of Texas at El Paso

28. **Thermal Characteristics of the Moon** (1972)
John W. Lucas
Jet Propulsion Laboratory

29. **Fundamentals of Spacecraft Thermal Design** (1972)
John W. Lucas
Jet Propulsion Laboratory

30. **Solar Activity Observations and Predictions** (1972)
Patrick S. McIntosh
Murray Dryer
Environmental Research Laboratories, National Oceanic and Atmospheric Administration

31. **Thermal Control and Radiation** (1973)
Chang-Lin Tien
University of California at Berkeley

32. **Communications Satellite Systems** (1974)
P.L. Bargellini
COMSAT Laboratories

33. **Communications Satellite Technology** (1974)
P.L. Bargellini
COMSAT Laboratories

34. **Instrumentation for Airbreathing Propulsion** (1974)
Allen E. Fuhs
Naval Postgraduate School
Marshall Kingery
Arnold Engineering Development Center

35. **Thermophysics and Spacecraft Thermal Control** (1974)
Robert G. Hering
University of Iowa

36. **Thermal Pollution Analysis** (1975)
Joseph A. Schetz
Virginia Polytechnic Institute

37. **Aeroacoustics: Jet and Combustion Noise; Duct Acoustics** (1975)
Henry T. Nagamatsu, Editor
General Electric Research and Development Center
Jack V. O'Keefe, Associate Editor
The Boeing Company
Ira R. Schwartz, Associate Editor
NASA Ames Research Center

38. **Aeroacoustics: Fan, STOL, and Boundary Layer Noise; Sonic Boom; Aeroacoustic Instrumentation** (1975)
Henry T. Nagamatsu, Editor
General Electric Research and Development Center
Jack V. O'Keefe, Associate Editor
The Boeing Company
Ira R. Schwartz, Associate Editor
NASA Ames Research Center

SERIES LISTING

39. **Heat Transfer with Thermal Control Applications** (1975)
M. Michael Yovanovich
University of Waterloo

40. **Aerodynamics of Base Combustion** (1976)
S.N.B. Murthy, Editor
Purdue University
J.R. Osborn, Associate Editor
Purdue University
A.W. Barrows
J.R. Ward, Associate Editors
Ballistics Research Laboratories

41. **Communications Satellite Developments: Systems** (1976)
Gilbert E. LaVean
Defense Communications Agency
William G. Schmidt
CML Satellite Corporation

42. **Communications Satellite Developments: Technology** (1976)
William G. Schmidt
CML Satellite Corporation
Gilbert E. LaVean
Defense Communications Agency

43. **Aeroacoustics: Jet Noise, Combustion and Core Engine Noise** (1976)
Ira R. Schwartz, Editor
NASA Ames Research Center
Henry T. Nagamatsu, Associate Editor
General Electric Research and Development Center
Warren C. Strahle, Associate Editor
Georgia Institute of Technology

44. **Aeroacoustics: Fan Noise and Control; Duct Acoustics; Rotor Noise** (1976)
Ira R. Schwartz, Editor
NASA Ames Research Center
Henry T. Nagamatsu, Associate Editor
General Electric Research and Development Center
Warren C. Strahle, Associate Editor
Georgia Institute of Technology

45. **Aeroacoustics: STOL Noise; Airframe and Airfoil Noise** (1976)
Ira R. Schwartz, Editor
NASA Ames Research Center
Henry T. Nagamatsu, Associate Editor
General Electric Research and Development Center
Warren C. Strahle, Associate Editor
Georgia Institute of Technology

46. **Aeroacoustics: Acoustic Wave Propagation; Aircraft Noise Prediction; Aeroacoustic Instrumentation** (1976)
Ira R. Schwartz, Editor
NASA Ames Research Center
Henry T. Nagamatsu, Associate Editor
General Electric Research and Development Center
Warren C. Strahle, Associate Editor
Georgia Institute of Technology

47. **Spacecraft Charging by Magnetospheric Plasmas** (1976)
Alan Rosen
TRW Inc.

48. **Scientific Investigations on the Skylab Satellite** (1976)
Marion I. Kent
Ernst Stuhlinger
NASA George C. Marshall Space Flight Center
Shi-Tsan Wu
The University of Alabama

49. **Radiative Transfer and Thermal Control** (1976)
Allie M. Smith
ARO Inc.

50. **Exploration of the Outer Solar System** (1976)
Eugene W. Greenstadt
TRW Inc.
Murray Dryer
National Oceanic and Atmospheric Administration
Devrie S. Intriligator
University of Southern California

51. **Rarefied Gas Dynamics, Parts I and II (two volumes)** (1977)
J. Leith Potter
ARO Inc.

52. **Materials Sciences in Space with Application to Space Processing** (1977)
Leo Steg
General Electric Company

53. **Experimental Diagnostics in Gas Phase Combustion Systems** (1977)
Ben T. Zinn, Editor
Georgia Institute of Technology
Craig T. Bowman, Associate Editor
Stanford University
Daniel L. Hartley, Associate Editor
Sandia Laboratories
Edward W. Price, Associate Editor
Georgia Institute of Technology
James G. Skifstad, Associate Editor
Purdue University

54. **Satellite Communications: Future Systems** (1977)
David Jarett
TRW Inc.

55. **Satellite Communications: Advanced Technologies** (1977)
David Jarett
TRW Inc.

56. **Thermophysics of Spacecraft and Outer Planet Entry Probes** (1977)
Allie M. Smith
ARO Inc.

57. **Space-Based Manufacturing from Nonterrestrial Materials** (1977)
Gerard K. O'Neill, Editor
Princeton University
Brian O'Leary, Assistant Editor
Princeton University

58. **Turbulent Combustion** (1978)
Lawrence A. Kennedy
State University of New York at Buffalo

59. **Aerodynamic Heating and Thermal Protection Systems** (1978)
Leroy S. Fletcher
University of Virginia

60. **Heat Transfer and Thermal Control Systems** (1978)
Leroy S. Fletcher
University of Virginia

61. **Radiation Energy Conversion in Space** (1978)
Kenneth W. Billman
NASA Ames Research Center

62. **Alternative Hydrocarbon Fuels: Combustion and Chemical Kinetics** (1978)
Craig T. Bowman
Stanford University
Jorgen Birkeland
Department of Energy

63. **Experimental Diagnostics in Combustion of Solids** (1978)
Thomas L. Boggs
Naval Weapons Center
Ben T. Zinn
Georgia Institute of Technology

64. **Outer Planet Entry Heating and Thermal Protection** (1979)
Raymond Viskanta
Purdue University

65. **Thermophysics and Thermal Control** (1979)
Raymond Viskanta
Purdue University

66. **Interior Ballistics of Guns** (1979)
Herman Krier
University of Illinois at Urbana-Champaign
Martin Summerfield
New York University

*67. **Remote Sensing of Earth from Space: Role of "Smart Sensors"** (1979)
Roger A. Breckenridge
NASA Langley Research Center

68. **Injection and Mixing in Turbulent Flow** (1980)
Joseph A. Schetz
Virginia Polytechnic Institute and State University

69. **Entry Heating and Thermal Protection** (1980)
Walter B. Olstad
NASA Headquarters

70. **Heat Transfer, Thermal Control, and Heat Pipes** (1980)
Walter B. Olstad
NASA Headquarters

71. **Space Systems and Their Interactions with Earth's Space Environment** (1980)
Henry B. Garrett
Charles P. Pike
Hanscom Air Force Base

72. **Viscous Flow Drag Reduction** (1980)
Gary R. Hough
Vought Advanced Technology Center

73. **Combustion Experiments in a Zero-Gravity Laboratory** (1981)
Thomas H. Cochran
NASA Lewis Research Center

74. **Rarefied Gas Dynamics, Parts I and II (two volumes)** (1981)
Sam S. Fisher
University of Virginia at Charlottesville

75. **Gasdynamics of Detonations and Explosions** (1981)
J.R. Bowen
University of Wisconsin at Madison
N. Manson
Université de Poitiers
A.K. Oppenheim
University of California at Berkeley
R.I. Soloukhin
Institute of Heat and Mass Transfer, BSSR Academy of Sciences

76. **Combustion in Reactive Systems** (1981)
J.R. Bowen
University of Wisconsin at Madison
N. Manson
Université de Poitiers
A.K. Oppenheim
University of California at Berkeley
R.I. Soloukhin
Institute of Heat and Mass Transfer, BSSR Academy of Sciences

77. **Aerothermodynamics and Planetary Entry** (1981)
A.L. Crosbie
University of Missouri-Rolla

78. **Heat Transfer and Thermal Control** (1981)
A.L. Crosbie
University of Missouri-Rolla

79. **Electric Propulsion and Its Applications to Space Missions** (1981)
Robert C. Finke
NASA Lewis Research Center

80. **Aero-Optical Phenomena** (1982)
Keith G. Gilbert
Leonard J. Otten
Air Force Weapons Laboratory

81. **Transonic Aerodynamics** (1982)
David Nixon
Nielsen Engineering & Research, Inc.

82. **Thermophysics of Atmospheric Entry** (1982)
T.E. Horton
The University of Mississippi

83. **Spacecraft Radiative Transfer and Temperature Control** (1982)
T.E. Horton
The University of Mississippi

84. **Liquid-Metal Flows and Magnetohydrodynamics** (1983)
H. Branover
Ben-Gurion University of the Negev
P.S. Lykoudis
Purdue University
A. Yakhot
Ben-Gurion University of the Negev

85. **Entry Vehicle Heating and Thermal Protection Systems: Space Shuttle, Solar Starprobe, Jupiter Galileo Probe** (1983)
Paul E. Bauer
McDonnell Douglas Astronautics Company
Howard E. Collicott
The Boeing Company

86. **Spacecraft Thermal Control, Design, and Operation** (1983)
Howard E. Collicott
The Boeing Company
Paul E. Bauer
McDonnell Douglas Astronautics Company

87. **Shock Waves, Explosions, and Detonations** (1983)
J.R. Bowen
University of Washington
N. Manson
Université de Poitiers
A.K. Oppenheim
University of California at Berkeley
R.I. Soloukhin
Institute of Heat and Mass Transfer, BSSR Academy of Sciences

88. **Flames, Lasers, and Reactive Systems** (1983)
J.R. Bowen
University of Washington
N. Manson
Université de Poitiers
A.K. Oppenheim
University of California at Berkeley
R.I. Soloukhin
Institute of Heat and Mass Transfer, BSSR Academy of Sciences

89. **Orbit-Raising and Maneuvering Propulsion: Research Status and Needs** (1984)
Leonard H. Caveny
Air Force Office of Scientific Research

90. **Fundamentals of Solid-Propellant Combustion** (1984)
Kenneth K. Kuo
The Pennsylvania State University
Martin Summerfield
Princeton Combustion Research Laboratories, Inc.

91. **Spacecraft Contamination: Sources and Prevention** (1984)
J.A. Roux
The University of Mississippi
T.D. McCay
NASA Marshall Space Flight Center

92. **Combustion Diagnostics by Nonintrusive Methods** (1984)
T.D. McCay
NASA Marshall Space Flight Center
J.A. Roux
The University of Mississippi

93. **The INTELSAT Global Satellite System** (1984)
Joel Alper
COMSAT Corporation
Joseph Pelton
INTELSAT

94. **Dynamics of Shock Waves, Explosions, and Detonations** (1984)
J.R. Bowen
University of Washington
N. Manson
Université de Poitiers
A.K. Oppenheim
University of California
R.I. Soloukhin
Institute of Heat and Mass Transfer, BSSR Academy of Sciences

95. **Dynamics of Flames and Reactive Systems** (1984)
J.R. Bowen
University of Washington
N. Manson
Université de Poitiers
A.K. Oppenheim
University of California
R.I. Soloukhin
Institute of Heat and Mass Transfer, BSSR Academy of Sciences

96. **Thermal Design of Aeroassisted Orbital Transfer Vehicles** (1985)
H.F. Nelson
University of Missouri-Rolla

97. **Monitoring Earth's Ocean, Land, and Atmosphere from Space — Sensors, Systems, and Applications** (1985)
Abraham Schnapf
Aerospace Systems Engineering

98. **Thrust and Drag: Its Prediction and Verification** (1985)
Eugene E. Covert
Massachusetts Institute of Technology
C.R. James
Vought Corporation
William F. Kimzey
Sverdrup Technology AEDC Group
George K. Richey
U.S. Air Force
Eugene C. Rooney
U.S. Navy Department of Defense

99. **Space Stations and Space Platforms — Concepts, Design, Infrastructure, and Uses** (1985)
Ivan Bekey
Daniel Herman
NASA Headquarters

100. **Single- and Multi-Phase Flows in an Electromagnetic Field Energy, Metallurgical, and Solar Applications** (1985)
Herman Branover
Ben-Gurion University of the Negev
Paul S. Lykoudis
Purdue University
Michael Mond
Ben-Gurion University of the Negev

101. **MHD Energy Conversion: Physiotechnical Problems** (1986)
V.A. Kirillin
A.E. Sheyndlin
Soviet Academy of Sciences

102. **Numerical Methods for Engine-Airframe Integration** (1986)
S.N.B. Murthy
Purdue University
Gerald C. Paynter
Boeing Airplane Company

103. **Thermophysical Aspects of Re-Entry Flows** (1986)
James N. Moss
NASA Langley Research Center
Carl D. Scott
NASA Johnson Space Center

104. **Tactical Missile Aerodynamics** (1986)
M.J. Hemsch
PRC Kentron, Inc.
J.N. Nielsen
NASA Ames Research Center

105. **Dynamics of Reactive Systems Part I: Flames and Configurations; Part II: Modeling and Heterogeneous Combustion** (1988)
J.R. Bowen
University of Washington
J.-C. Leyer
Université de Poitiers
R.I. Soloukhin
Institute of Heat and Mass Transfer, BSSR Academy of Sciences

106. **Dynamics of Explosions** (1986)
J.R. Bowen
University of Washington
J.-C. Leyer
Université de Poitiers
R.I. Soloukhin
Institute of Heat and Mass Transfer, BSSR Academy of Sciences

107. **Spacecraft Dielectric Material Properties and Spacecraft Charging** (1986)
A.R. Frederickson
U.S. Air Force Rome Air Development Center
D.B. Cotts
SRI International
J.A. Wall
U.S. Air Force Rome Air Development Center
F.L. Bouquet
Jet Propulsion Laboratory, California Institute of Technology

108. **Opportunities for Academic Research in a Low-Gravity Environment** (1986)
George A. Hazelrigg
National Science Foundation
Joseph M. Reynolds
Louisiana State University

109. **Gun Propulsion Technology** (1988)
Ludwig Stiefel
U.S. Army Armament Research, Development and Engineering Center

110. **Commercial Opportunities in Space** (1988)
F. Shahrokhi
K.E. Harwell
University of Tennessee Space Institute
C.C. Chao
National Cheng Kung University

111. **Liquid-Metal Flows: Magnetohydrodynamics and Applications** (1988)
Herman Branover, Michael Mond, and Yeshajahu Unger
Ben-Gurion University of the Negev

112. **Current Trends in Turbulence Research** (1988)
Herman Branover, Michael Mond, and Yeshajahu Unger
Ben-Gurion University of the Negev

113. **Dynamics of Reactive Systems Part I: Flames; Part II: Heterogeneous Combustion and Applications** (1988)
A.L. Kuhl
R & D Associates
J.R. Bowen
University of Washington
J.-C. Leyer
Université de Poitiers
A. Borisov
USSR Academy of Sciences

114. **Dynamics of Explosions** (1988)
A.L. Kuhl
R & D Associates
J.R. Bowen
University of Washington
J.-C. Leyer
Université de Poitiers
A. Borisov
USSR Academy of Sciences

115. **Machine Intelligence and Autonomy for Aerospace** (1988)
E. Heer
Heer Associates, Inc.
H. Lum
NASA Ames Research Center

116. **Rarefied Gas Dynamics: Space-Related Studies** (1989)
E.P. Muntz
University of Southern California
D.P. Weaver
U.S. Air Force Astronautics Laboratory (AFSC)
D.H. Campbell
The University of Dayton Research Institute

117. **Rarefied Gas Dynamics: Physical Phenomena** (1989)
E.P. Muntz
University of Southern California
D.P. Weaver
U.S. Air Force Astronautics Laboratory (AFSC)
D.H. Campbell
The University of Dayton Research Institute

118. **Rarefied Gas Dynamics: Theoretical and Computational Techniques** (1989)
E.P. Muntz
University of Southern California
D.P. Weaver
U.S. Air Force Astronautics Laboratory (AFSC)
D.H. Campbell
The University of Dayton Research Institute

(Other Volumes are planned.)

Other Volumes in the Rarefied Gas Dynamics Series*

Rarefied Gas Dynamics, Proceedings of the 1st International Symposium, edited by F. M. Devienne, Pergamon Press, Paris, France, 1960.

Rarefied Gas Dynamics, Proceedings of the 2nd International Symposium, edited by L. Talbot, Academic Press, New York, 1961.

Rarefied Gas Dynamics, Proceedings of the 3rd International Symposium, Vols. I and II, edited by J. A. Laurmann, Academic Press, New York, 1963.

Rarefied Gas Dynamics, Proceedings of the 4th International Symposium, Vols. I and II, edited by J. H. deLeeuw, Academic Press, New York, 1965.

Rarefied Gas Dynamics, Proceedings of the 5th International Symposium, Vols. I and II, edited by C. L. Brundin, Academic Press, London, England, 1967.

Rarefied Gas Dynamics, Proceedings of the 6th International Symposium, Vols. I and II, edited by L. Trilling and H. Y. Wachman, Academic Press, New York, 1969.

Rarefied Gas Dynamics, Proceedings of the 7th International Symposium, Vols. I and II, edited by D. Dini, C. Cercignani, and S. Nocilla, Editrice Tecnico Scientifica, Pisa, Italy, 1971.

Rarefied Gas Dynamics, Proceedings of the 8th International Symposium, Edited by K. Karamcheti, Academic Press, New York, 1974.

Rarefied Gas Dynamics, Proceedings of the 9th International Symposium, Vols. I and II, edited by M. Becker and M. Fiebig, DFVLR-Press, Porz-Wahn, Germany, 1974.

Rarefied Gas Dynamics, Proceedings of the 10th International Symposium, Parts I and II of Vol. 51 of *Progress in Astronautics and Aeronautics,* edited by L. Potter, AIAA, New York, 1977.

Rarefied Gas Dynamics, Proceedings of the 11th International Symposium, Vols. I and II, edited by R. Campargue, Commissiriat a l'Energie Atomique, Paris, France, 1979.

Rarefied Gas Dynamics, Proceedings of the 12th International Symposium, Parts I and II of Vol. 74 of *Progress in Astronautics and Aeronautics,* edited by S. Fisher, AIAA, New York, 1981.

Rarefied Gas Dynamics, Proceedings of the 13th International Symposium, Vols. I and II, edited by O. M. Belotserkovskii, M. N. Kogan, C. S. Kutateladze, and A. K. Rebrov, Plenum Press, New York, 1985.

Rarefied Gas Dynamics, Proceedings of the 14th International Symposium, Vols. I and II, edited by H. Oguchi, University of Tokyo Press, Tokyo, 1984.

Rarefied Gas Dynamics, Proceedings of the 15th International Symposium, Vols. I and II, edited by V. Boffi and C. Cercignani, B. G. Teubner, Stuttgart, 1986.

Rarefied Gas Dynamics, Proceedings of the 16th International Symposium, Vols. 116, 117, and 118 in *Progress in Astronautics and Aeronautics,* edited by E. P. Muntz, D. Weaver, and D. Campbell, AIAA, Washington, DC, 1989.

*Copies may be purchased directly from the publisher in each case. The AIAA cannot fill orders for volumes published by other publishers. This list is provided for information only.